D0900163

EUCARYOTIC GENE REGULATION

Academic Press Rapid Manuscript Reproduction

Proceedings of the 1979 ICN–UCLA Symposia on
Molecular and Cellular Biology held in Keystone,
March 4–9,1979

ICN-UCLA Symposia on Molecular and Cellular Biology
Volume XIV, 1979

EUCARYOTIC GENE REGULATION

edited by

RICHARD AXEL
College of Physicians and Surgeons
Columbia University
New York, New York

TOM MANIATIS
Division of Biology
California Institute of Technology
Pasadena, California

C. FRED FOX
Department of Micobiology and Molecular Biology Institute
University of California, Los Angeles
Los Angeles, California

ACADEMIC PRESS 1979

A Subsidiary of Harcourt Brace Jovanovich, Publishers

New York London Toronto Sydney San Francisco

ACADEMIC PRESS, INC.
111 Fifth Avenue, New York, New York 10003

United Kingdom Edition published by
ACADEMIC PRESS, INC. (LONDON) LTD.
24/28 Oval Road, London NW1 7DX

ISBN 0–12–068350–4

PRINTED IN THE UNITED STATES OF AMERICA

79 80 81 82 9 8 7 6 5 4 3 2 1

CONTENTS

CONTRIBUTORS

Numbers in parentheses refer to chapter numbers.

ABELSON, JOHN (7), Department of Chemistry, B-017, University of California, San Diego, La Jolla, California 92093

ALT, FREDERICK W. (35), Center for Cancer Research, E17-517, Massachusetts Institute of Technology, Cambridge, Massachusetts 02139

ASHBURNER, M. (9), Department of Genetics, University of Cambridge, Cambridge, England

AXEL, RICHARD (39), College of Physicians and Surgeons, Columbia University, New York, New York 10032

BALTIMORE, DAVID (35), Center for Cancer Research, E17-517, Massachusetts Institute of Technology, Cambridge, Massachusetts 02139

BANK, ARTHUR (31), Hammer Health Sciences Center, Room 1602, Columbia University, New York, New York 10032

BEGGS, JEAN D. (40), Plant Breeding Institute, Cambridge, England

BENDER, A. (24), Department of Biology, 16-437, Massachusetts Institute of Technology, Cambridge Massachusetts 02139

BENOIST, C. (26), Institut de Chimie Biologique, 11 Rue Humann, Strasbourg 67085, France

BERG, JOHAN VAN DEN (40), Gist-Brocades, Delft, Holland

BERG, PAUL (52), Department of Biochemistry, Stanford University Medical Center, Stanford, California 94305

BERK, A. J. (48), Department of Microbiology, University of California, Los Angeles, Los Angeles, California 90024

BERNARDS, R. (30), University of Amsterdam, Box 60.000, 1005 GA, Amsterdam, Netherlands

BIRKENMEIER, E. (42), Carnegie Institute of Washington, 115 West University Parkway, Baltimore, Maryland 21210

BLAIR, LINDLEY C. (2), Institute of Molecular Biology, University of Oregon, Eugene, Oregon 97403

BOLL, WERNER (40), Institut für Molekularbiol I, Universität Zurich, 8093 Zurich, Switzerland

BOTHWELL, ALFRED L. M. (35), Center for Cancer Research, E17-517, Massachusetts Institute of Technology, Cambridge, Massachusetts 02139

BOTSTEIN, DAVID (8), Department of Biology, Building 56-721, Massachusetts Institute of Technology, Cambridge, Massachusetts 02139

BREATHNACH, R. (26), Institut de Chimie Biologie, 11 Rue Humann, Strasbourg 67085, France

BRITTEN, R. J. (22), Kerckhoff Marine Laboratory, California Institute of Technology, Corona del Mar, California 92625

BROKER, THOMAS R. (50), Cold Spring Harbor Laboratory, P. O. Box 100, Cold Spring Harbor, New York 11724

BROOME, STEPHANIE (28), Biological Laboratory, Harvard University, Cambridge, Massachusetts, 02138

BROWN, D. D. (42), Carnegie Institute, 115 West University Parkway, Baltimore, Maryland 21210

BROWN, PETER C. (41), Department of Biological Science, Stanford University, Stanford, California 94305

BURNS, ALEXANDER L. (31), Columbia University, Room 1602, 701 West 168th Street, New York, New York 10032

BUTLER, E. T. (29), Division of Biology, California Institute of Technology, Pasadena, California 91125

CALAME, K. (34), California Institute of Technology, 156-29, Pasadena, California 91125

CAMFIELD, R. (9), Genetics Research Unit, University of Nottingham Medical School, Nottingham, England

CARBON, J. (6), Department of Biological Sciences, University of California, Santa Barbara, California 93106

CATTERALL, JAMES F. (27), Baylor College of Medicine, 1200 Mourshund Avenue, Houston, Texas 77030

CAVALLESCO, C. (32), Department of Medicine and Human Genetics, Yale University School of Medicine, New Haven, Connecticut 06510

CHAMBON, PIERRE (26), Institut de Chimie Biologique, 11 Rue Humann, Strasbourg 67085, France

CHICK, WILLIAM (28), Joslin Research Laboratory, Harvard Medical School, Boston, Massachusetts 02115

CHINAULT, A. C. (6), Department of Medicine, Baylor College of Medicine, Houston, Texas 77025

CHOI, EDMUND (36), Molecular Biology Institute, Room 540, University of California, Los Angeles, California 90024

CHOUDARY, P. V. (32), Department of Medicine and Human Genetics, Yale University School of Medicine, New Haven, Connecticut 06510

CHOVNICK, ARTHUR (11), Genetics and Cell Biology Section, University of Connecticut, Storrs, Connecticut 06268

CHOW, LOUISE T. (50), Cold Spring Harbor Laboratory, P. O. Box 100, Cold Spring Harbor, New York 11724

CLARK, ELLEN M. (47), Department of Biology, University of Virginia, Charlottesville, Virginia 22901

CLARK, STEPHEN H. (11), Genetics and Cell Biology Section, University of Connecticut, Storrs, Connecticut 06268

CLARKE, B. (9), Genetics Research Unit, University of Nottingham Medical School, Nottingham, England

CLARKE, PAT (36), Molecular Biology Institute, University of California, Los Angeles, Los Angeles, California 90024

COCHET, M. (26), Institut de Chimie Biologique, 11 Rue Humann, Strasbourg 67085, France

DAVIDSON, ERIC H. (22), Division of Biology, California Institute of Technology, Pasadena, California 91125

DAVIDSON, NORMAN (13), Department of Chemistry 164-30, California Institute of Technology, Pasadena, California 91125

DAVIS, M. (34), California Institute of Technology, 156-29, Pasadena, California 91125

DAVIS RONALD W. (5), Department of Biochemistry, Stanford Medical Center, Stanford, California 94305

DAWID, I. B. (15), National Institute of Health, Building 37, Room 4D-07, Bethesda, Maryland 20205

DE BOER, E. (30), University of Amsterdam, Box 60.000, 1005 GA, Amsterdam, Netherlands

DE RIEL, J. K. (32), Department of Medicine and Human Genetics, Yale University School of Medicine, New Haven, Connecticut 06510

DIGAN, MARY ELLEN (18), Department of Biology, Indiana University, Bloomington, Indiana 47401

DODGSON, JERRY B. (33), Department of Chemistry, 164-30, California Institute of Technology, Pasadena, California 91125

DUGAICZYK, ACHILLES (27), 1200 Mourshund Avenue, Baylor College of Medicine, Houston, Texas 77030

EARLY, P. (34), California Institute of Technology, 156-29, Pasadena, California 91225

EARNSHAW, WILLIAM C. (45), Medical Research Council, Laboratory of Molecular Biology, Hills Road, Cambridge, England

EFSTRATIADIS, ARGIRIS (28), Department of Biological Chemistry, Harvard Medical School, Boston, Massachusetts 02115

ELGIN, S. C. R. (46), Harvard University, 16 Divinity Avenue, Cambridge, Massachusetts 02138

EMMONS, SCOTT W. (21), Department of Molecular, Cellular and Developmental Biology, University of Colorado, Boulder, Colorado 80309

ENEA, VINCENZO (35), Center for Cancer Research, E17-517, Massachusetts Institute of Technology, Cambridge, Massachusetts 02139

ENGEL, J. D. (33), Department of Biochemistry and Molecular Biology, Northwestern University, Evanston, Illinois 60201

ENGELKE, DAVID R. (43), Washington University Medical School, 660 South Euclid Avenue, St. Louis, Missouri 63110

FARABAUGH, PHILIP J. (4), Department of Botany, Genetics and Development, Cornell University, Ithaca, New York 14853

FELDENZER, JOHN (31), Columbia University, Room 1602, 701 West 168th Street, New York, New York 10032

FELSENFELD, G. (44), National Institute of Arthritis, Metabolism and Digestive Diseases, Bethesda, Maryland 20205

FILES, JAMES G. (21), Department of Molecular, Cellular, and Developmental Biology, University of Colorado, Boulder, Colorado 80309

FINK, GERALD, R. (4), Department of Botany, Genetics, and Development, Cornell University, Ithaca, New York 14853

FIRTEL, RICHARD A. (20), Department of Biology, University of California, San Diego, La Jolla, California 92093

FLAVELL, R. A. (30), University of Amsterdam, Box 60.000, 1005 GA, Amsterdam, Netherlands

FORGET, BERNARD G. (32), Department of Internal Medicine, Yale University School of Medicine, New Haven, Connecticut 06510

FRIESEN, JAMES (4), York University, Toronto, Ontario, Canada

FRITSCH, E. F. (29), Division of Biology, California Institute of Technology, Pasadena, California 91125

GANNON, F. (26), Institut de Chimie Biologique, 11 Rue Humann, Strasbourg 67085, France

GAREN, ALAN (16), Department of Molecular Biophysics and Biochemistry, Yale University, New Haven, Connecticut 06520

GERLINGER, P. (26), Institut de Chimie Biologique, 11 Rue Humann, Strasbourg 67085, France

GEFTER, MALCOLM L. (49), Department of Biology, Massachusetts Institute of Technology, Cambridge, Massachusetts 02139

GILBERT, WALTER (1, 28), Biological Laboratories, Harvard University, Cambridge, Massachusetts 02138

GROSVELD, G. C. (30), University of Amsterdam, Box 60.000, 1005 GA, Amsterdam, Netherlands

HALL, T. J. (22), Kerckhoff Marine Laboratory, California Institute of Technology, Corona del Mar, California 92625

HARDISON, R. C. (29), Division of Biology, California Institute of Technology, Pasadena, California 91125

HARLAND, RICHARD M. (45), Medical Research Council, Laboratory of Molecular Biology, Hills Road, Cambridge, England

HARRISON, T. (48), Medical Research Council, Virology Institute, Cambridge, England

HARSA-KING, MARY LOU (24), Department of Biology, 16-437, Massachusetts Institute of Technology, Cambridge, Massachusetts 02139

HEILIG, R. (26), Institut de Chimie Biologique, 11 Rue Humann, Strasbourg 67085, France

HENIKOFF, STEVEN (12), Department of Zoology, University of Washington, Seattle, Washington 98195

HERSHEY, N. DAVIS (13), Department of Chemistry, California Institute of Technology, Pasadena, California 91125

HERSKOWITZ, IRA (2), Institute of Molecular Biology, University of Oregon, Eugene, Oregon 97403

HICKS, JAMES B. (3), Cold Spring Harbor Laboratory, P. O. Box 100, Cold Spring Harbor, New York 11724

HILLIKER, ARTHUR J. (11), Plant Industry, CSIRO, Canberra City, Act 2601, Australia

HINNEN, ALBERT (4), F. Mischer Institute, Ciba-Geigy, Basel, Switzerland

HIRSH, DAVID I. (21), Department of Molecular, Cellular, and Developmental Biology, University of Colorado, Boulder, Colorado 80309

HITZEMAN, R. A. (6), Department of Biological Sciences, University of California, Santa Barbara, California 93106

HOEIJMAKERS-VAN DOMMELEN, H. A. M. (30), University of Amsterdam, Box 60.000, 1005 GA, Amsterdam, Netherlands

HONDA, BARRY M. (45), Medical Research Council, Laboratory of Molecular Biology, Hills Road, Cambridge, England

HOOD, LEROY E. (34), California Institute of Technology, 156-29, Pasadena, California 91125

HOWARD, BRUCE H. (52), Department of Biochemistry, Stanford University Medical Center, Stanford, California 94305

ILGEN, CHRISTINE (4), Department of Botany, Genetics, and Development, Cornell University, Ithaca, New York 14853

JAMRICH, MILAN A. (47), Department of Biology, University of Virginia, Charlottesville, Virginia 22901

JUDD, B. H. (10), Department of Zoology, University of Texas, Austin, Texas

KANG, HYEN S. (7), Department of Microbiology, Seoul National University College of National Science, Seoul 151, Korea

KAUFMAN, RANDAL J. (41), Department of Biological Sciences, Stanford University, Stanford, California 94305

KEJZLAROVA-LEPESANT, JANA (16), CNRS Institut de Recherche, Universite Paris VII, Paris, France

KIMMEL, ALAN R. (20), Department of Biology, University of California, San Diego, La Jolla, California 92093

KINGSMAN, A. J. (6), Department of Biological Sciences, University of California, Santa Barbara, California 93106

KLAR, AMAR J. S. (3), Cold Spring Harbor Laboratory, Box 100, Cold Spring Harbor, New York 11724

KLASS, MICHAEL R. (21), Department of Molecular, Cellular, and Developmental Biology, University of Colorado, Boulder, Colorado 80309

KLOBUTCHER, LAWRENCE (38), Yale University School of Medicine, New Haven, Connecticut 06520

KNAPP, GAYLE (7), Department of Chemistry, University of California, San Diego, La Jolla California 92093

KOLODNER, RICHARD (28), Sidney Farber Cancer Institute, Boston, Massachusetts 02115

KOMAROMY, MICHAEL (36), Molecular Biology Institute, University of California, Los Angeles, California 90024

KOOTER, J. M. (30), University of Amsterdam, Box 60.000, 1005 GA, Amsterdam, Netherlands

KORN, L. J. (42), Carnegie Institute of Washington, 115 West University Parkway, Baltimore, Maryland 21210

KRUST, A. (26) Institut de Chimie Biologique, 11 Rue Human Strasbourg 67085, France

KUEHL, MICHAEL (36), Molecular Biology Institute, University of California, Los Angeles, California 90024

KUSHNER, PETER J. (2), Institute of Molecular Biology, University of Oregon, Eugene, Oregon 97403

LACY, E. (29, 39), Division of Biology, California Institute of Technology, Pasadena, California 91125

LAI, CARY (20), Department of Biology, University of California, San Diego, La Jolla, California 92093

LAI, EUGENE C. (27), Baylor College of Medicine,1200 Mourshund Avenue, Houston, Texas 77030

LANE, D. (51), Cold Spring Harbor Laboratory, Cold Spring Harbor, New York 11724

LASKEY, RONALD A. (45), Medical Research Council, Laboratory of Molecular Biology, Hills Road, Cambridge, England

LAWN, R. M. (29), Division of Biology, California Institute of Technology, Pasadena, California 91125

LAWRENCE, PETER A. (17), Medical Research Council, Laboratory of Molecular Biology, Hills Road, Cambridge, England

LEE, A. S. (22), Division of Biology, California Institute of Technology, Pasadena, California 91125

LEE, F. (48), Department of Pharmacology, Stanford University, Stanford, California 94305

LE MEUR, M. (26), Institut de Chimie Biologique, 11 Rue Humann, Strasbourg 67085, France

LE PENNEC, J. P. (26), Institut de Chimie Biologique, 11 Rue Humann, Strasbourg 67085, France

LEPESANT, JEAN-ANTOINE (16), CNRS Institut de Recherche, Université Paris VII, Paris, France

LEV, Z. (22), Division of Biology, California Institute of Technology, Pasadena, California 91125

LIVANT, D. (34), California Institute of Technology, Pasadena, California 91125

LODISH, H. F. (24), Department of Biology, Massachusetts Institute of Technology, Cambridge, Massachusetts 02139

LOMEDICO, PETER (28), Biological Laboratory, Harvard University, Cambridge, Massachusetts 02138

LONG, ERIC (15), National Institutes of Health, National Cancer Institute, Bethesda, Maryland 20205

MAHOWALD, ANTHONY P. (18), Department of Biology, Indiana University, Bloomington, Indiana 47401

MANDEL, J. L. (26), Institut de Chimie Biologique, 11 Rue Humann, Strasbourg 67085, France

MANIATIS, TOM (29, 39), Division of Biology, California Institute of Technology, Pasadena, California 91125

MANLEY, JAMES L. (49), Department of Biology, Massachusetts Institute of Technology, Cambridge, Massachusetts 02139

MANTEI, NED (40), Institut für Molekularbiol I, Universität Zurich, Zurich 8093, Switzerland

MARSH, J. LAWRENCE (19), Department of Biology, University of Virginia, Charlottesville, Virginia 22901

MASCHAT, FLORENCE (16), CNRS Institut de Recherche, Université Paris VII, Paris, France

McDONELL, M. (5), Department of Biochemistry, Stanford Medical Center, Stanford, California 94305

McGHEE, J. (44), National Institute of Arthritis, Metabolism and Digestive Diseases, Bethesda, Maryland 20205

MEARS, J. GREGORY (31), Columbia University, Room 1602, 701 West 168th Street, New York, New York 10032

MILLER JR., O. L. (47), Department of Biology, University of Virginia, Charlottesville, Virginia

MILLS, ANTHONY D. (45), Medical Research Council, Laboratory of Molecular Biology, Hills Road, Cambridge, England

MULLIGAN, RICHARD C. (52), Department of Biochemistry, Stanford Medical Center, Stanford California 94305

MINTZ, BEATRICE (25), Institute for Cancer Research, 7701 Burholme Avenue, Philadelphia, Pennsylvania 19111

MOIR, DONALD (8), Department of Biology, Massachusetts Institute of Technology, Cambridge, Massachusetts 02139

MOORE, G. P. (22), Division of Biology, California Institute of Technology, Pasadena, California 91125

MORATA, GINES (17), CSIC Centro de Biologia Molecular, Universidad Autonoma de Madrid, Canto Blance, Madrid, Spain

NABER STEPHEN (28), Joslin Research Laboratory, Harvard Medical School, Boston, Massachusetts 02115

NEWLON, CAROL S. (3), Department of Zoology, University of Iowa, Iowa City, Iowa 52240

NG, SUN YU (43), Washington University Medical School, 660 South Euclid Avenue, St. Louis, Missouri 63110

NUNBERG, JACK H. (41), Department of Biological Science, Stanford University, Stanford California 94305

OGDEN, RICHARD C. (7), Department of Chemistry, University of California, San Diego, La Jolla, California 92093

O'HARE, K. (26), Institut de Chimie Biologique, 11 Rue Humann, Strasbourg 67085, France

O'MALLEY, BERT W. (27), Department of Cell Biology, Baylor College of Medicine, Houston, Texas 77030

OOYEN, ALBERT VAN (40), Institut für Molekularbiol I, Universität Zurich, Zurich 8093, Switzerland

PARKER, R. C. (29), Division of Biology, California Institute of Technology, Pasadena, California 91125

PEEBLES, CRAIG L. (7), Department of Chemistry, University of California, San Diego, La Jolla, California 92093

PELLICER, A. (39), Institute of Cancer Research, 701 West 168th Street, New York, New York 10032

PERLER, FRANCINE (28), Department of Biological Chemistry, Harvard Medical School, Boston, Massachusetts 02115

PERRIN, F. (26), Institut de Chimie Biologique, 11 Rue Humann, Strasbourg 67085, France

PETERSON, R. (42), Carnegie Institute of Washington, 115 West University Parkway, Baltimore, Maryland 21210

RAMIREZ, FRANCESCO (31), Columbia University, Room 1602, 701 West 168th Street, New York, New York 10032

RAT, LUCE (16), CNRS Institut de Recherche, Université Paris VII, Paris, France

REDDY, V. B. (32), Department of Medicine and Human Genetics, Yale University School of Medicine, New Haven, Connecticut 06510

ROBBINS, A. (51), Department of Biochemistry, University of California, Berkeley, California 94720

ROBINSON, RANDY R. (13), Department of Chemistry, California Institute of Technology, Pasadena, California 91125

ROEDER, ROBERT G. (43), Department of Biological Chemistry, Washington University Medical School, St. Louis, Missouri 63110

ROSENTHAL, NADIA (28), Department of Biological Chemistry, Harvard Medical School, Boston, Massachusetts 02115

RUDDLE, FRANK H. (38), Department of Human Genetics, Yale University School of Medicine, New Haven, Connecticut 06520

ST. JOHN, T. (5), Department of Biochemistry, Stanford Medical Center, Stanford, California 94305

SAKONJU, S. (42), Carnegie Institute of Washington, 115 West Unviersity Parkway, Baltimore, Maryland 21210

SCANGOS, GEORGE (38), Kline Biology Tower, Yale University, New Haven, Connecticut 06520

SCHERER, S. (5), Department of Biochemistry, Stanford Medical Center, Stanford, California 94305

SCHIMKE, ROBERT T. (41), Department of Biological Science, Stanford University, Stanford California 94305

SEGALL, JACQUELINE (43), Washington University Medical School, 660 South Euclid Avenue, St. Louis, Missouri 63110

SHARP, PHILLIP A. (48, 49), Department of Biology, Massachusetts Institute of Technology, Cambridge, Massachusetts 02139

SHASTRY, BARKUR (43), Washington University Medical School, 660 South Euclid Avenue, St. Louis, Missouri 63110

SHEN, C.-K. J. (29), Division of Biology, California Institute of Technology, Pasadena, California 91125

SILVERSTEIN, S. (39), Institute of Cancer Research, 701 West 168th Street, New York, New York 10032

SIM, G. K. (39), Institute of Cancer Research, 701 West 168th Street, New York, New York 10032

SIMINOVITCH, LOUIS (37), Department of Medical Genetics, University of Toronto, Toronto, Ontario, Canada

SMITH, L. DENNIS (23), Department of Biological Sciences, Purdue University, West Lafayette, Indiana 47907

SOLLNER-WEBB, B. (44), National Institute of Arthritis, Metabolism and Digestive Diseases, Bethesda, Maryland 20205

SPRADLING, ALLAN C. (18), Department of Biology, Indiana University, Bloomington, Indiana 47401

SPRITZ, R. A. (32), Department of Medicine and Human Genetics, Yale University School of Medicine, New Haven, Connecticut 06510

STEIN, JOSEPH P. (27), Department of Cell Biology, Baylor College of Medicine, Houston, Texas 77030

STEWART, SUE ELLEN (8), Department of Biology, Massachusetts Institute of Technology, Cambridge, Massachusetts 02139

STINCHCOMB, D. (5), Department of Biochemistry, Stanford Medical Center, Stanford, California 94305

STRATHERN, JEFFREY N. (3), Cold Spring Harbor Laboratory, Box 100, Cold Spring Harbor, New York 11724

STRAUSBAUGH, LINDA D. (14), Biology Department, The Johns Hopkins University, Baltimore, Maryland 21218

STROMMER, J. (33), Molecular Biology Institute, University of California, Los Angeles, California 90024

STRUHL, K. (5), Department of Biochemistry, Stanford Medical Center, Stanford, California 94305

SWEET, R. (39), Institute of Cancer Research, 701 West 168th Street, New York, New York 10032

THATCHER, D. (9), Department of Molecular Biology, University of Edinburgh, Edinburgh, Scotland

THOMAS, T. L. (22), Division of Biology, California Institute of Technology, Pasadena, California 91125

TIZARD, RICHARD (28), Biology Laboratory, Harvard University, Cambridge, Massachusetts 02115

TJIAN, ROBERT (51), Department of Biochemistry, University of California, Berkeley, California 94720

VILLA-KOMAROFF, LYDIA (28), Department of Microbiology, University of Massachusetts, Worcester, Massachusetts 01605

WALL, RANDOLPH (36), Molecular Biology Institute, University of California, Los Angeles, California 90024

WARING, GAIL L. (18), Department of Biology, Marquette University, Milwaukee, Wisconsin 53233

WASSERMAN, WILLIAM J. (23), Department of Biological Science, Purdue University, West Layfayette, Indiana 47907

WEAVER, ROBERT F. (40), Institut für Molekularbiol I, Universität Zurich, Zurich 8093, Switzerland

WEIL, PETER A. (43), Washington University Medical School, 660 South Euclid Avenue, St. Louis, Missouri 63110

WEINBERG, E. S. (14), Biology Department, The Johns Hopkins University, Baltimore, Maryland 21218

WEISSMAN, S.M. (32), Department of Medicine and Human Genetics, Yale University School of Medicine, New Haven, Connecticut 06510

WEISSMAN, CHARLES (40), Institut für Molekularbiol I, Universität Zurich, Zurich 8093, Switzerland

WIGLER, M. (39), Institute of Cancer Research, 701 West 168th Street, New York, New York 10032

WILLIAMS, J. (48), Mellon Institute, Carnegie–Mellon University, Baltimore, Maryland

WILLIAMSON, P. (44), National Institute of Arthritis, Metabolism and Digestive Diseases, Bethesda, Maryland 20205

WILSON, J. T. (32), Department of Medicine and Human Genetics, Yale University School of Medicine, New Haven, Connecticut 06510

WILSON, L. B. (32), Department of Medicine and Human Genetics, Yale University School of Medicine, New Haven, Connecticut 06510

WOLD, B. (39), Institute of Cancer Research, 701 West 168th Street, New York, New York 10032

WOO, SAVIO L. C. (27), Baylor College of Medicine, 100 Mourshund Avenue, Houston, Texas 77030

WOODRUFF, R. (9), Department of Genetics, University of Cambridge, Cambridge, England

WRIGHT, THEODORE R. F. (19), Department of Biology, University of Virginia, Charlottesville, Virginia 22901

WU, CARL (46), Biological Laboratories, Harvard University, Cambridge, Massachusetts 02138

YEN, PAULINE (13), Department of Chemistry, California Institute of Technology, Pasadena, California 91125

INTRONS AND EXONS: PLAYGROUNDS OF EVOLUTION

Walter Gilbert[1]

Department of Biochemistry and Molecular Biology
Harvard University, Cambridge, Massachusetts 02138

ABSTRACT Eukaryotic genes are mosaic structures: regions of DNA bearing the coding information separated by stretches that have no apparent function. I speculate that this interspersion plays an important evolutionary role.

The structure of genes in higher cells is not simple but complex, characterized by the coding information lying in noncontiguous segments along the DNA. In order for such a gene to function, the RNA polymerase first makes a long transcription product, then a splicing enzyme removes portions of that transcript and splices together the remaining segments to form a mature message, which will be translated in the cytoplasm. This genetic structure and process was first detected in adenovirus (1,2), and since then it has been observed for a large number of eukaryotic genes: the various genes for globins (3,4,5), for the immunoglobulin light chains (6,7), for ovalbumin (8,9), for lysozyme (10), for the immunoglobulin heavy chains (11,12), for the cytochrome b gene of yeast mitochondria (13), for some yeast transfer RNAs (14,15), and, as we shall hear later in this meeting, for the genes for preproinsulin (16). However, such a structure was not seen in the analysis of histone genes (17). How can we understand this profound difference in structure between higher cell genes and those to which we are accustomed in bacteria?

The new picture conceives of a genetic region of DNA expressed as a long transcription unit, which through splicing can rearrange and eliminate some of its sequences, in one or several ways, to produce one or several final gene products. That a single protein chain may not lie on a contiguous region of DNA will show up in a genetic analysis. The map positions of mutations that lie in the protein will cluster, the clusters will lie at a distance from each other, and in the

[1]American Cancer Society Professor of Molecular Biology, supported by NIH Grant GM09541.

ISBN 0-12-068350-4

spaces between may fall mutations that block RNA transcription or the necessary processing. Benzer's notion of the cistron, while still having an operational validity, has now become far more complex. Those regions of DNA, that lie within the final gene as defined genetically, but that lie between segments that encode parts of the mature RNA product, we call introns (for intra-cistronic regions) (18,6). The DNA regions that correspond to sequences expressed in the mature message, we call exons, these being both the coding regions translated into protein as well as non-translated regions which appear in the mature RNA gene product. The gene is thus an alternating series of exons and introns, the introns being eliminated from the transcript by splicing. This language is neutral and does not presume that the regions called introns, also known as intervening sequences, have been inserted into the coding region of some preexisting gene. The basic question, as one looks at these genes lying in pieces, is: Were the genes once whole but became separated by the insertion of elements whose purpose is to keep the exons apart? Or were the parts of the genes always separated and then put together by the introns to create the evolving genetic element?

Some properties of introns have emerged over the past year. The transcription unit very often has multiple introns, whose lengths range from as little as ten bases to ten thousand bases, both within the coding region and within untranslated parts of the ultimate message. No convincing structural feature has been seen at the intron-exon boundaries which would in itself completely characterize the splice point. The most general homology identified so far is to a short repeated sequence at the two ends of the intron, a CAGG tetranucleotide, and as observed by Chambon (19), the greatest consistency is that there is a GT at the left boundary of the intron and an AG at the right boundary, although even this has an exception. The intron sequence is generally pyrimidine rich. Although these elements of sequence are suggestive, they are not sufficient for recognition. (See the review by Crick (20) for a discussion of splicing.)

In general the average length of the intron is much greater than that of the exon; the genes are spread across the chromosomal DNA to a much greater extent than the final coding capacity would require. In large part this explains the excess DNA in higher organisms: their genes are not trimmed down efficient structures but are instead long transcription units, five to ten times larger than the aggregate of the exons. In addition, these genes lie apart at distances comparable to their lengths. In principle, a single transcription unit can be read out in different ways, a different subset of sequences, a different set of exons, spliced together to make a gene product. Introns for one reading may be exons for another.

What is the role of the introns? The first possibility that comes to mind is that they are regulatory elements, sequences which must be removed in order for specific gene products to be made. One such notion is that a specific, specialized, splicing enzyme could be used as a control element for a unique gene. Nuclear transcription would run all the time, but, when needed, the control gene product, a splicing enzyme, would call for and create a specific functional sequence. In this picture, of course, the sequences of an intron used as a recognition element should be different from all of the others; one should not expect to find homologies in the intron-exon boundaries. Viruses are the best candidates for such roles: eliminating host splicing and providing a new virus splicing pattern would be an effective way of taking over the cell. An alternative regulatory use might be an obligatory removal property: the introns could be elements which must be removed in order for any message to move from the nucleus to cytoplasm, a role certainly better played by these elements when they occur in the untranslated regions. This last model would envisage both a control and a structural role for the intron RNA seqences. As far as control is concerned, a major lesson of molecular biology has been that if any biochemical process can be used for control, it will be. Furthermore, the relationship between biochemistry and molecular biology is not a one-to-one mapping of biological processes on unique biochemical solutions. There is not one way of replicating DNA but many, differing in the details of the enzymology. There is not one way of controlling genes in bacteria, such as that imagined when only repressors were thought to play a role, but many, all having in common only the property that the recognition of structure at a point of possible biochemical control can permit a genetic element to exercize specific control. Thus I would not be surprised to find that some introns are used as control functions.

A different hypothesis is that the introns are simply adventitious structural features. The fact that they often lie between repeated sequences is reminiscent of the transposing elements of bacteria, either the insertion sequences or the transposons, because when these sequences enter new DNA they create a short repeating seqence, nine bases long in the case of IS1 (21,22), five long for IS2 (23), flanking the intrusion. Although we recognize these insertion elements in bacteria through their inactivation of genes, if the cell had a mechanism to splice between the terminal redundancies of these seqences as they appeared in a message, the DNA element would be invisible in terms of gene expression. Certainly the role of the omega sequence in the mitochondrial RNA (24,25,26) is reminiscent of such an element, and there are suggestions of transposable elements in yeast, if not yet in mammals.

However, introns that are added elements might well appear in different places in duplicate genes, if they add after the original events that caused the gene duplication. Secondly, if the introns are transposable elements, one would expect to find homology between them, yet in families of genes the intronic sequences drift faster than the exonic sequences, with the exception of the regions near the intron-exon borders. In the few cases that have been examined the intronic sequences represent unique DNA (27), consistent with their playing a role that does not depend on the specific nature of their sequences.

Let us pursue the view that the introns exist and are preserved because they play some critical role in evolution. One hypothesizes that the splicing patterns are induced by general enzymes. Given the existence of such preexisting enzymes in the cell, evolutionary arguments propose that the extra DNA carried in the introns is of sufficient use that it would be preserved over long periods. The lack of introns in histone genes, which occur as multiple repeats, would be a consequence of the need for a very high rate of synthesis: under extreme evolutionary pressure extra baggage should be dropped. In the general situation one would imagine a balance between the energy lost in maintaining excess DNA and the freedom to do certain things over evolutionary times. But what are those things?

One possibility, suggested by Philip Leder, is that the introns represent sequences added to structural genes to freeze duplications in place. The role of these sequences in such a model is that after a gene is duplicated, the diversification of DNA sequence permitted by the rapid drift of both intronic and flanking seqences would cut down recombination between the copies, which recombination could lead, by unequal crossing over, to the loss of one of those genes. Such a role for introns would suggest that they should only occur in genes which have been duplicated, members of multi-gene families. This is true for the hemoglobins and the immunoglobulins but not so clearly for ovalbumin and other proteins. Furthermore, to block recombination best and to stabilize a duplicated gene, one might expect introns to arise in different places in such genes rather than in the same places throughout the family, since this last fact argues that they antedate the duplications. This model suggests a negative role, that the introns block recombination in order to slow down gene loss and exchange. The alternative view is that introns play a positive role, that they speed up the search for new gene products either by opening new pathways for gene evolution or by enhancing recombination, which shuffles the parts of genes.

A general, non-specific mechanism involving splicing out long segments of RNA can be of strong evolutionary benefit. Alterations of the splicing pattern, mutations leading to incomplete splicing at a high frequency, provide ways to read out the same DNA region in many different patterns simultaneously. In this manner gene functions can drift without abolishing entirely the function of the preexisting gene. New gene products can be sought while the old gene product is still made at a reasonable rate. There is no necessity for the genes to duplicate before they drift to new functions; the extra DNA required for the new uses, on the intron-exon picture, is imbedded within the DNA of the genes rather than being carried as extra copies. Evolution's constant exploration can take place through subtle modifications of the splicing pattern while maintaining the underlying, still useful structure of the previous gene.

The other evolutionary thought is the realization that expanding the gene in size moves the exons apart along the DNA and thus will permit a higher rate of both legitimate and illegitimate recombination between them. The illegitimate recombination will shuffle the exons from different genes and bring them together in new combinations. It will permit gene duplications, or exon duplications, to fuse, not by an exact match of expressed sequence at the point of fusion but through a more random crossover that causes a duplication of the previous exon anywhere within several thousand bases and puts it on the same transcription unit. Such an event is likely to lead to a low production of the protein containing the duplicated exon sequence and this provides a handle, if this duplication of exon sequence is a benefit, for natural selection to polish the system.

Legitimate recombination between homologous sequences within introns will also play a role. Consider the evolution of a gene made up of introns and exons. A mutation in one exon might produce a better gene product, and that gene would begin to spread by selection through the population. At the same time, a mutation to better performance in another exon might also be spreading through the population. Can these two better exons get together in the same gene? They can by recombination -- and the separation of the two exons by a long intron will increase the amount of recombination between these mutations and thus the ease by which both changes can be brought within a single gene. This process will work best, and thus will select for, those cases in which the exons represent separate functional elements in the final gene product -- since then the mutations to better function will always complement in cis.

Thus, if the introns are used to assort the exons, then the exons should represent useful solutions to the structure/function problem. They should not only represent domains, in the sense of complete structures, but also functional elements of the protein, components that can be used again and again in different proteins, to solve parts of the problem of overall function. This picture arose naturally from the first structures worked out at the DNA level for the light chain of the immunoglobulins (6,7). Here the coding sequence for the final gene product is interrupted by two introns, a short one separating the hydrophobic signal sequence at the amino terminus of the protein between amino acids -4 and -3 and a longer one dividing the gene into two halves which correspond to the structural domains of the immunoglobulin light chain. The immunoglobulin molecule consists of a series of repeated domains derived from a common ancestor. The light chain has two of these; the heavy chain of IgG or IgA has four, while that of IgM contains five. Figure 1 shows the position of the intron on an outline of the X-ray structure of the immunoglobulin light chain. Clearly it separates the two domains.

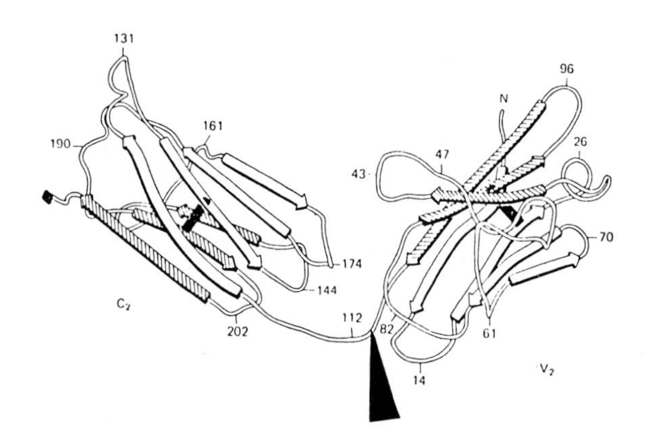

FIGURE 1. Schematic course of the backbone of an immunoglobulin light chain (28).

The initial nineteen amino acids in the coding seqence for the immunoglobulin light chain are a hydrophobic leader sequence. The first sixteen lie on a separate exon. One might wonder why only sixteen amino acids lie separated. Is this a violation of the idea that a functional unit must be carried on the exon? Israel Schechter and his coworkers (29) have shown that these first sixteen amino acids carry a complete hydrophobic signal and the cutting signal, the cutting occurring three amino acids over from the last glycine, irrespective of the intervening amino acids. Thus the region coded on this first exon will serve as a signal sequence if moved to other proteins.

Very recent work in the laboratories of Tonegawa (11) and Lee Hood (12) has demonstrated that the separation of domains by introns is a general property of immunoglobulin heavy chains. Introns separate the three constant region domains CH1, CH2, and CH3 in the cases of the gamma 1 and the alpha heavy chains. Tonegawa's laboratory has gone further and shown by DNA seqencing that a separate exon encodes the hinge region, a sequence of fifteen amino acids that lies between CH1 and CH2 to provide a flexible point in the molecule and to carry those cysteines that form the disulfide connections between the two heavy chains. Thus the immunoglobulins provide a complete example in which the exons code domains and functional elements. In these cases, when we examine the DNA, we find written upon it, in an exon-intron code, an insight into the functional and physical structure of the protein, which we could otherwise only obtain from biochemical and crystallographic analysis.

There are a number of proteins now being studied which have clearly repeating structures, such as serum albumin or ovomucoid; one awaits the correlation of the exons with those structures.

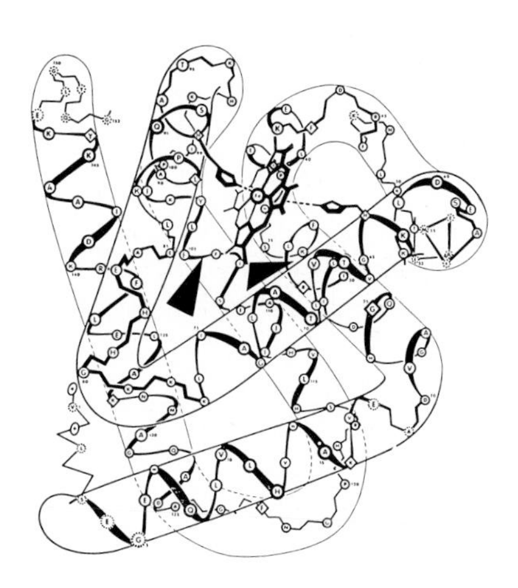

FIGURE 2. Schematic outline (30) of the structure of myoglobin, showing the positions of the introns.

What about the globins? These proteins have a simpler structure; there are no obvious domains to be represented by the three exons into which the genes for the alpha, the beta and the other globins are separated (3,4,5). However, figure 2 shows that the introns lie in alpha-helical runs in the globin molecule. This suggests that the "functional elements," those elements that might evolve separately, are the corners, the turns in the protein structure, exposed to the medium. The two flanking exons code for elements that wrap the product of the central exon, which carries the contacts that determine the heme binding site: the two histidines that coordinate to the iron and a total of 18 out of the 21 amino acids that contact the heme in the beta chain, 15 out of 19 in the alpha chain of globin (31). Figure 3 dissects the molecule to show a view of the central section and the two wings. One might conjecture that the mini-globin encoded in the central exon might provide a sufficient structure to bind the heme in a functional way, it may be a remnant of an early heme carrier.

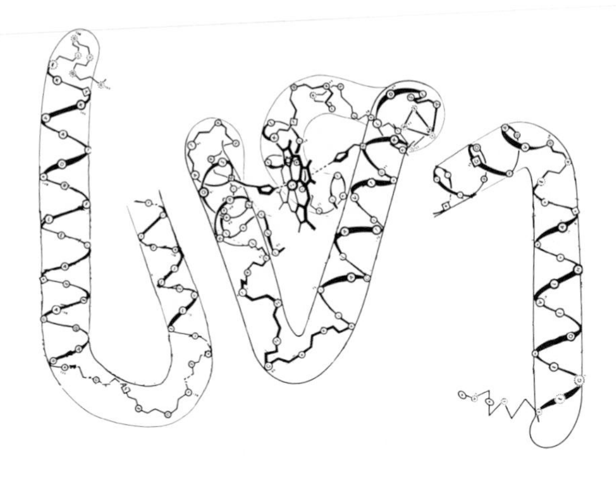

FIGURE 3. The products of the central and flanking exons of globin.

As we will report elsewhere in this meeting, we recently examined the two genes for insulin in the rat (16). Insulin is a small two-chain peptide hormone, whose two chains are connected together by an intermediate peptide in the precursor molecule, proinsulin. The purpose of the intermediate peptide (the C peptide) is simply to hold the chains together so that the disulfide bonds can form easily; at the last stage of maturation, a tryptic-like activity cleaves off the C peptide from proinsulin. Proinsulin itself is also the result of a cleavage; the direct gene product is preproinsulin, a molecule that carries an amino-terminal hydrophobic signal sequence, cleaved as the peptide chain passes through the membrane. The rat has two non-allelic genes for preproinsulin. One of these genes has two introns, one in the C peptide and the other in the untranslated region immediately before the gene. Although we do not see any element that distinguishes the hydrophobic presequence, the intron in the C peptide may divide the gene into portions that indicate the way in which two ancestral units came together. Figure 4 shows the position of the intron on a schematic structure for insulin (32). The inslin molecule consists of two chains lying on top of each other (cross-linked by the disulfides). The C peptide probably lies as a flexible loop down one side of the molecule, exposed to enzymatic attack at its ends. Thus, the intron in the C peptide divides the molecule into its two essential halves.

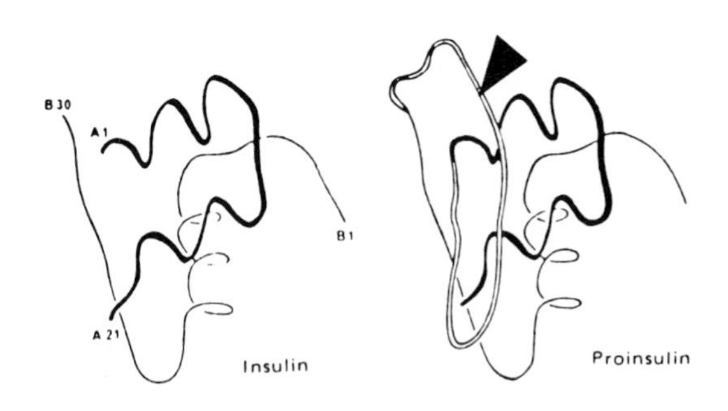

FIGURE 4. The position of the intron in proinsulin.

The second gene for insulin in the rat does not contain the intron in the coding sequence. It is entirely missing; the DNA sequence there running colinearily with the protein sequence. Both genes, however, share the first intron in the 5' untranslated region. What does this tell us? First, since the two genes in the rat function at similar rates, the intron that divides the coding region can not be essential, in cis, for the functioning of one of the genes. Second, this pair of genes demonstrates that in some sense introns come and go. Either the intron in gene II has been added after the gene was duplicated in the rodent family some 40 million years ago, or a precise deletion removed the intron from gene I in the same time span. By examining the insulin genes in other species, we should soon be able to decide which of these alternatives is correct, since these genes will have evolved from, and thus will reflect the structure of, a common ancestor. One hypothesizes that introns will be found to be lost occasionally, as they may have been between the hydrophobic region and the coding sequence of the insulins, and the gene with many introns will be the older form.

How likely is the loss of an intron? If there is no selection pressure, i.e. if the extra DNA in an intron is not an observable hindrance to growth, then an intron should be lost only through neutral drift. For neutral point mutations the rate of fixations is about 1% in 1 to 2 million years (33). Thus a specific base will mutate in 100 to 200 million years. If we examine the introns in globin (34). small additions and deletions occur at about one sixth the rate of base changes: an addition or a deletion at a specific nucleotide every 10^9 years. The frequency of a deletion covering a specific base should be the sum of all possible deletions covering (or ending at) that base, and so the frequency of one deletion, of specified endpoints, would be perhaps 100 to 1000 fold less than that of all deletions. Thus the neutral loss of an intron, by exact deletion, would occur over times of the order of 10^{11} to 10^{12} years, a rare event. Introns should breath in size, as a result of additions and deletions, but not easily vanish.

One might wonder how the splicing enzyme arose. Certainly there would be an evolutionary force for the co-evolution of the splicing enzyme and a spliced gene that provides an essential function. If at the moment that the gene that will make the splicing enzyme first mutates to that function there is in the cell a second gene, spliced by the first, that makes a product selected by evolution, then the two will be selected together. As soon as other genes mutate to produce products formed using the same splicing mechanism, the splicing enzyme would become so essential for the cell that it could not be lost. However, there is an even more general model that endows a strong evolutionary advantage on the splicing mechanism. Suppose a higher cell, in striving to make a number of proteins, has difficulty creating promotors, ribosome sites and all of the paraphernalia of exact protein synthesis. It might transcribe its DNA into long messages, but only manage to read out some limited regions of these messages into proteins. Suddenly a new ability emerges, a weak splicing activity that makes a number of connections and splices within the long transcripts. This provides a combinatorial mixture of expressed regions. The many newly spliced messengers enable the cell in a single mutational step to explore a large number of possibilities.

The ideas discussed here present the gene not as a coding structure interrupted by introns but as an intron background into which the exons have been either inserted by recombination or called forth by mutation. Like bubbles, the exons float on the sea of introns.

REFERENCES

1. Berget, S. M., Moore, C., and Sharp, P.A. (1977). Proc. Nat. Acad. Sci., U.S.A., 74, 3171.
2. Chow, L. I., Gelinas, R. E., Broker, T. R., and Roberts, R. J. (1977). Cell 12, 1.
3. Tilghman, S. M., Tiemeier, D. C., Seidman, J. G., Peterlin, B. M., Sullivan, M., Maizel, J. V., and Leder, P. (1978). Proc. Nat. Acad. Sci., U.S.A. 75, 725.
4. Jeffreys, A. J., and Flavell, R. A. (1977). Cell 12, 1097.
5. Leder, A., Miller, H. I., Hamer, D. H., Seidman, J. G., Norman, B., Sullivan, M., and Leder, P. (1978). Proc. Nat. Acad. Sci., U.S.A. 75, 6187.
6. Tonegawa, S., Maxam, A. M., Tizard, R., Bernard, O., and Gilbert, W. (1978). Proc. Nat. Acad. Sci., U.S.A. 74, 3518.
7. Brack, C., and Tonegawa, S. (1977). Proc. Nat. Acad. Sci., U.S.A. 74, 5652.
8. Breathnach, R., Mandel, J. L., and Chambon, P. (1977). Nature 270, 314.
9. Dugaiczyk, A., Woo, S. L. C., Lai, C. C., Mase, M. L., McReynolds, L., and O'Malley, B. W. (1978). Nature 274, 328.
10. Nguyen-Huu, M. C., Strathmann, M., Groner, B., Wurtz, T., Land, H., Giesecke, K., Sippel, A. E., and Schuetz, G. (1979). Proc. Nat. Acad. Sci., U.S.A. 76, 76.
11. Sakano, H., Rogers, J. H., Hueppi, K., Brack, C., Traunecker, A., Maki, R., Wall, R. and Tonegawa, S. (1979). Nature 277, 627.
12. Early, P. W., Davis, M. M., Kaback, D. B., Davidson, N., and Hood, L. (1979). Proc. Nat. Acad. Sci., U.S.A. 76, 857.
13. Slonimski, P. P., Claisse, M. L., Foucher, M., Jacq, C., Kochko, A., Lamouroux, A., Pajot, P., Perrodin, G., Spyridakis, A., and Wambier-Kluppel, M. L. (1978). In "Biochemistry and Genetics of Yeast" (M. Bacila, B. L. Horecker, and O. M. Steffani, eds.), pp. 339-401. Academic Press, New York.
14. Goodman, H. M., Olson, M. V., and Hall, B. D. (1977). Proc. Nat. Acad. Sci., U.S.A. 74, 5433.
15. Valenzuela, P., Vanegas, A., Weinberg, F., Bishop, R., and Rutter, W. J. (1978). Proc. Nat. Acad. Sci., U.S.A. 75, 190
16. Efstratiadis, A., Lomedico, P., Rosenthal, N., Kolodner, R., Tizard, R., Perler, F., Villa-Komaroff, L., Naber, S., Chick, W., Broome, S., and Gilbert, W. (1979). This

volume.
17. Schaffner, W., Kunz, G., Daetwyler, H., Telford, J., Smith, H. O., and Birnsteil, M. L. (1978). Cell 14, 655.
18. Gilbert, W. (1978). Nature 271, 501.
19. Breathnach, R., Benoist, C., O'Hare, K., Gannon, F., and Chambon, P. (1978). Proc. Nat. Acad. Sci., U.S.A. 75, 4853-4857.
20. Crick, F. (1979). Science 204, 264.
21. Calos, M. P., Johnsrud, L., and Miller, J. H. (1978). Cell 13, 411.
22. Grindley, N. D. F. (1978). Cell 13, 419.
23. Rosenberg, M., Court, D., Shimatake, H., Brady, C., and Wulff, D. L. (1978). Nature 272, 414.
24. Bos, J. L., Heyting, C., Borst, P., Arnberg, A. C., and von Bruggen, E. F. (178). Nature 275, 336-338.
25. Faye, G., Dennebouy, N., Kujawa, C., and Jacq, C. (1979). Molec. Gen. Genetics 168, 101-109.
26. Dujon, B., and Morimoto, R., private communication.
27. Miller, H. I., Konkel, D. A., and Leder, P. (1978). Nature 275, 772.
28. Schiffer, M., Girling, R. L., Ely, K. R., and Edmundson, A. B. (1973). Biochemistry 12, 4620.
29. Schechter, I., Wolf, O., Zemell, R., and Burstein, V. (1979). Fed. Proc. 38, 1839.
30. Dickerson, R. E. (1964). In "The Proteins, Vol. II" (H. Neurath, ed.), p. 634. Academic Press, New York.
31. Perutz, M. F., Muirhead, H., Cox, J. M., and Goaman, L. C. G. (1968). Nature 219, 131.
32. Blundell, T. L., Bedarkar, E., Rinderknecht, E., and Humbel, R. E. (1978). Proc. Nat. Acad. Sci., U.S.A. 75, 180-184.
33. Brown, W. M., George, M., and Wilson, A. C. (1979). Proc. Nat. Acad. Sci., U.S.A. 76, 1967.
34. van den Berg, J., van Ooyen, A., Mantei, N., Schamboeck, A., Grosveld, G., Flavell, K. A., and Weissmann, C. (1978). Nature 275, 37.

MUTATIONS OF THE HMa AND HMα LOCI AND THEIR BEARING ON THE CASSETTE MODEL OF MATING TYPE INTERCONVERSION IN YEAST

Lindley C. Blair, Peter J. Kushner, and Ira Herskowitz

Department of Biology and Institute of Molecular Biology
University of Oregon, Eugene, Oregon 97403

ABSTRACT We have isolated two types of mutations of the HMa and HMα loci of the yeast *Saccharomyces cerevisiae*. One type appears to be a change from HMa to hma (interpreted as a change from silent α to silent a) or from HMα to hmα (interpreted as a change from silent a to silent α). These "mutations" may occur by intrachromosomal recombination or by an error in the interconversion process. The second type of mutations, possibly point mutations in HMa and HMα, allows a test of a key feature of the cassette model of mating type interconversion--that silent information from HMa and HMα is moved to the mating type locus, where it is expressed. We find this prediction to be satisfied--mutations of HMa and HMα are observed to move to the mating type locus, where they exhibit phenotypes like those of known mating type locus mutations.

INTRODUCTION

Yeast haploid cell types, mating types 'a' and 'α', are stable in some strains and unstable in others. In the latter, "homothallic" strains, the mating type of a cell changes as often as every cell division cycle and occurs in a specific pattern [1,2]. As a result, homothallic spores produce clones containing both 'a' and 'α' cells, which mate with each other--a process called "diploidization " [see 3]). Mating type can also change in strains with stable mating type, "heterothallic" strains, doing so at a frequency of approximately 10^{-6} [see 1]. In both homothallic and heterothallic mating type interconversion, the change in mating type of a cell is due to a change in the mating type locus (MAT), which has alleles a (MATa) or α (MATα) [4,5,1]. Homothallic strains differ from standard heterothallic strains in carrying the HO allele [6], which promotes high frequency of mating type interconversion but which is not necessary to maintain a cell's mating type [5,1]. Mating type interconversion is thus an example of a genetic switch promoted by some action of the HO gene.

ISBN 0-12-068350-4

In the cassette model of mating type interconversion [7], switches between 'a' and 'α' occur by inserting genetic blocs, "cassettes", of a or α regulatory information into the mating type locus. All cells contain silent, "library" copies of a and α information (at loci HMα and HMa, discussed below), which become activated when copies of this information are "plugged" into the mating type locus via action of HO (see Figure 1). Cell type in yeast is thus proposed to be determined by the position of the cassette--silent at HMa and HMα, active at MAT. Experiments described elsewhere suggest that the cassettes at HMa and HMα are kept silent by action of Sir and Mar functions [8,9,10,11] and that transposition of information to the mating type locus removes the information from Sir-Mar control.

Two sets of experimental observations figured prominently in formulating the cassette model. First, mutations of the mating type locus ($\alpha 1^-$, $\alpha 2^-$, $a1^-$) are efficiently "healed" by

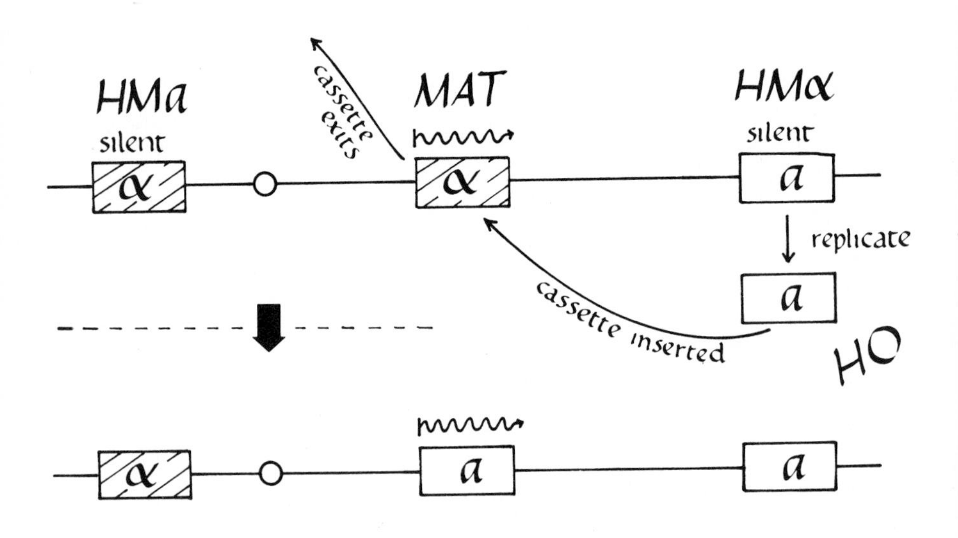

Figure 1. The cassette model. Mating type information, a or α, is expressed at the mating type locus (MAT) near the centromere on chromosome III. Silent (library) copies of a (HMα) and α (HMa) information are situated on the arms of chromosome III. To switch mating type, a replica of a silent copy can be inserted into MAT with concomitant removal of the previous resident information. The new information at MAT can then be expressed. Here a switch from mating type α to mating type a is depicted.

mating type interconversion [12,13,14]. For example, a spore that is α1$^-$ switches efficiently to α+ within three generations of germination; its mating type locus is no longer defective. These results provide the experimental basis for invoking the existence of silent copies of a and α information, which are the sources of functional a and α information in the healing process.

Secondly, genetic analysis by Oshima and colleagues [15] and by Santa Maria and Vidal [16] identified the HMa and HMα loci, which are required for switching from MATa → MATα and MATα → MATa, respectively. HMa was mapped to the left arm of chromosome III, and HMα was mapped to the right arm of chromosome III [17,18]. The natural variant allele hma prevents the MATa → MATα switch, while the natural variant allele hmα prevents the MATα → MATa switch. Subsequent analysis revealed that hma is functionally equivalent to HMα in its ability to switch MATa → MATα, and that hmα is functionally equivalent to HMa in its ability to switch MATa → MATα [17,19,20,21]. Having both HMa and HMα, or both hma and hmα, enables a strain containing HO to switch efficiently in either direction; hma and HMα allow switching only to MATa, while HMa and hmα allow switching only to MATα. These observations support the idea that the hma and hmα alleles are not simply point mutations of HMa and HMα. In summary, as proposed in the cassette model, yeast cells contain three sites at which a or α information can reside: 1) the HMa locus containing silent α information (or silent a for hma) on the left arm of chromosome III; 2) the HMα locus containing silent information (or silent α for hmα); and 3) the mating type locus, either MATaor MATα, located near the centromere on chromosome III, the only site where a or α information is normally expressed. The HMa and HMα loci generally act only as sources for the mating type cassettes while the MAT locus is normally the only recipient for these cassettes.

HMa and HMα loci are thus key protagonists on the mating type interconversion stage. By isolating mutants defective in diploidization, we have addressed two questions concerning HMa, HMα, and the cassette model: (1) How did hma and hmα variant alleles arise? We describe here "mutations" to hma and hmα that we have isolated, and propose that they occur by an error in mating type interconversion or by intrachromosomal recombination. (2) Does information stored at HMa and HMα move to the mating type locus, as predicted in the cassette model? We describe here apparent point mutations in HMa and HMα, and experiments which indicate that the defective informating is transmitted to the mating type locus.

Two methods are used here to identify HMa and HMα alterations. The first is a general screen for mutants defective in diploidization and is expected to identify mutations affecting any of the functions necessary for mating type interconversion (e.g., HO) or mating itself (e.g., genes necessary for mating between siblings). The second is a screen which is expected to identify alterations of HMa and HMα specifically.

ISOLATION OF MUTANTS DEFECTIVE IN DIPLOIDIZATION (METHOD I)

Within a few generations after germination, a clone derived from an a or α HO HMa HMα spore contains both a and α cells, which mate to form a/α diploid cells. Because mating type interconversion is turned off in a/α cells [see 13], these colonies have the phenotype of an a/α cell. In particular, most cells in the colony are unable to mate with either a or α tester cells. Mating proficiency of colonies is determined by ability of the a/α cells and the tester cells to complement each other's nutritional requirements and is done utilizing replica-plating techniques. In contrast, mutants defective in diploidization, e.g. due to a mutation in HO or in HMa or HMα, will maintain an 'a' or 'α' mating type and be identified by their ability to form prototrophs upon mating with the appropriate tester strain.

Approximately 40 mutants which exhibit 'a' or 'α' mating type have been obtained from 15,000 colonies grown from lightly UV-irradiated a and α HO HMa HMα spores. Several classes of mutants have been obtained:

1. Mutations in HO. When crossed to HO HMa HMα spores of opposite mating type, 2 diploidization-proficient (Dip^+) and 2 diploidization-deficient (Dip^-) segregants are obtained in each tetrad. The Dip^- segregants are both a and α. When the mutants are crossed to ho HMa HMα cells of opposite mating type, no diploidizing segregants are observed in a total of 90 tetrads. These results indicate clearly that this class of Dip^- mutants carries alterations in the HO gene. Two of the three mutants of this class do, in fact, exhibit a low frequency of mating type interconversion (higher than in ho strains) and presumably have "leaky" mutations in HO. Further analysis of these mutations will involve determining the number of complementation groups at the HO locus and determining whether the wildtype alleles of the HO mutations are present in standard ho strains.

2. Mutations in genes other than HO necessary for switching by both a and α cells. When the mutant 6B34 is

crossed to HO HMa HMα spores of the opposite mating type, segregants of both mating type result. However, when 6B34 is crossed to ho HMa HMα, HO is found still to be present in 6B34 with a separate mutation, affecting switching in either direction, segregating independently of HO. A mutation in this strain thus appears to identify a gene similar to HO in that it is necessary for both a and α cells to switch mating types. The mutation has also been shown to be in a different complementation group than HO. Diploids not turned off for HO (e.g. a*/α) and heterozygous for HO and the mutation can switch mating type loci to become a/α. A mutation with similar properties (swi1-1) has been reported by Garvik and Haber [22], but it is not known if the mutation in 6B34 and swi1-1 are in the same gene.

3. Mutations affecting HMa and HMα. Strain 2B17 has an 'a' mating behavior despite carrying HO. When crossed to an α hma HMα HO spore, 29 of 30 tetrads gave 2 Dip$^+$ and 2 Dip$^-$ 'a' segregants. (The exceptional tetrad may be due to gene conversion.) These results indicate that 2B17 has an alteration of the HMa locus. Several different kinds of alterations can be imagined: i) a "mutation" to hma, ii) a point mutation in the HMa structural information, iii) a mutation of HMa such that its information cannot be moved to the mating type locus (e.g., because it is defective in a site necessary for copying HMα). The distinctive feature of a "mutation" to hma is that hma is not simply inactivation of HMa; as discussed above, hma appears to be an alteration of the HMa locus such that is is now equivalent to HMα. We have therefore determined whether the HMa-2B17 allele allows mating type interconversion by α HO hmα spores. 2B17 was thus crossed to an α hma hmα HO spore. If HMa-2B17 is hma, all α segregants should be able to diploidize regardless of whether they carry hmα or HMα since all spores are hma. If HMa-2B17 is not hma, then half of the α spores (those receiving hmα and HMa-2B17) should be unable to diploidize and should presumably exhibit an α mating type. No segregants with α mating ability were observed in 55 tetrads, in which approximately 55 'α' mating segregants would be expected if the HMa-2B17 mutations were of type ii or iii. In other words, all segregants which initially carried an α mating type locus were able to diploidize. These results are strong presumptive evidence that HMa-2B17 is hma. A similar analysis has been performed on mutant strain 3B31 which indicates that it has sustained a change from HMα to hmα. The production of hma and hmα alleles by "mutation" has also been reported by Kozhina [23].

An additional type of analysis can distinguish between,

for example, hma and HMa⁻ (type ii). As described more fully below, strains which have a point mutation in HMa are predicted by the cassette model to switch between cells with an a mating type locus and cells with a defective α mating type locus. A clone of a HO HMa⁻ HMα cells would thus contain both a and α⁻ cells, whereas a clone of a HO hma HMα would contain only a cells. We observe that all cells in 2B17 clones exhibit an 'a' phenotype (response to α-factor, see below). Since known point mutations of the α mating type locus do not lead to response to α-factor [24], this observation is consistent with HMa-2B17 being hma and is inconsistent with it being HMa⁻. (The α-factor response of an a HO HMa⁻ HMα strain is described below).

Although we have not carried out analysis to the same extent as for 2B17, presumptive hma mutants appear to be rather common. In a screen of approximately 5000 colonies grown from UV irradiated spores, 16 mutants were found to be "hma-like". These mutants carry HO, grow into colonies giving an 'a' mating response, and all cells in the colony respond to α-factor. No "hma-like" mutants were found among 3000 unirradiated colonies. Because the irradiated spores were grown for one generation before plating, the 16 mutants represent at least 8 independent mutational events. To explain the production of hma "mutations", for example, we propose that a information from either MATa or HMα could replace α information at HMa (see Figure 2). The mechanism might involve either recombin-

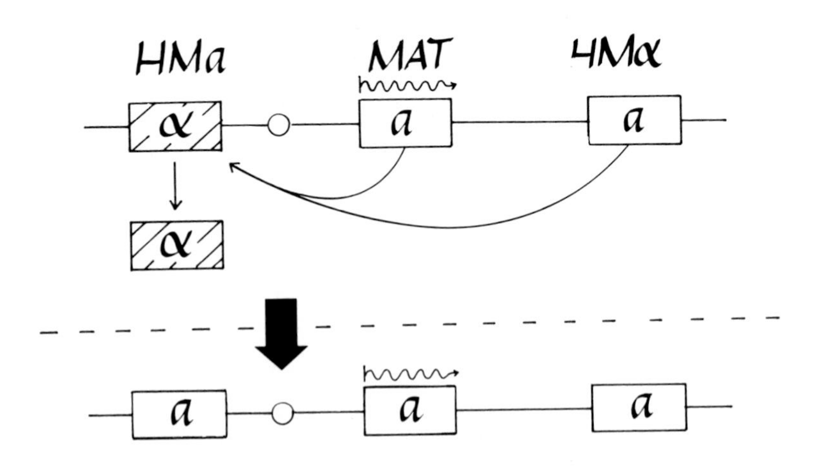

Figure 2. Production of hma "mutations". Information from MATa or from HMα might replace α information at HMa by recombination, aberrant interconversion, or a combination of both.

ation or an error in the mating type interconversion process or both. "Mutations" to hmα could evolve similarly.

ISOLATION OF MUTANTS DEFECTIVE IN HMa AND HMα (METHOD II)

A. Mutations in HMa

1. Isolation and preliminary characterization. We have utilized the natural variant allele of the HMa locus, hma, to develop an isolation scheme for mutations specifically in HMa. The mutants have been isolated in a parental strain which is ho HMa HMα and carries the α1-5 mutation, which reduces mating efficiency [24]. When this strain is mated (by a selected mating) to an a HO hma HMα strain, the resultant diploid segregates HO HMa spores, which give rise to α^+ progeny cells. In contrast, a mutant defective in HMa forms a diploid with the a HO hma HMα strain that should segregate only HMa^- or hma spores, which will not generate α^+ progeny cells.

We have isolated one mutant, strain E66 (carrying mutation HMa66), by looking for derivatives of α1-5 ho HMa HMα with the above behavior. Several lines of evidence indicate that E66 has an alteration of HMa: (1) Spores which are a HO HMa66 HMα are defective in diploidization, giving rise to colonies with 'a' mating behavior; α HO HMa66 HMα spores diploidize normally. (2) The mutation maps to HMa--no HMa^+ recombinants segregate from HMa66/hma diploids in about 100 tetrads. (It is clear that HMa66 is not simply a mutation affecting mating per se since ho HMa66 α or a strains mate normally.) These observations indicate that HMa66 is a mutation affecting HMa function but leave open whether, for example, HMa66 is a change to hma or whether it is a mutation within HMa.

2. Switching pattern of HO HMa66 cells--the "homothallic wounding experiment". The cassette model predicts that strains with mutations within HMa (i.e., within the silent α information) should switch from a to α^-. Indeed, strains which are initially α should switch to a and then to α^-. We tested this prediction by following the cell lineage of a and α spores produced by a diploid homozygous for HO and HMa66 and found that this prediction was satisfied. The a spores gave rise to cells which were resistant to α-factor (and hence no longer a) but which were unable to mate--the phenotype of α^- mutants. Likewise, the α^+ spores gave rise to a cells which subsequently switched to cells with an 'α^-' phenotype: each of the 20 lineages of α^+ cells displayed

"wounding"--$\alpha^+ \rightarrow$ a $\rightarrow \alpha^-$. In no instance did α^+ cells return after the initial switch to a. We note that the α^+ genotype of the initial spores in some of these cell lineages was confirmed by mating unswitched progeny with a tester cells (see Figure 3). The 'α^-' cells were shown to have a mutation at the α mating type locus by crosses (selected matings) to an ho a strain, which yielded ho α^- segregants. All of the α^- mutations recovered in this way (denoted as α-66) exhibit the same phenotype--a temperature-sensitive mating defect, support of sporulation of diploids formed with a ho cells, and healability to α^+ (like other α^- mutations).

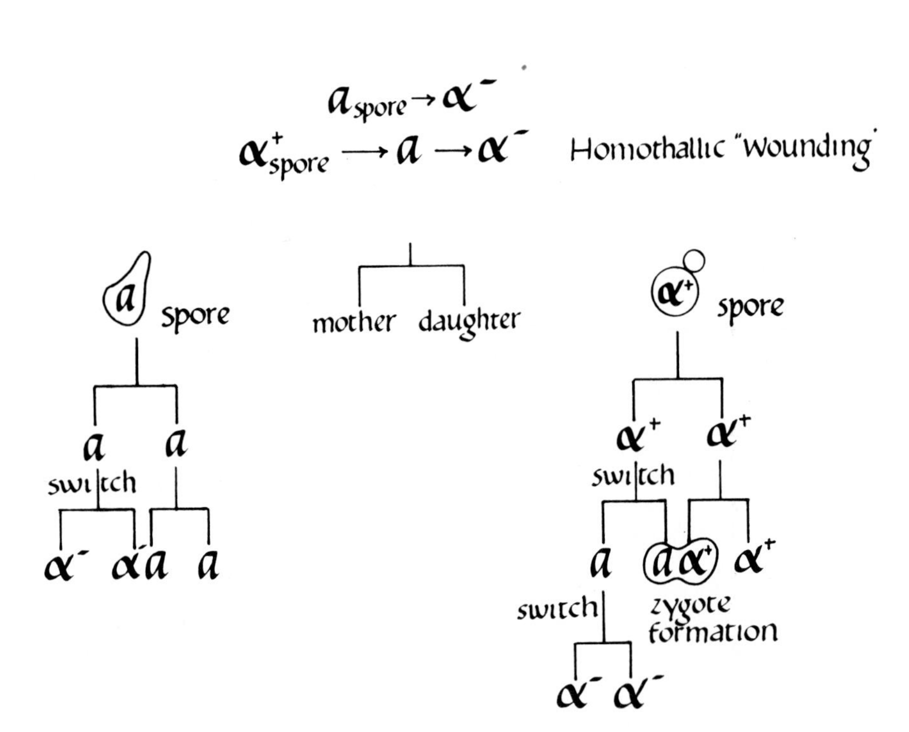

Figure 3. Lineage of HO HMa-66 (silent α^-) spores. See text for explanation of experiments and additional information. In terms of the cassette model, these pedigrees resulting in a $\rightarrow \alpha^-$ and $\alpha^+ \rightarrow$ a $\rightarrow \alpha^-$ switching sequences can be most simply explained by the insertion of defective α information into the mating type locus. The original mutation maps to the HMa locus proposed to contain silent α information. See also Figure 1.

In summary, a HO HMa66 HMα and α HO HMa66 HMα spores can give rise to cells of α^- genotype. The new mutation, α-66, arising in this manner can be segregated into ho backgrounds where it behaves like other α^- mutations. Thus a mutation originally in the HMa locus can be "transmitted" to the mating type locus by action of the mating type interconversion system. The observation that α HO HMa66 HMα spores switch to a and then to α^- but not to α^+ supports the prediction of the cassette model that the information transfer from HMa to the mating type locus is unidirectional. The failure to observe the return of α^+ information also indicates that the exiting cassette is "lost"; it is not stored in the cell in a recoverable form.

B. Mutations in HMα

1. Isolation and preliminary characterization. In order to isolate mutations of HMα, we have utilized the natural variant allele hmα. Mutants have been isolated in an ho HMa HMα strain carrying a mutation of the a mating type locus, a^*, which affects mating little if at all [25]. a^*/α diploids, however, in contrast to a/α cells, are unable to sporulate and do not turn off mating or mating type interconversion [24,12,14]. When an a^* ho HMa HMα strain is crossed to an α HO HMa hmα strain, the resulting diploid switches from a^*/α to a/α and thus to sporulation proficiency [13,14]. A mutant defective in HMα ("$HM\alpha^-$") is expected to be unable to convert to a/α and hence would not give rise to cells able to sporulate. We have isolated several mutant strains using this Spo^- phenotype and here describe two, C11 and C42. The latter mutant exhibits a distinctive "leaky" phenotype after mating with the α HO HMa hmα strain--approximately 1% of the cells sporulate, which indicates that the mutant maintains at least some residual HMα function. In contrast, the parent strain gives approximately 80% sporulation, and the C11 mutant<0.1% sporulation. The mutation responsible for the block in switching from a^*/α to a/α behaves as an $HM\alpha^-$ mutation in that it is recessive to HMα (the diploid formed between C11 or C42 and an α HO HMa HMα strain sporulates), but is not complemented by hmα (the screening criterion--diploids formed between C11 and C42 and an α HO HMa hmα strain do not sporulate).

Two other lines of evidence indicate that C11 and C42 have mutations of HMα (denoted as HMα11 and HMα42). First, the mutations map to HMα: they segregate as alleles of HMα and map very near (approximately 3 centimorgans) to the MAL2 locus, which is on the right arm of chromosome III [3].

Secondly, a HO HMa HMα11 or HMα42 spores grow into clones containing a/α cells, whereas α HO HMa HMα11 or HMα42 spores grow into colonies which mate with both a and α tester strains and which contain few cells able to sporulate. The "bi-mating" ability of α HO HMa HMα11 or HMα42 suggests that the HMα mutations are not hmα, since α HO HMa hmα cells mate only as 'α' [16].

2. Switching pattern of HMα⁻ mutants--homothallic wounding once again. The cassette model predicts that strains with mutations within HMα should convert from α to a⁻. We have tested this prediction in two ways, by following the pattern of phenotypic changes in the cell lineages of HO HMα⁻ spores and by genetic analysis of the product of mating type interconversion by α ho HMa HMα⁻ cells.

Beginning with spores from an a/α diploid homozygous for HO, HMa, and HMα42, we determined the mating type of the spores by their response to α-factor and then removed the spores to agar lacking α-factor to observe their subsequent development. Both the α and a spores gave rise to clones in which zygotes were formed by mating of sibling cells (Figure 4). Thus, both kinds of spores can undergo interconversion, a spores to 'α' and α spores to 'a'. The key question concerning the latter interconversion events is whether or not the 'a' cells generated from α spores have a defective mating type locus. If the mating type locus is functional, then the resulting 'a'/α diploids would be stable, since mating type interconversion is turned off in a/α cells. If these 'a' cells have a defective mating type locus, with a defect like that of a* cells, then the mating type interconversion system would still be active in these cells [13]. Indeed, the zygotes which arose in the clones of the α spores invariably displayed an a⁻/α phenotype, i.e. they continued to produce zygotes formed by mating between siblings derived from the zygote. On the other hand, zygotes formed at the four-cell stage of a spore clones had the normal a/α phenotype (no zygotes formed between cells derived from the zygotes). In a separate experiment, a spores were allowed to switch to α, and zygotes from these α cells were analyzed. These zygotes behaved like the zygotes derived from α spores. These observations indicate that spores from the a/α HO/HO HMa/HMa HMα⁻/HMα⁻ diploid, whether initially a or α, can give rise to cells with a phenotype of an a⁻ mating type locus (Figure 4). We are currently attempting to segregate the mating type locus of such 'a⁻' strains into ho backgrounds to determine whether these strains have a defective a mating type locus. In summary, HO HMa HMα42 cells appear to undergo homothallic "wounding" of the mating type locus, switching from a → α → a⁻.

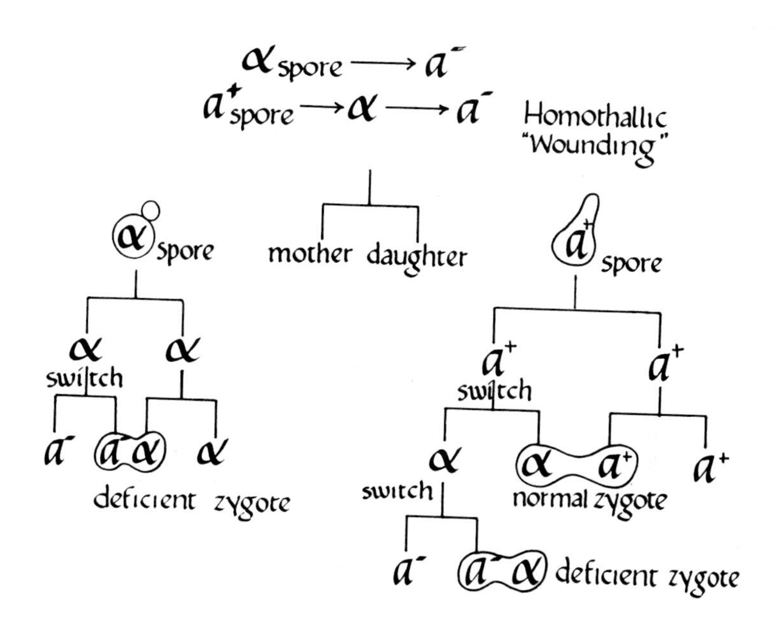

Figure 4. Lineage of HO HMα-42 (silent a⁻) spores. See text and Figure 1 for explanation of experiments and additional information. Analogous to Figure 3, the switching sequences of α → a⁻ and a⁺ → α → a⁻ can be explained by the insertion of defective a information into the mating type locus. Here the original mutation maps to HMα proposed to contain silent a information.

In order to determine whether the information containing the HMα42 mutation can be moved into the mating type locus, we have examined the mating type locus of strains derived from an α ho HMa HMα42 strain which have switched at low frequency to 'a'. In each of three independent switches from α to 'a', we find that the resultant strain when mated to an α ho strain has phenotypic characteristics similar to those of the a* mutation, i.e. failure to support sporulation, turn off mating, and turn off mating type interconversion (when mated to α HO cells). The determinant for this phenotype maps to the mating type locus and is designated as a-42. Each of the α to 'a' interconvertants gives 3-4% sporulation when mated with α ho cells, indicating that this mutation is "leaky" and suggesting that the same mutation is present at the mating type locus in each of the independent interconvertants. In contrast, an ho α HMa HMα11 strain switches to an 'a' strain that is unable to support sporulation when mated with an α ho strain. These studies indicate, at least for HMα42, that mutations within the HMα locus are introduced into the mating type locus as a result of mating type interconversion.

CONCLUSIONS AND THE FUTURE

The experiments described here provide support for a key feature of the cassette model of mating type interconversion—that information coded by the HMa and HMα loci is moved to the mating type locus [7,26]. Another key feature of the cassette model, the equivalence of HMa and HMα information with α and a mating type loci respectively, receives support from studies of mutations that allow *in situ* expression of HMa and HMα [9,10], and from studies of chromosomal rearrangements that appear to allow *in situ* expression of HMa and HMα [7,27]. Cell type in yeast thus appears to be determined by a genetic element, a cassette, whose position in the genome determines whether or not it is expressed. Among the many specific questions that remain, we would like to understand the mechanism by which information from HMa and HMα is "transmitted" to the mating type locus and the mechanism which determines the competence of a cell to switch mating types [see 2]. We would also like to know whether, on a "local" (phylogenetic) scale, a cassette mechanism of gene control is general. For example, can such a mechanism explain mating type interconversion in other yeasts, such as *Schizosaccharomyces pombe*, for which a "flip-flop" model has been proposed [28]. And, finally, can a cassette mechanism account for production of multiple cell types in higher eukaryotic development? Due to recent technical advances (DNA transformation and cloning) and the power of genetics in general, and of yeast genetics in particular, answers to these questions should be forthcoming.

ACKNOWLEDGEMENTS

We thank Kerrie Rine for preparation of the figures and Judy Retherford for preparation of the manuscript. This work was supported by a Training Grant to the Institute of Molecular Biology and by a Research Grant and a Research Career Development Award from the Public Health Service.

REFERENCES

1. Hicks, J. B., and Herskowitz, I. (1976). Genetics 83, 245.
2. Strathern, J. N., and Herskowitz, I. (1979). Cell, in press.
3. Mortimer, R. K., and Hawthorne, D. C. (1969). In "Yeast Genetics" (A. H. Rose and J. S. Harrison, eds.), pp. 385. Academic Press, New York.
4. Hawthorne, D. C. (1963). Proc. 11th Intern. Congr. Genet. 1, 34 (Abstr.).
5. Takano, I., and Oshima, Y. (1970). Genetics 65, 421.
6. Winge, Ö., and Roberts, C. (1949). Compt. Rend. Trav. Lab. Carlsberg., Ser. Physiol. 24, 341.
7. Hicks, J. B., Strathern, J. N., and Herskowitz, I. (1977). In "DNA Insertion Elements, Plasmids, and Episomes" (A. I. Bukhari, J. A. Shapiro, and S. L. Adhya, eds.), Cold Spring Harbor Laboratory, Cold Spring Harbor, New York, pp. 457.
8. Herskowitz, I., Strathern, J. N., Hicks, J. B., and Rine, J. (1977). In "Proceedings of the 1977 ICN-UCLA Symposium" (G. Wilcox, J. Abelson, and C. F. Fox, eds.), pp. 193. Academic Press.
9. Rine, J., Strathern, J. N., Hicks, J. B., and Herskowitz, I. (1979). Genetics, submitted.
10. Klar, A. J. S., Fogel, S., and MacLeod, K. (1979). Genetics, in press.
11. Haber, J. E., and George, J. P. (1979). Genetics, in press.
12. Hicks, J. B., and Herskowitz, I. (1977). Genetics 85,373.
13. Strathern, J. N., Blair, L. C., and Herskowitz, I. (1979). Proc. Nat. Acad. Sci. USA, in press.
14. Klar, A. J. S., Fogel, S., and Radin, D. N. (1979). Genetics, in press.
15. Takano, I., and Oshima, Y. (1967). Genetics 57, 875.
16. Santa Maria, J., and Vidal, D. (1970). I. N. Invest. Agron. (Madrid) 30, 1.
17. Harashima, S., Nogi, Y., and Oshima, Y. (1974). Genetics 77, 639.
18. Harashima, S. and Oshima, Y. (1976). Genetics 84, 437.
19. Naumov, G. I., and Tolstorukov, I. I. (1973). Genetika 9, 82.
20. Klar, A. J. S., and Fogel, S. (1977). Genetics 85, 407.
21. Harashima, S., and Oshima, Y. (1978). Genetics 88, s37.
22. Haber, J. E., and Garvik, B. (1977). Genetics 87, 33.
23. Kozhina, T. N. (1977). 3rd Meeting of Genetical and Selection Society of the U.S.S.R. "Science", Leningrad, 222 (Abstr.).

24. MacKay, V., and Manney, T. R. (1974). Genetics 76, 255.
25. Kassir, Y., and Simchen, G. (1976). Genetics 82, 187.
26. Oshima, Y., and Takano, I. (1971). Genetics 67, 327.
27. Strathern, J. N., Newlon, C., Herskowitz, I., and Hicks, J. B., in preparation.
28. Egel, R. (1977). In "DNA Insertion Elements, Plasmids, and Episomes" (A. I. Bukhari, J. A. Shapiro, and S. L. Adhya, eds.), pp. 447. Cold Spring Harbor Laboratory, Cold Spring Harbor, New York.

CONTROL OF MATING TYPE IN YEAST[1]

James Hicks, Amar Klar, Jeffrey N. Strathern
and Carol S. Newlon[2]

Cold Spring Harbor Laboratory
Cold Spring Harbor, New York, 11724

ABSTRACT The mating behavior of Saccharomyces yeasts is controlled by the alleles of a single mating type locus (MAT). The efficient interconversion of mating types is the result of a reversible genetic alteration of that locus which directs the alternative expression of the MATa and MATα alleles. The molecular basis of the switching event is proposed to involve sequential transposition of DNA copies of MATa and MATα information into the MAT locus from unexpressed or "library" sites elsewhere in the genome. This proposal has been termed the "cassette model" for differentiation. We have documented the cassette mechanism by demonstrating the transposition of mutant genetic information from a silent site to the MAT locus during the interconversion process and by characterizing chromosomal rearrangements which are also associated with interconversion.

[1]This work was supported by NIH Research Grants GM25624 and GM25678, NSF Grant PCM78-07793 and postdoctoral fellowship DRG244 from the Damon Runyon-Walter Winchell Cancer Research Foundation (JNS).

[2]Present address: Department of Zoology, University of Iowa, Iowa City, Iowa 52240

ISBN 0-12-068350-4

INTRODUCTION

The differentiation of cell types during the developmental growth of an organism depends on the expression of a number of genes in a particular pattern organized in time and space. Once triggered, this process proceeds in a cascade of gene expression pre-programmed in the genome. The molecular basis of the regulating circuits which control the development of a complex higher organism is not known. However, simple examples of cell differentiation in lower eukaryotic organisms can provide an intellectual framework for the examination of such processes in higher cells.

The interconversion of mating types in yeast is an example of a simple, yet complete, developmental event in which one complex cell type gives rise to a population of two distinct cell types which then interact to form a third cell type. This novel behavior lead to the "cassette model" for the control of cell type in which it was proposed that switches between cell types were accomplished through the transposition of genetic information (1,2). In this report we present several lines of evidence confirming predictions of the cassette model as applied to the mating type genes of yeast thus verifying the hypothesis of activation of these genes occuring through transposition.

BACKGROUND OF THE CASSETTE HYPOTHESIS:

The mechanics of the mating type system in yeast and the cassette model have been described in detail elsewhere (3,4,5) and will be only summarized briefly here. The mating behavior of *Saccharomyces* is controlled by the mating type locus (*MAT*) which has two alleles *MATa* and *MATα*. Cells expressing *MATa* (mating type *a*) conjugate readily with cells expressing *MATα* (mating type *α*) to create a *MATa/MATα* diploid which does not mate but which is capable of meiosis and sporulation. Meiosis yields four haploid spores, two containing *MATa* and two containing *MATα*. *MATa* and *MATα* are codominant alleles. The phenotype of *MATa/MATα* strains is distinct from that of *MATa/MATa* or *MATα/MATα* homozygotes. Therefore, both alleles contain active genes. One is not simply the absence of the other.

The stability of the *MAT* alleles is influenced by another genetic locus, *HO*. Most laboratory strains carry the recessive *ho* allele. In such so-called heterothallic strains cells of one mating type can switch to the other mating type only at low frequency (approximately 10^{-6}). However in *HO* (homothallic) strains mating type switches may occur as often as every cell division. A haploid *HO* cell of one mating type can divide twice to yield four daughter cells two exhibiting the parental mating type and two of the opposite type. This switching event can be traced to specific cells in the pedigree and occurs in a defined pattern (3,6). The two resultant cell types then conjugate to form the diploid a/α cell type in which both mating and switching (*HO*) functions are shut off. Both types of mating type interconversion are due to alterations of the *MAT* locus and lead to the formulation of the "cassette model" in which changes in cell type were proposed to occur through transposition of *MATa* and *MATα* information to *MAT* from sites elsewhere in the genome.

The cassette model was proposed to account for the following observations:

1) Most laboratory yeast strains can reversibly change mating types either at high frequency (*HO* strains) or low frequency (*ho* strains). Therefore both *MATa* and *MATα* are present in such strains.

2) Mutant cells carrying defective alleles at the *MAT* locus can be "healed" by the switching process. In the presence of the *HO* gene *matα*$^-$ cells can switch to *MATa*$^+$ and thence to *MATα*$^+$ within three generations (7). This "healing" occurs with all *mata*$^-$ and *matα*$^-$ mutations so far tested, (8,9,10) and can also be shown to occur at low frequency in heterothallic (*ho*) strains. Therefore, cells capable of switching must contain *MATa* and *MATα* information in the genome independent of the allele currently expressed at *MAT*.

Thus, under the provisions of the cassette model a change of cell type is viewed as loss of the previously expressed DNA from *MAT* followed by insertion of a copy (cassette) of the *MATa* or *MATα* information stored elsewhere in the genome. The scheme, which is diagrammed in Figure 1 accounts for the reversible alternation of cell type and the "healing" of mutant alleles through switching. Furthermore, we have proposed that the sites of the stored mating type information are the *HMa* and *HMα* loci described below.

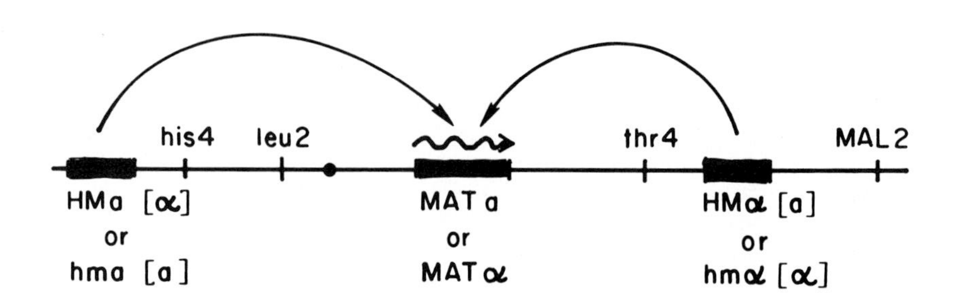

Figure 1. Map of yeast chromosome III and the cassette model for mating type interconversion. Brackets ([]) indicate silent copies of a and α genes.

The sites of the stored MAT information were proposed to be the HMa and HMα loci based on their pattern of involvement in homothallic interconversion. HMa and HMα have been mapped to the left and right arms, respectively, of chromosome III which also contains the MAT locus (11). However, the three loci are not closely linked to one another (Figure 1). There are two naturally occurring codominant variants of each locus: HMa and hma on the left arm; HMα and hmα on the right arm (12, 13). Oshima and co-workers described these loci as coding for "controlling elements" which served to determine the expression of the a and α alleles at MAT (13). However, in order to explain the "healing" results it was proposed by Hicks, Strathern, and Herskowitz (1) that, rather than controlling the expression of genes at MAT these loci actually contain the a and α structural genes.

Cells containing either HMα or hma are capable of switching from α to a (14). Therefore, it was proposed that HMα and hma are unexpressed MATa genes which are the sources of MATa cassettes (1,2). Similarly, because cells containing either HMa or hma can switch from a to α, HMa and hmα were proposed as sources of the MATα cassettes (1,2). This proposed arrangement can be summarized as follows: There are three sites at which MATa or MATα information can reside. A silent locus on the left arm of chromosome III (HMa), a silent locus on the right arm of chromosome III (HMα) and an active locus (MAT) between them at which either MATa or MATα can be expressed. Either MATa or MATα can occupy any of the three loci, although in most laboratory strains the source copy of MATα is located on the left arm (HMa) and the source copy of MATa is on the right arm (HMα) no matter which allele is present and active at the MAT locus.

In this paper we will provide a shorthand interpretation of the position of each cassette in parentheses following the formal genetic nomenclature for each strain to be discussed. For example, a normal laboratory haploid strain of mating type α might be described as: HMa MATα HMα ([α] α [a]). In this convention the cassettes are listed from left to right as they map on the chromosome with brackets indicating silent cassettes. Thus, strains of the genotypes HO HMa HMα ([α][a]) or HO hma hmα ([a] [α]) contain both silent MATα and silent MATa information in addition to the allele expressed at MAT and therefore can switch in both directions. HO hma MATa HMα ([a] a [a]) cells can not switch to α and HO HMa MATα hmα ([α] α [α]) cells can not switch to a because in each case they lack the alternate MAT allele.

In summary, in the cassette model for mating type interconversion the alternate expression of the MATa and MATα alleles is proposed to occur via transposition of normally silent genetic information to the MAT locus where it can be expressed. It should be noted that this transposition event does not alter the genetic content of the silent loci which are the sources of the transposed information. Therefore, we will define transposition in this system as the replication of the source information followed by the substitution of that replicated copy or cassette for the information previously expressed at MAT. The fate of the exiting cassette is not clear but it has not been shown to reappear in the genome and is presumably lost.

The cassette model makes several very strong genetic predictions. One particularly important prediction is that if mutant information were introduced into the silent cassettes at HMa and HMα that information should be faithfully copied into the MAT locus during the switching process and be expressed as a mutant mating type allele. The generation of mutations at the HMα locus and their behavior during switching is described in the following section.

THE MAR1 HYPOTHESIS AND THE ISOLATION OF MUTATIONS OF THE HMa AND HMα LOCI

Recently, a locus designated MAR1 (Mating type Regulator), situated about 27 cM away from trp1 on the left arm of linkage group IV, has been described by Klar, Fogel and MacLeod (15). The mating phenotype of a strain carrying the mar1⁻ allele depends on the arrangement of HMa, HMα and mating type alleles in the same fashion as does homothallic switching ability. That is, a strain of genotype HMa MATα HMα ([α] α [a]) or hma MATα hmα ([a] α [α]) can switch in the presence of HO and is a nonmater in the presence of mar1⁻. On the other hand a strain incapable of switching in the presence of HO, such as HMa MATα hmα ([α] α [α]) or hma MATa HMα ([a] a [a]), behaves as an α or a mater, respectively, in the presence of mar1. It has been suggested that in cells containing the mar1 mutation, all copies of the mating type alleles are expressed including the ordinarily silent information proposed to exist at HMa and HMα. A strain containing both MATa and MATα alleles in any combination thus exhibits the nonmating phenotype typical of MATa/MATα diploids. If only one type of MAT allele is present (e.g., α α α mar1) mating ability is unaffected (Fig. 2). Under the terms of the cassette hypothesis the nonmating phenotype of mar1⁻ HMa MATα HMα (α α a) strains should provide a novel selection for mutations in the a information at HMα since elimination of a should make such a cell mate as an α. Likewise, selection for mating type a cells from a mar1⁻ HMa MATa HMα (α a a) nonmater should yield mutations in the α information expressed at HMa. This scheme is summarized in Figure 2. Using this selection we have obtained a number of spontaneous and ethylmethane sulfonate [EMS] induced mutations at HMa and HMα including amber and ochre nonsense-suppressible mutations. These results are consistent with the interpretation that the silent mating type a and α regulatory information resides at HMα and HMa respectively and that a mar1 mutant allows the expression of this information.

STRAINS CARRYING HMα⁻ MUTATIONS GENERATE CORRESPONDINGLY DEFECTIVE mata⁻ ALLELES:

A critical prediction of the specific cassette model is that strains with mutations at the "storage" loci HMa and HMα should generate correspondingly defective MAT alleles during the switching process. Several spontaneously induced mutations of the HMα locus obtained by the procedure described above have been used to test this prediction. The

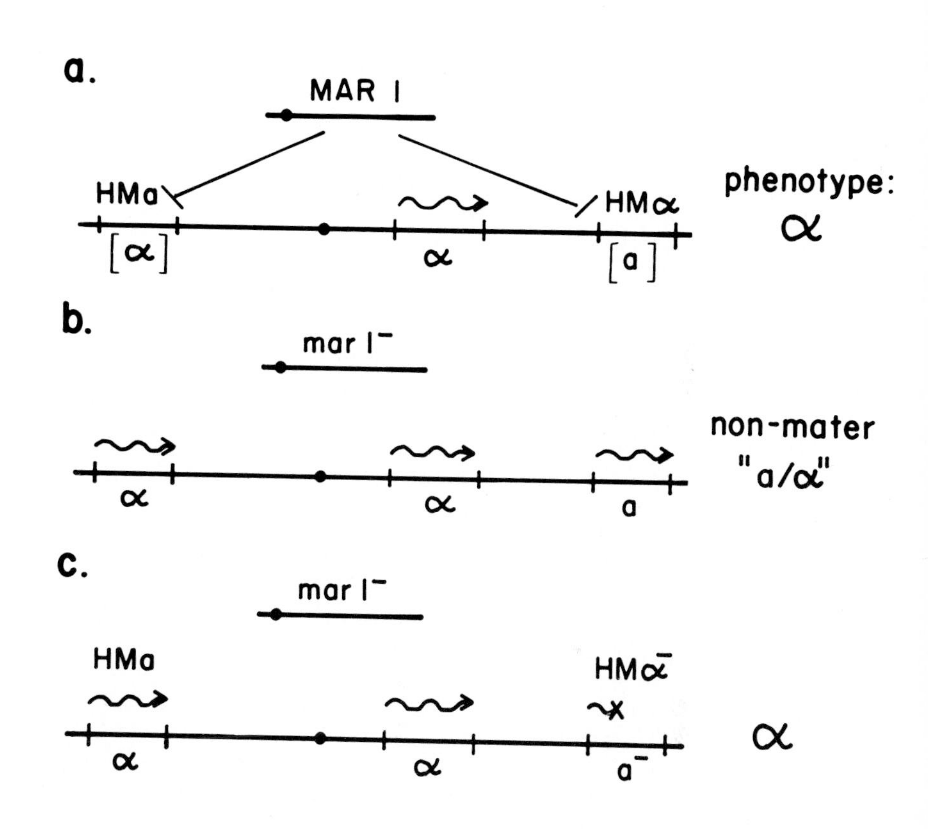

Figure 2. Proposed function of the MAR1 locus and a scheme for selection of mutations in HMα ([a]). The top panel shows the normal control of HMa ([a]) and HMα ([a]). In the middle panel a mutation to mar1 allows expression of HMa and HMα yielding the non-mating "a/α" phenotype. In the bottom panel the mar1⁻ strain has reverted to the "α" phenotype by a mutation in HMα (a⁻).

results indicate that defective $mata^-$ alleles, mapping at MAT, are obtained by switching the MATα locus in $HM\alpha^-$ mutants (8). Blair, Kushner and Herskowitz (16 and this volume) similarly observed that a strain carrying a defective allele at HMa yields defective $mat\alpha^-$ alleles by the interconversion process. We have extended this analysis to strains carrying EMS induced amber and ochre mutations at HMα.

Heterothallic (ho) strains can effect switches at MAT with a low frequency of about 10^{-6} (17,18). These infrequent events are recovered by "rare mating" strains which are otherwise incapable of mating (e.g., a x a, α x α). Strains to be mated in this fashion carry multiple complementary auxotrophic markers and only the hybrids resulting from rare fusions grow on the selective media. The resulting diploids are sporulated and the spontaneous switches of MAT are recovered and identified by tetrad analysis. We employed this technique to recover cells in which MATα switched to MATa in strains carrying nonsense-suppressible mutations at HMα.

In a typical experiment hybrids between standard α ho ([α] α [a]) and α ho strains with $HM\alpha^-$ amber or ochre mutations ([α] α $[a^-_{am/oc}]$) were selected by the rare mating technique. The rare mated hybrids were sporulated and subjected to tetrad analysis. These strains are heterozygous for cryptopleurine resistance (CRY^s/cry^r) and since this marker is closely linked to MAT (19), segregation of this determinant provides an assay for which of the parents had experienced a switch. It was observed that functional MATa alleles are generated only by switches in the standard α strain and that the strains with $HM\alpha^-$ amber and ochre mutations yielded correspondingly amber and ochre $mata^-$ alleles while retaining the nonsense mutation at HMα. The amber and the ochre defects at the $HM\alpha^-$ locus and in the recovered mata alleles were defined by suppression of the mutations with known translational suppressors. A schematic representation of such an experiment is present in Figure 3.

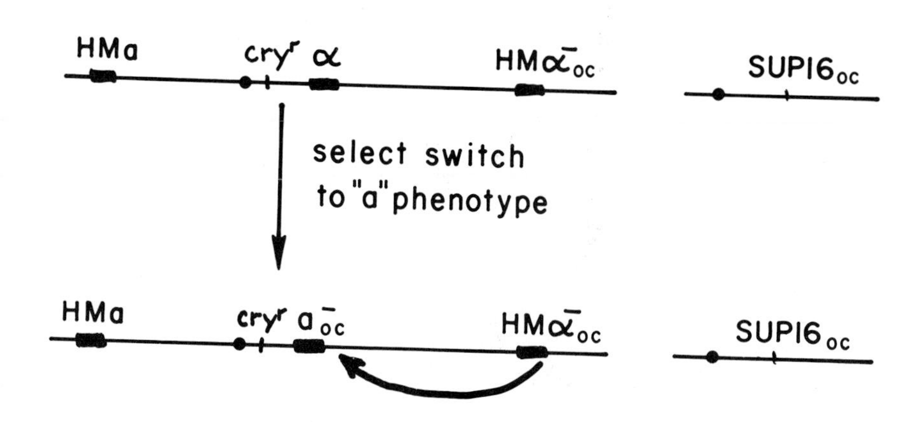

Figure 3. Selection of a mating type switch from α→a in a strain containing an $HM\alpha^-_{am/oc}$ ([$a^-_{am/oc}$]) mutation. After the switch the nonsense version of a is found at both MAT and HMα.

Since the specific mutational defect of the $HM\alpha^-$ allele is reflected in the $mata^-$ allele, it was concluded that as predicted by the cassette hypothesis the information originating from HMα is transposed to the site of the mating type locus due to the process of switching. Identical results were obtained when $HM\alpha^-_{am/oc}$ alleles were introduced into homothallic (HO) strains. Within three divisions spores of genotype HO HMa MATa $HM\alpha^-$ ([α] a [$a^-_{am/oc}$]) gave rise to cells of the HO HMa $mata^-_{am/oc}$ $HM\alpha^-_{am/oc}$ [α] [$a^-_{am/oc}$] [$a^-_{am/oc}$] genotype. These cells then alternated between $MAT\alpha^+$ and $mata^-$ in the subsequent generations and had lost the $MATa^+$ allele (Klar, unpublished results).

ACTIVATION OF SILENT INFORMATION AT HMa AND HMα BY CHROMOSOMAL REARRANGEMENTS

In addition to heterothallic transposition of MAT alleles as described in the previous section and high frequency transposition mediated by the HO gene, it appears that mating type interconversion can also occur by chromosomal rearrangement. This mechanism is clearly distinct from normal heterothallic and homothallic interconversion because it does not involve replication, transposition and substitution of genetic information but rather the reorganization of the chromosome such that the MAT locus is disrupted and the normally silent mating type information is expressed. Evidence supporting such a mechanism is described in this section.

Activation of HMα [a]. It was first observed that cells could switch from α to a by a deletion extending distal from MAT to between thr4 and mal2 by Hawthorne (17). This observation had to be incorporated into any model for the structure of the mating type locus and the mechanism of interconversion and, in fact, led directly to the cassette hypothesis (1,2). The Hawthorne deletion can be routinely obtained from α x α rare matings. These a/α diploids sporulate and yield uniformly two α and two inviable a spores in each tetrad. Thus, the lethal lesion is tightly linked to the MAT locus. The other end of the deletion has been mapped to within 1 centimorgan of the HMα locus (20). In terms of the cassette hypothesis we interpret the mating type switch associated with this deletion as the removal of the MATα information originally present at MAT and activation of the normally silent MATa at HMα by fusion to MAT as diagrammed in Figure 4.

The cassette fusion interpretation has been verified using the amber and ochre nonsense mutants of HMα described in a previous section. Rare matings were performed between a standard α cry^r strain ([α] cry^r α [a]) and α CRY^+ strains carrying $HM\alpha^-$ amber [a^-_{am}] or $HM\alpha^-$ ochre [a^-_{oc}] mutations along with corresponding nonsense suppressors. The segregation of the cry^r marker in these crosses identifies which parent experienced a mating type switch during the formation of the hybrid. Switching events occurring in the $HM\alpha^-$ CRY^+ parent lead to two classes of diploids capable of sporulation. One class yields four viable spores and is the result of a normal interconversion event activating a^- information as described in a previous section. The remaining diploids segregate 2α:2a-lethal spores characteristic of the Hawthorne deletion. In each of the diploids showing lethal segregation it was determined that the lethal event was associated with the activation of the mutant (amber or ochre) a phenotype originally present at HMα. This confirms that the deletion mediated switch is the result of activation of the MATa information located at HMα.

Activation of HMa [α]. We have recently chacterized a to α-lethal switches analogous to Hawthorne's a-lethal which support the identification of HMa as the source of α cassettes. These recessive α-lethal switches were obtained in a x a rare matings which can be selected between strains with complementary requirements. Diploids obtained in this fashion fall into two main categories: 1) normal a/α diploids in which one parent has undergone a switch similar to that observed in homothallic strains; 2) aberrant a/α diploids in which the switch from a to α is accompanied by a recessive

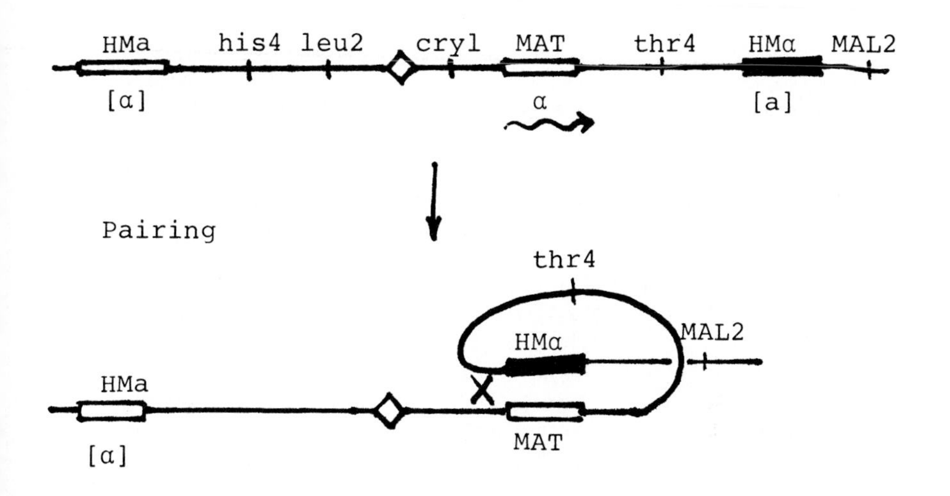

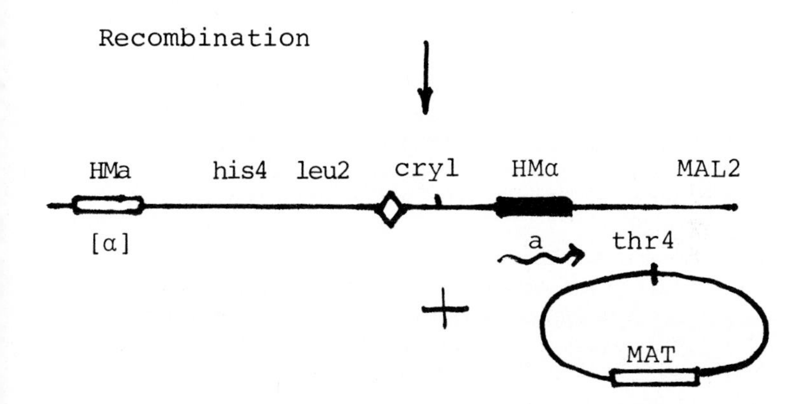

Figure 4. Proposed mechanism for the generation of an α→a switch by the Hawthorne deletion. The wavy arrow denotes an active *MAT* allele but is not meant to imply transcriptional control. Silent mating type information is enclosed in brackets ([ ]) and the centromere is represented by ◆ . Not drawn to physical or genetic scale.

α-lethal inseparable from MAT. Such diploids segregate two viable a spores and two inviable α-lethal spores which can be rescued by mating to a cells.

Mortimer and Hawthorne (personal communication) characterized one such α-lethal as the loss of chromosome III markers to the right of MAT associated with decreased recombination in the his4-leu2-MAT region. In the cassette model this event is viewed as the fusion of MAT and HMa [α] creating a ring chromosome III which simultaneously deletes MATa, activates the α information stored in HMa and results in the loss of the acentric fragments distal of MAT and HMa. Such an event is diagrammed in Figure 5. This interpretation is consistent with several characteristics of strains containing the aberrant chromosome.

1) The aberrant chromosome is unstable and frequently lost during mitotic growth.

2) There is a high frequency of meiotic non-disjunction for chromosome III in these strains giving rise to nonmating (a/α) aneuploid spores.

3) Recombination in the his4 leu2, MAT interval is reduced.

The circular rearrangement hypothesis is based on a recombination event between HMa [α] and the MAT locus. This implies that both HMa and MAT should be absolutely linked to the lethal and should not be rescued by recombination with the linear chromosome. Analyses of such diploids show that while his4, leu2 and cryl alleles can be rescued from the aberrant chromosome, MATα and the MATα [α] information from HMa is not recovered. Therefore, the lethal lesion is tightly linked to both MAT and HMa consistent with the ring chromosome hypothesis.

Physical studies have borne out the prediction of the genetic results. The rearranged version of chromosome III associated with the a→α-lethal switch has been visualized as a 63 μm supercoiled molecule after purification of closed covalent circular (ccc) DNA from an ethidium bromide-cesium chloride density gradient. Furthermore, these molecules have been shown to contain DNA sequences unique to chromosome III by sequence homology with cloned yeast DNA fragments containing the his4 and leu2 structural genes (21).

These results provide the first evidence of a physical interaction between the MAT locus and the silent information at HMa. It is possible that the recombination events which lead to the ring chromosome in the absence of the HO gene are due to sequence homology between portions of the MAT locus and HMa. There may be recognition sequences common to both loci which are important in the homothallic interconversion process. The circular rearrangement may represent the aberrant resolution of an intermediate in the replication-

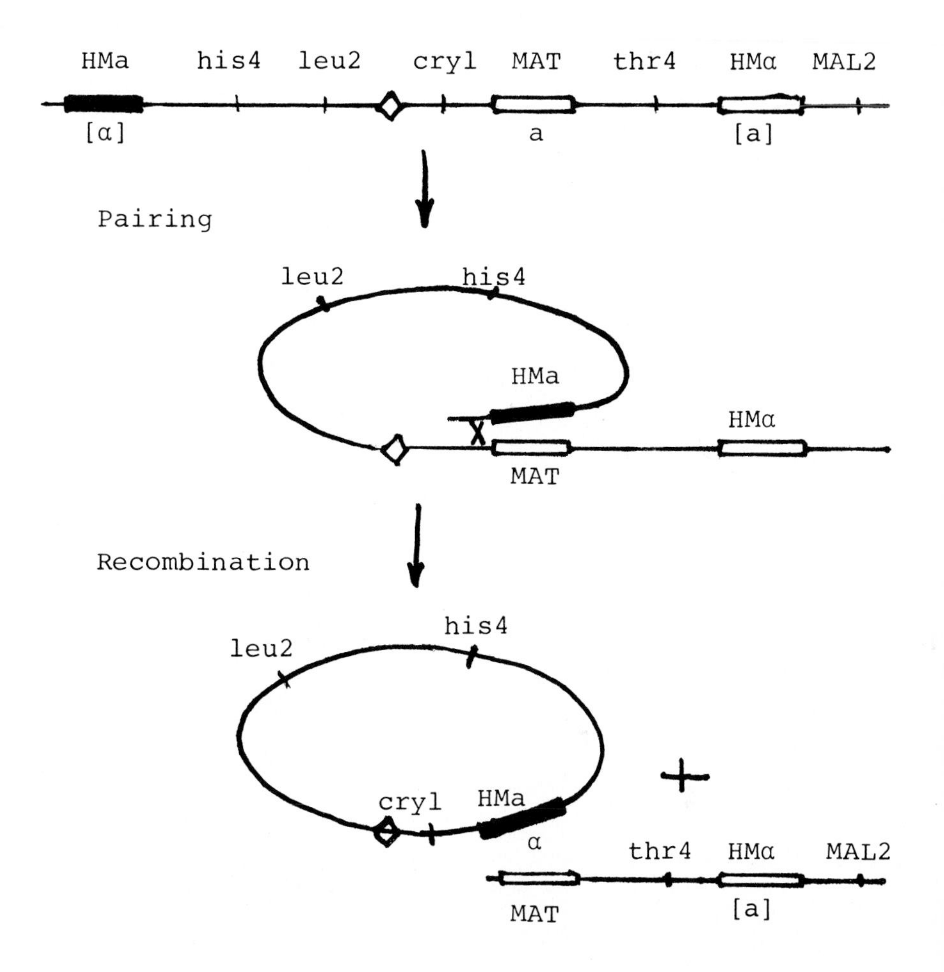

Figure 5. Proposed mechanism for the generation of a ring chromosome concomitant with an a→α switch. After recombination the acentric fragment is rapidly lost. Symbols are as described in the legend to Figure 3.

translocation-substitution process which normally leads to heterothallic switching. In addition, enrichment of the ring chromosome in the ccc DNA fraction provides a means by which a large segment of a yeast chromosome can be isolated intact for analysis of gene organization and provides insight into the structure of the chromosome. The supercoiled structure indicates that the chromosome contains a continuous double-helical DNA molecule extending through the centromere.

CONCLUSIONS

The observations reviewed in this paper represent strong evidence in support of the basic concept of the cassette model, that is, the control of gene expression by transposition of structural gene sequences from storage loci to active sites. The functions normally associated with the alleles of the *MAT* locus (*MATa* and *MATα*) can be expressed at the proposed storage loci (*HMa* and *HMα*) in the *mar1* background. This expression permits the isolation of nonsense-suppressible mutations in the structural genes at these loci and the eventual transposition of this mutant information to the *MAT* locus both at low frequency in heterothallic (*ho*) strains and at high frequency in the presence of *HO*. The normally silent copies of a and α information can also be expressed by chromosomal rearrangements with endpoints mapping at *HMα* and *HMa*, respectively. The molecular details of the cassette mechanism, however, remain to be determined. What is the structure of the mating type locus? How do the active *MATa* and *MATα* alleles differ from their silent counterparts? What is the difference between *MATa* and *MATα*? What is the function of the *HO* gene? How is the silent information duplicated and substituted into *MAT*? What is the fate of the exiting cassette? To answer these questions we have begun to isolate the components of the mating type interconversion system with the eventual goal of reconstructing the switching system *in vitro*. To this end we have cloned a yeast DNA fragment which complements *matα* mutations and used this fragment as a hybridization probe to identify the *MAT* locus on agarose gels. These results will be reported elsewhere.

ACKNOWLEDGEMENTS

We thank Ms. Jean McIndoo for expert technical assistance and Louisa Dalessandro for preparation of the manuscript.

REFERENCES

1. Hicks, J.B., Strathern, J. and Herskowitz, I. (1977). In "DNA Insertion Elements, Plasmids and Episomes." (A.I. Bukhari, J.A. Shapiro and S.L. Adhya, eds.), pp. 457-462. Cold Spring Harbor Laboratory, New York.
2. Herskowitz, I., Strathern, J., Hicks, J. and Rine, J. (1977). In "Proceedings of the 1977 ICN-UCLA Symposium. (A. Wilcox, J. Abelson and C.F. Fox, eds.), Academic Press, New York.
3. Hicks, J.B. and Herskowitz, I. (1976). Genetics 83, 245.
4. Hicks, J., Strathern, J., and Herskowitz, I. (1977). Genetics 85, 393.
5. Crandall, M., Egel, R. and MacKay, V. (1976). In "Advances in Microbial Physiology." (A.H. Rose and D.W. Davenport, eds.), Academic Press, New York.
6. Strathern, J., and Herskowitz, I. (1979). Proc. Nat. Acad. Sci. (USA) (in press).
7. Hicks, J. and Herskowitz, I. (1977). Genetics 85, 373.
8. Klar, A.J.S. and Fogel, S. (1979). Proc. Nat. Acad. Sci. (USA) (in press).
9. Mascioli, D. and Haber, J. (personal communication).
10. Strathern, J., Blair, B., and Herskowitz, I. (1979). Proc. Nat. Acad. Sci. (USA) (in press).
11. Harashima, S. and Oshima, Y. (1976). Genetics 84, 437.
12. Naumov, G.I. and Tolstorukov, I.I. (1973). Genetika 9, 82.
13. Harashima, S., Nogi, Y. and Oshima, Y. (1974). Genetics 77, 639.
14. Klar, A.J.S. and Fogel, S. (1977). Genetics 85, 407.
15. Klar, A.J.S. and Fogel, S., and MacLeod, K. (1979). Genetics (in press).
16. Kushner, P., Blair, L. and Herskowitz, I. (1979). "Control of Yeast Cell Type by Mobile Genes - a test." (in preparation).
17. Hawthorne, D.C. (1963). Genetics 48, 1727.
18. Rabin, M. (1970). Mating type mutations obtained from "rare matings" of cells of like mating type. M.S. Thesis, University of Washington, Seattle, Washington.
19. Grant, P., Sanchez, L., and Jimenez, A. (1974). J. Bacteriol. 120, 1308.
20. Strathern, J.N. (1977). Control of Cell Type in Yeast. Ph.D. Thesis, University of Oregon, Eugene, Oregon.
21. Strathern, J.N., Newlon, C.W., Herskowitz, I., and Hicks, J.B. (in preparation).

ISOLATION OF A YEAST GENE (HIS4) BY TRANSFORMATION OF YEAST[1]

Albert Hinnen,[2] Philip Farabaugh, Christine Ilgen, and Gerald R. Fink

Department of Botany, Genetics and Development, Cornell University, Ithaca, New York 14853

and

James Friesen

York University, Toronto Ontario, Canada

ABSTRACT The his4 gene of yeast was isolated from a collection of pBR313 plasmids containing Bam H1 fragments of yeast DNA by transformation into a yeast his4 deletion. Mutants devoid of his4A, B and C. activities can be transformed to HIS4+ with pYeHis4. The transforming sequences integrate on either side of the recipient his4− region on chromosome III.

INTRODUCTION

The isolation of yeast genes has been greatly facilitated by recent advances in recombinant DNA technology. Most attempts at gene isolation have begun with the construction of hybrid molecules containing yeast DNA fragments inserted into *E. coli* plasmids or bacteriophages (1, 2, 3).

Three approaches have been used for the detection of specific yeast genes in the pool of hybrid plasmids.

1. Complementation of *E. coli* Mutations. Several yeast genes have been isolated because, upon introduction into *E. coli* by transformation, they provide the function missing in the *E. coli* host. Among those yeast genes identified in this way are HIS3+ (1) and LEU2+ (2). Other functioning genes may be detected by immunochemical procedures.

[1]This work was supported by NIH GM15408 and CA23441

[2]Present address: F. Mischer Institute, Ciba-Geigy, Basel, Switzerland.

ISBN 0-12-068350-4

2. RNA-DNA Colony Hybridization. Genes which express abundant RNA's in yeast can be detected in *E. coli* using the purified RNA as a probe. Among the genes identified in this way are the genes for ribosomal RNA (3), tRNA (4, 5), the galactose enzymes etc. Many alternative approaches are possible if the complementary RNA is available. For example, the reverse transcript of a synthetic, partial mRNA for the cytochrome c of yeast has been used as a probe to isolate the CYC1 gene (6).

3. Transformation of Yeast. Genes which fail to function in *E. coli* and do not make abundant transcripts should be expressed when introduced back into yeast by transformation (7).

We used our recently described yeast transformation protocol (7, 8) for the direct isolation of yeast genes by yeast transformation. Specifically, we report here on the isolation of the *his4* gene cluster (9) using the combination of plasmid technology and yeast transformation. A bank of *E. coli*-yeast hybrids consisting of yeast Bam H1 restriction fragments introduced into the *E. coli* plasmid pBR313 has been used to transform a *his4* deletion of yeast.

The method described here is straightforward since it allows the use of already existing plasmid or phage banks. It should be generally applicable for the isolation of any yeast genes for which a strong selection system can be devised.

RESULTS

Isolation of a Plasmid Which Complements a *his4* Deletion. Our isolation of a plasmid containing the *HIS4*$^+$ gene required a stable *his4* mutant which could be transformed at high frequency. Many standard yeast strains transform at very low frequencies. In order to be able to check the transformability of the yeast recipient, we constructed a strain (6657-9B) which contained a *his4* deletion (*his4-34*) in a background containing the *leu2-33 leu2-112* double mutation. This construction allowed us to monitor the transformation frequency by transforming the strain to Leu$^+$ with plasmid pYe*leu*10 (2, 4). The results showed that the transformation frequency of this strain was high enough to permit a selection for *HIS4*$^+$ using the plasmid bank. The chance of encountering a Bam H1 restriction site within the deleted DNA sequence was reduced by selecting a deletion which was shown by genetic analysis to cover only a very small part of the *his4A* region (10).

We used a plasmid bank of yeast Bam H1 restriction fragments constructed by J. Friesen. The Bam H1 yeast fragments from yeast strain S288C were cloned into the Tet^r gene of the E. coli plasmid pBR313. Yeast Bam H1 fragments have an average size of 9500 kb. Thus, a bank of 4800 colonies, at least statistically, covers about 94% of the yeast genome. In addition, the structure of the plasmid allowed us to cut out the inserted DNA from the plasmid and to transform with linear yeast DNA. We have reported recently that linear DNA molecules with free yeast ends transform at a frequency which is about 5- to 20-fold higher than circular DNA (8).

Since the yeast transformation frequency is not high enough to transform with a mixture of plasmid DNA of the whole bank (50-200 transformants per μg linear DNA) we arranged the 4800 clones into one hundred separate pools. The E. coli strain containing the appropriate yeast DNA had to be identified by transforming 6657-9B with each of these sub-pools and looking for the one which gave $HIS4^+$ transformants. The appropriate pool was then sub-divided to find the correct pYeHis4 clone.

The 4800 strains in the clone bank were grown up separately in microtiter dishes to reduce selective growth advantages of individual clones. After overnight growth in L-broth containing ampicillin (500 μg/ml), 48 cultures were pooled, grown up batch-wise, and plasmids isolated from each pool. Each of the 100 plasmid preparations was then used to transform recipient 6657-9B to $HIS4^+$. Once the pool which transformed our recipient to $HIS4^+$ was identified, the 48 clones were sub-divided into smaller pools and eventually a plasmid containing purified $HIS4^+$ was isolated.

pYeHis4 Carries the $HIS4^+$ Gene of Yeast. Two lines of evidence show that plasmid pYeHis4 carries the $HIS4^+$ region of yeast. Previous work (7, 8) established that yeast transformation involves a recombination event between the yeast sequence on the hybrid plasmid and the homologous sequence on the recipient chromosome. That is to say, the yeast sequences on the plasmid "find" the corresponding sequences on the recipient chromosome. This event creates an insertion of the bacterial plasmid sequences flanked by direct repeats of the yeast sequences (see Fig. 1). Thus, one line of evidence that the appropriate segment has been cloned is that it is inserted by transformation at the homologous locus on the yeast chromosome.

FIGURE 1

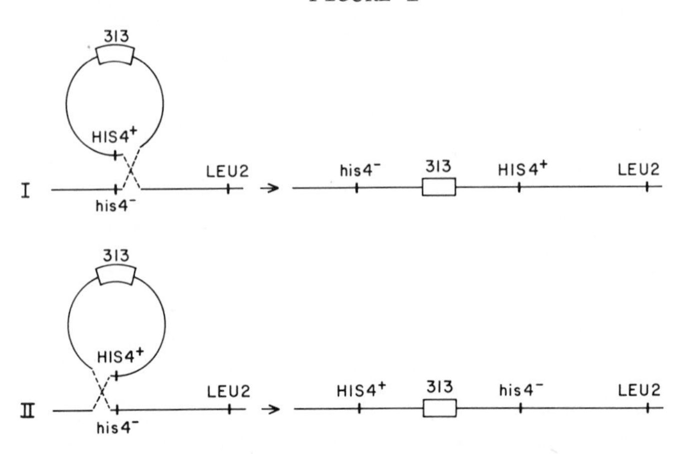

FIGURE 1. Results of transformation by integration of pYeHis4. In orientation (I) the crossover responsible for integration occurs to the right of his4 (between his4 and leu2). In orientation (II) the crossover occurs to the left of the resident his4 region.

Our results show that transformation with pYeHis4 involves the insertion of $HIS4^+$ and pBR313 sequences into the chromosome adjacent to the his4-34 region of the recipient. The $HIS4^+$ region in the transformants can be mapped by standard genetic crosses (11). Table 1 summarizes the results with pYeHis4. Clearly, pYeHis4 inserts at the his4 locus on chromosome III.

TABLE 1
TRANSFORMANTS OBTAINED WITH pYeHis4

% with pBR	% at HIS4	ORIENTATION[a]	
		I	II
78	100	21	14

[a]Orientation I and orientation II refer to the location of the $HIS4^+$ and pBR313 sequences with respect to the recipient $his4^-$ and $leu2^-$ regions as shown in Figure 1. There were 98 transformants examined for chromosome III linkage and pBR sequences. In 35 of these the orientation was determined.

The second line of evidence comes from hybridization experiments using ^{32}P-labeled pYeHis4 as a probe. DNA's from strains carrying various deletions of his4 and as a control, DNA from wild type S288C were cut with a restriction enzyme and the restriction fragments from each DNA preparation were separated on an agarose gel. The DNA was transferred to nitrocellulose by the method of Southern (12) and the resulting filter was hybridized to the labeled pYeHis4. The deletion strains showed predictable alterations in the hybridization pattern as compared to wild type, indicating that various portions of the his4 region were physically missing in the deletion strains. The presence of these sequences in pYeHis4 and S288C and their absence in his4 deletion strains is convincing evidence that pYeHis4 contains the $HIS4^+$ DNA of yeast.

pYeHis4 Contains His4ABC. The his4 region catalyzes three steps in histidine biosynthesis (13), the 3rd, the 2nd, and the 10th encoded by his4A, B and C respectively. To determine whether pYeHis4 contains all three regions, transformation of yeast strains containing various mutations within the his4 region was carried out. Strains carrying each of the mutations shown in Fig. 2 can be transformed to $HIS4^+$ using pYeHis4. His4-912 is a polar mutation which destroys all three his4 activities. By transformation we have obtained a duplication containing 912 pBR313 HIS^+. This strain has a normal doubling time on minimal medium. The transformation of his4-912 to $HIS4^+$ shows that pYeHis4 contains all of the information encoded at $HIS4^+$.

FIGURE 2

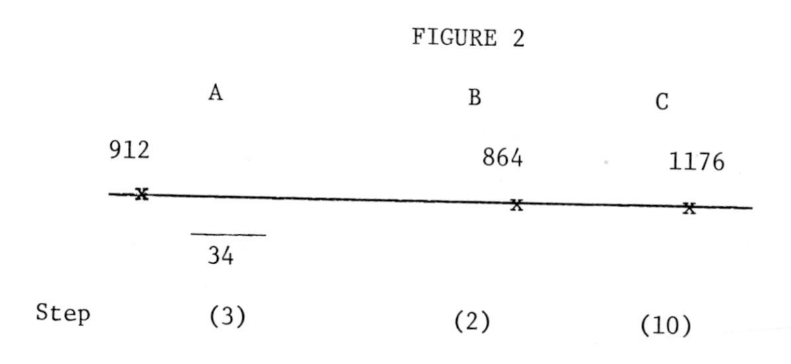

Figure 2. The his4 region of yeast. The sites of mutation and his4-912 ($A^-B^-C^-$), 864 (C^-), 1176 (C^-), and 34 (an A^-) deletion.

Integration of pYeHis4 Occurs on Both Sides of His4. There are two possible orientations of the transforming sequences with respect to the recipient sequences (see Fig. 1). The orientation will be determined by the site of the crossover event which is responsible for the integration of the transforming sequences.

We sought to determine the orientation of the $HIS4^+$ region in a large number of transformants. Using $leu2^-$ as an outside marker the orientation of the two histidine regions can be determined. The order is either I) $his4^-$ $HIS4^+$ $leu2^-$ or II) $HIS4^+$ $his4^-$ $leu2^-$. By examining meiotic tetrads from diploids heterozygous for the duplication and leu2 we could identify crossovers between the $HIS4^+$ and $his4^-$ regions. If the order in I) were obtained, the recombinants would be $leu2^-$; whereas if the order in II) were obtained, the recombinants would be $LEU2^+$. As can be seen in Table 1 there are roughly as many transformants which have orientation I) as those which have orientation II).

DISCUSSION

We have shown that it is possible by transformation into yeast to clone a yeast gene which fails to function in *E. coli*. The reagents necessary for success in this experiment are a recipient yeast strain with a stable, selectable mutation in the desired gene and a DNA bank containing restriction fragments of total yeast DNA cloned into a suitable vector. Suitable vectors are either a bacterial plasmid like pBR313 or 322 or various *E. coli*-yeast hybrid plasmids containing portions of the yeast 2μ plasmid or the yeast trp1 gene.

TABLE 2
DIFFERENT YEAST TRANSFORMATION SYSTEMS

Plasmid	transformants/μg DNA	Stable	Reference
E. coli	1 - 10	+	6
E. coli + 2μ	10,000	-	14, 15, 16
E. coli + trp1	500 - 2000	-	15

As shown in Table 2, the choice of the vector will determine the nature of the transformants. If the vector is only a bacterial plasmid, then the transformation of yeast genes into yeast occurs by homologous recombination. The event usually integrates both the plasmid and the yeast sequences adjacent to the resident sequence (Fig. 1). However, if the transforming DNA has either a piece of the yeast 2μ plasmid or the trp1 gene on it, the transformants do not have the sequences on the transforming DNA integrated into their chromosomes. Rather, these transformants contain the circular plasmid as an autonomously replicating element (14, 15, 16). Such transformants show instability for the transformed marker, and at high frequency segregate clones which have lost the plasmid.

In our experiment we used an *E. coli* clone bank and assayed it by looking for integrative transformation of yeast. It is also possible to use non-integrating plasmids (lines 2 and 3, Table 2) to obtain his4. If the non-integrating plasmids are used, the frequencies of transformation are high enough so that one can transform directly from the reaction mix in which yeast restriction fragments have been joined to the plasmid. Yeast transformants obtained with non-integrating plasmids are unstable since the gene conferring prototrophy remains associated with the unintegrated plasmid. Using these methods virtually any gene from yeast can be isolated.

REFERENCES

1. Struhl, K., Cameron, J. R., and Davis, R. W. (1976). Proc. Natl. Acad. Sci. U.S.A. 73, 1471.
2. Ratzkin, B., and Carbon J. (1977). Proc. Natl. Acad. Sci. U.S.A. 74, 487.
3. Kramer, R. A., Cameron, J. R. and Davis, R. W. (1976). Cell 8, 227.
4. Beckmann, J. S., Johnson, P. F., and Abelson, J. (1977). Science 196, 205.
5. Petes, T., Broach, J., Wensink, P., Herefore, L., Fink, G. and Botstein, D. (1978). Gene 4, 37.
6. Montgomery, D. L., Hall, B. D., Gillam, S., and Smith, M. (1978). Cell 14, 673.
7. Hinnen, A., Hicks, J. B., and Fink, G. R. (1978). Proc. Natl. Acad. Sci. U.S.A. 75, 1929.
8. Ilgen, C., Farabaugh, P. J., Hinnen, A., Walsh, J. M., and Fink, G. R. (1979). Genetic Engineering 1, 117.

9. Bigelis, R., Keesey, J., and Fink, G. R. (1977). ICN-UCLA Symposia on Molecular and Cell Biology, Vol VII, pp. 179-187.
10. Fink, G. R. and Styles, C. (1974). Genetics 77, 231.
11. Sherman, F., Fink, G. R. and Lawrence, C. W. (1974). Methods in Yeast Genetics (Cold Spring Harbor Laboratory, Cold Spring Harbor, N.Y.) pp. 13-14.
12. Southern, E. M. (1975). J. Mol. Biol. 98, 503.
13. Fink, G. R. (1966). Genetics 53, 445.
14. Hicks, J., Hinnen, A., and Fink, G. R. (1978). Cold Spring Harbor Symp. (in press).
15. Struhl, K., Stinchcomb, D., Scherer, S. and Davis, R. W. (1979). Proc. Natl. Acad. Sci. 76, 1035.
16. Beggs, J. (1978). Nature 275, 104.

STRUCTURAL AND FUNCTIONAL ANALYSIS OF THE HIS3 GENE AND GALACTOSE INDUCIBLE SEQUENCES IN YEAST

Ronald Davis, Tom St. John, Kevin Struhl,
Dan Stinchcomb, Stewart Scherer, and Mike McDonell

Department of Biochemistry, Stanford Medical Center,
Stanford, California 94305

ABSTRACT: The His3 gene and galactose inducible sequences from Saccharomyces cerevisiae have been isolated by molecular cloning in E. coli. The regions homologous to the poly-A containing RNA have been mapped for these genes and for the surrounding region. RNA for the his3 gene was found to be inducible by histidine starvation. The galactose inducible region was homologous to three poly-A containing RNAs. The induction was at the transcriptional level and was specific for galactose. The amount of RNA homologous to the regions surrounding the his3 gene were unaffected by histidine starvation. The amount of RNA homologous to the regions surrounding the three galactose inducible sequences were also unaffected by the presence of galactose. Deletions have been isolated in and near the his3 gene which cause loss of expression of the his3 gene in yeast.

INTRODUCTION

Saccharomyces cerevisiae is a simple eukaryotic micro-organism. Numerous genes have been isolated by molecular cloning in E. coli by virtue of their complementation of E. coli auxotrophs (1,2,3,4). However, the number of genes that can be isolated by their expression in E. coli is limited. Molecular clones can also be identified by in situ hybridization with a purified RNA (5,6,7,8). We have recently developed an extension of this approach called differential plaque filter hybridization (9). RNA prepared from two different states of a cell can be used to isolate differentially expressed genomic sequences. By using RNA purified from cells grown on galactose and other carbon sources, we have isolated sequences which are specifically induced with galactose. One of these regions is homologous to three poly-A containing RNAs, and thus is likely to be the gal1,7 and 10 gene cluster on chromosome 2 (10).

ISBN 0-12-068350-4

RESULTS AND DISCUSSION

Structural Analysis of a Galactose Induced Region. λgt-Sc481 DNA contains three EcoRI DNA fragments from a partial digest and is homologous to differentially expressed RNA from yeast (9). These EcoRI fragments make up a total of 7 kilobases of yeast DNA and contain two central SalI sites. Sequences flanking this region were also cloned. The region to the left to the next SalI fragment was cloned in λgt30, as was the region to the right to the next XhoI fragment. The total region that has been cloned is about 24 kilobases. The whole 24 kilobase region has been broken up into a large number of sub-clones. These fragments comprising 1-4 kb intervals were inserted into pBR322 DNA. The RNA in yeast that is homologous to each portion of the 24 kb region can now be identified. The method of characterization is to electrophorese total or poly-A containing RNA on a glyoxal agarose gel (11). The RNA is then transferred and covalently coupled to diazotized paper (12). Subclones which are ^{32}P labeled by nick translation can then be hybridized to these immobilized RNA strips (12). The paper is reusable and the numerous clones can be hybridized in succession to the same strip. Each interval over the entire 24 kb was hybridized to RNA that was extracted from either glucose, galactose or acetate grown cells. This region was found to contain 3 RNAs which are specifically induced for galactose. These 3 RNAs are closely clustered in the central region of the 24 kb and correspond to those sequences cloned in λgt-Sc481. Region on both sides of this cluster of about 2-3 kb are not homologous to any RNA under any of the growth conditions checked. Adjacent to these apparently silent regions are poly-A containing RNAs that are not specifically induced with galactose. There are at least three RNAs to the left and one RNA to the right. The two RNAs immediately to the left and the RNA to the right appear to be constitutive. The RNA furthest to the left is not present in glucose grown cells, and may be under some form of catabolite repression. Because there are three galactose-induced RNAs that are closely clustered, it is likely that this locus is the gal1,7 and 10 gene cluster on chromosome 2. This interpretation is consistent with the observation that sequences homologous to these three RNAs are absent in a strain genetically characterized as a deletion that removes the three galactose-induced genes 1, 7 and 10. The DNA fragment containing the end points of the deleted region has been cloned. The endpoints of the deletion were found to be just outside the galactose-induced sequences. This observation makes it very likely that the cloned region does correspond to gal1,7 and 10. Positive identification of this locus will require transformation of mutant yeast strains as described below.

Structural and Functional Characterization of the His3 region. The his3 gene was cloned in λgt1 by complementation of E. coli his-B mutants (1). The initial clone in λgt-Sc2601 was an EcoR1 DNA fragment of 10 kb. This was divided by subcloning small fragments into pBR322 DNA. RNA homologous to the 10 kb region was identified by the methods described above. It was found that there were 5 polyA containing RNAs homologous to this region. The distance between the RNAs is on the order of a few hundred bases. The his3 RNA was clearly identified by hybridization of a small 1.7 kb subcloned region. This small region is functional in E. coli and yeast and is homologous to a 700 base pair poly-A containing RNA from yeast. Deletion mutants of λgt4-Sc2601 with at least one end-point in the yeast DNA were isolated (13). The end-points of the deletions were mapped by electron microscopy or by hybridization and gel techniques. Deletions with endpoints near the his3 gene were collected for further analysis. It is hoped that deletions can be isolated that remove part or all of the promoter for his3 but not the structural his3 gene. Such deletions would likely have endpoints near the 5' end of the RNA but do not delete past the 5' end of the RNA. A more precise map of the RNAs was obtained by the methods of Berk and Sharp (14). Deletions which were within a few hundred base pairs of the 5' end of the RNA were used for a functional analysis in yeast. By use of the transformation of yeast as developed by Hinnen, et al. (15), the functional expression of these mutants could be assessed. However, homologous integration of the altered sequences into the yeast chromosome could complicate the functional analysis. Recombination could occur between the mutation in the structural gene and a promoter mutation which could give a functional his3 gene. It was therefore essential to develop an autonomously replicating yeast vector. The simplest approach is to find sections of the yeast genome that, when cloned into an *E. coli* vector, will allow the hybrid molecule to autonomously replicate in the yeast cell (16). Such vectors would generally contain an *E. coli* replicator, an *E. coli* selectable marker, a yeast replicator, and a yeast selectable marker. A 1.4 kb *Eco*RI fragment was found that contains both the yeast *trp1* gene and a yeast replicator. When this 1.4 kb fragment is introduced into a plasmid and transformed into *trp1*$^-$ yeast cells the resulting Trp$^+$ cells contain small covalently closed circular DNA molecules homologous to the hybrid DNA. This yeast vector, *YRp7*, was then used for the cloning of the altered *his3* sequences, (as *Bam*H1 fragments). *His3*$^-$ *trp1*$^-$ cells were then transformed with these clones selecting for Trp$^+$. Several of the deletions were found to be *his3*$^-$. It seems likely that these deletions have removed part or all of the *his3* promoter. It is essential to assess whether these altered forms express an RNA. This is complicated because of the mutant chromosomal copy of *his3*. Although this gene is nonfunctional it does produce an RNA molecule homologous to *his3*. Therefore a method was developed for deleting the chromosomal *his3* gene. This method, called transplacement (17), was used to delete the central region of the *his3* gene. It will now be possible to determine whether the deletions with end points near the 5' end of the *his3* RNA do not transcribe the *his3* sequence. If they are defective in the promoter of *his3*, presumably new mutants could be isolated from these altered sequences which re-establish *his3* function. By sequencing the wild-type gene and the resulting mutant forms, it may be possible to achieve a functional analysis of a eukaryotic promoter. Some of the mutant forms may also have altered regulation patterns. A similar analysis is in progress for the galactose-induced sequences.

REFERENCES

1. Struhl, K., Cameron, J. R., and Davis, R. W. (1976). Proc. Natl. Acad. Sci. USA 73, 1471.
2. Ratzkin, B. and Carbon, J. (1977). Proc. Natl. Acad. Sci. USA 74, 487.
3. Hicks, J. B. and Fink, G. R. (1977). Nature 269, 265.
4. Struhl, K. and Davis, R. W. (1977). Proc. Natl. Acad. Sci. USA 74, 5255.

5. Grunstein, M. and Hogness, D. (1975). Proc. Natl. Acad. Sci. USA 72, 3961.
6. Jones, K. and Murray, K. (1975). J. Mol. Biol. 96, 455.
7. Kramer, R. A., Cameron, J. R., and Davis, R. W. (1976). Cell 8, 227.
8. Benton, W. D. and Davis, R. W. (1977). Science 196, 180.
9. St. John, T. and Davis, R. W. (1979). Cell, in press.
10. Bassel, J. and Mortimer, R. (1971). J. Bact. 108, 179.
11. Carmichael, G. C. and McMaster, G. K. (1977). Proc. Natl. Acad. Sci. USA 74, 4835.
12. Alwine, J. C., Kemp, D. J. and Stark, G. R. (1977). Proc. Natl. Acad. Sci. USA 74, 5350.
13. Parkinson, J. S. and Huskey, R. J. (1971). J. Mol. Biol. 56, 369.
14. Berk, A. J. and Sharp, P. A. (1977). Cell 12, 721.
15. Hinnen, A., Hicks, J. B. and Fink, G. R. (1978). Proc. Natl. Acad. Sci. USA 75, 1929.
16. Struhl, K., Stinchcomb, D. T., Scherer, S. and Davis, R. W. Proc. Natl. Acad. Sci. USA, in press.
17. Scherer, S. and Davis, R. W., in preparation.

DETECTION OF *E. COLI* CLONES CONTAINING SPECIFIC YEAST GENES BY IMMUNOLOGICAL SCREENING[1]

Ronald A. Hitzeman, A. Craig Chinault,[2]
Alan J. Kingsman, and John Carbon

Department of Biological Sciences, University of California,
Santa Barbara, California 93106

ABSTRACT Immunological screening techniques have been developed for the identification of specific antigen-producing clones in banks of bacterial colonies containing hybrid plasmids. Method I (indirect) involves covalent attachment of antiserum to cyanogen bromide-activated paper discs, contact of this paper with lysed colonies on agar plates, and finally detection of the bound antigen with ^{125}I-labeled antibody. In Method II (direct), the antigen is picked up directly by binding to CNBr-activated paper discs, and detected by binding to ^{125}I-labeled antibody. The latter method does not require multivalent antigenic determinants and could be useful for the detection of small polypeptide fragments. Using Method I, we have identified several *E. coli* colonies, containing yeast DNA inserts in plasmid ColE1, that produce antigen which combines with antibody directed against yeast hexokinase or against yeast 3-phosphoglycerate kinase (PGK). The 13 kbp yeast DNA insert in plasmid pYe57E2 from a clone producing PGK cross-reacting material has been shown to originate from chromosome III of yeast, by using quantitative Southern hybridizations to DNA from yeast strains aneuploid for this chromosome. This is consistent with previous genetic results (9), which have placed the *pgk* locus near the centromere on chromosome III.

[1]This work was supported by an NIH grant (CA-11034) from the National Cancer Institute and by fellowship support from the NIH (A.C.C.), Damon Runyon-Walter Winchell Cancer Fund (R.A.H.), and Abbott Laboratories (A.J.K.).
[2]Present address: Department of Medicine, Baylor College of Medicine, Houston, Texas 77025.

ISBN 0-12-068350-4

INTRODUCTION

Several methods have been developed for the screening of *E. coli* colony banks to identify specific segments of cloned foreign DNA. If a purified mRNA (or cDNA copy) is available for use as labeled probe, colony hybridization can be used (1). Other methods depend upon expression of the cloned foreign gene, either to produce a functional enzyme which can complement a mutation in the *E. coli* host (2,3,4), or an antigenically competent polypeptide which can be detected by screening the clones with labeled antibody (5,6,7). The latter approach requires only the antiserum made against the purified protein, and, in contrast to complementation selection, does not depend upon synthesis in the bacterial host cell of a complete and functional enzyme.

We have previously described an antibody screening method based on the use of antibody-coated plastic microtiter dishes, to which were added crude extracts of recombinant DNA-containing clones (5,8). Antigen (Ag) present in the extract binds to unlabeled antibody (Ab) adsorbed to the plastic well surfaces, and this bound Ag is then detected by reaction with ^{125}I-labeled antibody. This screening procedure has now been modified to permit the direct detection of antigen produced from bacterial colonies growing on agar plates. This is accomplished using CNBr-activated paper discs to which either unlabeled Ab is covalently attached or the colony protein is covalently attached directly. After lysis of colonies in a top agar layer, Ag producing colonies are detected by either the indirect array of paper-Ab-Ag-(^{125}I)Ab (Method I) or by the direct array of paper-Ag-(^{125}I)Ab (Method II). Although Method I depends upon the synthesis of a multi-valent Ag in the host cell, Method II has the advantage that monovalent polypeptide fragments could be detected.

The application of these methods has enabled us to detect *E. coli* colonies containing cloned yeast DNA segments which produce antigens reactive to antibodies made against yeast hexokinase and 3-phosphoglycerate kinase (PGK). The cloning of the yeast *pgk* gene, which is centromere-linked on chromosome III (9), is an essential step in our studies on yeast centromeric DNA (5,10).

METHODS

Proteins Used. Yeast hexokinase (D-hexo-6-phosphotransferase) was obtained from Sigma as type C-300 containing 495 units/mg. Yeast phosphoglycerate kinase (3-phospho-D-glycerate-1-phosphotransferase) was obtained from

Sigma as type IV containing 2640 units/mg protein. Yeast PGK was further purified by Sephadex G100 chromatography, and gave a single sharp band on SDS-acrylamide gel electrophoresis.

Yeast Bank Construction. The yeast bank was made in *E. coli* strain JA221 (*hsdM*$^+$ *hsdR*$^-$ *lacY leuB6 ΔtrpE5 recA1*) by transformation with annealed recombinant plasmids consisting of ColE1 vector (*Eco*RI cut) with poly (dT) "tails" and randomly sheared segments of yeast DNA with poly(dA) "tails" (5,10,11). About 8000 individual colicin E1-resistant colonies were picked, of which 70-75% contained hybrid plasmids (15 kbp average insert). These were stored in 96-well microtest II dishes (Falcon Plastics) at -80°C in the presence of 8% dimethylsulfoxide.

Antibody Preparation. Antiserum was made to the above yeast hexokinase and PGK preparations by standard procedures. A portion (1-3 ml) of the antiserum (50% $(NH_4)_2SO_4$ cut) was further purified by affinity chromatography as described by Shapiro *et al.* (12) to obtain purified antibody for labeling with $Na^{125}I$ (0.5 mCi per 10 μg antibody) to a specific activity of 10^6-10^7 cpm/μg by the lactoperoxidase method of David and Reisfeld (13).

Preparation of CNBr-Activated Paper. This paper was prepared for the covalent attachment of protein by a modification of the method used by Axén and Ernback (14) to produce cyanogen bromide-activated agarose or cellulose powder for the covalent attachment of enzymes. Twenty Whatman #40 (ashless), 9 cm diameter paper discs were washed in a 1 liter beaker with 400 ml of 0.1 M $NaHCO_3$ followed by a H_2O wash. The wet discs were placed in a 2 liter beaker containing 12 g of cyanogen bromide (all work with this compound must be done in a fume hood) dissolved in 480 ml of H_2O at 23-25°C. The reaction was started by the addition of 2 N NaOH to obtain a pH of about 11. The pH should be maintained near 11 by the occasional addition of 2 N NaOH while swirling the beaker vigorously by hand for 8 min. The reaction was stopped by decanting the CNBr solution and the discs were washed with two 500 ml portions of 0.1 M $NaHCO_3$. Five filters at a time were then washed by suction using 250 ml of 0.1 M $NaHCO_3$, 250 ml of H_2O, 200 ml of 50% acetone in H_2O, and 50 ml of acetone. The discs were vacuum dried and stored over dessicant at 4°C.

Binding of Antibody to Treated Paper Discs. Thirty to 50 μl of the $(NH_4)_2SO_4$ fraction of antiserum in PBS (10 mM sodium phosphate, 0.14 M NaCl, pH 7.0) was mixed with 3 ml of

20 mM sodium phosphate (pH 6.8) in a glass Petri dish. A single CNBr-activated paper disc was put in each dish and left at 4°C for at least 18 h. The paper discs were then removed and put into a deaerated BSA/glycine solution containing 0.1 M glycine, 0.5% BSA, 0.14 M NaCl, and 10 mM sodium phosphate at pH 7.0 (5 ml/disc) for 6 h at 37°C. The discs can be used at this point, left overnight at 4°C, or frozen at -20°C for later use. Immediately before use, the discs were soaked with a solution (5 ml/disc) containing 5% calf serum in PBS (CS/PBS) for 30 min, followed by suction wash of each disc with 25 ml of CS/PBS. Finally, they were kept in a solution (4 ml/disc) of CS/PBS until use.

RESULTS

Detection of Clones Producing Antigen that Reacts with Antibody Made to Yeast Hexokinase. Antigen-producing clones, from an *E. coli* bank containing ColE1 with sheared and "tailed" yeast DNA inserts, were detected using yeast hexokinase-antiserum covalently attached to CNBr-activated paper (Fig. 1). Positive spots are dark and are produced by the array of paper-Ab-Ag-(^{125}I)Ab. Fifteen such positives were obtained screening a bank of 6240 colonies. The relative intensities of the spots were consistently the same upon rescreening.

Cell lysis using chloroform vapors followed by application of top agar containing lysozyme and SDS was used to release protein from colonies. All three components contribute to this release of protein. Without chloroform, this release was decreased about 5-fold as detected by the antibody-containing paper. The absence of SDS resulted in a 2-fold decrease in sensitivity; while higher concentrations (3x and 5x) showed slight inhibition. SDS may be important for the release of antigen that is associated with membranes or other proteins, thus affecting its diffusion through the top agar. The agar concentration can be decreased to about 0.3% if the antigen does not diffuse well through the top agar (however, the protein made in *E. coli* might be quite different than protein made from the gene in its normal background). The absence of lysozyme also resulted in a 2-fold decrease in sensitivity, and concentrations 5x that used in Fig. 1 did not improve the sensitivity.

The plasmid DNAs were isolated from nine hexokinase-positive clones. Transformation of strain JA221 with these plasmid DNAs resulted in colicin-resistant strains that carry the plasmids. Five of each of these plasmid transformants were screened for Ag production and all were positive. Since

JA221 and JA221/ColE1 do not produce this Ag, it can be concluded that the effect is mediated by the yeast DNA inserts in the plasmids.

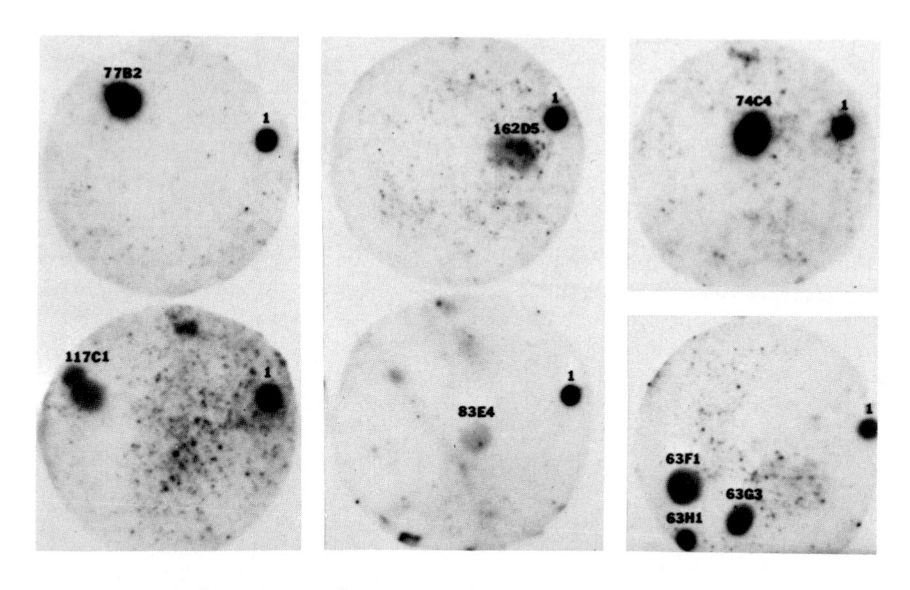

FIGURE 1. Detection of colonies producing antigen that reacts with antibody made to purified yeast hexokinase (screening Method I). Forty-eight bank colonies were transferred from a microtest dish onto each S-agar plate (32 g tryptone, 5 g NaCl, 15 g Difco agar, and 0.2 g NaOH in 1 liter of solution). After two days growth at 37°C, the plates were inverted for 30 min over 7 cm 3 MM Whatman paper discs each containing 1 ml of chloroform. One hour after removing the plates, a top agar layer was applied to each plate, containing 1.5 ml top agar (1 g tryptone, 0.8 g Difco purified agar, and 0.8 g NaCl per 100 ml, pH 6.8), 100 μl 5 mg/ml SDS, and 100 μl 12.5 mg/ml lysozyme in PBS (added immediately before plating). Each plate was spotted with a 1 μl control containing 1 ng of purified antigen in CS/PBS. The paper discs containing covalently bound antiserum were blotted on paper towels and applied to the plates for 9 h at room temperature. Four were removed at a time and stirred gently for 1 h in 100 ml of CS/PBS in a one liter beaker. Each disc was washed by suction with 200 ml PBS and four discs were again put into each beaker with 50 ml CS/PBS. The CS/PBS was then replaced with 4 x 10^6 cpm of ^{125}I-labeled antibody in 10 ml of a solution containing 25% calf serum, 10 mM Tris (pH 7.5), and 0.14 M NaCl. After swirling for 5 h at room temperature, the ^{125}I-containing solution was replaced with 50 ml of CS/PBS. After

Comparison of the Indirect (Method I) and Direct (Method II) Screening Procedures. Since Method I detects the Ag by an indirect "sandwich" (paper-Ab-Ag-(^{125}I)Ab) and requires at least divalent antigenic determinants, it would be desirable to modify the system to permit detection of monovalent antigens. This would increase the applicability of the method, since expression of a foreign gene in *E. coli* may be altered to the extent that only relatively small polypeptides with single antigenic determinants are produced.

The direct method (II) was developed using the hexokinase-positive clones detected by Method I. Method II detects positive colonies by the array of paper-Ag-(^{125}I)Ab. This method is made possible by the high binding potential of CNBr-activated paper (20% protein by weight (15)) and again depends on the diffusion of protein through a top agar layer. Fig. 2 shows a direct comparison of the two methods for the same hexokinase-positive colonies. It can be concluded that Method I (2a) and Method II (2b) have similar sensitivities.

The greatest efficiency of release of colony protein required all three lysis components. The chloroform treatment is absolutely essential, without it the intensity of spots decreased about 10-fold. If both SDS and lysozyme were left out of the top agar, intensity dropped 3- to 5-fold. The concentrations of SDS and lysozyme used in Fig. 2 were optimal; however, much higher concentrations of each did not appear to be inhibitory. Finally, it should be noted that NaCl was left out of the top agar, due to its inhibitory effect on the intensity of spots.

Detection of a Clone That Reacts with Antibody Made to Yeast Phosphoglycerate Kinase. Another glycolytic enzyme from yeast, 3-phosphoglycerate kinase, with a molecular weight of about 50,000 (16), has been used to do immunological screening by Method I. The genetic locus specifying PGK is of particular interest since yeast mutations resulting in lack of this enzyme map very close to the centromere on chromosome III (9, and Mortimer, R.K., unpublished results). Fig. 3 shows one clone that was detected by Method I, using antibody made to purified PGK, from a screening of 4800 bank

---

30 min the discs were each washed by suction using 200 ml of PBS and finally soaked for 30 min in a pan containing PBS. The discs were exposed for two days at -70°C to Kodak XRP1 x-ray film using a DuPont Cronex Lighting-Plus intensifying screen. The numeral 1 refers to 1 ng of hexokinase spotted on each plate as control.

nitrocellulose strip used in the hybridization. Note the change of scale at the dotted line.

Restriction Map of pYe57E2. A number of restriction endonuclease cleavage sites have been mapped on pYe57E2 by analysis of agarose gel electrophoresis patterns generated from partial restriction enzyme digests (Fig. 5). Experiments involving recloning of restriction fragments to localize the region coding for the polypeptide which is antigenically-active with antibody directed against yeast PGK are in progress. In addition, pYe57E2 was labeled with ^{32}P by nick translation (21) and, after removing the common ColEl vector sequences, was used as a probe for colony hybridization with approximately 3000 bacterial colonies from the bank to search for overlapping fragments. One of the plasmids identified from this screening (pYe75H2) has also been partially characterized (Fig. 5).

A DNA sequence which is repeated several times in the yeast genome has been localized to the region around the 1.5 kbp *Bgl*II-*Bgl*II fragment on pYe75H2 (a portion of this sequence is also present on pYe57E2). Other cloned yeast DNA segments have been isolated in which this sequence is contiguous to a repeated sequence discovered previously near the *leu2* locus (10), which maps on the opposite side of the centromere on chromosome III. Additional experiments are in progress to further characterize these sequences and to examine the possibility that they are relevant to the structure of the centromere.

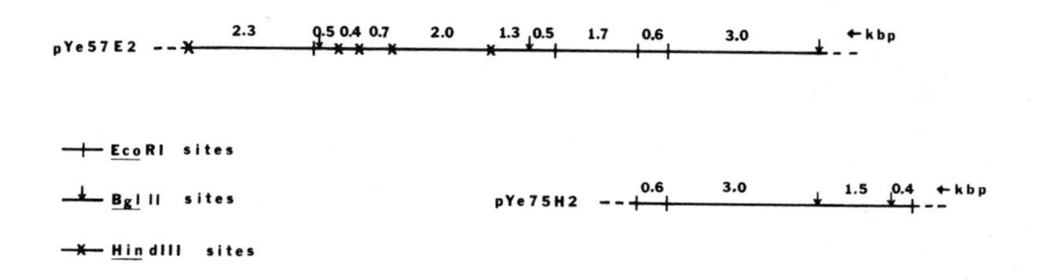

FIGURE 5. Preliminary restriction maps of plasmid pYe57E2 and an overlapping plasmid pYe75H2.

REFERENCES

1. Grunstein, M., and Hogness, D. S. (1975). Proc. Natl. Acad. Sci. U.S.A. 72, 3961.
2. Struhl, K., Cameron, J. R., and Davis, R. W. (1976). Proc. Natl. Acad. Sci. U.S.A. 73, 1471.
3. Ratzkin, B., and Carbon, J. (1977). Proc. Natl. Acad. Sci. U.S.A. 74, 487.
4. Vapnek, D., Hantala, J. A. Jacobson, J. W., Giles, N.H., and Kushner, S. R. (1977). Proc. Natl. Acad. Sci. U.S.A. 74, 3508.
5. Carbon, J., Clarke, L., Chinault, C., Ratzkin, B., and Walz, A. (1978). In "Biochemistry and Genetics of Yeast" (M. Bacila, B. L. Horecker, and A. O. M. Stoppani, eds.), pp. 425-443. Academic Press, New York.
6. Erlich, H. A., Cohen, S. N., and McDevitt, H. O. (1978). Cell 13, 681.
7. Broome, S., and Gilbert, W. (1978). Proc. Natl. Acad. Sci. U.S.A. 75, 2746.
8. Clarke, L., Hitzeman, R., and Carbon, J. (1979). Methods in Enzymol., "Recombinant DNA", in press.
9. Lam, K., and Marmur, J. (1977). J. Bacteriol. 130, 746.
10. Chinault, A. C., and Carbon, J. (1979). Gene 5, 111.
11. Carbon, J., Ratzkin, B., Clarke, L., and Richardson, D. (1977). Brookhaven Symposia in Biology 29, 277.
12. Shapiro, D. J., Taylor, J. M., McKnight, G. S., Palacios, R., Gonzalez, C., Kiely, M. L., and Schimke, R. T. (1974). J. Biol. Chem. 249, 3665.
13. David, G. S., and Reisfeld, R. A. (1974). Biochem. 13, 1014.
14. Axén, R., and Ernback, S. (1971). Eur. J. Biochem. 18, 351.
15. Miles, L. E. M., and Hales, C. N. (1968). Biochem. J. 108, 611.
16. Scopes, R. K. (1971). Biochem. J. 122, 89.
17. Chang, A. C. Y., Nunberg, J. H., Kaufman, R. J., Erlich, H. A., Schimke, R. T., and Cohen, S. N. (1978). Nature 275, 617.
18. Southern, E. M. (1975). J. Mol. Biol. 98, 503.
19. Hicks, J., and Fink, G. R. (1977). Nature 269, 265.
20. Walz, A., Ratzkin, B., and Carbon, J. (1978). Proc. Natl. Acad. Sci. U.S.A. 75, 6172.
21. Rigby, P. W. J., Dieckmann, M., Rhodes, C., and Berg, P. (1977). J. Mol. Biol. 113, 237.

STRUCTURE OF YEAST tRNA PRECURSORS CONTAINING INTERVENING SEQUENCES[1]

Hyen S. Kang[2], Richard C. Ogden, Gayle Knapp, Craig L. Peebles and John Abelson

Department of Chemistry
University of California, San Diego, La Jolla, CA 92093

ABSTRACT. The sequences of five identified tRNA precursors from Saccharomyces cerevisiae are presented. Comparison of these sequences indicates certain similarities in the primary and probable secondary structures. These similarities are discussed. In addition, a tertiary structure model is proposed.

INTRODUCTION

The discovery that precursors to certain tRNAs accumulate in a temperature-sensitive Saccharomyces cerevisiae mutant (1), ts-136, has considerably facilitated the study of eukaryotic tRNA biosynthesis. This mutant defines the rnal gene of yeast (2). It is presumed to be defective in a step in RNA transport from nucleus to cytoplasm. At the nonpermissive temperature, the 35S ribosomal RNA precursor accumulates (1), the appearance of messenger RNA in the cytoplasm is halted and poly A-containing RNA accumulates in the nucleus (3), and a particular subset of tRNA precursors accumulates (4). The set includes pre-$tRNA^{Phe}$, pre-$tRNA^{Tyr}$, pre-$tRNA^{Trp}$, pre-$tRNA^{Leu}_{3}$ and pre-$tRNA^{Ser}_{UCG}$ as well as at least four other RNAs which hybridize to E. coli recombinant plasmid clones containing yeast tRNA genes of unknown specificity.

The investigation of this subset of tRNA precursors by a variety of methods indicates that all nine precursors share common features. Each of these RNAs contains a unique, or closely related, set of sequences, located near the center of

[1]This work was supported by grants from the National Institutes of Health (CA 10984, GM 05518 and GM 07199).

[2]Present address: Department of Microbiology, Seoul National University, College of Natural Sciences, Seoul 151 Korea.

ISBN 0-12-068350-4

the molecule, which are not found in the mature tRNA. These sequences range in size from 14 nucleotides (pre-$tRNA^{Tyr}$) to approximately 60 nucleotides (precursor 19) and, where evidence has been obtained, they have been shown to be faithful transcripts of the intervening sequence in the gene. These RNAs are not primary transcripts, but have already been partially matured. The 5' and 3' termini are likely to be the same as those of the mature tRNA however the intervening sequence has not been excised and only certain base modifications are present in molar yields. All nine of the RNAs serve as substrates for *in vitro* splicing in soluble extracts of *Saccharomyces cerevisiae* and give rise in the absence of ATP to two half-tRNA sized molecules and the excised intervening sequence (5, unpublished results). These structural and functional similarities are discussed below.

RESULTS

Isolation of tRNA Precursors. *In vivo* ^{32}P-labelling of a diploid strain, homozygous for the temperature-sensitive *rnal* locus, has been described extensively (4,6). Figure 1 shows the separation by two-dimensional polyacrylamide gel electrophoresis of the RNAs accumulating at the nonpermissive temperature in this mutant. The tRNA precursors were originally identified by hybridization to a set of *E. coli* recombinant plasmid clones which contain one or more yeast tRNA genes (7). The four numbered RNA species hybridize to unique sets of these clones containing genes of unknown tRNA specificity. Some of the RNAs appear as doublet spots on the two-dimensional gel (*e.g.*, precursor 13). This precursor heterogeneity may be due to 3' terminal differences or, in some cases, to intervening sequence heterogeneity.

Sequence Analysis of tRNA Precursors. The sequences of genes for four yeast tRNAs containing intervening sequences have been determined. The sequence analyses of three $tRNA^{Tyr}$ genes and an ochre allele of one of them (*SUP4-O*) have been reported (8). All contain intervening sequences of 14 base pairs interrupting the anticodon region of the mature tRNA sequence. Three $tRNA^{Phe}$ gene sequences have been reported (9) and intervening sequences of 18 or 19 base pairs were found in the same region. In neither case could the exact location of the intervening sequences be determined due to the presence of repeated sequences at the junction positions. One $tRNA^{Trp}$ gene and one $tRNA_3^{Leu}$ gene, contained on independent pBR313-yeast DNA recombinant plasmids, have been sequenced (Kang, Ogden and Abelson, unpublished results). Classical RNA fingerprint analysis (10) has indicated that the precursors to

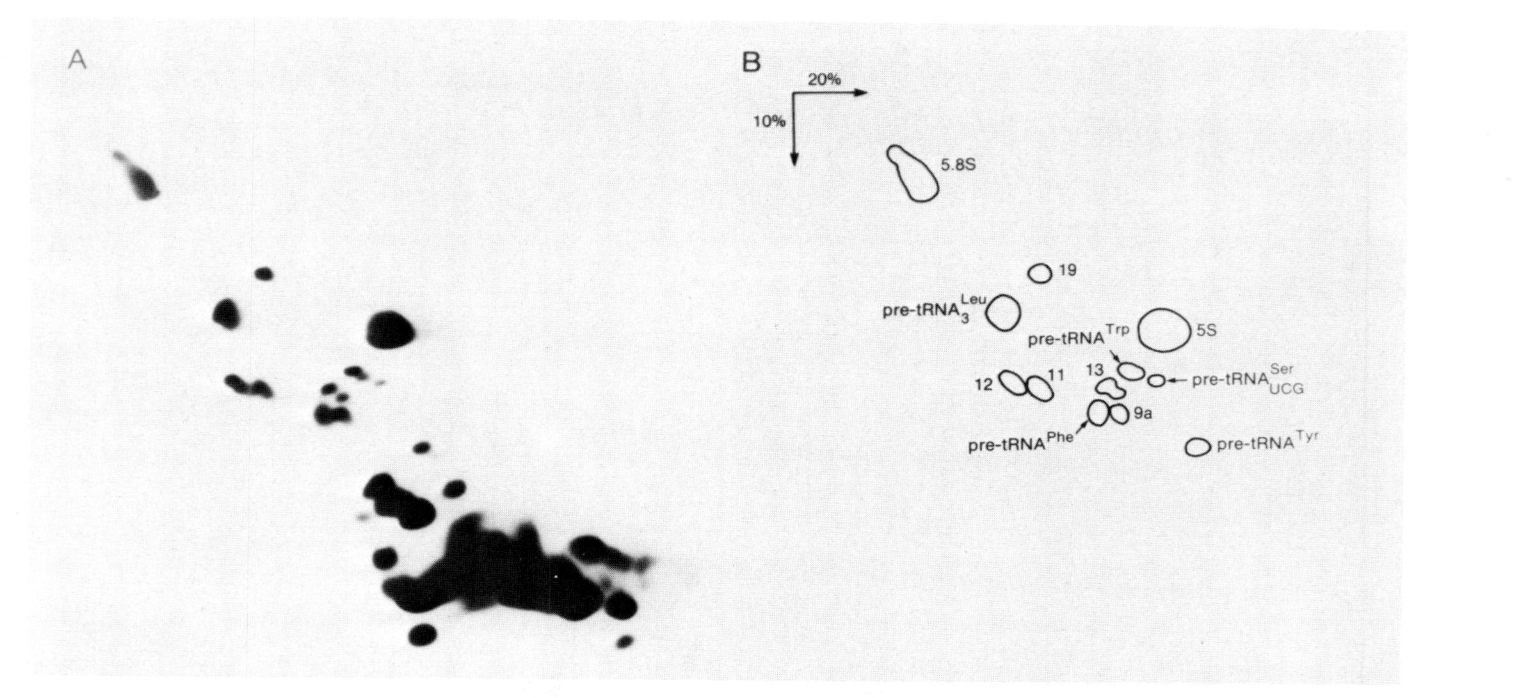

Figure 1. Two-dimensional polyacylamide gel electrophoresis of ^{32}P-RNA isolated from the yeast <u>rna1</u> mutant (4). (A) Autoradiogram of the gel separations. (B) Diagrammatic representation and identification of the ^{32}P-pre-tRNAs.

these tRNAs accumulating in the temperature-sensitive mutant contain intervening sequences in the anticodon region (6, unpublished results). The $tRNA^{Trp}$ gene contains an intervening sequence of 34 base pairs; the $tRNA_3^{Leu}$ gene contains one of 32 base pairs. Again, the DNA sequence did not indicate the precise location of the intervening sequences with respect to the mature tRNA.

The precursor tRNA sequences have been compared to the sequences of the genes in all four cases. In the cases of pre-$tRNA^{Tyr}$ and pre-$tRNA^{Phe}$ (4,11) the fingerprints indicated the presence of mature 5' and 3' termini and demonstrated that the precursors were faithful transcripts of the heterologous population of genes. The predicted heterogeneities in the intervening sequence were observed. In a similar manner, the characterization of pre-$tRNA^{Trp}$ (6) and pre-$tRNA_3^{Leu}$ (Kang, Ogden and Abelson, unpublished results) demonstrated that the RNase T1 and RNase A oligonucleotides predicted by the DNA sequences are found in the precursors and that the precursor-specific intervening sequence is located in the middle of the mature tRNA sequence. The primary sequence of the pre-$tRNA_{UCG}^{Ser}$ intervening sequence has been determined (12). It has a length of 19 nucleotides and is unambiguously located at the 3' side of the nucleotide that is often hypermodified in the anticodon loop. The post-transcriptional modification of the precursor tRNAs is discussed below.

The sequences of the five identified tRNA precursors are shown in Figure 2. The secondary structure models of pre-$tRNA^{Tyr}$ and pre-$tRNA^{Phe}$ are suggested from calculation of the most favorable free energy using the rules derived by Tinoco *et al*. (13) and Borer *et al*. (14). The secondary structures of pre-$tRNA^{Tyr}$ and pre-$tRNA_{UCG}^{Ser}$ are consistent with studies of the nuclease sensitivity of the precursors (11,12). The secondary structures of pre-$tRNA^{Trp}$ and pre-$tRNA_3^{Leu}$ have been drawn to maximize conventional base pairing within the larger intervening sequences and between the intervening sequence and tRNA. In addition, a generalized secondary structure model showing constant features of pre-tRNA structure is shown in Figure 2. This model and a proposed tertiary structure model are discussed later.

Post-transcriptional Modification of tRNA Precursors. These tRNA precursors represent intermediates in the yeast tRNA biosynthetic pathway. A variety of steps are involved in the maturation of tRNA and in these precursors some steps are evidently complete whereas others are incomplete or, in some cases, have not occurred to any detectable extent. The five known precursors have the same 5' and 3' termini as the mature tRNAs indicating that terminal processing is completed. In no case is the 3' CCA_{OH} terminus coded for in the gene indicating that tRNA nucleotidyl transferase can add these nucleotides to the precursor prior to splicing.

A catalogue of the nucleoside modifications found at the various positions of the precursors is presented in Table 1. As can be seen, none of the ribose modifications (e.g., 2'-O-methyl G or 2'-O-methyl C) are observed and none of the hypermodified nucleotides (e.g., Y or 6-(Δ^2-isopentenyl)-A) occuring immediately adjacent to the 5' side of the splice point are found. The extent of a particular modification (e.g., ψ) varies with its position in the precursor. In general, the extent of pyrimidine modification is greater than that of purine modification. In no case have modified nucleotides been detected in the intervening sequence.

Splicing of tRNA Precursors. All nine precursors serve as substrates for the tRNA splicing activities in yeast soluble extracts (4,5, unpublished results). The kinetics of the reaction, distribution of products in the presence and absence of ATP and inhibition by mature tRNA are similar for each precursor. At the present degree of purification there is no discrimination by the activities between the precursors. Figure 3 demonstrates the ATP-independent generation of half-tRNA sized molecules and intervening sequences from each precursor in an incubation with yeast ribosomal wash (15). In most cases, the RNA products are well resolved. However, the halves of 11, 12 and pre-$tRNA^{Trp}$ co-migrate in this system. They can be resolved by inclusion of CTP in the incubation (probably due to the reconstruction of the CCA_{OH} terminus by tRNA nucleotidyl transferase activity in the extract). This general method has enabled a detailed investigation of the intermediates of in vitro RNA splicing (15).

The structures of the half-tRNA sized molecules for the five known precursors are shown in Figure 4. The sequences adjacent to the splice point (in bold lettering) have been determined by fingerprint analyses of the isolated halves. These structures demonstrate the similarities between the substrate RNAs after excision of the intervening sequence. The fact that mature tRNA inhibits only ligation suggests that the

---

Figure 2. Nucleotide Sequences of Five Yeast tRNA Precursors. The arrows indicate the splice points. A general secondary structure model is also shown summarizing constant and variable features of the structures. Variable nucleotides are indicated by (-O-) in the mature portion of the precursor and (-X-) in the intervening sequence. Pu and Py indicate positions where a purine or pyrimidine is conserved in the structures. Regions with variable numbers of nucleotides are indicated adjacent to the structure.

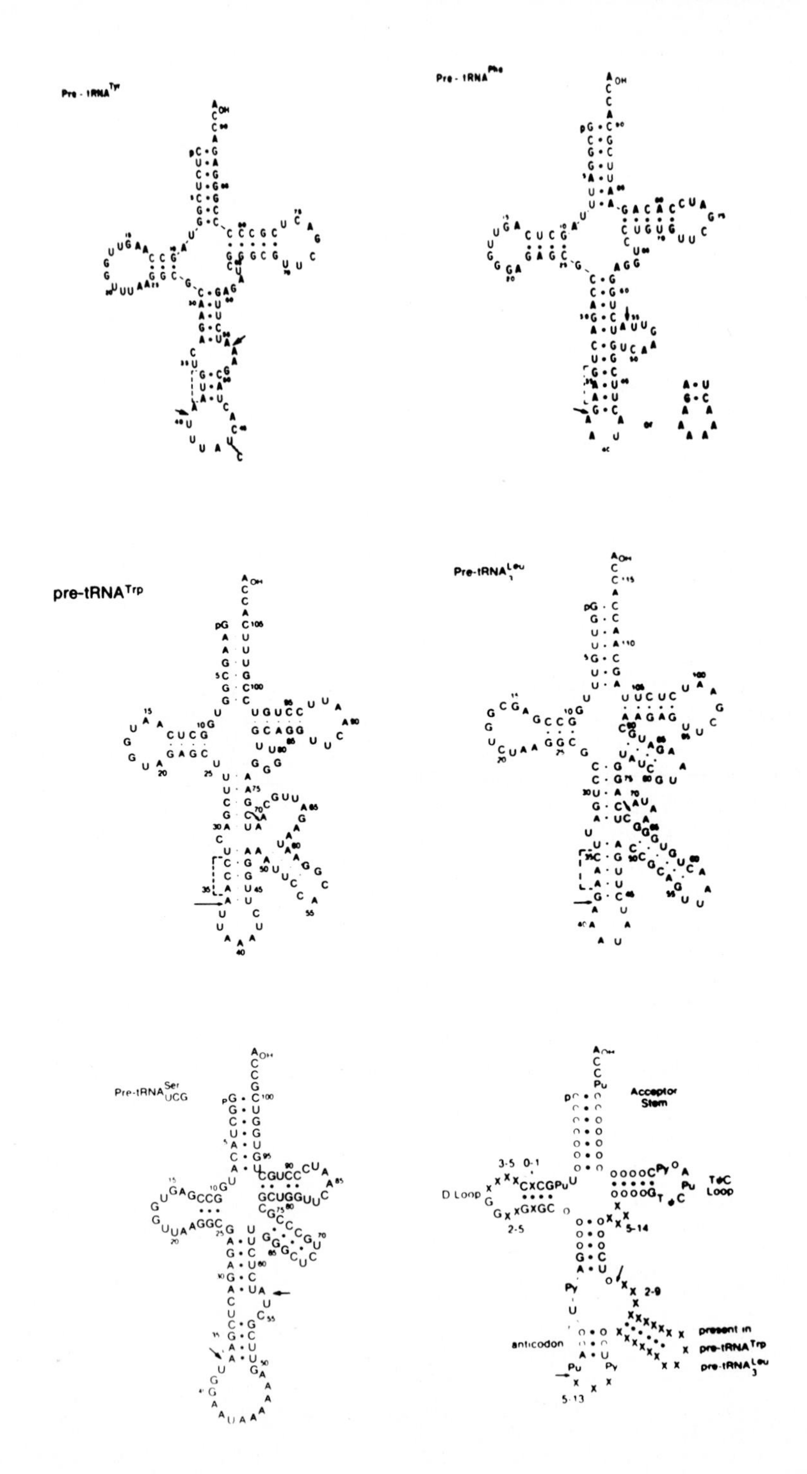
Pre-tRNA^Tyr
Pre-tRNA^Phe
pre-tRNA^Trp
Pre-tRNA^Leu_3
Pre-tRNA^Ser_UCG
Acceptor Stem
TψC Loop
D Loop
anticodon
present in pre-tRNA^Trp pre-tRNA^Leu_3

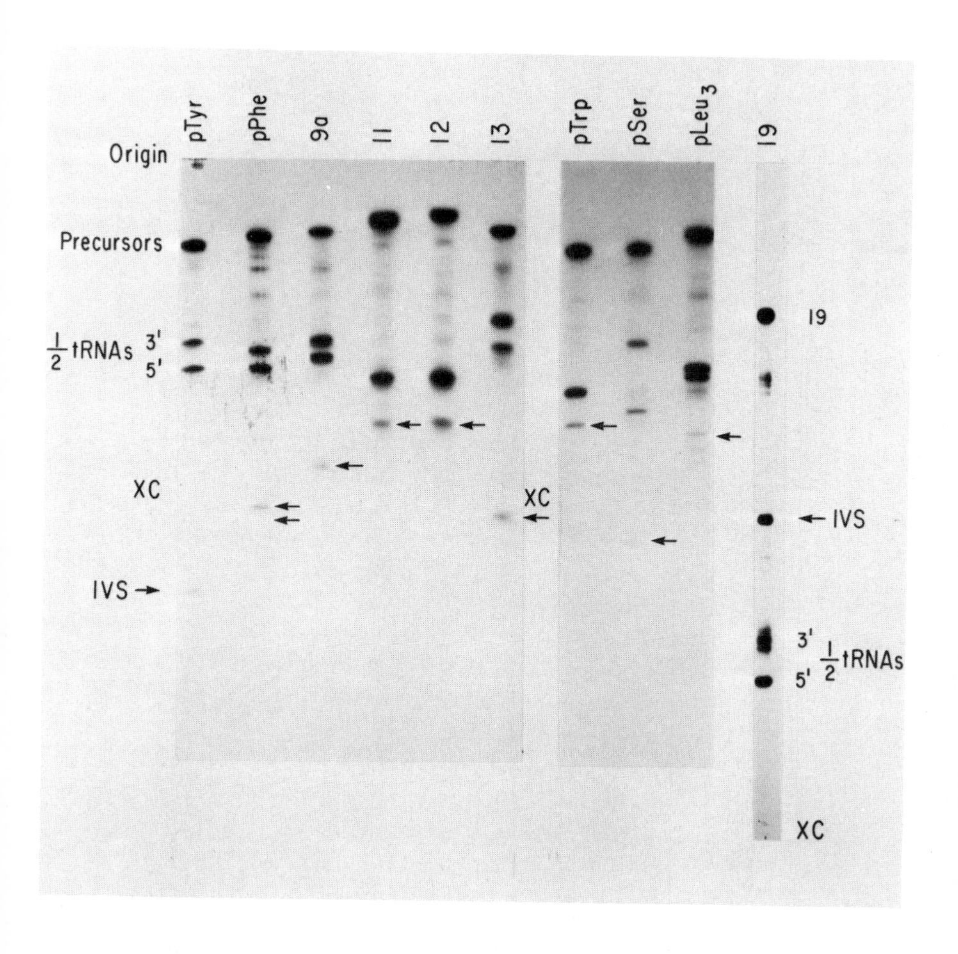

Figure 3. Preparative cleavage of tRNA precursors. Cleavage of the nine tRNA precursors was performed using conditions described in (15). The tRNA halves and intervening sequence were identified by sequence analysis, mobility and/or minor nucleotide analysis. Arrows indicate the intervening sequences, the 5' half always migrates faster than the 3' half except for precursors 11, 12 and pre-tRNA^{Trp} (pTrp) where the halves comigrate in this system. XC indicates the position of xylene cyanol dye marker.

halves assume a structure similar to that of the mature tRNA (5, 15). The common structure of the ligase substrates is a direct consequence of the constant position of the intervening sequence in the five known precursors.

TABLE 1

MODIFIED NUCLEOTIDES DETECTED IN tRNA PRECURSORS

Modification	Precursor (Position[a])	Approximate Extent[b] (Position)
Ψ	pre-tRNA^{Tyr} (71,55,37)	>90% (71); <10% (55,37)
	pre-tRNA^{Phe} (73)	>90% (73)
	pre-tRNA$^{Ser}_{UCG}$ (83,58)	80-100% (83); 50-60% (58)
	pre-tRNA^{Trp} (25,26,27,72,88,98)	>90% (25,26,27[c]); 80% (72); 30% (88); 60% (98)
	pre-tRNA$^{Leu}_{3}$ (72,96)	>90% (72,96)
D	pre-tRNA^{Tyr} (16,17,20,21,22,63)	50% (all positions[c])
	pre-tRNA^{Phe} (16,17)	50% (16,17)
	pre-tRNA$^{Ser}_{UCG}$ (16,19,20)	70-100% (16); 100% (19,20[c])
	pre-tRNA^{Trp} (16,19,80)	50% (16); >90% (19); <10% (80)
	pre-tRNA$^{Leu}_{3}$ (19,21)	60% (19,21[c])

TABLE 1 (continued)

T	pre-$tRNA^{Tyr}$ (70)	
	pre-$tRNA^{Phe}$ (72)	
	pre-$tRNA^{Ser}_{UCG}$ (82)	>90% (all cases)
	pre-$tRNA^{Trp}$ (87)	
	pre-$tRNA^{Leu}_{3}$ (95)	
m^5C	pre-$tRNA^{Phe}$ (67)	50% (67)
	pre-$tRNA^{Ser}_{UCG}$ (76)	90% (76)
	pre-$tRNA^{Leu}_{3}$ (89)	90% (89)
ac^4C	pre-$tRNA^{Ser}_{UCG}$ (12)	75–100% (12)
	pre-$tRNA^{Leu}_{3}$ (12)	>70% (12)

TABLE 1 (continued)

m^1A [d]	pre-$tRNA^{Tyr}$ (74)	
	pre-$tRNA^{Phe}$ (76)	>50% (all cases)
	pre-$tRNA^{Trp}$ (91)	
$\overset{*}{A}$ [e]	pre-$tRNA^{Ser}_{UCG}$ (85 or 86)	100%
m^2_2G	pre-$tRNA^{Tyr}$ (28)	>80% (28)
	pre-$tRNA^{Phe}$ (26)	>90% (26)
	pre-$tRNA^{Ser}_{UCG}$ (26)	90% (26)
	pre-$tRNA^{Leu}_{3}$ (27)	>80% (27)
m^7G	pre-$tRNA^{Phe}$ (64)	>70% (64)
	pre-$tRNA^{Trp}$ (79)	>70% (79)

TABLE 1 (continued)

---

[a] The position number refers to the numbering in Figure 2.

[b] The extent of modification was estimated in two ways: (1) when an oligonucleotide containing a modified residue separated from the unmodified one during electrophoresis scintillation counting of the oligonucleotides provided a quantitative measurement; (2) Total enzymatic digestion and separation by two-dimensional thin layer chromatography as described previously (6) enabled a visual comparison of the amount of radioactivity in modified and unmodified residues for any RNase T1 or RNase A oligonucleotide.

[c] The extent of modification at individual positions has not been determined.

[d] m^1A was in some cases detected as a rearranged nucleotide (m^6A) and/or as a depurinated product.

[e] This uncharacterized modification is described by Etcheverry, Colby and Guthrie (12).

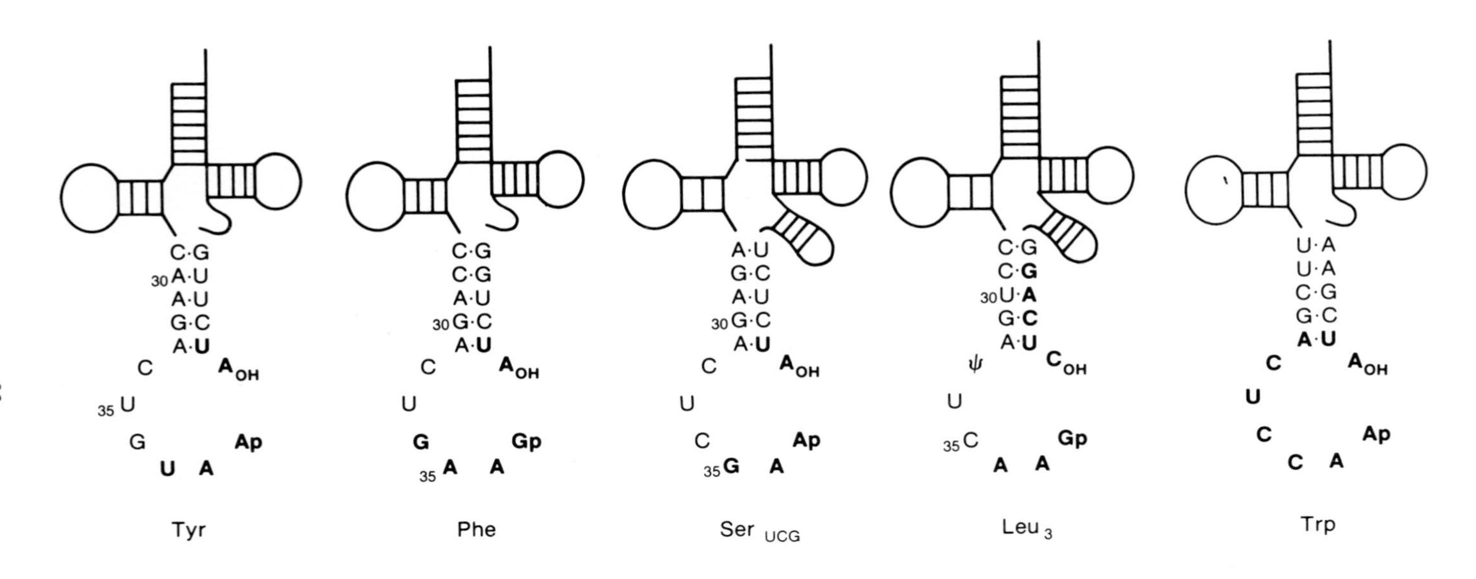

Figure 4. Half-tRNA molecules arising from excision of the intervening sequence. The terminal sequences (bold lettering) were ascertained by fingerprint analysis.

DISCUSSION

The detailed investigation of five tRNA precursors containing intervening sequences has led to the discovery of several unifying features which may prove significant in the mechanism of tRNA splicing and possibly of RNA splicing in general. There is very little primary sequence homology among the tRNA precursors either within the intervening sequence or at the splice points. However, all of the intervening sequences do have high A + U base compositions (A + U:G + C = 2:1). On the other hand, the tRNA portion of the precursors shows two possibly significant regions of sequence similarities. First, the last two base pairs in the mature anticodon stem are the same in all five tRNAs and, second, the D stems of the tRNAs are very similar (Figure 2).

There are also some interesting structural similarities. The position of the intervening sequence, as determined by analyses of the excised intervening sequences and the half-tRNA molecules, is the same in all five precursors. It is immediately adjacent to the nucleotide that is usually hypermodified in the mature tRNA. This implies that the cleavage sites in all the precursors must occur in exactly the same position relative to the tRNA portion of the precursor. Furthermore, the base pairing of the anticodon with its complement in the intervening sequence (Figure 2) can occur in all five cases--producing additional common structure near the cleavage sites. In all cases, except that of pre-$tRNA^{Tyr}$, nucleotides adjoining the anticodon can also be base paired. This extends the anticodon helix almost to the helix of the anticodon stem. Regions of variation among the intervening sequences are found in loops at either end of the anticodon helix as is indicated in the generalized structure (Figure 2). At one end of this helix, there is a hairpin loop varying from 5 to 13 nucleotides. At the other end is an interior loop located between the helices of the anticodon and anticodon stem. Depending on the size of the anticodon helix, this interior loop includes zero, one or two nucleotides between the anticodon and the anticodon stem and includes from three (pre-$tRNA^{Ser}_{UCG}$) to twenty-two or twenty-three extra nucleotides in the form of a large helical stem and loop (pre-$tRNA^{Leu}_{3}$ and pre-$tRNA^{Trp}$) between the complement of the anticodon and the anticodon stem. It is not clear how a large interior loop (*e.g.*, in pre-$tRNA^{Trp}$ or pre-$tRNA^{Leu}_{3}$) might disrupt a continuous helix that might be formed by the anticodon helix and the anticodon stem. In the case of pre-tRNAs with shorter intervening sequences, the number of bases in the interior loop is small. For example, in pre-$tRNA^{Phe}$ there might be no discontinuity between the helices. The anticodon stem and anticodon helices are adjacent on the anticodon side and

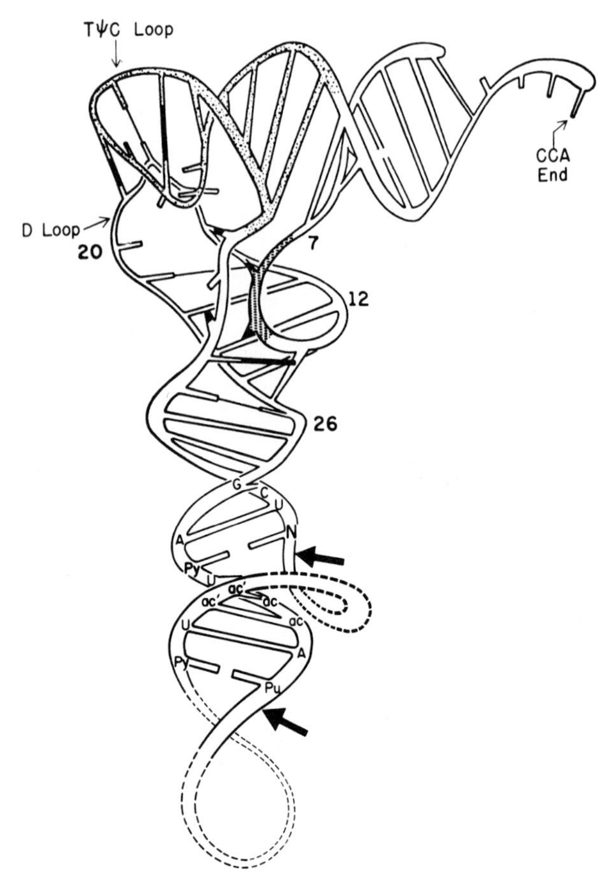

Figure 5. Proposed model for the tertiary structure of the tRNA precursors containing intervening sequences. This preliminary model has been drawn utilizing the conserved elements of secondary structure shown in Figure 2. The tertiary structure of the tRNA has been retained except for the foreshortened anticodon stem. This stem has been extended in the form of the RNA double helix (A form). The arrows indicate the splice points. Invariant or semi-invariant positions are indicated in the anticodon stem and loop area and in the intervening sequence. The anticodon and its complementary sequence in the intervening sequence are denoted by ac-ac-A and U-ac'-ac', respectively. N indicates a position of a variable nucleotide. The numbers refer to the nucleotide positions in the mature sequence. The broken areas in the phosphate-ribose chain indicate the positions of variable size in the intervening sequence.

interrupted by a moderate-sized bulge on the other (Figure 2). The proposed secondary structures for pre-tRNA$^{Ser}_{UCG}$ and pre-tRNA^{Tyr} (Figure 2) only disrupt the continuity of the helices by small interior loops. In these two precursors, there are two unpaired nucleotides near the anticodon which are opposed by three and four unpaired nucleotides, respectively. Thus, it is possible that there could be a continuous helix formed by the anticodon stem and anticodon helices.

A preliminary tertiary structure model is proposed in Figure 5. From this structure it may be seen that the two cut sites, separated by a distance equivalent to 6 base pairs, would be situated conveniently on the same side of the helix. Also, the conserved region of the D stem is located on the same side of the precursor as the cut sites. In order for the five precursors to all conform to this structure the interior loop would have to fold out to one side of the helix or possibly be accommodated in the groove of the helix formed by the anticodon and its complement. These are, of course, speculations and knowledge of the true structure of the tRNA precursor will require further investigation using physical techniques.

ACKNOWLEDGMENTS

We would like to thank Tina Etcheverry, Diane Colby and Christine Guthrie for permission to quote their work prior to publication.

REFERENCES

1. Hopper, A.K., Banks, F., and Evangelidis, V. (1978). Cell. 14, 211.
2. Hutchison, H.T., Hartwell, L.H., and McLaughlin, C.S. (1969). J. Bact. 99, 807.
3. Shiokawa, K., and Pogo, A.O. (1974). Proc. Nat. Acad. Sci. 71, 2658.
4. Knapp, G., Beckmann, J.S., Johnson, P.F., Fuhrman, S.A. and Abelson, J. (1978). Cell. 14, 221.
5. Peebles, C.L., Ogden, R.C., Knapp, G., and Abelson, J. (1979). Cell. (submitted).
6. Ogden, R.C., Beckmann, J.S., Abelson, J., Kang, H.S. Söll, D., and Schmidt, O. (1979). Cell. (in press).
7. Beckmann, J.S., Johnson, F.P., Abelson, J., and Fuhrman, S.A. (1977). In Molecular Approaches to Eukaryotic Systems, G. Wilcox, J. Abelson and C.F. Fox, eds. (New York: Academic Press), 213.
8. Goodman, H.M., Olson, M.V., and Hall, B.D. (1977). Proc. Nat. Acad. Sci. USA 74, 5453

9. Valenzuela, P., Venegas, A., Weinberg, F., Bishop, R., and Rutter, W.J. (1978). Proc. Nat. Acad. Sci. USA 75, 190.
10. Brownlee, G.G. (1972). In Laboratory Techniques in Biochemistry and Molecular Biology, T.S. Work and E. Work, eds. (New York: American Elsevier), 67.
11. O'Farrell, P.Z., Cordell, B., Valenzuela, P., Rutter, W.J., and Goodman, H.M. (1978). Nature 274, 438.
12. Etcheverry, T., Colby, D., and Guthrie, C. (1979). Cell. (submitted).
13. Tinoco, I., Jr., Borer, P.N., Dengler, B., Levine, M.D., Uhlenbeck, O.C., Crothers, D.M., and Gralla, J. (1973). Nature New Biol. 246, 40.
14. Borer, P.N., Dengler, B., Tinoco, I., Jr., and Uhlenbeck, O.C. (1974) J. Mol. Biol. 86, 843.
15. Knapp, G., Ogden, R.C., Peebles, C.L. and Abelson, J. (1979). Cell. (submitted).

ANALYSIS OF YEAST CELL-CYCLE MUTATIONS BY PSEUDO-REVERSION

Don Moir, Sue Ellen Stewart, and David Botstein

Department of Biology, Massachusetts Institute of Technology,
Cambridge, Massachusetts 02139

ABSTRACT Both extragenic and apparent intragenic suppressors which exhibit heat-sensitivity were isolated from revertants of cold-sensitive cell-division-cycle mutants of *Saccharomyces cerevisiae*. In some cases the extragenic suppressor is also a cell-division-cycle mutation, and could specify a protein which interacts with the original cold-sensitive gene product. These results have certain implications for models of control of the cell cycle.

INTRODUCTION

The eukaryotic cell division cycle consists of an ordered series of morphological and chemical events which occur in a regular sequence during each cell division. In the yeast *Saccharomyces cerevisiae* the cell cycle has proved amenable to analysis by genetic methods, especially the isolation and characterization of conditional-lethal mutations (mainly temperature-sensitive) which result in arrest at the restrictive temperature of all cells at a single morphological state characteristic of a particular point in the cell cycle (1,2). Such mutations (*cdc* for cell-division-cycle) are therefore thought to affect the cell cycle specifically, and are a relatively rare subset (about ten percent) of all conditional-lethals.

In analyzing mutations which affect specifically this ordered series of events, it is relevant to consider whether the mutations affect products made sequentially or continuously during the cycle. On the one extreme, it might be supposed that cell-cycle defects occur primarily in proteins made once per cycle, used up in the process of their activity, and resynthesized only when required in the next cycle. At the other extreme, one might take the view that the relevant products (presumably proteins) are synthesized continuously throughout the cycle, and mutants exhibit specific arrest at some point in the cell cycle because of a failure in some protein assembly which is sequentially regulated morpho-

ISBN 0-12-068350-4

genetically, as in the case of bacteriophage T4 tails (3,4).

In the former view, many cell cycle mutations might interrupt the cell cycle by a failure in regulation causing a gene not to be expressed. In the latter view, all cell cycle defects are failures of protein function. The results presented below suggest that cell cycle genes often specify interacting proteins. While these results cannot themselves distinguish the extreme models of cell cycle regulation, they lend plausibility to the idea that protein assembly plays a substantial role.

When two proteins interact, they may affect each other's structure. In fact, the expression of a mutation in a protein's conformation and function can often be altered by a second change in an interacting protein. Such phenomena are frequently observed genetically: mutations affecting the structure of a protein can be "suppressed" by mutations in a gene specifying an interacting protein (5,6). This kind of suppression is commonly observed in protein assembly processes. In one study of phage morphogenesis, it was observed that cold-sensitive mutations can be used to identify mutations in genes specifying interacting proteins by selecting for pseudo-revertants which are cold-independent but which have acquired simultaneously a second phenotype, e.g., heat-sensitivity (7). Such suppressors usually were found to retain their new phenotype when separated from the original mutation they suppressed. Conversely, heat-sensitive mutations could also be used to generate cold-sensitive lesions in interacting proteins by pseudo-reversion.

In this paper, we describe the application of this pseudo-reversion scheme to heat-sensitive and cold-sensitive cell-division-cycle mutations of *Saccharomyces cerevisiae*.

RESULTS

Isolation and Characterization of Revertants. Revertants of seven different cold-sensitive cell cycle mutants were selected at the cold restrictive temperature (17°C) and screened for concomitant acquisition of temperature-sensitivity at 35°C. As shown in Table I, four of the cold-sensitive mutants, all defective in the same step (medial nuclear division) at 17°, yielded temperature-sensitive revertants. These revertants were analyzed by crosses to wild type in order to assess linkage of the suppressor with the newly acquired temperature-sensitivity and with the original cold-sensitive mutation. In virtually all cases examined (exceptions are noted below), the two phenotypes, suppression of cold-sensitivity and temperature-sensitivity for growth, are tightly linked to each other, suggesting that they are the

TABLE I

TEMPERATURE-SENSITIVE REVERTANTS OF COLD-SENSITIVE CELL-DIVISION-CYCLE MUTANTS

_cs_ mutants	Independent revertants examined	_ts_ mutants found	Linkage to _cs_ marker	Cell cycle terminal phenotype at 35°	Efficiency of suppression
_cs_A18	100	rA18-50, rA18-69, rA18-74, rA18-95 [a]	None	Yes	Efficient
		rA18-66	None	No	Weak
		rA18-79	None	No	Weak
_cs_E24	143	rE24-6, rE24-15, rE24-23 [a]	Tightly linked	Yes	Efficient
_cs_H80	161	rH80-16	None	No	Weak
		rH80-38	None	Yes	Not detectable
		rH80-119	None	No	Weak
_cs_CC30	128	reCC30-45[b]	None	No	Weak
		reCC30-93[b]	None	Yes	Not detectable

[a] Bracketed mutants are members of a single complementation group.
[b] "re" designates a revertant isolated from EMS-mutagenized stock.

result of the same mutational event. Each revertant was further examined in the microscope to determine whether it exhibits a typical cell-division-cycle phenotype at 35° (8). Three of the four cold-sensitive mutants yielded two types of revertants, those with cdc terminal phenotypes at 35° and those without.

Revertants of a cs-cdc Mutant Yield ts Alleles of an Unlinked cdc Gene. The cold-sensitive cdc mutant csA18 yielded six independently derived temperature-sensitive revertants (see Table I). The revertants fall into two general classes, those which efficiently suppress cold-sensitivity and exhibit a medial nuclear division terminal phenotype at 35°, and those which suppress cold-sensitivity only weakly and do not display a cell-division-cycle phenotype. Mutants in the former class are in a single complementation group not genetically linked to the original cold-sensitive marker (PD:NPD:T ratio of 10:3:24 for segregation of the cs and ts markers). Spores which have segregated a ts marker free of the cs background are themselves well-behaved temperature-sensitive cdc mutants. Such segregants complement all of the heretofore described temperature-sensitive cdc mutants (2). Sporulation of the diploid, derived from a segregant spore containing only the ts marker crossed to csA18, yields many recombinant cs-ts spores which display temperature-sensitivity but not cold-sensitivity. Furthermore, sporulation of a cross of such a recombinant spore back to wild type of opposite mating type proves that the cs marker was present in the reconstructed cs-ts strain, since cs spores frequently segregate free of the ts suppressor. Revertants in the second class (see Table I), weak suppressors with no cdc terminal phenotype, produced a high proportion of inviable spores (∿60%); however, cold-sensitive spores were found frequently (∿15%), indicating the ts suppressor is not linked to the cs mutation. These revertants do complement revertants from the first class at 35°.

The results of reversion of csA18 thus provide an example of a cold-sensitive cdc mutation which can be suppressed by mutations in another cell-cycle gene, as judged from the heat-sensitive phenotypes of the suppressors when free of the original cold-sensitive mutation. Two points deserve emphasis. First, the phenotype of the original mutations and the suppressors is very similar: both arrest at the same point in the cycle at non-permissive temperature. Second, complementation analysis of the heat-sensitive mutations showed that they belong to new cell-division-cycle complementation group (gene) not previsouly described (2).

Revertants of a cs-cdc Mutant Yield ts Alleles of the Same cdc Gene. Reversion of the cold-sensitive cdc mutant csE24 produced three independently derived temperature-sensitive revertants (see Table I). All three contain efficient suppressors of cold-sensitivity and each fails to complement the other for growth at 35°. In addition, the ts suppressors are tightly linked to the cold-sensitive marker which they suppress; no cold-sensitive spores were observed in 33 tetrads examined from a cross of rE24-15 with wild type (PD:NPD:T ratio of 33:0:0) although temperature-sensitivity segregated 2:2 in all cases. Since no cold-sensitive spores were observed, tetrads from the cross rE24-15 x csE24 were analyzed to ensure that the apparent linkage was not due to two mutational events--intragenic reversion of cold-sensitivity and acquisition of an unlinked ts mutation. Again, only PD tetrads were observed (8:0:0) indicating the ts suppressor is tightly linked to the csE24 mutation. These revertants exhibit a medial nuclear division terminal phenotype at 35° and they complement all of the previously described temperature-sensitive cdc mutants (2), including rA18-95.

The results of reversion of csE24 provide an example of a second expected class of pseudo-revertants: those with compensating lesions in the same gene. Such second-site revertants are also frequently found in the phage assembly systems (7).

Some More Complicated Cases in which Reversion of a cdc Mutant Results in the Appearance of New cdc Mutations. Three independent temperature-sensitive revertants of csH80 were analyzed for linkage of the ts and cs markers by crosses to wild type (see Table I). Two revertants, rH80-16 and rH90-119, harbor ts suppressors which are not linked to the original cs locus; furthermore, in both cases suppression is weak and the mutants do not exhibit a cell cycle terminal phenotype at 35°. Both suppressors also exhibit temperature-sensitivity when removed from the cs background by segregation in crosses. Reconstruction of the cs-ts strain yielded the expected results for both revertants--cs-ts spores are temperature-sensitive but not cold-sensitive.

The third revertant of csH80, rH80-38, yielded no cs spores upon analysis of the sporulated diploid rH80-38 crossed to wild type (PD:NPD:T ratio of 19:0:0), suggesting that the ts suppressor is tightly linked to the cs mutation. However, when the rH80-38 x csH80 diploid was sporulated and analyzed, both the ts and cs markers segregated 2:2 completely independently of one another. In other words, the ts marker is not linked to csH80 and is not a suppressor of csH80; reconstruction of a cs-ts spore results in a strain which exhibits both conditional-lethal defects. Apparently, two mutational events have occurred to produce rH80-38, an intragenic reversion of the cold-sensitivity and acquisition of an unlinked ts marker. The ts mutation in the pseudo-revertant rH80-38 fails to complement mutations in cdc2.

An observation related to this unexpected result has emerged from efforts to obtain cs revertants of tsFT20, a new ts allele of cdc 2. Although spontaneous reversion of tsFT20 failed to produce revertants which exhibit cold-sensitivity, reversion induced by the mutagen EMS did result in one cs revertant out of 20 examined (reFT20-78, not shown in Table I). No ts spores were found upon analysis of a cross between the revertant and wild type (PD:NPD:T ratio of 8:0:0), but dissection of tetrads from a cross of the revertant with tsFT20 demonstrated that the cs and ts markers are unlinked. Analogous to the rH80-38 x csH80 results, in the case the cs marker does not suppress the temperature-sensitivity and the ts marker is apparently not present in the reFT20-78 revertant. In addition, the revertant exhibits a cdc terminal phenotype at 17° and the cold-sensitive marker is allelic to csH80. Thus, reciprocal events have apparently occurred — reversion of csH80 coupled with acquisition of a ts allel of cdc 2, and reversion of a different ts cdc 2 allele coupled with acquisition of a new cs allele of csH80.

Since selection for spontaneous revertants of another cold-sensitive cdc mutant (csCC30) failed to produce any temperature-sensitive ones (50 examined), revertants were selected from EMS-mutagenized stocks. Two temperature-sensitive revertants were obtained (see Table I). One of these reCC30-45 was shown, by means of analysis of the cross to wild type to harbor a weak temperature-sensitive suppressor of the cold-sensitivity. The ts suppressor is unlinked to the cs marker which it suppresses (PD:NPD:T ratio of 4:4:13) and does not exhibit a cdc terminal phenotype at 35°. The suppressor also displays temperature-sensitivity when segregated free of the cs marker, and a reconstructed cs-ts strain is temperature-sensitive but not cold-sensitive. The second temperature-sensitive revertant obtained from EMS-mutagenized stocks, reCC30-93, yielded no cs spores after a cross to the wild type parent, suggesting tight linkage of the ts suppressor to the cs marker. However, analysis of tetrads from a cross of the revertant back to the original cold-sensitive strain revealed that the cs and ts markers are unlinked. As observed for rH80-38, the ts marker fails to suppress the cold-sensitivity, and reconstructed cs-ts spores are both temperature-sensitive and cold-sensitive. The ts marker in reCC30-93 exhibits a medial nuclear division phenotype at 37°, and it complements all of the previously described ts cdc mutants (2).

These results show that one can obtain cdc mutants in our pseudo-reversion scheme even when suppression cannot be demonstrated in a reconstructed cs-ts recombinant. While we cannot show specificity of the reversion by suppression, we are persuaded that some specific interaction must occur because of the coincidence: a cs mutant (csH80) yielded upon reversion a ts mutation in cdc 2 and a pseudo-reversion of a ts allele of cdc 2 yielded a cs allele of the same gene as csH80. Such cases seem not to be rare.

Dominance Tests of Suppressors. The temperature-sensitive suppressors of cold-sensitivity exhibit two phenotypes: temperature-sensitivity for growth and suppression of the cold-sensitive cell cycle defect. In cases where the suppressor can be segregated free of the cs mutation, it is possible to analyze the dominance of both phenotypes in relation to the wild type allele. An extensive analysis of this type was carried out for the temperature-sensitive suppressor of the revertant rA18-95, and the results are shown in Table II. Diploid B indicates clearly that the temperature-sensitive suppressor is recessive to the wild type allele for growth at 35°, while diploid F demonstrates that the suppressor is dominant to the wild type allele for suppression of cold-sensitivity. An identical analysis was performed on the suppressor from the revertant reCC30-45; the results (not shown here) are the same as those for the rA18-95 suppressor.

These results are consistent with protein-interactions as the basis of suppression. The suppressor is dominant because it can repair the cs-defect by binding, but recessive as a ts-mutation since it has an essential function.

DISCUSSION

We found that several cold-sensitive mutations defective in the progress of the yeast cell cycle could be suppressed by mutations in other cell-cycle genes. The ease with which we found such suppression suggests that many cell-cycle genes specify proteins which interact with one another during the course of cell division. Pseudo-reversion of three different cold-sensitive cdc mutants resulted in strains with heat-sensitive mutations in cdc genes unlinked to the original cold-sensitive mutation. This implies that the inherent defect in many cdc mutants is a failure in function of an existing cell-cycle-specific protein and that this altered protein may be "corrected" by a compensating interaction with a second mutant cdc protein. In the case of phage assembly, such protein interactions could be documented directly (7).

TABLE II

DOMINANCE TESTS ON THE TEMPERATURE-SENSITIVE SUPPRESSOR IN rA18-95

	Strain	Genotype		Phenotype Growth	
		cs locus	ts locus	17°	35°
haploid control	rA18-9503C	cs	ts	+	-
A	csA1824D x 640	$\frac{+}{cs}$	$\frac{+}{+}$	+	+
B	rA18-9521D x 640	$\frac{+}{+}$	$\frac{+}{ts}$	+	+
C	rA18-9521D x csA1824B	$\frac{+}{cs}$	$\frac{+}{ts}$	+	+
D	rA18-9521D x rA18-9517C	$\frac{+}{+}$	$\frac{ts}{ts}$	+	-
E	csA1824 x csA1824B	$\frac{cs}{cs}$	$\frac{+}{+}$	-	+
F	rA18-9503C x csA1824B	$\frac{cs}{cs}$	$\frac{+}{ts}$	+	+
G	rA18-9503C x rA18-9503D	$\frac{cs}{cs}$	$\frac{ts}{ts}$	+	-
H	rA18-9521D x rA18-9503D	$\frac{+}{cs}$	$\frac{ts}{ts}$	+	-

Supporting evidence for this view has recently come from a different kind of experiment, cleverly devised by L. Sowder and B. Byers at the U. of Washington in Seattle, which involves construction of single cells composed of a wild type cytoplasm and a mutant cdc nucleus (9). They studied twenty different cdc mutations in this way. In all cases except one, a shift to restrictive temperature resulted in one to four extra division cycles before arrest at the terminal phenotype. These same cdc mutants normally arrest within a single division cycle. Apparently, the wild type cytoplasm supplied sufficient functional cdc protein for more than one generation, and this protein is gradually diluted by newly synthesized mutant protein. This result stands in opposition to the view that cell cycle gene products are necessarily synthesized anew during each cycle.

Although our results do not eliminate a "regulation of gene expression" model for control of the cell cycle, the finding that apparent interactions between two different cell-cycle-specific proteins can compensate for some cdc defects suggests that in a substantial number of cases, control may be exerted at the level of protein interaction.

ACKNOWLEDGEMENTS

This research was supported by a grant from NIH (GM21253). DM was supported by an ACS postdoctoral fellowship.

REFERENCES

1. Hartwell, L.H. (1974). Bacteriol. Rev. 38, 164.
2. Hartwell, L.H., Mortimer, R.K., Culotti, J., and Culotti, M. (1973). Genetics 74, 267.
3. Edgar, R.S., and Wood, W.B. (1966). Proc. Nat. Acad. Sci. USA 55, 498.
4. Kikuchi, Y., and King, J. (1975). J. Mol. Biol. 99, 695.
5. Apirion, D., and Schlessinger, D. (1967). J. Bacteriol. 94, 1275.
6. Tomizawa, J. (1971). In "The Bacteriophage Lambda" (A.D. Hershey, ed.), pp. 549-552. Cold Spring Harbor Press, Cold Spring Harbor, N.Y.
7. Jarvik, J., and Botstein, D. (1975). Proc. Nat. Acad. Sci. USA 72, 2738.
8. Hartwell, L.H., Culotti, J., and Reid, B. (1970). Proc. Nat. Acad. Sci. USA 66, 352.
9. Sowder, L., and Byers, B., personal communication.

A GENETIC ANALYSIS OF THE LOCUS CODING FOR ALCOHOL DEHYDROGENASE, AND ITS ADJACENT CHROMOSOME REGION, IN *DROSOPHILA MELANOGASTER*[1]

M. Ashburner[2], R. Camfield[3], B. Clarke[3], D. Thatcher[4] and R. Woodruff[2]

[2]Department of Genetics, University of Cambridge, Cambridge, [3]Genetics Research Unit, University of Nottingham Medical School, Nottingham and [4]Department of Molecular Biology, University of Edinburgh, Edinburgh, Great Britain.

ABSTRACT Alcohol dehydrogenase (ADH) in *Drosophila melanogaster* is a homodimeric protein of subunit molecular weight 24,000 daltons. Several electrophoretic and enzyme negative alleles have been mapped within the locus and the map orientated on the chromosome by determining the amino acid substitutions in electrophoretic variants. The structural gene for ADH is located within polytene chromosome bands 35B2 or 35B3. We discuss the results of attempts to 'saturate' genetically the chromosome region around *Adh*.

INTRODUCTION

Despite the degree of sophistication of many genetic studies with *Drosophila melanogaster* very few genes have been described by a combination of both genetic and molecular techniques. Indeed, although the structural genes for over 50 *Drosophila* proteins have now been mapped, and the coding regions for the ribosomal and many tRNAs identified, it is only for the *rudimentary* and *rosy* loci that we have both a detailed genetic map and some knowledge of the protein products. Even in these two well studied cases no protein sequence data are available. Thus it is impossible to distinguish rigorously between amorphic mutations that result from coding region changes and those that result from 'control' region changes. With the enzyme alcohol dehydrogenase (E.C.1.1.1.1.) we have a preliminary fine structure map and an almost complete primary protein sequence. We can now begin to consider the important question

[1]Supported by grants from the Medical Research Council to M.A., B.C. and Dr. R. Ambler (University of Edinburgh) and by grants from the Science Research Council to D.T.

ISBN 0-12-068350-4

whether the amorphic *Adh* mutations occur only in the coding region or whether they occur in both coding and 'control' regions. The purpose of this review is to present the 'state of the art' of studies on the *Adh* gene and its protein, drawing from the work we have done in the last few years and from the work of other groups, and particularly that of Dr. Sofer of The Johns Hopkins University.

RESULTS AND DISCUSSION

The Protein. The alcohol dehydrogenase (ADH) of *D.melanogaster* is a homodimeric protein of molecular weight 48,000 daltons, the two polypetide chains being 225 amino acid residues long. *In vivo* the substrate for ADH is assumed to be ethanol and other short chain (up to 7 or 8 carbon) primary and secondary alcohols. Since the habitat of *D.melanogaster* includes fermenting substrates, the bulk of these alcohols presumably comes from exogenous, rather than endogenous sources. The enzyme may, therefore, be considered to play a role in detoxification of ingested alcohols.

In genetically homozygous strains of *D.melanogaster* ADH may exist in up to three different forms (isozymes) separable by gel electrophoresis. These forms are inter-convertible and result from the binding to the enzyme of an NAD-carbonyl addition complex (1,2). This complex not only affects the electrophoretic mobility (i.e. net charge) of the enzyme, but also its activity and stability. Sofer and his colleagues have recently shown that feeding flies with acetone produces a dramatic reduction in the specific activity of ADH, presumably by driving all the enzyme into the form with two bound acetone-NAD molecules (3).

The question is often asked whether or not ADH activity is 'controlled'. In the sense that different tissues of the fly display very different specific activities and that the amount of ADH per fly varies with developmental age the answer to this question is 'yes'. In larvae the great bulk of an animal's ADH is found in one tissue, the fat body (4). Other tissues, such as the larval salivary gland (and all permanent *Drosophila* cell lines so far assayed) have little or no ADH activity.

Genetic Variation for ADH. Genetic variation affecting ADH was discovered soon after the first crude extracts of *Drosophila* were stained histochemically for the enzyme (5). In wild and laboratory populations of *D.melanogaster* five classes of electrophoretic variant are now known, although only two of them, the so-called fast (*F*) and slow (*S*) forms, are at all common. It is probable that this figure underestimates the degree of genetic polymorphism for ADH since there are no very strong reasons to believe that, for example, an *F* allele isolated from a Texan population codes for a protein of identical primary sequence to an *F* allele from a population found in a vineyard just outside Macon. Doubtless this question, which is not without interest to population geneticists, will soon be answered by studies of protein or nucleic acid sequence.

Electrophoretic variants of ADH may also be generated by mutagenesis. The Adh^D allele of Grell (6) was induced by ethylmethane sulphonate from an Adh^F allele and has proved to be a substitution of Glu for Gly. Some ADH negative (i.e. Adh^n) mutations produce a protein which, as well as being enzymically inactive, has an altered electrophoretic mobility compared to the enzyme coded by the parental allele. For example Adh^{n11} was induced with EMS from an Adh^F allele yet Adh^{n11}/Adh^F heterozygotes show a heterodimer with a mobility apparently due to the Adh^{n11} protein having a net charge similar to that of the Adh^{uf} enzyme (7). Our sequence studies show the amino acid substitution of Adh^{n11} to be an Asp for Gly change.

A second class of variation known to affect ADH alters the enzyme's stability without, necessarily, affecting its net charge or activity. There are several reports (e.g. 8,9) of thermostability variants for ADH and some of them have been mapped to the region of the *Adh* gene. However the absence of purified enzyme from such strains makes it difficult to know how much of the variation in thermostability is due to changes in ADH *per se*, and how much of it is due to other genetic differences between stocks. One EMS induced Adh^n (Adh^{n5}) has been shown to be a leaky allele whose enzyme is very thermolabile *in vivo* and in partially purified preparations (10).

An obvious class of genetic variation, if there is an interest in gene control, is that affecting the level of the gene's expression. In fact the analysis of differences in ADH specific activity between different strains of *Drosophila* proves to be more difficult than at first sight may appear. The reason is that the quantity of ADH per fly (or per unit of soluble protein or fly weight) is very susceptible to changes in nutritional conditions. Under some circumstances the assumption that the amount of enzyme is related to the first power of total protein (an assumption used in calculating "specific activity"), does not hold. In fact the amount of ADH per fly is related to total protein (or wet weight) by an allometric constant that varies between strains. Nevertheless, there is one example, so far, of a strain that does appear to be a true under-producer of ADH. The genetic difference responsible for this low activity has been mapped at, or very close to, *Adh* and it is cis-dominant (11). We must not, however, jump to the conclusion that this represents a 'control' mutation, since we do not yet know whether it affects the protein's primary sequence. In principle we can imagine that the low activities of ADH in strains carrying this variant might be due to an amino acid substitution decreasing the enzyme's stability. This would recombine with substitutions at other sites within the gene and be cis-dominant to them. Since we do not know whether the DNA coding for ADH is contiguous or interrupted, map distances between activity variants and *bona-fide* structural gene substitutions are a poor guide. This argument applies equally to variants affecting xanthine dehydrogenase activity, and mapping some distance proximal to the limits of the XDH structural gene as defined by the amorphic *rosy* mutations (12).

Nevertheless the search for, and analysis of, activity variants must continue, because the genetic and molecular tools for their rigorous analysis are now becoming available. For *Adh* several selection techniques for the recovery of such variants can be designed, but we know of none that has produced clear results.

Mutations deficient in ADH are very easy to recover after mutagenesis of *D.melanogaster*. They can be selected chemically (see below) or by the brute method of histochemically testing progeny of a mutagenised father and a mother carrying an *Adh* deficiency. Although chemical selection has been widely used, it probably yields only a proportion of induced ADH negative (Adh^n) mutations. For example Sofer and Hatkoff (13) recovered one Adh^n per 8,000 chromosomes after EMS mutagenesis, selecting adults with 1-pentene- 3-ol. By histochemical screening our recovery of Adh^n mutations after EMS is an order of magnitude larger and, indeed, corresponds well with the frequency of recovery of lethal mutations at the 'average' vital locus in *Drosophila*.

Aaron (14) has recently completed a large study of X-ray mutagenesis at *Adh* using 1-pentene-3-ol selection. The frequency of "point" Adh^n mutants was found to be 3×10^{-8} per rad. Fully two thirds of the mutants recovered were obvious deficiencies (that is included one or more neighboring lethal loci). In fact we have found that even some of the apparent 'point' nulls recovered by Aaron in these experiments are deficiencies, since they include a wing locus adjacent to *Adh* (*outspread*) and enhance the phenotype of *Scutoid* (Tsubota, unpublished).

An important question of new Adh^n mutations is whether or not they code for an enzymically inactive, but immunologically cross-reacting protein. Of sixteen different Adh^n's studied by Schwartz and Sofer (7) seven failed to precipitate an anti-ADH antibody in the Ouchterlony test. One of these (Adh^{nA}) did show an active enzyme on gels when heterozygous with Adh^{n11} and must therefore, produce a protein. With another (Adh^{n4}) we can detect CRM by radial immunodiffusion using either anti-ADH^F or anti-ADH^{n11} sera. Thus at least two thirds (11/16) of the nulls analysed are CRM positive. The levels of cross reacting material may equal the levels of protein in wild type strains, or they may be lower. For example, by radial immunodiffusion against an anti-ADH^F we estimate that there are 11.40+3.40μg of ADH per fly in the Adh^D strain. In the leaky Adh^{n5} and the non leaky Adh^{n4} mutants derived from Adh^D there are about 10.40$\pm$0.01μg and 4.3$\pm$0.01μg of ADH protein respectively.

The fact that an Adh^n mutant strain may have no detectable cross-reacting material is not, of course, a firm criterion for the genetic inactivity of the allele. The protein coded by the allele may be immunochemically non-competent or it may be very unstable.

Selective Techniques for The Analysis of the Adh Gene

We have already mentioned that chemical selection may be used to recover mutations affecting ADH activity. Adh^n homozygotes are normally viable, and are morphologically indistinguishable from wild type flies. Yet, as Grell (6) discovered, they are conditional lethals. They are killed on exposure to ethanol. Quite literally Adh^n flies become drunk and incapable in the presence of moderate (e.g. 8%) concentrations of ethanol. They lose motor co-ordination, roll over and eventually die. Flies with an active enzyme are able to withstand ethanol and, even to use it as an energy source (15). Thus ethanol selection allows, in principle, the recovery of rare Adh^+ flies from a large population of Adh^n flies. Sofer and his colleagues (13,16) have discovered techniques for the reciprocal selection, the recovery of rare Adh^n flies from a large population of Adh^+ flies. This is based on the fact that various unsaturated secondary alcohols (for example 1-pentene-3-ol or 1-pentyne-3-ol), at low concentrations are not toxic to *Drosophila*, but can be converted, by ADH, to very toxic ketones. Thus flies lacking ADH can survive under conditions that kill all wild type flies.

In practice the efficiency of ethanol selection is not great enough to allow its sole use for the efficient screening of rare *Adh*$^+$ recombinants. For this reason we have made strains in which various *Adh*^{n} chromosomes have been marked with lethals mapping at loci very close to *Adh*.

One of the aims of our work is to make a genetic map of the *Adh* locus using amorphic mutants and then to determine how much of this map is occupied by the coding region itself. In principle one of several different answers might be expected: for example the coding region might be co-extensive with the map based upon amorphs, or it might occupy only a limited region of the map. Following the work by Chovnick and his colleagues on the *rosy* gene we have attempted to define the coding region itself by the mapping of electrophoretic variants with respect to the amorphs. The general procedure for this is simple: to induce *Adh*^{n}'s on different electrophoretic alleles and then to progeny-test for their electrophoretic allele all *Adh*$^+$ chromosomes recovered as recombinants between different nulls. In principle, were an electrophoretic site to map between two nulls then the *Adh*$^+$ reciprocal recombinants between them would fall into two electrophoretic classes, whose frequencies would depend upon the distance between each null and the electrophoretic substitution. If, on the other hand, the electrophoretic substitution was either distal or proximal to both nulls then the recombinants would all be of one electrophoretic class. Of course, matters are not quite so simple when mapping over small distances, because many of the *Adh*$^+$ chromosomes come from non-reciprocal, rather than reciprocal events. Very close substitutions co-convert. In the *Adh* gene, which is about 1 x 10^{-3} map units long, recombination is due both to reciprocal exchange (or a process that mimics reciprocal exchange, i.e. is accompanied by exchange of outside markers) and gene conversion. The present map, which we stress is provisional and in need of verification, is shown in figure 1. It includes three different electrophoretic alleles (*Adh*^{D}, *Adh*^{Uf3} and *Adh*^{S}) and four nulls (*Adh*^{n2}, *Adh*^{n4}, *Adh*^{n11} and *Adh*^{nC1}). It is interesting that the 'size' of the gene, in terms of recombination, is about one-fifth that of *rosy* and that the sizes of the ADH and XDH polypeptides are in a similar proportion.

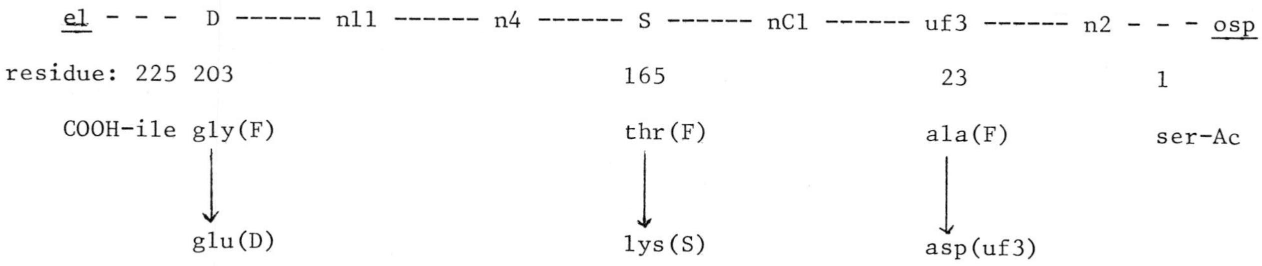

```
el - - -  D ------ n11 ------ n4 ------ S ------ nC1 ------ uf3 ------ n2 - - - osp

residue: 225 203                        165                 23         1

    COOH-ile gly(F)                     thr(F)              ala(F)     ser-Ac
              |                           |                   |
              v                           v                   v
             glu(D)                     lys(S)              asp(uf3)
```

Figure 1. A preliminary fine structure map of the Adh gene of *D.melanogaster* and its correlation to the amino acid sequence of the ADH protein. Adh^{D}, Adh^{S} and Adh^{uf3} are electrophoretic alleles and Adh^{n11}, Adh^{nC1} and Adh^{n2} are all EMS induced ADH negative mutants. The amino acid positions shown immediately below the map are those of the protein from a *b* Adh^{F} *pr* stock. The substitutions and their positions found in the three electrophoretic variants are shown. The substitutions in the Adh^{S}, Adh^{D} and Adh^{uf3} enzymes have previously been published (16,22,23).

The amino acid substitution of the three electrophoretic alleles are also shown on this map, as well as the absolute orientation of the gene, with respect to the centromere (which is to the right). We have also determined the amino acid substitution of a so far unmapped electrophoretic allele (F', glu for ala at residue 29 in a protein with the *Adh*^{S} lys substitution). The *Adh*^{n11} protein contains an asp for gly in a peptide for which our data do not yet allow a precise localisation.

It will be seen that there is, as yet, no evidence suggesting that the coding region is anything but co-extensive with the region within which mutation can abolish ADH activity. There remains, it is true a possibility that the CRM^{-} *Adh*^{n2} maps distal to the amino terminus but the data are inadequate to prove this.

The Genetic Environs of Adh. More than forty deficienices that include *Adh* have been recovered by the joint efforts of Sofer's group (20), Aaron (14) and ourselves (18, 21). They all show a dosage effect for ADH, the specific activity of deficiency heterozygotes being about 50% that of their *Adh*$^{+}$ sibs. Cytological study of these deficiencies locates *Adh* to bands 35B2 or 35B3 of chromosome arm 2L, a position fully consistent with that obtained from conventional genetic mapping.

The polytene chromosome region within which *Adh* is located is atypical in several respects. Region 34-35 has been known since the days of Calvin Bridges to be notorious for the extent to which it may undergo ectopic pairing with non-homologous chromosome regions. These ectopic pairing sites are, in addition, late replicating and show some evidence for local under-replication in the form of chromosome "weak points". These features of chromosome structure may all reflect the occurence of interstitial heterochromatin in region 34-35. An additional feature is shown by genetic studies. Figure 2 is a map of the small region immediately surrounding *Adh*. Of the deletions shown here 22 were selected, after X-ray or formaldehyde mutagenesis, on the basis of the fact that they delete *Adh*. Of the 44 independant break points 19 are within the small (8 band) region shown in figure 2. It is clear from the figure that these break points are distributed non-randomly in relation to the eight loci shown. Between *ScoR* and the lethal *br3* there are nine break-points, between the lethal *br29* and *Adh* five, yet there are none between *osp* and *ScoR* and only one between *br3* and the locus to its immediate right (*br4*, not shown on this figure). The basis for this clustering of deficiency end points is unclear.

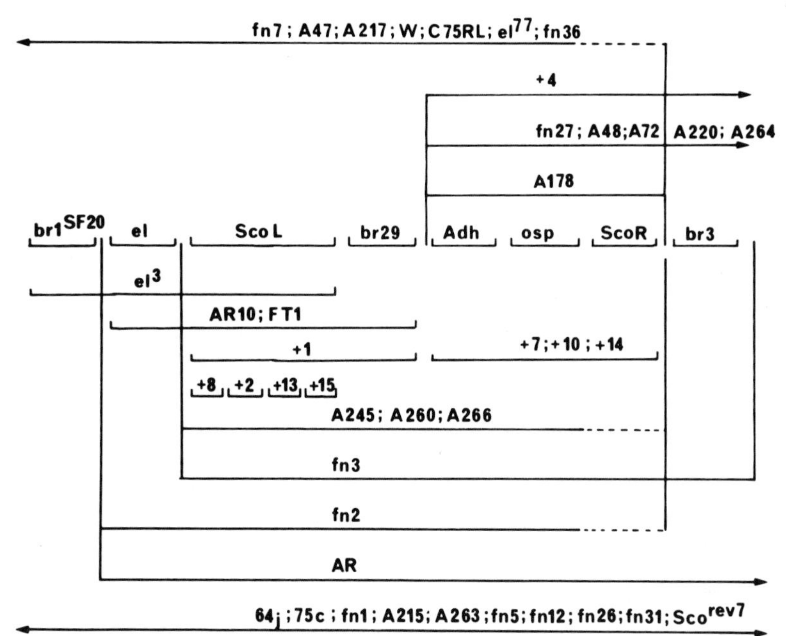

Figure 2. A genetic map of the immediate environs of *Adh*, representing approximately bands 35A1 to 35B4 of chromosome arm 2L (8 bands). Eight loci are shown, these are the lethals *br1*, *br29* and *br3*, two loci affecting wing phenotype, elbow (*el*) and outspread (*osp*), the two components of the Scutoid (*Sco*) complex which affect macrochaetae and *Adh* itself. The region is approximately 0.4 map units long, *br1* mapping 0.20 map units to the left of *Adh* and *br3* 0.18 map units to the right. The genetic break-points of 27 deficiencies within the region are indicated, those shown by arrows to the left or right being outside this immediate region. The mutants symbolised as *+1* to *+15* are induced revertants of the dominant Scutoid. Of these, five (*+1*, *+4*, *+7*, *+10* and *+14*) behave as deficiencies, the others as alleles of *ScoL* (see ref. 21). Deficiencies with a *fn* prefix were induced with formaldehyde, those with an *A* prefix by X-rays (except *A260* which was spontaneous). *el3* was induced with EMS, *W* was spontaneous and the remainder were induced with X-rays except that the right hand limit of *C75RL* comes from a spontaneous aberration (see ref. 17 for further details). We are grateful to Sidney Aaron, Ed Grell and Bill Sofer for deficiency stocks.

One possible reason is that breakage is occuring, in the sperm chromosome (the mutagen target), in sequences that are both genetically silent and are under-replicated during the process of polytenisation. Thus the mutational target in these regions may be much larger than either the genetic data or the polytene chromosomes would indicate.

Apart from the clustering of break-points the cytological features of the region 34-35 do not appear to be reflected in its genetic organisation. For example within *Df(2L)64j*, a large deficiency lacking about 34 chromosome bands, both we (18,19) and O'Donnell et al (20) have isolated a large number of lethal and visible mutations. These map, by complementation tests, to about 30 complementation groups (the uncertainty is due to examples of complex complementation pattern whose genetic basis remains to be discovered). The gross numerical relationship between 'genes' and 'bands' is not, therefore unusual. The region covered by *Df(2L)64j* is approximately 0.6% of the total cytological map (in terms of band number) and represents about 0.76% of the total genetic map. There is about 0.07% recombination per band, and adjacent complementation groups are of the order of 0.10% apart. These figures may be compared with those for the zeste-white region, cytologically more typical, where there is about 0.054% exchange per band and where adjacent complementation groups are about 0.08% apart.

Since flies that are homozygous for an Adh^n allele are both viable and fertile (at least in the absence of ethanol) we might expect homozygotes for an *Adh* deficiency also to be viable. Indeed they are, though the situation is rather more complex than might be expected. It can be seen from the map (fig. 2) that there are three loci (*Adh*, *osp* and *ScoR*) included within a region unbroken by any deficiency. We know from genetic data that *Adh* and *osp* are very closely linked indeed, *osp* being less than 0.01% to the right of *Adh*. The position of *ScoR* relative to these loci is uncertain, since *Sco* severely decreases the exchange frequency around *Adh*, but the data suggest that it is to the right of *Adh* (20,21). Some homozygous viable Adh^n mutants behave as if they were deficient for *Adh* (since they are ADH negative), for *osp* (since they shown an extreme *osp* phenotype over recessive *osp* alleles), and for *ScoR* (since they enhance the phenotype of *Sco*). Flies of a very similar phenotype can be readily generated by various trans heterozygous combinations of long deficiencies: for example *Df(2L)fn7* is broken just proximal to *Adh* and extends distally over at least 12 loci, on the other hand *Df(2L)A48* is broken just distal to *Adh* and extends proximally over at least 12 loci. Heterozygotes between these two deficiencies are viable and express an *Adh*⁻ *osp*⁻ phenotype. The cytological data, though ambiguous because of difficulties in interpreting the banding pattern indicate that these flies are homozygously deleted for the adjacent bands 35B2 and 35B3.

The closestknown locus to *Adh* on its proximal side is *osp*. Although we have so far failed to separate *Adh* from *osp* by a deficieny break-point we think that this reflects only their closeness, and not a functional relationship. Many *Adh*^{n} alleles are *osp*$^{+}$ and, conversely, we have *osp* alleles that are *Adh*$^{+}$. All *Adh* alleles that also have an *osp* phenotype behave as deficiencies by other genetic criteria. The situtation distal to *Adh* is less clear. By exchange the closest established locus is *elbow*, this is 0.16% distal to *Adh* and maps cytologically to bands 35A1.2, about three bands to the left of *Adh* (the map position of *elbow* is based upon the break-point of a *T(Y;2)el* chromosome and deficiency mapping). The analysis of the interaction of deficiencies for this region with *Sco* and the analysis of a number of X-ray induced revertants of *Sco* indicate that the left hand component of the *Sco* complex (*ScoL*) and a lethal *(br29)* are located between *el* and *Adh* (21). However in the absence of point alleles of *br29* (all we have are small deficiencies such as *AR10*, *FT1* and *Sco*$^{R+1}$) this map must be regarded as hypothesis, rather than established fact.

In the long run we regard the opportunity to extend the fine structure analysis of *Adh* into its neighboring loci as an exciting one. We believe that detailed genetic analyses of this type, combined with a more direct approach to gene and chromosome organisation through recombinant DNA technology, is the recipe for an understanding of the rules that relate genetic organisation, and the control of gene expression, to chromosome structure.

ACKNOWLEDGEMENTS

We are indebted to Pam Angel, Alison Galvin, Trevor Littlewood, Christopher Pitts and Anne Tomsett for their technical assistance at all stages of our work. We are also very grateful to Drs. Sidney Aaron, Ed Grell and Bill Sofer for their generosity in sending us stocks and sharing their work with us in advance of its publication.

REFERENCES

1. Jacobson, K.B. (1968). Science 159, 324.
2. Schwartz, M., and Sofer, W. (1976). Nature, 263, 129.
3. Papel, I., Henderson, M., van Herrewege, J., David., J. and Sofer, W. (1979). Biochem. Genetics, In press.
4. Ursprung, H., Sofer, W. and Burroughs, N. (1970). Wilhelm Roux' Archiv. 164, 201.
5. Johnson, F., and Denniston, C. (1964). Nature 204, 906
6. Grell, E.H., Jacobson, K.B., and Murphy, J.R. (1968). Ann. N.Y. Acad. Sci. 151, 441.
7. Schwartz, M., and Sofer, W. (1976). Genetics 83, 125.
8. Thorig, G.W., Schoone, A.A., and Scharloo, W. (1975) Biochem. Genetics 13, 721.
9. Sampsell, B. (1977). Biochem. Genetics, 15, 971.
10. Vigue, C. and Sofer, W. (1974). Biochem. Genetics 11, 387.
11. Thompson, J., Ashburner, M., and Woodruff, R.C. (1977) Nature, 270, 363.
12. Chovnick, A., Gelbart, W. and McCarron, M. (1977). Cell, 11, 1.
13. Sofer, W., and Hatkoff, M.A. (1972). Genetics 72, 545.
14. Aaron, C.S. (1979). Mutation Res. In press.
15. van Herrewege, J., and David, J. (1974). C. rend. Acad. Sci. Paris, 279D, 335.
16. Fletcher, T.S., Ayala, F.J., Thatcher, D.R., and Chambers, G.K. (1978). Proc. nat. Acad. Sci. U.S.A., 75, 5609.
17. O'Donnell, J., Gerace, L., Leister, F., and Sofer W. (1975). Genetics 79, 73.
18. Woodruff, R.C., and Ashburner, M. (1979). Genetics, In press.
19. Woodruff, R.C., and Ashburner, M. (1979). Genetics, In press.
20. O'Donnell, J., Mandel, H.C., Krauss, N., and Sofer, W. (1977). Genetics 86, 553.
21. Ashburner, M., Woodruff, R.C., and Camfield, R, (1979). Genetics, In press.
22. Schwartz, M., and Jörnvall, H. (1976). Europ. J. Biochem. 68, 159.
23. Camfield, R.G. and Thatcher, D. (1977). Biochem. Soc. Trans. 5, 271.

ALLELIC COMPLEMENTATION AND TRANSVECTION IN *DROSOPHILA MELANOGASTER*

Burke H. Judd
Department of Zoology, The University of Texas,
at Austin, Austin, Texas 78712

ABSTRACT Interactions between alleles in homologous chromosomes may offer important clues to the function and regulation of eukaryotic genes. Complementation among alleles is usually attributed to the formation of a multimeric protein where subunits encoded by two different alleles can still form a functional complex. However, the large number of loci that exhibit allelic complementation in Drosophila compels alternative explanations. The study of several such complex loci leads to the suggestion that there may exist two general categories of regulatory mechanisms that can explain allelic complementation and a related phenomenon, transvection. The mechanisms are (i) multiple sites for the initiation of transcription of a gene, where each site responds to cell-type or stage specific activation signals and (ii) post-transcriptional recombination of RNA molecules. This latter set of mechanisms also explains transvection effects. Transvection is a position effect in Drosophila seen as an intensification of the mutant phenotype of alleles in trans-configuration by a heterozygous chromosomal rearrangement that upsets pairing between homologs. Examples of allelic complementation and transvection are given and some speculative models on the molecular basis for these phenomena are discussed.

INTRODUCTION

Although details are lacking, there are some general features of eukaryotic gene control that are taking shape. Some examples of genes that act in trans-configuration as regulators for other loci are known (1; 2), and more recently cis-acting control elements tightly linked to or integral parts of structural elements have been identified (3; 4; 5; 6). The very large average size of eukaryotic genes, the lack of an indication of polycistronic transcription and the small fraction of nuclear RNA that becomes messenger RNA all point clearly to a consideration that a large proportion of a locus may be involved in regulatory roles.

ISBN 0-12-068350-4

Complex loci in Drosophila exhibit some characteristics that I believe give valuable clues about some aspects of gene function and regulation. There are two phenomena in particular that command attention in this regard, namely allelic complementation and transvection. I would like to discuss these characteristics and offer some examples and at the same time speculate on the mechanisms that underlie these expressions of gene function and regulation.

As a framework for this speculation, I would like to mention two general categories of control mechanisms that the analysis brings to mind. The first is that there may be more than one initiation site for transcription of a gene. Differential activity of the gene could be controlled if each initiation site responds to a signal limited to specific cell types or developmental stages. Second, I suggest that an important aspect of regulation may involve the post-transcriptional processing, recombination and utilization of RNA transcripts.

ALLELIC COMPLEMENTATION

Typically allelic complementation occurs when a heterozygote, a_1/a_2, produces a wild-type phenotype while the combinations a_1/a_3 and a_2/a_3 are mutant. It is of course necessary that a_3 is recombinationally separable from a_1 and a_2 to rule out the possibility that it is a deletion that encroaches on two different units of function.

The conventional explanation for allelic complementation is that a polypeptide encoded by a single locus functions as a dimer or higher multimer. Modifications of the polypeptide structure may render it inactive in a homozygote while the two different chains produced by a heterozygote might be able to form a multimer with normal or nearly normal function.

There are no proven examples of allelic complementation that involve mechanisms other than the interaction of polypeptide chains, but as I have pointed out earlier (7), the rather high proportion of loci that exhibit this characteristic prompts me to examine alternative explanations. I have suggested that regulated loci that are active in several stages or cell types as development proceeds may respond to a number of stage or tissue specific activation signals. Mutations in the regulatory sequences of a locus may make it insensitive to one type of signal but leave other activation sites unimpaired. Different mutations of this type might be expected to complement each other while failing to complement alleles that modify the sequence encoding the polypeptide chain or those that totally inactivate the locus.

Cut Locus Complementation. The organization of the cut locus of Drosophila melanogaster makes it a candidate for regulation by a multiple activation-site system (8). A cluster of lethal alleles maps at the proximal end of the locus. Alleles causing modifications of wing structures are located in the central region of the gene while the most distal segment is occupied by alleles that affect leg structure. Analysis of a number of mutations, including a series deletions and other rearrangements, shows that those parts of the locus containing the morphological mutant sites cannot mutate to a lethal state. Lethality is not achieved even when the distal segment plus a part of the central segment are removed. The interpretation that best fits the genetic and cytological observations is that there is a single gene product encoded by the sequence near the proximal end of the cut gene. The product appears to be indispensible in some way so that mutations to the null state are lethal. Mutations that produce only morphological changes are thought to be lesions in regulatory sequences that result in reduced activity in particular tissues or at specific stages. Complementation is expected in individuals heterozygous for mutants in different activator units since the structural sequence is normal in both homologs and each defective activator has a normal copy in the homologous chromosome. This is precisely the pattern exhibited by cut alleles. Those affecting wing morphology complement with those affecting leg structures. Lethals, on the other hand, complement neither each other nor any of the alleles causing morphological changes. The former cases produce lethality, the latter give phenotypes typical of the morphological mutants.

Quite a number of complex loci in Drosophila show allelic complementation by morphological alleles and the absence of it for the most extreme (lethal) alleles. Also typical of complex loci is a pleiotropic phenotype where different complementing alleles often show rather uniquely different effects on developmental pattern. Some examples are the Notch locus (9, 10, 11, 12), the bithorax locus (13, 14), the dumpy locus (15, 16, 17) and the Nasobemia-Antennapedia-proboscipedia-Multiple sex comb-Extra sex comb-Humeral gene complex (18, 19, 20).

Other ways that allelic complementation can be explained are to invoke mechanisms for the recombination of RNA transcripts. These can be by recombining RNAs that function in RNA-protein complexes or in multimeric RNA complexes or as mRNAs that are covalently spliced together during processing. Schemes such as these will also explain the phenomenon known as transvection, described by E. B. Lewis (21).

TRANSVECTION

During the analysis of mutants of the bithorax complex, Lewis noted that the phenotype of the _bx/Ubx_ heterozygote was significantly more abnormal if a chromosome rearrangement with a breakpoint near the bithorax locus was introduced into one of the homologous chromosomes. Rearrangements that are most effective in producing the transvection effect have breakpoints proximal to the bx locus and upset the somatic pairing of homologs as viewed in polytene cells. Lewis interpreted this effect to mean that gene products can be transported between homologs and that the transport process is upset when homologs can no longer pair normally.

Lewis regarded the mutants exhibiting transvection as marking units that have separate and distinct functions. He pictured the product of one unit being utilized or modified by the next unit in a sequential reaction. Genetic blocks in different units could be overcome in trans-heterozygotes if products were available by diffusion to the functioning units of paired homologs.

Clearly it is not logical to assume that the products being exchanged by diffusion are polypeptides because they should be as readily available to functioning units on unpaired as on paired chromosomes. Transvection and allelic complementation could possibly be closely related effects if (i) there are functions that are carried out by RNAs in the nucleus and/or (ii) there are mechanisms for recombining segments from two different RNA transcripts either in a complex or actually splicing them together as occurs during some of the processing of mRNAs. One scheme of this type is diagrammed in Figure 1.

The figure pictures a large initial transcript that is then processed to mature mRNA by cutting and splicing together three different segments of the original transcript. The recombination mechanism suggested here is the exchange of fragments from different transcripts during the processing steps. The model assumes that there are specific sites at which the nicking and splicing occurs. Such sites demarcate boundaries of subunits that compose the mature message. The model predicts that allelic complementation and transvection will be possible only between alleles whose mutant sites are in different subunits. Mutants that fail to complement then might tend to cluster in particular places in the genetic map while those that complement would map to different clusters. Such a clustering is typical of several complex loci in Drosophila.

Transvection can now be understood if the transcripts from the genes on the two homologs can undergo recombination. In trans-heterozygotes a normal messenger might be spliced together by recombining the functional pieces of two RNA transcripts. The position of the two homologous chromosomes would then play an important role in gene expression in trans heterozygotes because the unpaired condition would make it very much less probable that recombination between products from two different transcripts could occur. The same would

23. Ashburner, M. (1969). Chromosoma (Berlin) 27, 156.
24. Korge, G. (1977). Chromosoma (Berlin) 62, 155.
25. Jack, J.W. and Judd, B.H. (1979). Proc. Natl. Acad. Sci. USA 76, 1368.
26. Judd, B.H. (1961). Proc. Natl. Acad. Sci. USA 47, 545.
27. Pontecorvo, G. (1952). Symp. Soc. Exptl. Biol. 6, 218.
28. Lewis, D. (1961). Genet. Res. 2, 141.
29. Apirion, D. (1966). Genetics 53, 935.
30. Burton, E.G. and Metzenberg, R.L. (1972). J. Bacteriol. 109, 140.
31. Dietrich, P.S. and Metzenberg, R.L. (1973). Biochem. Genet. 8, 73.
32. Broker, T.R., Chou, L.T., Dunn, A.R., Glinas, R.E., Hassell, J.A., Klessig, D.F., Lewis, J.B., Roberts, R.J. and Zain, B.S. (1977). Cold Spring Harbor Symp. Quant. Biol. 42, 531.
33. Berget, S.M., Berk, A.J., Harrison, T. and Sharp, P.A. (1977). Cold Spring Harbor Symp. Quant. Biol. 42, 523.
34. Leder, P., Tilghman, S.M., Tiemer, D.C., Polsky, F.I., Seidman, J.G., Edgell, M.H., Enquist, L.W., Leder, A. and Norman, B. (1977). Cold Spring Harbor Symp. Quant. Biol. 42, 915.
35. Tonegawa, S., Brack, C., Hozumi, N. and Pirrotta, V. (1977). Cold Spring Harbor Symp. Quant. Biol. 42, 921.

GENETIC ORGANIZATION OF THE ROSY LOCUS IN *Drosophila melanogaster*: PILOT STUDIES ON THE INDUCTION AND ANALYSIS OF LOW XDH ACTIVITY MUTANTS OF THE ROSY LOCUS AS PUTATIVE CONTROL MUTANTS*

Stephen H. Clark, Arthur J. Hilliker,† and Arthur Chovnick

Genetics and Cell Biology Section, The Biological Sciences Group, University of Connecticut, Storrs, Connecticut 06268

ABSTRACT. The rosy locus (3-52.0) defines a segment of DNA associated with the biogenesis of the enzyme xanthine dehydrogenase (XDH). In addition to containing amino acid sequence coding information, we assume that the locus possesses transcription and translation regulating sequences. Previous reports from this laboratory have described experiments which define the left border of the XDH structural element as well as a *cis*-acting control element immediately to the left of the structural element. In an effort to further elaborate the organization of this genetic unit, a series of mutagenesis and selection experiments were conducted to produce an array of low XDH activity variants (XDH lows). One selection protocol involved an F_2 screening system which employed purine as a selective agent (flies with low levels of XDH activity die on purine supplemented media). This procedure did produce an array of XDH lows. However, the labor involved precludes large-scale mutagenesis experiments. A second screening system was developed which involves F_1 selection of XDH lows by virtue of the fact that flies with low levels of XDH have a rosy mutant eye phenotype when reared on standard medium supplemented with the XDH inhibitor 4-hydroxypyrazolo(3,4-d) pyrimidine (HPP or allopurinol). The results of biochemical and genetic analysis on several induced XDH lows is discussed.

*This investigation was supported by research grant GM09886 from the Public Health Service and a postdoctoral fellowship to S.H.C., GM05260, from the Public Health Service.

†Present address: Plant Industry, CSIRO, P. O. Box 1600, Canberra City, A.C.T. 2601 Australia

ISBN 0-12-068350-4

INTRODUCTION

In recent years much attention has been focused on questions concerning the regulation of gene activity in higher organisms. Numerous experimental approaches have been employed to this end and the results of such studies have been amply reviewed (*i.e.* Cold Spring Harbor Symposium, 1977). Despite the considerable increase in our knowledge of gene organization in recent years, our understanding of the functional organization of the eucaryotic gene remains primitive. Genetic studies suggest that the eucaryotic gene is a bipartite entity consisting of a non-coding, control element which lies adjacent to a coding or structural element (6, 2). To fully understand the control mechanisms associated with such a genetic unit one must undertake a coordinated genetic and biochemical analysis of individual genetic loci. The rosy locus in *Drosophila melanogaster* provides an ideal system for this type of experimental approach. The rosy locus and surrounding region has been well defined genetically (manuscript in preparation) and a *cis*-acting control element has been described (1, 5). The present work describes a series of experiments designed to produce an array of rosy locus variants which reduce the level of XDH activity. In this experimental system XDH lows can be further characterized as to the type of genetic lesion(s) that caused the reduced levels of XDH activity. The production and analysis of such variants serves a dual function in providing insight into the genetic mechanisms associated with XDH biogenesis, as well as to produce XDH activity variants on an isoallelic background. The latter is of considerable import for future molecular analysis of the rosy locus since the currently available control variants were isolated from different wild type alleles found in natural populations. Given the high degree of genetic variability to be found in such isolates from natural populations, identification of DNA sequences related to specific control variants may be impossible. In contrast, control variants induced on a specific wild type isoallele should be ideal material for such analysis.

RESULTS AND DISCUSSION

Two mutation and selection protocols have been designed to recover rosy locus variants with reduced levels of XDH activity. The first system makes use of the fact that flies with low levels of XDH activity will possess a wild type eye color, and like XDH null mutants, will die during larval development (or exhibit dramatically delayed development) when reared on purine supplemented medium. The details of this selective system have already been published (3). Basically, the approach involves the treatment of males bearing a defined wild type ry^+ isoallele with EMS and mating such males to females carrying appropriate balancer third chromosomes. F_1 male progeny from this cross are selected and pair mated to appropriately marked rosy females. Two brood vials of such pair matings are established, one vial containing purine medium and a second vial containing only standard medium. Selection of potentially interesting variants is determined by larval death or delayed development in the purine treated culture. These purine sensitive chromosomes are then isolated from the untreated vial and a stock established for further analysis. Several purine sensitive alleles were isolated in this fashion in a selective screen which examined over 5,000 mutagenized chromosomes.

Utilizing a series of biochemical tests the XDH low activity variants (XDH lows) were subdivided into several classes (details of this biochemical analysis will appear elsewhere).

One class of mutants has normal levels of XDH CRM but reduced levels of XDH activity indicative of XDH structural modification that lowers the enzyme's specific activity. A second type of alteration apparently alters the *in vivo* stability of the XDH monomer. (Active XDH exists as a homodimer.) This class exhibits both reduced levels of XDH activity and XDH CRM as a homozygote, but normal levels of hybrid dimer when heterozygous with a normal XDH isoallele. By these criteria the two above classes of XDH lows are classed as alterations in the coding or structural element of the gene, and not particularly interesting for an analysis of genetic mechanisms associated with the regulation of XDH biogenesis. However, to confirm the accuracy of the biochemical analysis several of these structural variants were subject to genetic fine structural analysis. All such biochemically defined structural variants mapped within the XDH structural element of the rosy locus. A third class of XDH lows consisted of two different variants which had a concomitant reduction of XDH CRM and XDH activity and low levels of hybrid dimer when heterozygous with a normal XDH isoallele. These variant alleles were designated as putative control variants since none of the tests could detect a structural alteration in the XDH peptide. These putative control variants were then subject to extensive fine structure analysis to determine their map position relative to the previously defined control and structural regions of the rosy locus. (Details of this genetic analysis will appear elsewhere.) Both of the mutants in this class mapped within the defined structural region of the rosy locus. The simplest interpretation is that these variants represent alterations in the amino acid sequence of the XDH peptide which reduce peptide stability (*i.e.*, half life) *in vivo*, and hence, the low CRM, low monomer and low XDH activity.

It became obvious from this study that there may be many sites within the XDH structural element that, upon mutation, can lead to mutant phenotypes that simulate putative "underproducer" control variants. Moreover, the number of control element sites capable of mutation to yield a clearly identifiable underproduction of XDH may be few. That the control element is capable of such mutation is proven (5).

It becomes apparent, then, that a large number of XDH lows must be induced and analyzed in order to insure success in the recovery of induced control variants. Unfortunately, the purine selection scheme (described above) cannot be used in a large-scale mutagenesis experiment. It requires two generations before detection of exceptions, and each treated chromosome is screened in pair-matings, and in two broods (with and without purine supplementation). The labor requirement of this protocol for a large-scale experiment is enormous. Consequently, a second mutagenesis and screening procedure was developed that will permit such large-scale screening. The new system makes use of the XDH inhibitor, 4-hydroxypyrazolo(3,4-d) pyrimidine (HPP or allopurinol). When added to standard fly medium in appropriate concentration, HPP will produce rosy eye color phenocopies in a genotype with low levels of XDH, while flies with normal levels of enzyme will have a wild type eye phenotype on the same HPP supplemented medium (4). Thus, mutagenized males can be mated to appropriate rosy tester females and F_1 progeny screened directly on HPP supplemented media. Flies with a rosy eye phenotype are easily selected, and the chromosome isolated for further analysis. Not only does this procedure permit massive screening, but it also permits recovery of still another class of interesting variant that would not have been detected by the purine selection system. In wild type strains, XDH activity increases during larval development reaching a peak at the prepupal stage. During the early stages of pupal development, XDH activity declines and subsequently rises again during the later stages of pupal differentiation (2). We believe that it is during these later stages, when eye pigments are being synthesized, that the HPP inhibition of XDH activity leads to the production of rosy phenocopies. Thus, mutants identified by the HPP screening procedure might be associated with a general decrease in XDH activity throughout development, or result from a decrease in activity late in development when eye pigment is being synthesized. Such a variant might reflect an alteration in the temporal control of XDH biogenesis. The former class of allele (*i.e.*, lower levels of XDH activity throughout development) would be both purine sensitive and exhibit a rosy eye color when reared on HPP supplemented media. The latter class (normal larval levels of XDH activity but lowered levels during the later stages of pupal development) would not be sensitive to purine in the medium but would phenocopy on HPP supplemented food. A preliminary experiment utilizing the HPP selection protocol did produce the types of variants described above. One class of variant is both purine sensitive and produces a rosy eye phenotype when reared on HPP supplemented media, and a second variant class is not sensitive to purine but does phenocopy on HPP supplemented media. Further biochemical and genetic analysis of these variants is currently in progress and details of these experiments will be forthcoming.

REFERENCES

1. Chovnick, A., Gelbart, W., McCarron, M., Osmond, B., Candido, E.P.M. and Baillie, D.L. (1976) *Genetics* 84, 233.
2. Chovnick, A., McCarron, M., Hilliker, A., O'Donnell, J., Gelbart, W. and Clark, S. (1977) *Cold Spring Harbor Symp. Quant. Biol.* 42, 1011.
3. Gelbart, W., McCarron, M. and Chovnick, A. (1976) *Genetics* 84, 211.
4. Keller, E.C. and Glassman, E. (1965) *Nature* 208, 202.
5. McCarron, M., O'Donnell, J., Chovnick, A., Bhuller, B.S., Hewitt, J. and Candido, E.P.M. (1979) *Genetics* 91, in press.
6. McClintock, B. (1951) *Cold Spring Harbor Symp. Quant. Biol.* 16, 13. _____ (1956) *Cold Spring Harbor Symp. Quant. Biol.* 21, 197.

DOES POSITION-EFFECT VARIEGATION IN DROSOPHILA RESULT FROM SOMATIC GENE LOSS?[1]

Steven Henikoff

Department of Zoology, University of Washington
Seattle, Washington 98195

ABSTRACT Chromosome aberrations that place a euchromatic gene locus near heterochromatin often result in a variegated phenotype in which the gene is fully active in some cells and apparently inactive in others. Since the gene itself is not permanently altered in the germ line, the possibility exists that mutant tissue results from physical loss of the gene in some somatic cells. In order to test this possibility, polytene chromosomes were examined from heat-shocked Drosophila larvae heterozygous for a normal sequence chromosome and an aberrant homolog with a puff-site placed near heterochromatin. Occasionally, the heat-shock locus did not puff on the mutant homolog in nuclei where the normal chromosome puff was fully active. ^{3}H-labeled RNA probes were used to detect puff-site sequences by hybridization in situ. In most cases, specific labeling near the euchromatin-heterochromatin junction of the aberrant homolog was apparent, even though no puff could be seen. This suggests that variegation can occur without the loss of a gene's DNA sequence. The level of labeling observed was usually much less than expected, as if variegation might cause under-replication or partial gene loss. Using a different approach, a particular model of gene excision was tested by examining eye pigment of flies carrying a "white-mottled" chromosome. When this chromosome was removed from dividing cells by mitotic recombination, the resulting clones lacked pigmented spots that would have revealed the presence of excised fragments.

[1]This work was supported by a fellowship grant from the Leukemia Society of America and by Grant GM 19179-05 from the National Institute of Health.

ISBN 0-12-068350-4

INTRODUCTION

Position-effect variegation was first observed by Muller fifty years ago as the mutant expression of a *Drosophila* gene in some cells of a tissue but not in others (reviewed by Spofford (1)). Variegation can occur when a gene lies near a chromosome aberration breakpoint in a heterochromatic region. In spite of an impressive accumulation of diverse and interesting observations on the phenomenon, its mechanism is still a subject of speculation. Mutant expression is generally thought to be caused by inactivation of the gene (1), although physical loss of the gene from somatic cells has not been ruled out. More than forty years ago, Schultz (2) proposed that excision of rings by intrachromosomal recombination in somatic cells might cause variegation. Schultz' hypothesis seems particularly attractive today in view of recent findings that somatic recombination is involved in the production of antibody diversity (3), and that non-chromosomal ribosomal DNA fragments exist in *Drosophila* (4). Also, based on their studies of *Drosophila* ribosomal genes in flies with rearranged X-chromosomes, Zuchowski-Berg (5) and Procunier and Tartof (6) have independently suggested that variegation is due to gene loss.

Baker (7) has suggested that many observations are irrelevant to the underlying mechanism of variegation. Therefore, he recommended looking at variegation of salivary gland puffs in order to study the phenomenon at the level of the chromosome itself. Previously, Schultz (8) had reported an example of variegation for a particular developmental puff lying next to a breakpoint in heterochromatin. Compared to the homologous normal puff, the puff on the aberrant chromosome was reduced in size. Rudkin's photograph of an example of this puff-site indicated that it was not distinguishable from the heterochromatic regions of the chromocenter (9). Since the normal euchromatic banding pattern was not preserved where the puff was absent, nothing could be determined concerning the presence or absence of the DNA sequences for the puff gene.

To test for somatic gene loss in variegated tissues, I have taken two quite different approaches. The first asks whether cells that are mutant for a gene in variegating tissue still carry that gene. This type of experiment theoretically is capable of ruling out gene loss as causing variegation. The second approach asks whether expression of a variegating gene can be detected in a cell from which the aberrant chromosome has been removed. This type of experiment is theoretically capable of demonstrating that extrachromosomal genes, perhaps generated by excision, exist in affected cells.

RESULTS AND DISCUSSION

Puff Variegation. Advances in the study of *Drosophila melanogaster* heat shock puffs (10-14) have made a molecular analysis of puff variegation possible. Probes are now available so that the presence or absence of DNA sequences at puff-sites can be determined using the technique of hybridization *in situ*. Therefore, an attempt was made to find variegated expression of a heat-shock puff. Unfortunately, there is presently no direct selection scheme for this. Instead, an existing stock was used. The translocation T(Y;3)A78, constructed by Lindsley, Sandler, and co-workers (15), is broken in the mostly heterochromatic long arm of a Y chromosome marked with *y*+ and $\underline{B}^s$, and in chromosome 3 in subdivision 87B. (Mutations are described by Lindsley and Grell (16)). One of the most prominent heat shock puffs, at band 87C1, lies very near the third chromosome breakpoint. When salivary chromosomes from heat-shocked individuals heterozygous for the translocation and a wild-type third chromosome are examined, nearly all of the nuclei show puffs of approximately equal size at region 87C of both homologs. Very rarely, however, a striking difference is seen between the two puff sites as shown in Figure 1. While the normal homolog shows a prominent puff at 87C, the translocated puff-site shows neither a puff nor a clear band. Such a difference in behavior of homologs has never been reported except for the example of variegation mentioned earlier. It seems likely that the occasional absence of puffing at the heat-shock locus is due to a position effect induced by nearby heterochromatin.

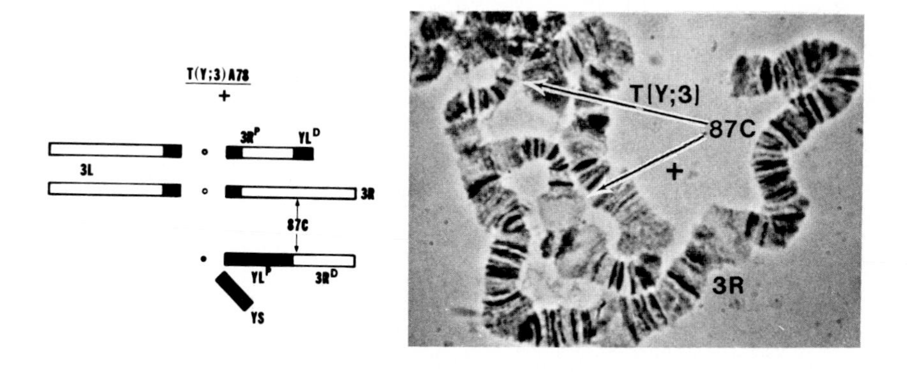

FIGURE 1. Schematic diagram of T(Y;3)A78/+ and polytene chromosome example of a position effect on the 87C puff-site.

Because the heat-shock puff-site on the aberrant chromosome was so seldom inactive, I hoped to increase the frequency by using genetic and environmental modifiers which enhance variegation. It has been known for decades that removal of heterochromatic regions from the chromosome complement of an individual as well as growth at low temperature usually enhance variegation (1). Therefore, larvae of the genotype C(1)DX,*y f*;+/T(Y;3)A78,*y*+B^S; M(2)$S2^{10}$/+ were used because they probably lack most of the X-heterochromatin and nearly all of the proximal heterochromatin of chromosome 2R in one homolog (16). When these larvae were grown at 18° and heat-shocked, an increase in the frequency of inactive puff-sites on the aberrant homolog was noted, reaching 20% for some individuals. Although it is possible that some unknown factor led to this increase, it seems likely that enhancement of variegation occurred because of these modifiers.

Hybridization in situ to a Variegated Puff-site. With a locus that fails to puff because of its proximity to heterochromatin, it should be possible to ask whether the DNA sequences normally present at the site are always missing when the puff is absent. Several molecular probes are available that are complementary to DNA sequences transcribed at 87C during heat shock. The major 2.6kb heat shock message has been previously purified to near homogeneity (12). When hybridized *in situ* to polytene chromosomes, this message specifically labels DNA at both 87C and 87A, another heat shock puff-site (10-12). Cloned DNA, complementary to this message and to neighboring sequences at 87C1 (14), has advantages over RNA labeled *in vivo* in its unquestioned purity and in the higher specific activity of copy RNA made to it *in vitro*. A second quite distinct molecular species of RNA is produced at 87C1 (12-14), for which copy RNA to a cloned DNA sequence is available. For the experiments described here, I have used only ^{3}H copy RNA made to pPW229 DNA, which is complementary to nearly all of the 2.6kb message as well as to some adjacent sequences (14). Other experiments using 2.6kb message ^{3}H-labeled *in vivo* (12), copy RNA to pPW232.1, a cloned fragment complementary only to the message (14), and copy RNA to a purified *Pst*I restriction fragment of the 1.5kb "non-message" repeat of pPW232 (14) gave essentially the same results. Cloned DNA and copy RNA were either generously supplied by Drs. K. Livak and M. Meselson or were prepared in collaboration with them.

Salivary glands from third instar larvae heat-shocked at 37° for 20 min. were squashed in acetic-orcein stain and examined for puff variegation. All clear examples in which 87C was apparently unpuffed in the aberrant homolog and puffed in the normal homolog were photographed. Following removal of the coverglass and fixation in 3 parts ethanol:1 part acetic acid which removed the orcein stain, slides were hybridized *in situ* as previously described (12) to ^{3}H copy RNA made from pPW229 plasmid DNA, autoradiographed and restained with Giemsa. Then the examples of variegation chosen in the fresh squash were rephotographed. Photographing beforehand not only permits analysis of the chromosomes under conditions of ideal morphology unobstructed by photographic emulsion, but also allows one to choose examples of variegation purely on the basis of morphology since no grains are present to influence the choice. One example is shown in Figure 2. Here the aberrant homolog is stretched in the 87C region. The autoradiograph shows a high concentration of grains over the 87C puff on the normal homolog and grains over the puff at 87A. There is also a fairly high concentration of grains over a small region on the aberrant homolog. As can be seen from the left-hand photograph, this particular region is stained and is clearly not puffed. As judged from the identifiable bands distal to it, this region could include bands 87C1-3, although its appearance is quite different from those bands. This region resembles centric heterochromatin in its lack of clear banding. (Some clear bands, probably derived from the Y-chromosome used in the construction of the translocation, can be seen proximal to the labeled region.)

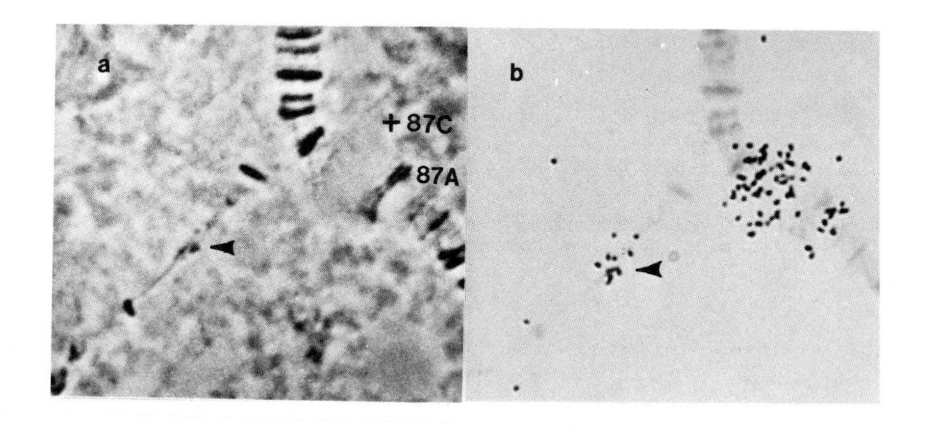

FIGURE 2. Hybridization of copy RNA made from pPW229 DNA to a variegation-enhanced T(Y;3)A78 chromosome. a. An example of a chromosome from an acetic-orcein squashed salivary gland. b. Autoradiograph of the same example after hybridization *in situ* and Giemsa staining. The puffs at 87A and 87C are marked on the normal sequence chromosome. The arrow points to the region that is labeled on the aberrant homolog.

In this example it is clear that absence of a puff can occur without complete loss of the DNA sequence that is transcribed in the puff. The possibility that transcription still occurs at the unpuffed site is being investigated.

Other examples of 87C puff inactivation were examined and compared to examples in which the puff locus was clearly active on the aberrant chromosome. Whereas approximately equal numbers of grains were consistently seen over active 87C puffs on the aberrant chromosomes and their normal homologs, a quite different pattern was seen in nuclei in which the aberrant chromosome puffs were inactive. This pattern can be seen in Figure 3 which shows the results from a single slide. Eleven clear examples of puff inactivation were photographed prior to fixation and the autoradiographic grains were counted after development. For comparison, 11 randomly chosen "fully active" nuclei were similarly scored. Background labeling was negligible. Although most examples in which the puff was inactive show significant labeling, the number of grains is greatly reduced compared to controls. In a few cases, no significant labeling is seen, although the normal homolog and the control nuclei show very significant labeling over all 87C puffs. The cause of reduced labeling could be under-replication of the "heterochromatized" gene sequence. This and other possibilities are currently under investigation.

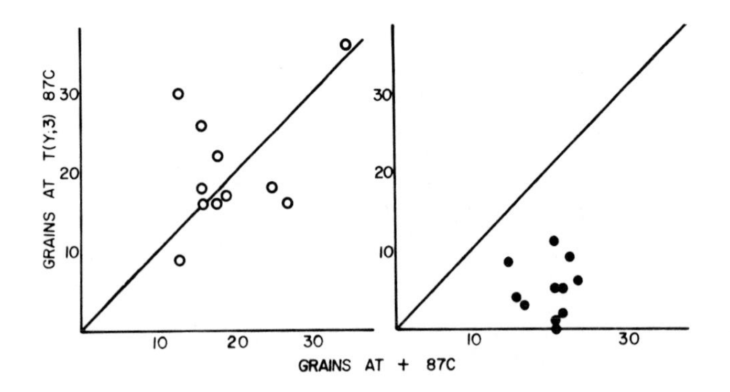

FIGURE 3. Grain counts over 87C on normal and aberrant homologs from single salivary nuclei of the glands of one heat-shocked individual. Open circles (left) are for single nuclei where 87C is puffed on the translocation chromosome and closed circles (right) where 87C is unpuffed. A total of 3 background grains were counted over 93D, an unlabeled puff, for all 22 nuclei.

Does Gene Loss by Excision Occur? The above results lead to the tentative conclusion that gene loss alone cannot account for puff variegation. Of course this does not mean that a gene sequence is never lost during the development of a tissue that will show a variegated phenotype for that gene. If such loss does occur, then an obvious prediction is that reversion to the wild phenotype should never be seen. In the case of the white (*w*) locus of *Drosophila melanogaster*, however, apparent reversion is quite common in so-called "pepper-and-salt" variegated eyes such as in white-mottled-4 (In(1)w^{m4}) adults. Here, an occasional fully pigmented cell will appear on a completely mutant background as if this cell has reversed its prior inactivation (1). Other white-variegated mutants show a quite different "clonal" pattern in which large groups of cells are all either *w+* or *w*. This pattern has been shown to result from a "decision" made prior to the end of the first larval instar (17) whereas the pepper-and-salt mottling must have been caused by a much later event. These two patterns are not exclusively confined to each type of mutant because clonal variegated eyes sometimes show small spots, and also an occasional w^{m4} fly will have a large pigmented clone. These and other observations have been interpreted as indicating that a decision made early is customarily confirmed or reversed at a much later stage (1). A gene loss model would seem to have serious difficulties explaining this two step process, since loss during the first step would not permit later gene activation.

Gene loss could give rise to the observed pattern if excision of the gene occurred as the first step of the process with actual loss of the excised fragment occurring later. That is, the relatively small fragment would sometimes replicate and segregate, while other times it would be lost. Presumably, genes on a surviving fragment would be active. This particular model is testable in that a segregating fragment can be separated from the aberrant chromosome that generated it by mitotic recombination in somatic cells. The scheme outlined in Figure 4 describes a test adapted from a study by Janning (17). Late first and early second instar female larvae of genotype In(1)w^{m4},w^{m4} *rux*/Y^S *y* w^a *rb* lz^{50e} were X-irradiated with 1-2 krad in order to induce somatic crossing-over, and their eyes were examined after eclosion for the presence of an unfaceted glassy clone indicative of homozygous lz^{50e} cells. As shown in Figure 4, crossing-over, which can occur between the centromere and the inversion breakpoint, will give rise to a "twin spot" (half of the time) in which the maternal chromosome is effectively homozygous in one member of the twin and the paternal chromosome is effectively homozygous in the other. The heterozygous background is apricot-colored and peppered with wild-type pigment spots and has a normally faceted surface. One clone is recognizable

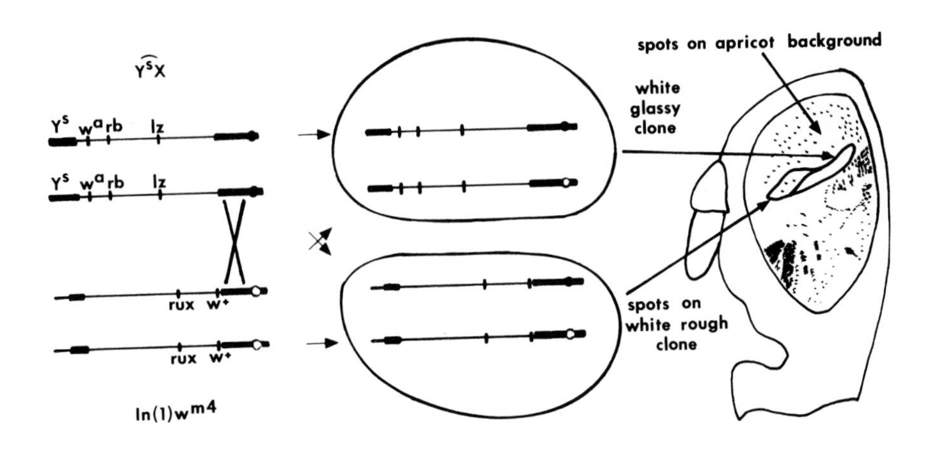

FIGURE 4. Mitotic recombination in In(1)w^{m4},w^{m4} rux^2/Y^S y w^a rb lz^{50e} heterozygote. Larvae were irradiated approximately 39-48 hours after egg deposition. A typical example of an eye that shows a clonal "twin-spot" is depicted.

because the w^{m4} rux^2 chromosome causes pigment spots on a white background and disarranged facets. The other clone is expected to have a glassy lz^{50e} surface on a white background because the combination w^a rb results in the absence of pigment. This clone of cells should not be peppered with w+ pigment spots because the w^{m4} chromosome is not present. However, if the early "decision" had resulted in excision of the w+ gene prior to irradiation, then the excised fragment would be expected to end up in either daughter cell and not just in the one that receives the two w^{m4} rux^2 chromatids.

The experiment is to look for pigment spots within the glassy twin-spot clone, where without excision and eventual survival of a fragment, no spots would be expected. The probability of pigment spots appearing should be increased by the presence of Y^S, a cell autonomous variegation suppressor (17) on the end of the chromosome carrying lz^{50e}. Table 1a shows the results of one series of observations. Not a single example of a pigment spot in a lz^{50e} clone was seen in this or in any other test. That this failure is not due to some effect of the lz^{50e} gene on formation or detection of pigment is shown in Table 1b for the control cross by the presence of spots in the lz^{50e} clones when the lz^{50e} gene is carried by the w^{m4} chromosome, and by the absence of pigment in the rux clones. A typical w^{m4} rux^2/$Y^S$$w^a$rb lz^{50e} twin spot

TABLE 1

ANALYSIS OF EYE CLONES IN PROGENY FEMALES[a]

Cross	Number with pigment spots/total: lz clones (twins + singles)	Number with pigment spots/total: rux clones (twins only)
a. $\frac{w^{m4}\ rux^{2}}{\overline{\quad\quad}/}$ x $\frac{Y^{s}yw^{a}rb\ lz^{50e}}{Y^{s}yw^{a}rb\ lz^{50e}}$	0/25	15/16
b. $\frac{w^{m4}\ lz^{50e}}{\overline{\quad\quad}/}$ x $\frac{Y^{s}yw^{a}rb\ rux^{2}}{Y\ yw^{a}rb\ rux^{2}}$	10/15	0/10

[a]Screened for scorable lz^{50e} clones and examined under water using a compound microscope for location of pigment.

is sketched in Figure 4. Since pigment spots are seen only where the w^{m4} chromosome is present, the excision model above is evidently wrong.

Conclusions. The experiments described here were aimed at testing the possibility that position-effect variegation of a *Drosophila* gene is a result of its loss from some cells of a tissue but not from others. In salivary gland cells, a variegating heat-shock puff locus generally was found to include heat-shock gene sequences in cases where the puff-site was inactive. Variable reduction in the amount of hybridizable DNA in these cases may have been due to under-replication of the sequences. Frequent autoradiographic labeling of the junction between heterochromatin and the puff-site region argues against the hypothesis that inactivity of the puff is simply physical absence of the gene. Furthermore, a particular model for gene loss involving excision of a fragment carrying the gene appears to be wrong when tested in w^{m4} eye pigment cells.

In spite of the fact that position-effect variegation remains an unexplained phenomenon, the molecular approach described here may be helpful in elucidating the normal chromosomal process that is interrupted when breaks occur in heterochromatic regions.

ACKNOWLEDGMENTS

I am grateful to C. Laird for advice and support, to S. Hawley for helpful discussions, and to M. Hammond, L. Wilkinson and M. Lamb for comments on the manuscript.

REFERENCES

1. Spofford, J. (1976). In "The Genetics and Biology of Drosophila" (M. Ashburner and E. Novitski, eds.), pp. 955-1018. Academic Press, New York.
2. Schultz, J. (1936). Proc. Nat. Acad. Sci. USA. 22, 27.
3. Hozumi, N. and Tonegawa, S. (1976). Proc. Nat. Acad. Sci. USA 73, 3628.
4. Zuchowski, C. and Harford, A. (1976). Chromosoma 58, 219.
5. Zuchowski-Berg, C. (1978). Nature 271, 60.
6. Procunier, J. and Tartof, K. (1978). Genetics 88,67.
7. Baker, W. (1968). Adv. Genet. 14 133.
8. Schultz, J. (1965). Brookhaven Symp. Biol. 18, 116.
9. Rudkin, G. (1965). Proc. 11th Intern. Congr. Genet. 2, pp. 359-374. Pergamon Press, Oxford.
10. McKenzie, S., Henikoff, S. and Meselson, M. (1975). Proc. Nat. Acad. Sci. USA 72, 1117.
11. Spradling, A., Pardue, M. and Penman, S. (1977). J. Mol. Biol. 109, 559.
12. Henikoff, S. and Meselson, M. (1977). Cell 12, 441.
13. Lis, J., Prestidge, L. and Hogness, D. (1978). Cell 14, 901.
14. Livak, K., Freund, R., Schweber, M., Wensink, P. and Meselson, M. (1978). Proc. Nat. Acad. Sci. USA 75, 5613.
15. Lindsley, D., Sandler, L., Baker, B., Carpenter, A., Dennell, R., Hall, J., Jacobs, P., Gabor Miklos, G., Davis, B., Gethmann, R., Hardy, R., Hessler, A., Miller, S., Nozawa, H., Parry, D. and Gould-Somera, M. (1972). Genetics 71, 157.
16. Lindsley, D. and Grell, E. (1968). "Genetic Variations of Drosophila melanogaster." Carnegie Institution, Washington.
17. Janning, W. (1970). Molec. Gen. Genet. 107, 128.

SEQUENCE ORGANIZATION OF DROSOPHILA tRNA GENES[1]

Pauline Yen, N. Davis Hershey,
Randy R. Robinson, and Norman Davidson

Department of Chemistry, California Institute of
Technology Pasadena, CA 91125

ABSTRACT The tRNA gene cluster at region 42A of chromosome arm 2R of the Drosophila melanogaster genome has been mapped on a number of recombinant DNA molecules with overlapping Drosophila inserts. The cluster extends over a region of 46 kb and contains at least 18 tRNA genes. It contains several $tRNA_2^{lys}$, $tRNA_2^{arg}$, and $tRNA^{asn}$ genes that are mutually interspersed and irregularly spaced. A preliminary account of experience with Benton-Davis screening for λ plaques with inserts carrying tRNA genes is given.

INTRODUCTION

Saturation hybridization experiments indicate that there are 600-800 tRNA genes in the haploid genome of Drosophila melanogaster. (1,2,3) There are about 90 resolvable peaks when total Dm tRNA is fractionated by RPC-5 chromatography(4). From an analysis of the kinetics of tRNA-genomic DNA hybridization, Weber and Berger estimate that Dm tRNA is made up of about 59 kinetic families (3). These data suggest that there are approximately 60-90 different tRNA sequences (*i.e.*, species) in Drosophila and an average reiteration frequency for any one species of six to 13.

Steffensen and Wimber (5) and later Elder, Szabo and Uhlenbeck (6) have mapped the sites on polytene chromosomes to which total labeled Dm tRNA hybridizes. In the latter study, 54 sites were mapped by hybridization of ^{125}I-labeled tRNA to polytene chromosomes. Of these 26 were estimated by grain counts to be strong sites with 4 or more gene copies, and 28 were weak with 1-3 gene copies. These sites are listed in Table 1.

The authors noted that other sites, such as those with small numbers of tRNA genes, those with tRNA genes of low

[1]This work was supported by a grant GM 10991 from the United States Public Health Service.

ISBN 0-12-068350-4

abundance, or those on portions of the genome that are under-replicated in polytene chromosomes (*e.g.*, the Y chromosome) may not have been detected.

Several laboratories have carried out *in situ* studies on purified tRNAs. Some of these results are also reported in Table 1 (7,8,9,10).

The overall conclusions from these studies are that most of the tRNA genes of Drosophila are clustered at a limited

Table 1

Drosophila 4S RNA Sites[a]

X	2L	2R	3L	3R	4
3D	22DE	41CD	61D Met3	84A Met2	none
3F	23EF	*42A Asn Arg2, Lys2	*62A Lys2 Glu4	84D Val3b	
5F6A	24DE	42E Lys2	63A	84F Arg2	
11A	_*27D_	_*43A_	_63B_	87BC	
*12E	28D	44EF	64DE Val3a	87F88A	
	28F29A	46A Met3	66B	89BC Val4	
	29D Asp2	48AB Met2	67B	90C Val3b	
	29E	48CD	69F70A	90DE	
	_*34A_	49AB	70BC Val4	_*91C_	
	35AB	49F50A	70DE	92AB Val3b	
		*50BC Lys2	72F*73A Met2	93A	
		52F53A Glu4	79F	95F96A	
		54A		97CD	
		54D		99EF	
		55EF			
		56D Val4			
		*56F Glu4			
		57D			
		58AB			
		60E			

[a]The locus assignments in heavy type were made by Elder et al. (6) using total labeled tRNA. Sites that were heavily labeled are underlined. Entries to the side of the main columns of the table are assignments made using purified tRNA's (5,6,7,8,9,10). Asterisks indicate assignments for recombinant Charon plaques that give strong tRNA signals, isolated in our laboratory. Entries in italics are *in situ* assignments that may be different from those of Elder *et al.* (6).

number of sites, perhaps 54 to 60, on the polytene chromosomes; that some of these clusters contain more than one species of tRNA gene; and that for some tRNA species there are genes at 2 or more sites.

We have previously described our initial studies of the sequence arrangement of the tRNA genes on a cloned fragment of Dm tRNA of length 9 kb (11). This fragment was shown by *in situ* hybridization to map at region 42A of chromosome 2R. *In situ* studies with labeled total Dm tRNA had shown that 42A is one of the richest tRNA gene clusters in the Drosophila melanogaster genome (5).

In the present communication, we describe the present status of our further studies of the sequence organization of tRNA genes in cloned DNA derived from this region of the chromosome. We also give a preliminary account of our experience in screening for recombinant λ plaques with Drosophila tRNA genes.

EXPERIMENTAL

A recombinant "library" of nuclear Dm DNA inserted into the modified λ bacteriophage Charon 4 (12) has been constructed. High molecular weight nuclear Dm DNA was partially digested with EcoR1 to varying extents (5%, 7.5%, 14%, 20%, 31% and 57%) of total digestion. Fractions of length 22 ± 4 kb were isolated from each digest by sucrose gradient velocity sedimentation. Equal masses of sized DNA from each digest were combined and then ligated to annealed EcoR1 ends of Charon 4 DNA. The ligation products were packaged *in vitro* (13,14) to yield 2×10^6 recombinant plaque forming units.

The library contains inserts ranging in size from 12 to 20 kb, and contains less than 1% Charon 4 contamination. Library Dm DNA is representative of the single copy Dm genome as demonstrated by its ability to drive 95% of single copy tracer into duplex with a $cot_{1/2}$ of 280, compared to 270 for nuclear Dm DNA driver.

The other experimental methods used here are all standard, and are referred to where necessary in the Results section.

RESULTS

a) The Gene Cluster at 42 A. The plasmid pCIT12 described in our initial study (11) contains a Dm DNA segment of length 9.3 kb fused to the vector ColE1. Fig. 1 includes a map of the Eco R1 cleavage sites and identified positions of tRNA genes on this insert. The methods by which the tRNA genes have been mapped and identified will be discussed later.

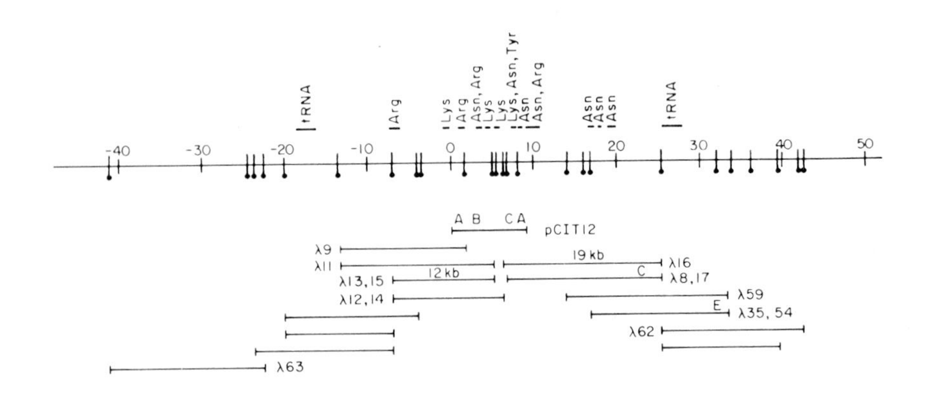

FIGURE 1. Dm inserts in Charon 4 phage covering the region 42A. The scale is in kilobases (kb). ⬇ denote EcoR1 clevage sites. Capital letters denote EcoR1 restriction fragments referred to in the text. Note that fragment A of pCIT12 contains the Col E1 vector as well as segments on both ends of the insert. Positions of tRNA genes were determined by blotting experiments with labeled total or purified tRNA probes as described in the text.

It seemed probable from the strong *in situ* hybridization of total Dm tRNA to locus 42A that the 9.3 kb segment of pCIT12 does not include the entire gene cluster at locus 42A. This inference was confirmed by our initial studies of Charon 4 recombinant molecules with inserts that overlapped with pCIT12, and by the studies of P. Gergan and P. Wensink who have independently isolated a Dm plasmid which partially overlaps with pCIT12 but extends further to the right (given the arbitrary orientation of the map of Fig. 1) and includes additional tRNA genes. We therefore wished to determine the total extent of the tRNA gene cluster of band 42A which includes pCIT12.

For this purpose we have engaged in the procedure of "walking down the chromosome". Initial attempts to isolate recombinant molecules with adjacent sequences from the Charon 4 library using pCIT12 DNA as a probe gave many plaques with inserts derived from other regions of the genome. This was due to the presence of repeated sequences in the EcoR1 fragment C of pCIT12. Therefore, a DNA segment consisting of the 2 R1 segments A and B of pCIT12 (which in addition to the Dm segment of the A fragment indicated in Fig. 1 includes the

6.7 kb of the vector ColEl) was subcloned. This DNA was labeled by nick translation and used as a probe in a Benton-Davis screen (15) of the Charon 4 library. Note that this probe includes segments from the two ends of pCIT12 but excludes the internal segments. With the probe used, the phages λ16, λ8, λ17, λ9, λ11, λ13, λ15, λ12 and λ14 of Fig. 1 were isolated. It was shown from the restriction patterns with several enzymes (data not shown in Fig. 1) that these inserts overlapped with pCIT12 as indicated in Fig. 1. λ17 has a length of 18.8 kb, extends 16 kb beyond pCIT12, and overlaps it for about 2.8 kb. The restriction map of λ17 was determined. The rightmost EcoRl fragment, 17C, of length 8.7 kb was shown by chromosomal blots to contain mainly single copy sequences. It was subcloned and used as a probe to isolate an additional group of phage from the library. The restriction patterns of these phage were determined. Those that did not contain the EcoRl fragments 17D and 17F were selected for further study. These include λ59, λ35, and λ54. These three phage all end at a point 8 kb to the right of λ17. The process was repeated using the 1.6 kb fragment 35E and phage extending to the right selected. The phage λ62 extends 9 kb beyond λ35.

In the same way phage extending to the left of pCIT12 were selected. The phage λ11 has a Dm DNA insert of length 18.4 kb and extends 13.3 kb beyond pCIT12. In a series of two additional steps which are obvious by inspection of Fig. 1, the phage λ63 was isolated. Its end point is about 41 kb beyond the left end of pCIT12. Thus, the set of recombinant molecules cover a region of length 83 kb at locus 42A.

Preliminary tRNA gene assignments on some of the cloned segments have been made by Southern blotting experiments, using several different kinds of probes. Total Dm 4S RNA has been purified from Drosophila pupae as previously described (11) and labelled with ^{32}P either at the 5' end by T4 polynucleotide kinase or at the 3' end as (5'-^{32}P)pCp with T4 RNA ligase (16). The specific activities achieved either way are approximately 2×10^7 cpm/μg tRNA. However, the tRNA is less degraded by the 3' end labeling procedure.

Specific genes have been identified with probes prepared by labeling purified Dm tRNA species. These latter preparations were generous gifts from the laboratory of Dr. Dieter Söll. Since the tRNAs are not always completely purified and contain other tRNA components as minor contaminants, and since the hybridizations are carried out in vast tRNA excess, care must be taken to carry out hybridizations for times such that only the major species gives strong spots in the autoradiographs. Weak signals which are probably due to minor tRNA impurities have been disregarded in the tRNA gene assignments

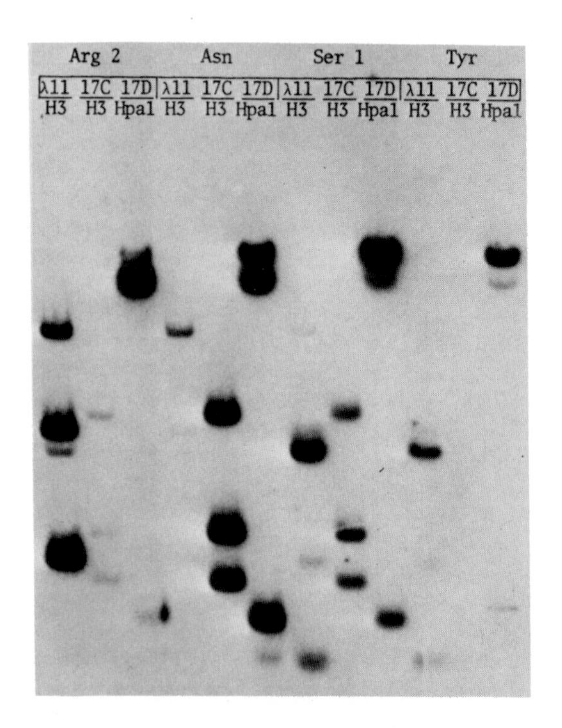

FIGURE 2. Blotting experiments with purified tRNA probes on appropriate restriction digests of the Eco Rl fragments C and D of λ17 (see Fig. 3) recloned in pBR322, as well as of a Hind III digest of λ11. The $tRNA^{ser}$ probe was heavily contaminated with $tRNA_2^{lys}$ and further experiments with purified $tRNA_2^{lys}$ show that the positive signals are due to this tRNA (D. Söll, personal communication). Because of impurities in the tRNAs, only strong autoradiographic signals are interpreted as positive for the main tRNA component.

in Figs. 1 and 3.

An example of such a Southern blot experiment with several purified tRNA species is given in Fig. 2. The tRNA gene assignments given in Figs. 1 and 3 have been made, on the basis of experiments like those of Fig. 2 in our laboratory, in that of Dr. Dieter Söll (17), and by P. Gergan and P. Wensink (personal communication).

Figs. 1 and 3 show that the gene cluster at region 42A on chromosome arm 2R of the Dm genome contains multiple

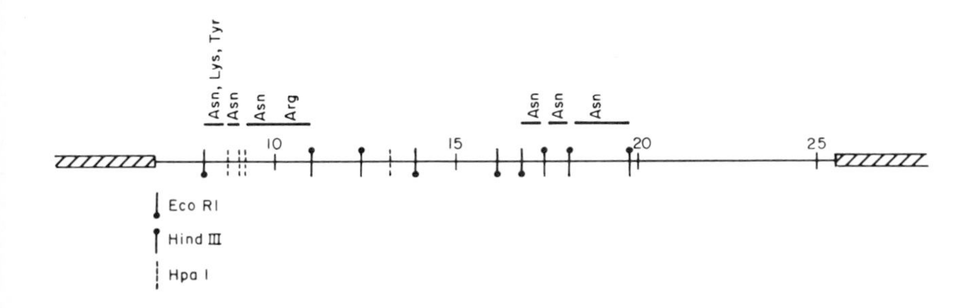

FIGURE 3. Restriction enzyme map and tRNA assignments for the DNA phage λ17 from blotting experiments such as illustrated in Fig. 2.

copies of tRNA genes for $tRNA_2^{arg}$, $tRNA_2^{lys}$, and $tRNA^{asn}$. At least one $tRNA^{tyr}$ gene seems to be present.

On the map in Fig. 1, there are two segments of DNA, one extending from 25 to 28 kb and one from -17 to -20 kb, which hybridize with a total RNA probe; experiments with specific tRNAs have not yet been performed. The hybridization experiments are not quantitative and the possibilities exist that: a) several genes for a particular species are present on a restriction fragment where only one has been assigned; b) there is a restriction endonuclease site sufficiently close to the center of a gene so that two restriction fragments hybridize to the tRNA probe even though there is only one gene; c) other as yet unidentified tRNA species, for which purified probes were not available, are also present.

The map of pCIT12 (11,17, P. Gergan and P. Wensink, personal communication) and the present mapping studies of λ17 (Fig. 3) show that the several species of genes are interspersed and irregularly spaced. At the level of resolution achieved at present, there is no evidence of extensive sequence homology in the regions flanking identical genes.

It is apparent from these studies that there is a segment of DNA of length 47 kb in the region 42A on chromosome 2R of the Dm genome which contains at least 18 tRNA genes. For a region of at least 14 kb on the right and 23 kb on the left there are no additional tRNA genes so that it is possible that this segment represents the complete tRNA gene cluster at region 42A.

By *in situ* hybridization experiments, Tener *et al.* (7) have estimated that there are approximately 18 $tRNA_2^{lys}$ genes in the haploid genome of Dm, and that these map at the sites 42A, 42E, 50B, 62A and 63B. The $tRNA_2^{arg}$ genes have been assigned to chromosomal locations of 42A and 84F.

It thus appears that at least for some of the repeated tRNA gene families of Drosophila, genes are located at several widely spaced loci; within each locus, the genes are irregularly spaced and interspersed with other tRNAs. The functional significance of this pattern remains an intriguing, unsolved problem.

b) Screening for tRNA genes. It may be useful to describe our preliminary experiences in screening a Charon 4 recombinant library for tRNA genes. Fig. 4 is an example of hybridizations of λ plaques adsorbed on to duplicate nitrocellulose filters by the Benton-Davis procedure (15). Hybridizations were carried out with 5' ^{32}P labeled total pupal tRNA. The plate in Fig. 4 contains about 1700 plaques. Of these, 29 gave apparent duplicate positive signals. These positive plaques were picked and replated. Seventeen gave

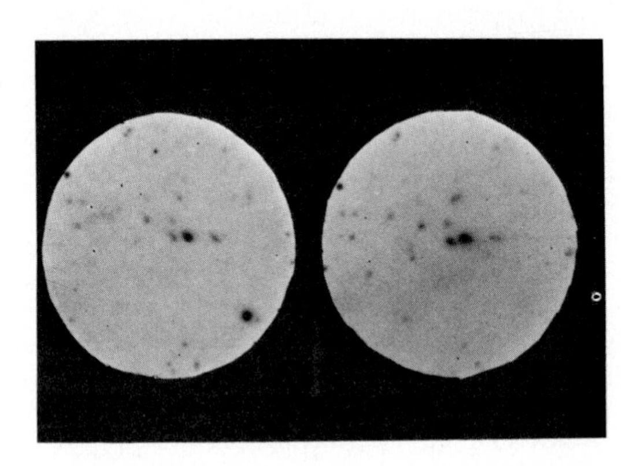

FIGURE 4. Duplicate Benton-Davis (15) screens of the Charon 4 Drosophila library. Each plate contained about 1700 plaques. Hybridizations with 5' ^{32}P labeled tRNA were conducted with 0.15 μg/ml tRNA ($\sim 2 \times 10^7$ cpm/μg) for 15 hrs at 68°C in 1.0 M NaCl, 0.050 M Tris, pH 8; 1 mM EDTA, 10 x Denhardt's solution, 0.1% SDS, 0.1% sodium pyrophosphate, and 20 μg/ml Dm rRNA.

positive plaques on at least one rescreen; thirteen were positive on duplicate rescreens. Nine of the strongest spots were selected for further study.

Similar results were obtained in screening the entire library (10 plates). Comparable results were obtained using ^{125}I labeled tRNA probes. A number of the plaques which gave strong signals were purified and characterized by hybridization, with the results recorded in Table 1.

At the present time, we do not know the sensitivity of the screening procedure. We suspect however that, with improvements in labeling and hybridization procedures, plaques with 3 or 4 tRNA genes can be identified.

ACKNOWLEDGMENTS

We are indebted to Drs. Dieter Söll, Olke Uhlenbeck, and Pieter Wensink for several personal communications. The purified tRNA's used were a generous gift from the laboratory of Dr. Dieter Söll. Department of Chemistry Contribution #5997.

REFERENCES

1. Ritossa, F.M., Atwood, K.C. and Spiegelman, S. (1966). Genetics. 54, 663.
2. Tartof, K.D. and Perry, R.P. (1970). J. Mol. Biol. 98, 503.
3. Weber, L. and Berger, E. (1976). Biochem. 15, 5511.
4. White, B.N., Tener, G.M., Holden, J. and Suzuki, D.T. (1973). Dev. Biol. 33, 185.
5. Steffensen, D.M. and Wimber, D.E. (1971). Genetics. 69, 163.
6. Elder, R., Szabo, P. and Uhlenbeck, O. (1979). Cold Spring Harbor Symposium on tRNA. in press.
7. Tener, G.M., Hayashi, S., Dunn, R., Delaney, A., Gillam, I.C., Grigliatti, T.A., Kaufman, T.C. and Suzuki, D.T. (1979). Cold Spring Harbor Symposium on tRNA. in press.
8. Dunn, R., Hayashi, S., Gillam, I.C., Delaney, A.D., Tener, G.M., Grigliatti, T.A., Kaufman, T.C. and Suzuki, D.T. (1979). J. Mol. Biol. in press.
9. Kubli, E. and Schmidt, T. (1978). Nucl. Acid Res. 5, 1465.
10. Schmidt, T., Egg, A.H. and Kubli, E. (1979). Molec. Gen. Genet. in press.
11. Yen, P.H., Sodja, A., Cohen, M., Conrad, S.E., Wu, M., Davidson, N. and Ilgen, C. (1977). Cell. 11, 763.
12. Blattner, F.R., Williams, B.G., Blechi, A.E., Thompson, K.D., Faber, H.E., Furlong, L.A., Grunwald, D.J., Kiefer, D.O., Moore, D.D., Schumm, J.W., Sheldon, E.L. and Smithies, O. (1977). Science. 196, 161.

13. Maniatis, T., Hardison, R.C., Lacy, E., Lauer, J., O'Connell, C., Quon, D., Sim, G.K. and Efstratiadis, A. (1978). Cell. 15, 687.
14. Sternberg, N., Tiemeier, D. and Enquist, L. (1977). Gene. 1, 255.
15. Benton, W.D. and Davis, R.W. (1977). Science. 196, 180.
16. England, T.E. and Uhlenbeck, O.C. (1978). Nature. 275, 560.
17. Schmidt, O., Mao, J.-I., Silverman, S., Hovemann, B. and Söll, D. (1978). Proc. Natl. Acad. Sci. USA. 75, 4819.

HETEROGENEITY IN HISTONE GENE ORGANIZATION IN *DROSOPHILA*[1]

Linda D. Strausbaugh and Eric S. Weinberg

Department of Biology, The Johns Hopkins University
Baltimore, Maryland 21218

ABSTRACT The organization of histone DNA in *Drosophila* has been analyzed using restriction enzyme digestion of genomic DNA and agarose gel electrophoresis, followed by transfer to nitrocellulose paper and hybridization with nick-translated cloned histone DNA probe. Our results fully confirm the findings of Lifton *et al.* (1) that the major portions of histone gene repeats are organized into 4.8 and 5.0 kb repeat units in *Drosophila melanogaster*. We have studied many strains of diverse geographic origin from this species and have found striking similarity in the size of the predominant histone DNA repeats, although there is heterogeneity in less abundant histone gene fragments between different strains. Individuals in a strain, as well as single chromosomes from a strain, also show a high degree of homogeneity of repeat units. In a similar analysis of DNA from other species in the genus, we have found conservation of a 5.0 kb repeat size, but some differences in other bands.

INTRODUCTION

The occurrence of repeated genes in eukaryotic genomes raises several interesting questions concerning the maintenance of reiteration frequency, sequence homogeneity, and organization of the genes with respect to non-coding sequences. We have chosen *Drosophila melanogaster* for our studies on the heterogeneity and stability of repeated genes because of the wealth of genetic information available and the ease of obtaining and rearing strains of diverse origin which have been maintained as reproductively isolated stocks over long periods of time. In addition, species within the genus *Drosophila* have well characterized evolutionary relationships established from studies on morphological and chromosomal relationships. Use of other species in addition to *melanogaster* allows characterization of changes in repeated genes over the large scale of time during which the species evolved.

[1]This work was supported by NSF grant SM177-12417 and PHS grant GM06318 to LDS and PHS grant GM22155 to ESW.

ISBN 0-12-068350-4

The histone genes provide a particularly interesting set of repeated genes because of their levels of organization. In both sea urchins and fruit flies, the genes coding for the five major histone types (H1, H2a, H2b, H3, H4) are clustered into a unit which is tandemly repeated many times per genome (1, reviewed in 2). Heterogeneity can occur, therefore, at several levels: 1) sequence heterogeneity might occur in either the genes or the spacers separating them; 2) changes might occur in the order of the genes within the repeating unit; or 3) changes may occur in the length of the repeat unit. The organization of histone DNA in *Drosophila melanogaster* has been extensively studied by Lifton, Goldberg, Karp, and Hogness (1). They have found that the majority of the 100 repeats per haploid genome in *Drosophila melanogaster* have lengths of 4.8 and 5.0 kb (the latter differs from the former by an extra 270 base pairs of spacer between the H1 and H3 genes) (1), and are clustered on the left arm of chromosome 2 at polytene chromosome position 39DE (1,3,4). The two size classes of repeats are non-randomly interspersed and are interpreted to be organized into blocks of tandem arrays separated from each other by non-histone DNA (1).

In this paper we report our initial investigations into the characteristics of repeat lengths in different strains of *melanogaster* and different species in the genus *Drosophila*.

METHODS

DNA from large quantities of flies (1 gram or more) is extracted by a proteinase K - phenol-chloroform method (Strausbaugh and Kiefer, submitted for publication). DNA from single flies or a few flies is extracted by a modification of the method used for bulk DNA preparation (Strausbaugh and Weinberg, manuscript in preparation). The DNA is then digested with the appropriate restriction enzyme and the fragments electrophoresed on agarose gels, transferred to nitrocellulose filters, and hybridized to nick-translated DNA as previously described (5). The probes used are one 4.8 kb repeat from *Drosophila melanogaster* cloned into the Hind III site of pBR313 (derived from cDm500 (1)) or one 6.3 kb repeat from *Strongylocentrotus purpuratus* cloned into the Hind III site of pBR313 (pCO2 (5)). All figures shown have been made from filters hybridized with homologous probe. Exposure for autoradiography for both bulk DNA (5-10 μg/slot) and single flies is routinely 48 hours, with longer exposures necessary for detection of some of the minor bands. The relative amounts of the major repeat classes are determined from densitometric scans of exposed films in which the band intensities are within the linear range of the film.

RESULTS

In agreement with Lifton *et al.* (1), we find that the majority of histone genes in *D. melanogaster* are organized into 4.8 and 5.0 kb repeats. We have studied DNA from 12 different strains, including wild type strains of diverse geographic origin and strains with mutagenized second chromosomes, and find that all have major repeat lengths of 4.8 and 5.0 kb. Figure 1 shows an Hind III digest of total *D. melanogaster* genomic DNA from several strains isolated from different parts of the world. Hind III is known to cut once in both the 4.8 and 5.0 kb repeat in the H1 genes (1). The doublet characteristic of the two major repeat lengths and the heterogeneity that occurs between strains are illustrated. First, we find that the relative amounts of 4.8 and 5.0 kb repeats can differ between the strains. Table 1 shows the relative amounts of the two major repeat classes in some of the strains. With the first three strains, we have digested the DNA with Hind III, Bam HI, and Hpa I in different experiments and all yield the 4.8 and 5.0 kb repeats. Within a strain there is good agreement for the distribution of histone genes between the 4.8 and 5.0 kb units. For example, for Amherst DNA the range is between 75% and 81% for the 5.0 kb unit with all three enzymes. The relative amounts of 4.8 and 5.0 kb repeats in the remaining strains have been determined with Bam HI.

The second kind of heterogeneity which occurs between the strains is in the minor bands which are of sizes other than 4.8 and 5.0 kb. As was first noted by Lifton *et al.* (1), these bands are present in lower frequency than the major repeat classes and are not easily explained by the loss or gain of restriction sites in the major repeats. Figure 1 shows some of the minor bands characteristic of a strain and a particular enzyme, such as the 4.35 kb Swedish and 3.9 kb Amherst fragments when Hind III was used. Other minor bands, such as the 3.6 and 1.85 kb bands occur in several of the strains. It is interesting that a fragment at 3.9 kb can occur in other strains on longer exposures (see Figure 3), indicating that some of the strain differences may be quantitative in nature. Another prominent strain difference can be seen in Hind III (not shown) and Xho I (Figure 4) digestions of Samarkand DNA, in the presence of a histone homologous band at 5.5 kb. This fragment, another potentially useful genetic marker, is demonstrable in both bulk and single fly preparations of Samarkand DNA and hybridizes with cloned sea urchin histone DNA.

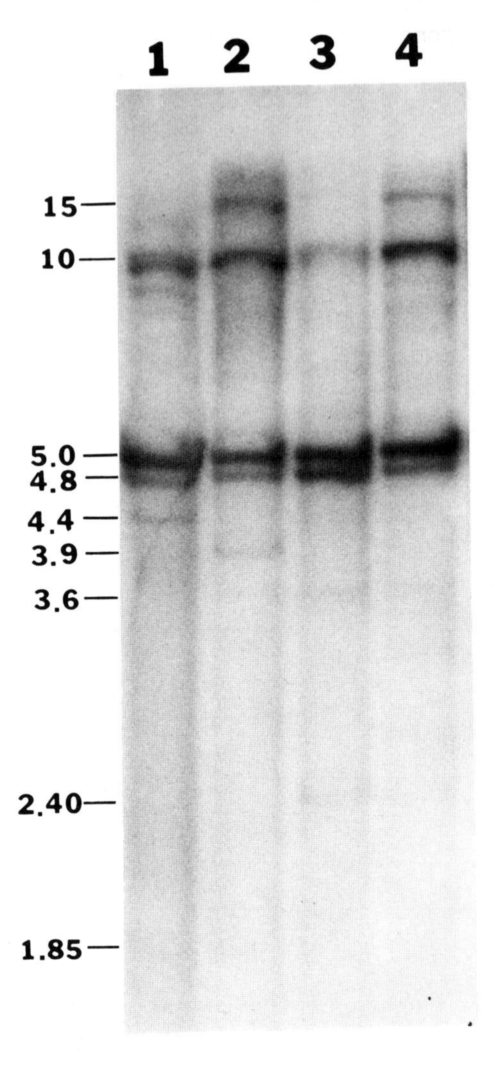

FIGURE 1. Hind III digestion of strains: 1 = Swedish, 2 = Amherst, 3 = Tunnelgaten, 4 = b Adhn^2 pr cn. Size in kb is given on the left.

TABLE 1
DISTRIBUTION OF HISTONE GENES BETWEEN THE TWO MAJOR REPEATS

Strain	%5.0 kb	%4.8 kb	# of exp.
Samarkand (USSR)	60	40	8
Amherst	78	22	6
Swedish	81	19	5
Kyoto (Japan)	74	26	2
Tunnelgaten (Germany)	59	41	2
Satto (Argentina)	76	24	1
b Adhn^2 pr cn	74	26	1

Based on Eco RI digestions of total genomic DNA, Lifton et al. (1) have concluded that 4.8 kb repeats are not found interspersed between two 5.0 kb repeats. Based on cloning results, however, it would appear that the 4.8 kb repeats do occur next to 5.0 kb repeats (1). We have used partial digestion of DNA from several of the strains to investigate whether there is different organization of the 4.8 and 5.0 kb repeats in DNA with different ratios of the two repeat types. Since Hind III cuts each kind of repeat once, by using partial digestion we should detect the occurrence of 4.8 kb repeats next to 5.0 kb repeats as a 9.8 kb band. Figure 2a shows

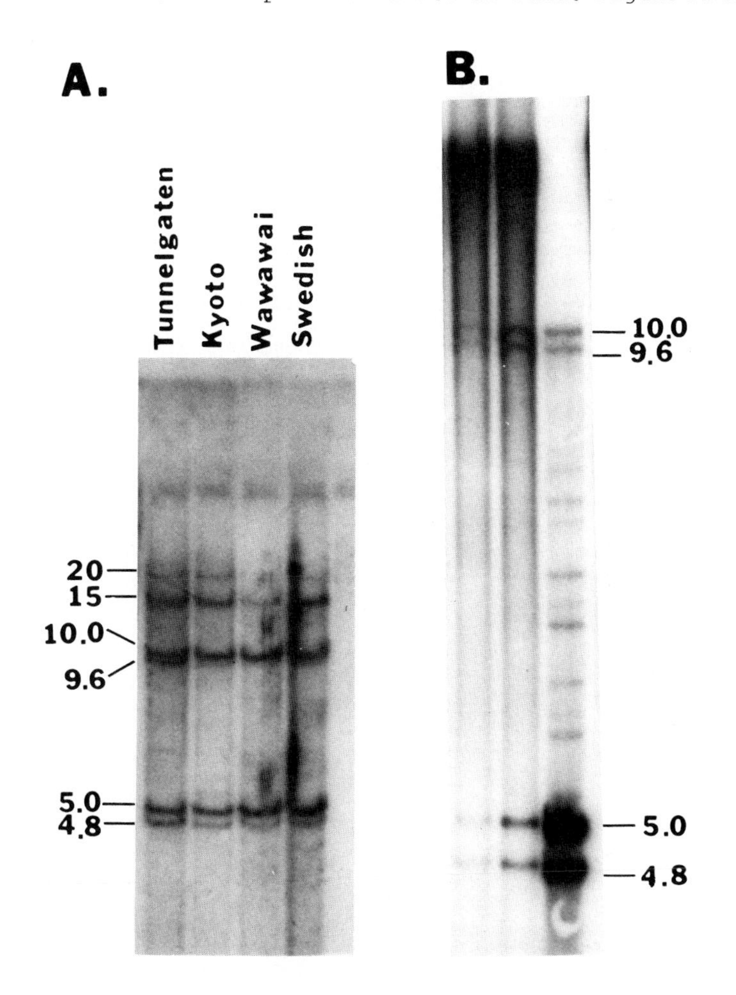

FIGURE 2. Partial Hind III digestions. A. Comparison of different strains. B. Tunnelgaten. Sizes in kb shown on the left (A) and right (B).

partial Hind III digestions of several strains . In none of the cases do we see a band corresponding to 9.8 kb. Additionally, the ratio of the 5.0 kb dimers to the 4.8 kb dimers reflects the distribution of the two major size classes in the strain. Figure 2b shows a progressive series of Hind III partials of Tunnelgaten DNA which has been electrophoresed for 80 cm to give better resolution in the region corresponding to two repeats. This film has also been overexposed to allow for visualization of any bands between the 9.6 and 10.0 kb bands. As can be seen, many additional bands become visible in the 5.0 - 9.6 kb range with this long exposure, but there is still no band evident which would correspond to 9.8 kb. Assuming that restriction enzymes have equal access to all repeats, these results indicate that the majority of the 5.0 and 4.8 kb repeats are each clustered monolithically and not interspersed, regardless of the relative amounts of the two size classes present.

Using a simple and rapid method of isolation of DNA from single flies, we have asked if the limited heterogeneity we observed in DNA extracted from large numbers of flies was due to differences between individuals or was a characteristic of each member of the population. Figure 3 shows DNA from individual flies from two different strains: Tunnelgaten which has almost equal amounts of the two repeat sizes in bulk DNA and Amherst which has more 5.0 than 4.8 kb repeat and has another characteristic band at 3.9 kb. As can be seen from a comparison of Figure 1 with Figure 3, DNA from both individual flies and bulk DNA preparations contain a distribution of the two size classes characteristic for each particular strain. The pattern of minor bands is also characteristic of each of the strains, although additional minor bands can also be seen in both strains with the single fly gel. The Amherst 3.9 kb fragment appears prominently in the Amherst individuals, but small amounts of a fragment of this size are also present in Tunnelgaten individuals. A further difference between individuals from the two populations occurs in the pattern of the minor bands larger than 5.0 kb. The heterogeneity within a population has been further investigated by constructing stocks which are isogenic for single second chromosomes from the Amherst strain. In studies of eight such stocks, the patterns are very similar, although some very minor differences which were obscured when DNA from the total population was analyzed can be seen (Strausbaugh and Weinberg, manuscript in preparation).

The results of the studies discussed so far indicate very little heterogeneity in major repeat classes at the level of populations isolated from widely separated geographical locations, individuals within a population, or single chromosomes within a population. At present we have not de-

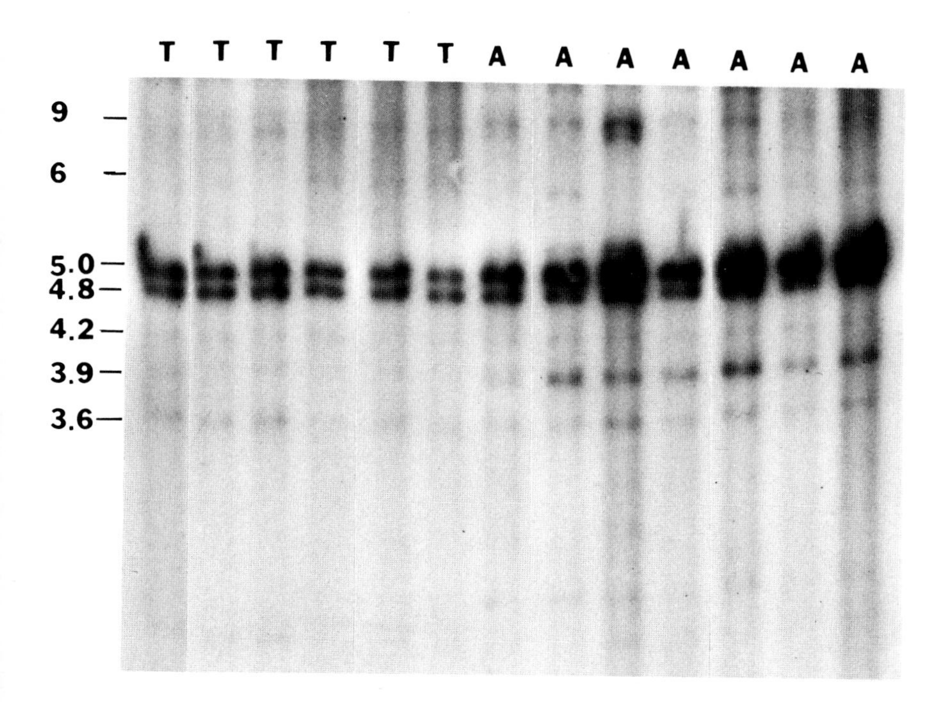

FIGURE 3. Hind III-digestion of DNA from single flies from Tunnelgaten (T) or Amherst (A) strains.

termined the relationship of the minor bands to the histone genes, but additional observations on the prominent 3.9 kb Amherst band support the idea that this band contains histone genes. First, this fragment hybridizes with the cloned 6.3 kb *S. purpuratus* histone repeat. Since the arrangement of the histone genes and spacers is very different between this sea urchin and *Drosophila melanogaster* (1), it is likely that histone gene sequences in the sea urchin repeat are hybridizing with histone gene sequences in the 3.9 kb Amherst band. Second, when the stocks isogenic for Amherst second chromosomes are analyzed, all have this characteristic band. Since the second chromosomes (where the histone genes are known to be located) is the only one selected for in these crosses, this result is consistent with the idea that the band is derived from DNA from the second chromosomes and not elsewhere in the genome. Third, this band is formed when DNA is digested with a number of other enzymes which are known to have one site in the *melanogaster* repeat, consistent with the suggestion that it could represent a set of sequences related to the main repeats. We are presently cloning this Amherst fragment to determine its organization.

We have been impressed by the overall conservation of the 4.8 and 5.0 kb repeats in *Drosophila melanogaster*. Unlike the cases in *S purpuratus* (5) and *L. pictus* (Overton, Donnelly, and Weinberg, unpublished results), where individuals in a population show both length and sequence heterogeneity in histone repeats, *D. melanogaster* strains and individuals show a remarkable maintenance of the basic repeat sizes. We have studied histone repeat length characteristics in other species in the genus *Drosophila*. Figure 4 shows the results of Xho I restriction analysis on DNA from a few flies of species which are close relatives of *melanogaster*. In these experiments, the *melanogaster* strain is Samarkand, a strain with a prominent band at 5.5 kb (discussed previously). All six of the species shown here are in the same subgenus, *Sophophora*. Three of the species, *D. simulans*, *D. mauritania* and *D. melanogaster* are in the malanogaster species group. *D. paulistorum* is in the willistoni species group and *D. pseudoobscura* and *D. affinis* are in different subgroups of the obscura species group. While all of the species have bands at 5.0 kb, only *D. melanogaster* has a prominent band at 4.8 kb. As can be seen, there are other minor band differences between the species. Preliminary results on re-

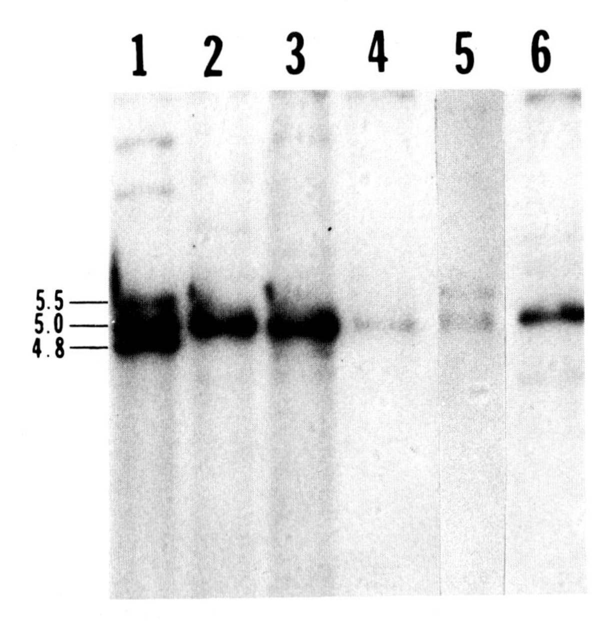

FIGURE 4. Xho I digestion of different species. 1 = *D. melanogaster* (Samarkand), 2 = *D. stimulans*, 3 = *D. mauritania*, 4 = *D. pseudoobscura*, 5 = *D. affinis*, 6 = *D. paulistorum*. Sizes in kb on the left.

striction of *simulans* DNA with additional enzymes indicate that this 5.0 kb repeat is similar to that of *melanogaster*. The DNA from *pseudoobscura* and *affinis* hybridize to a lesser degree to the *melanogaster* probe than does DNA of the other species, although ethidium bromide stains of the gels do not indicate any large differences in the amounts of DNA from flies of different species. *D. affinis* shows an additional, somewhat prominent band at about 5.5 kb which also appears when DNA is digested with Hind III. We are presently extending this analysis to other species in the genus which are more distantly related to *melanogaster*.

DISCUSSION

A variety of histone repeat size classes occur in a population in sea urchins and a very restricted number of these classes occur in any single individual (5, and Overton, Donnelly, and Weinberg, unpublished observations). In *D. melanogaster* we have not seen this extent of heterogeneity in the geographically isolated populations, in individual members of a population, or in single chromosomes. Heterogeneity exists in the different strains in the minor histone DNA homologous bands. Two possibilities are that the minor bands may represent the junction of histone and non-histone sequences (1) or they may represent sequence changes in the histone DNA present in the major repeat classes. The demonstration of exact relationships between such minor bands and the histone genes must await cloning of these sequences. It is interesting that heterogeneity in histone DNA is generated and distributed in sea urchins, but not in *Drosophila*. On one hand, it is possible that the different organization in the *Drosophila* repeat results in stronger selection pressures than in sea urchins, such that the majority of the repeats in *Drosophila* must have spacers of particular sequence and length for proper function. In this case, any chromosome with large proportions of unusual repeats would be eliminated. On the other hand, the conservation in *Drosophila* may be the result of some structural feature of the genes, such as chromosomal location near the centromere, which might affect the cellular mechanism(s) governing the stability of reiterated genes. In this case, heterogeneity might occur in any single repeat within a cluster, but an increase in the relative abundance in a chromosome of any changed sequence might not occur.

The results of a comparison of *melanogaster* with other species show some interesting differences. The most dramatic difference is the absence of 4.8 kb repeat sizes in species other than *melanogaster*. One possibility is that the other species do, in fact, have 4.8 kb repeats, but they are either present in such small amounts that they are not detectable in these experiments, or they are not tandemly arranged so that a restriction enzyme digestion would not produce a discrete fragment. Another possibility is that the other species do not have 4.8 kb size repeat classes and that repeats of this length are not required for the survival of the fly. The lower intensities of the bands in *pseudoobscura* and *affinis* could be due to low levels of sequence homology between these species and the *melanogaster* probe, or to the presence of fewer copies of the histone genes in these species, or to the possibility that some of the genes in these species may not be arranged tandemly.

We plan to continue our survey of strains and species and to use combined genetic and molecular approaches to study any differences in features of histone gene organization. Our ultimate goal is the use of different bands in *D. melanogaster* as genetic markers for studying recombination and stability of the histone gene sequences.

ACKNOWLEDGEMENTS

We thank R. Donnelly and C. Overton for technical advice and helpful discussions, and W. Sofer for many helpful discussions and for providing many of the strains and facilities for rearing flies. We also thank R. Richmond and S. Polivanov for supplying some of the species and J.G. Gall for sharing the plasmid subcloned from cDm500 Lifton *et al.* (1). We also thank P. Kuwabara and V. Murtif for excellent technical asistance.

REFERENCES

1. Lifton, R.P., Goldberg, M., Karp, R., and Hogness, D. (1978). Cold Spring Harbor Symposium on Quantitative Biology 42, 1047.
2. Kedes, L.H. (1979) Annual Review of Biochemistry 48, (in press).
3. Birnstiel, M., Weinberg, E.S., and Pardue, M.L. (1973) In "Molecular Cytogenetics" (B. Hamkalo and J. Papaconstantinou, eds.), pp. 75-93. Plenum Press, New York.
4. Pardue, M.L., Kedes, L.H., Weinberg, E.S., and Birnstiel, M.L. (1977) Chromosoma 63, 135.
5. Overton, G.C., and Weinberg, E.S., (1978) Cell 14, 247.

TRANSCRIPTION OF *DROSOPHILA MELANOGASTER* RIBOSOMAL GENES: ARE THE INSERTIONS TRANSCRIBED?[1]

Eric O. Long and Igor B. Dawid

Laboratory of Biochemistry, National Cancer Institute, Bethesda, Maryland 20205

ABSTRACT Some of the ribosomal genes in *D. melanogaster* are interrupted in the 28S RNA coding region by sequences called ribosomal insertions. The insertions are of two different types, not homologous to each other. This report deals with insertions of type 1, present in 50% of the ribosomal genes on the X chromosome and absent or rare on the Y chromosome. The majority of type 1 insertions are 5 kb long. It has also been shown that sequences homologous to type 1 insertions are present outside the rDNA locus. The presence of transcripts containing insertion sequences has been investigated by transferring nuclear RNA from *Drosophila* embryos to DBM[2]-paper and hybridizing it with cloned insertion DNA fragments. Several bands were detected ranging in size from 1 kb to over 8 kb. Molecules larger than the rRNA precursor (8 kb) are very rare. It is known that many thousand ribosomal transcripts are being synthesized in each embryo nucleus. No matter how fast splicing takes place, a single gene with a 5 kb insertion transcribed at a normal rate would produce more insertion transcripts than has been detected in our experiments. Therefore, unless a mechanism other than splicing is involved, genes with insertions can be considered not functional with respect to the production of 28S rRNA. This conclusion has also been reached analyzing RNA from ovaries, larvae, pupae and adult flies. Different models of transcription are discussed.

[1] This work was supported by a fellowship to E.O.L. from the Swiss Science Foundation.

[2] Abbreviations used: DBM: diazo-benzyloxy-methyl; kb: kilobase

ISBN 0-12-068350-4

INTRODUCTION

In *Drosophila melanogaster* there are about 150 to 200 tandemly repeated ribosomal genes in each nucleolus organizer. There is one nucleolus organizer on the X chromosome and one on the Y chromosome. It was first found by Glover and Hogness (1) that some of these genes contain an insertion in the sequence coding for the 28S rRNA. This was the first case reported of an intervening sequence in an eukaryotic gene. The structure of these genes and of their insertions has been studied in detail (2,3). There are two types of insertions, all occurring at the same place in the 28S rRNA coding sequence. Type 1 and type 2 insertions are not homologous. Type 1 insertions seem to occur only in the ribosomal genes on the X chromosome, while type 2 insertions are present at similar frequency among the ribosomal genes of the X and the Y chromosomes. Both types of insertions vary in size. Some of their characteristics are summarized in Table 1. This report will only deal with insertions of type 1; maps of such insertions are shown in Figure 1. It has been found that sequences homologous to type 1 insertions are present in the *Drosophila* genome outside the nucleolus organizer (4). These DNA regions are interspersed with sequences not homologous to ribosomal or insertion sequences.

Since the discovery of the *Drosophila* ribosomal insertions, other genes, viral and eukaryotic, have been found to contain intervening sequences. In some cases, as for instance the globin gene, it has been demonstrated that the intervening sequence is transcribed in the precursor RNA (5,6). That sequence is then removed from the RNA by a splicing mechanism which generates a continuous coding sequence in the mRNA. The question we have asked is whether *Drosophila* ribosomal genes with insertions are transcribed. If transcription of these genes occurs via a splicing mechanism there should be transcripts in nuclear RNA that contain insertion sequences. It is known from electron microscopic examination of spread nucleolus chromatin that active ribosomal genes are fully loaded with RNA polymerases and that many thousand ribosomal transcripts are being synthesized in every embryo nucleus (7). From these observations it can be deduced that a single gene with a 5 kb insertion (the most frequent insertion), transcribed at a normal rate, would produce 80 nascent transcripts containing insertion sequences. This is true no matter how fast splicing takes place since read-through of the whole sequence is necessary before splicing can occur. Furthermore, rRNA molecules up to 12 kb in size should exist provided their processing rate is comparable to normal 8 kb rRNA precursors. These predictions rely on the assumption that splicing takes

TABLE I

INSERTIONS IN *DROSOPHILA MELANOGASTER* rDNA

	Size in kb	% of genes on chromosome X	% of genes on chromosome Y	Characteristics
Type 1	5	35	0-1	Two *Sma* I restriction sites and two *Bam* HI restriction sites
	1	14	0-2	Two *Bam* HI restriction sites
	0.5			One *Bam* HI restriction site
	other sizes	rare		
				Short insertions are homologous to the rightmost part of large insertions
				Genes with type 1 insertions are interspersed with continuous genes
Type 2	1.5 - 4	16	13	Not homologous to type 1 insertions
				Several *Eco* RI restriction sites

place in one step. We have analyzed the transcription of insertion sequences using several different technical approaches.

RESULTS AND DISCUSSION

The materials and methods used, as well as the experimental data will be published elsewhere. We will summarize here the results obtained to date and discuss them.

As already reported (8), unlabeled nuclear RNA from Drosophila embryos was hybridized in excess to a 5 kb insertion probe labeled to a high specific activity. Even at Rot values that correspond to 1 RNA copy per nucleus, the probe could not be protected against the single-strand specific nuclease S1. We concluded then that ribosomal genes with insertions are either not transcribed at all or transcribed at a very low level. This conclusion was limited to the average embryonic cell.

To confirm and refine these results, nuclear RNA was electrophoresed on agarose gels after full denaturation with glyoxal, transferred to DBM-paper by the procedure developed by Alwine et al. (9), and hybridized with nick-translated DNA fragments. When such filters were hybridized with a cloned continuous rDNA repeat, the largest molecule detected was the known 8.0 kb rRNA precursor. This result is consistent with the liquid hybridizations described above and supports the conclusion that in embryos there are no frequent large rRNA precursors, as would be expected from transcription of a gene with its insertion. However, when nuclear RNA transferred to DBM-paper was hybridized with subcloned insertion fragments several RNA bands appeared. With the SmaI fragment as a probe (see Fig. 1), faint and broad bands appeared between 4 kb and 8 kb. These molecules are estimated (by comparison with diluted reference fragments run in parallel) to be rare, in the order of 1 or a few copies per nucleus. When the BamHI fragment (see Fig. 1) was used as a probe, stronger and more discrete bands appeared, ranging in size from 1 kb to about 8.5 kb. The two most prominent bands at 3 kb and 4 kb are certainly present in more than 10 copies per nucleus. We have repeated liquid hybridizations using the BamHI fragment as a probe. Kinetics of reassociation of nuclear RNA to this 1 kb insertion fragment showed that complementary transcripts are present at about 20 copies per nucleus. These transcripts were complementary to only a part of the BamHI fragment. This is probably the reason why they were not detected in the earlier liquid hybridizations with the entire 5 kb insertion as a probe.

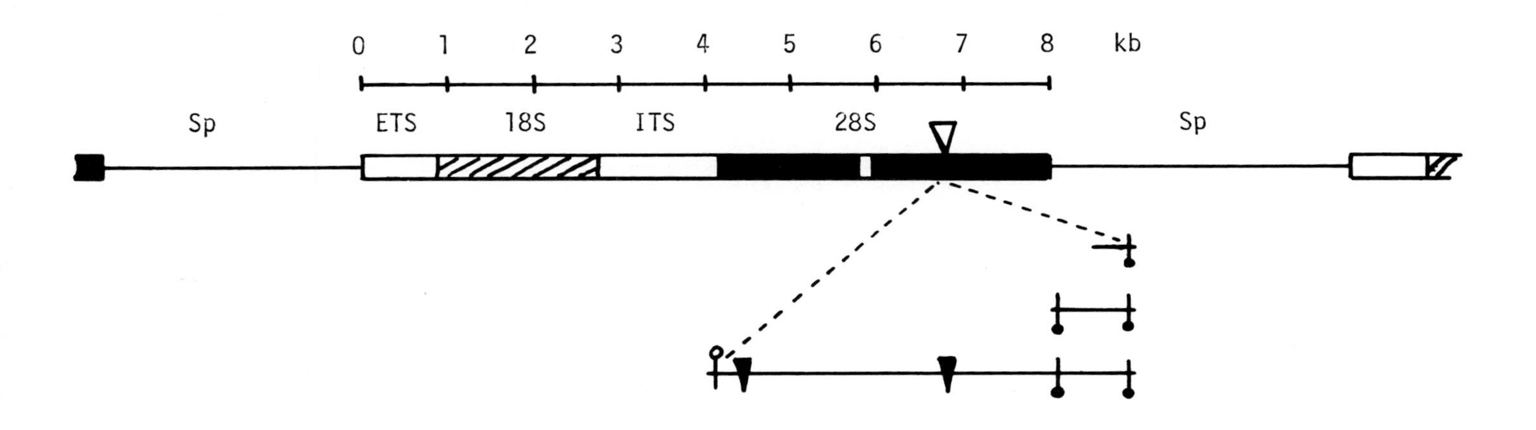

FIGURE 1. Map of a ribosomal gene and of type 1 insertions. A single continuous rDNA repeat is shown. The gene is separated from the neighbouring genes by nontranscribed spacer (Sp). The external and internal transcribed spacers (ETS and ITS) and the 18S and 28S rRNA coding regions are indicated. The point in the 28S rRNA coding sequence where insertions occur is indicated by a triangle (▽). Under the gene map are maps of the most frequent type 1 insertions. The sites recognized by the following restriction enzymes are shown : Hind III (⚲); Sma I (▼); Bam HI (↓). Some of the fragments generated by these restriction enzymes have been cloned in order to obtain pure insertion probes. The Sma I fragment has been cloned by direct ligation into the single Sma I site of the vector pMB9.(The isoschizomer Xma I has been used in that case to generate 5' overhanging ends). The Bam HI fragment has been cloned by direct ligation into the single Bam HI site of the vector pBR322. The Hind III-Bam HI fragment has been cloned by direct ligation into Hind III-Bam HI double digested pBR322.

Are any insertion transcripts derived from the ribosomal genes or do they derive from the insertion-homologous sequences outside the nucleolus organizer? We have tested this question by "sandwich" hybridization (10). Filter-bound insertion fragments were hybridized with unlabeled nuclear RNA and further hybridized with labeled rRNA coding regions. Some hybridization was observed that must have been due to annealing of RNA tails with the probe. Therefore RNA molecules exist which consist of both insertion sequences and rRNA sequences. We cannot say whether or not all RNA molecules that hybridize with insertion sequences also carry ribosomal regions. The strongest sandwich hybridization was observed with the BamHI insertion fragment (see Figure 1) suggesting that the RNA molecules responsible for this reaction may be derived from a ribosomal gene with a short insertion.

Compared to the number of complete and nascent rRNA precursors in embryo nuclei (up to 10,000 molecules) the total number of insertion transcripts large enough to be potential precursors to 28S rRNA is insignificant. Therefore, splicing of a precursor containing insertion sequences (Model 1a in Fig. 2) is not a relevant mechanism for the synthesis of rRNA in *Drosophila* embryos. We have not detected insertion transcripts large enough to contain a 5 kb insertion sequence (by transfers to DBM-paper). It is nevertheless possible that correct splicing takes place in rare ribosomal transcripts with short insertions. Genes with insertions can be considered not functional with respect to the production of 28S rRNA, unless the RNA polymerase can bypass the insertion. We consider the latter possibility unlikely (Model 1b in Fig. 2).

Genes with insertions may not be transcribed at all, except for a few copies (Model 2a). This interpretation is consistent with the observation by McKnight and Miller (7) that in nucleolar chromatin from *Drosophila* embryos there are long stretches of silent DNA interspersed with active transcription units. It is known that genes with type 1 insertions are interspersed with continuous genes (2). If the genes with insertions are transcribed then there must be either a strong termination signal at the point of the insertion (Model 2c) or abortive transcription into the insertion (Model 2d). Transcripts such as those predicted in Model 2d have indeed been detected in our experiments, but they are rare.

The major conclusion from this work is that genes with insertions do not contribute significantly to the synthesis of 28S rRNA. This is not only true in embryos, which were the source of RNA in all the experiments referred to sofar, but also in ovaries, larvae, pupae and adult flies. Quantitatively, as well as qualitatively, insertion transcripts were

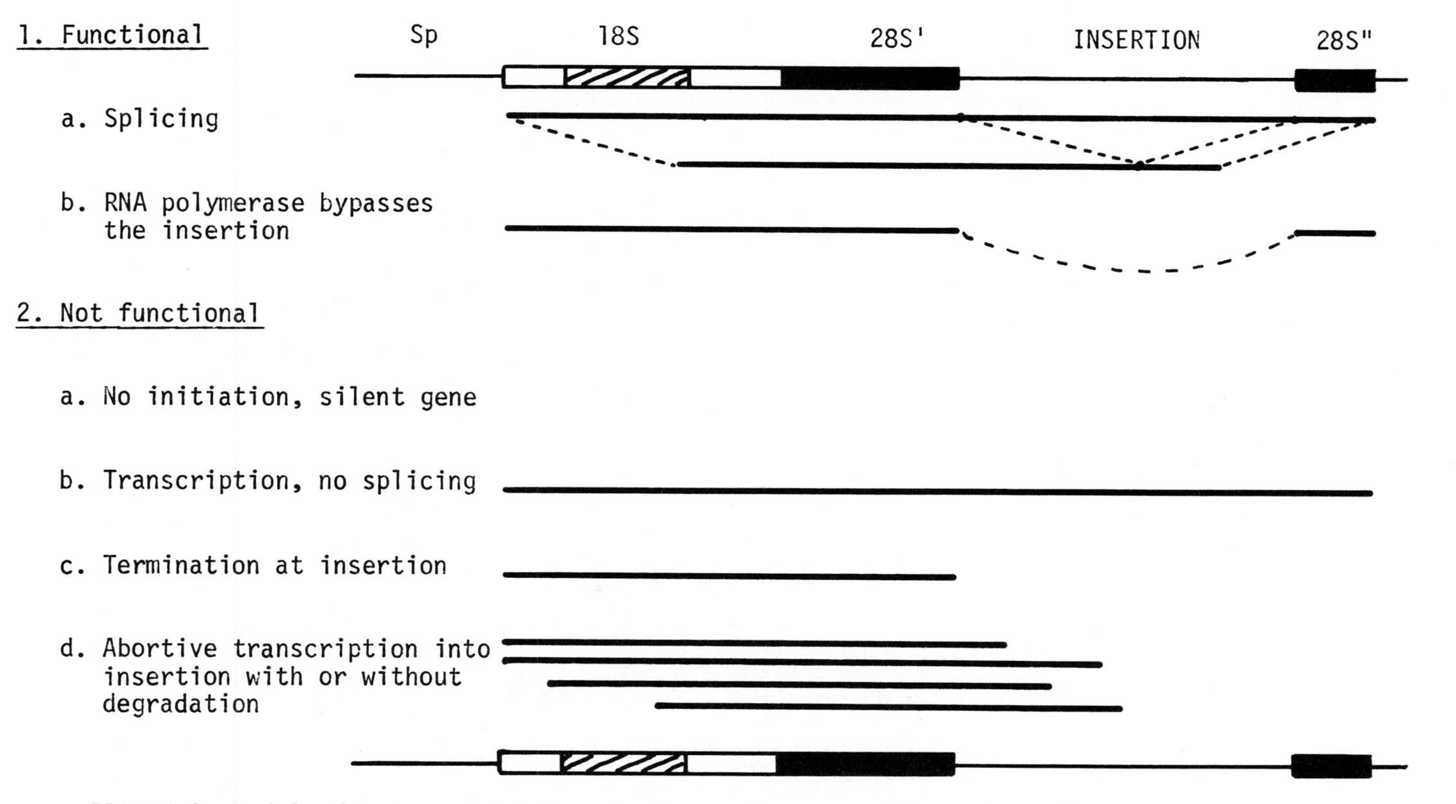

FIGURE 2. Models for transcription of ribosomal genes with an insertion. Predicted transcripts are shown between maps of a gene with a 5 kb insertion.

comparable in all stages examined by transfer to DBM-paper. It remains possible that ribosomal genes with insertions are functional only in specialized cells, but in general, it is clear that the expression of these genes differs from other eukaryotic genes with transcribed intervening sequences.

REFERENCES

1. Glover, D. M., and Hogness, D. S. (1977). Cell 10, 167.
2. Wellauer, P. K., and Dawid, I. B. (1977). Cell 10, 193.
3. Wellauer, P. K., Dawid, I. B., and Tartof, K. D. (1978). Cell 14, 269.
4. Dawid, I. B., and Botchan, P. (1977). Proc. Nat. Acad. Sci. USA 74, 4233.
5. Tilghman, S. M., Curtis, P. J., Tiemeier, D. C., Leder, P., and Weissmann, C. (1978). Proc. Nat. Acad. Sci. USA 75, 1309.
6. Kinniburgh, A. J., Mertz, J. E., and Ross, J. (1978). Cell 14, 681.
7. McKnight, S. L., and Miller, O. L. Jr. (1976). Cell 8, 305.
8. Dawid, I. B., and Long, E. O. (1978). In "Alfred Benzon Symposium XIII". In press, Munksgaard, Copenhagen.
9. Alwine, J. C., Kemp, D. J., and Stark, G. R. (1977). Proc. Nat. Acad. Sci. USA 74, 5350.
10. Dunn, A. R., and Hassell, J. A. (1977). Cell 12, 23.

ECDYSONE-INDUCIBLE FUNCTIONS OF LARVAL FAT BODIES IN *DROSOPHILA*[1]

Jean-Antoine Lepesant[2], Alan Garen[3],
Jana Kejzlarova-Lepesant[2], Florence Maschat[2] and Luce Rat[2]

CNRS Institut de Recherche en Biologie Moleculaire,
Universite Paris VII, Paris, France[2] and
Department of Molecular Biophysics and Biochemistry,
Yale University, New Haven, CT, USA[3]

In this report we describe a combined genetic and molecular approach to the problem of ecdysone action, involving the use of a temperature-sensitive mutant of *Drosophila melanogaster* called *ecd*[1], which becomes deficient for ecdysone after a temperature shift from 20°C to 29°C at various stages of development (1). The conditional ecdysone-deficient phenotype of the mutant provides a sensitive system for analyzing the control of specific developmental processes by a steroid hormone. We have focused on the late third instar larval stage of *Drosophila* development, when the ecdysone concentration increases sharply by a factor of about twenty within a few hours (1,2). This increase is blocked in *ecd*[1] larvae after a temperature shift to 29° earlier in the third instar (1); consequently, all of the late larval functions normally induced by the increase in ecdysone concentration should also be blocked after the shift. Such functions might be identified by an appropriate comparison between late third instar *ecd*[1] larvae that have an abnormally low concentration of ecdysone as a result of an earlier temperature shift and larvae that have a higher concentration of ecdysone, because ecdysone was added to the food after the shift or the temperature was kept at 20°C until the ecdysone titer had increased in the course of normal development. This procedure was used to test first for effects of ecdysone on the proteins of various larval tissues. The most pronounced effects were observed with the larval fat bodies, in which several major new proteins appeared as a result of two distinct responses induced by the increase in ecdysone concentration; one response was an incorporation back into the fat bodies of proteins synthesized earlier in the same tissue and subsequently secreted into the circulating haemolymph, and the other response was *de novo* synthesis of one major new protein called *P1* and

[1]This work was supported in part by grants to A. Garen from the American Cancer Society and National Institutes of Health, U.S.P.H.S.

ISBN 0-12-068350-4

two minor proteins (3). The ecdysone-induced synthesis of the P1 protein did not occur in four other larval tissues examined, namely the salivary glands, imaginal discs, brain with attached ganglion and discs, and male gonads, and therefore appears to be a tissue-specific response of the fat bodies. The response also appears to be stage-specific because synthesis of the P1 protein could not be induced prematurely in fat bodies of early third instar larvae, indicating that the capacity of the tissue to respond to the hormone signal depends on prior developmental steps.

The ecdysone-induced synthesis of the P1 protein was further analyzed by *in vitro* translation of RNA isolated from fat bodies of low and high ecdysone samples of late third instar ecd^1 larvae (3). The results indicate that ecdysone induces a major increase in the amount of translatable messenger RNA for the P1 protein. The RNA from the fat bodies was used as a probe for the isolation of a clone of λ bacteriophage carrying a segment of the *Drosophila* genome complementary to the P1 messenger RNA, as identified by specific inhibition of *in vitro* translation of the messenger RNA after hybridization with the cloned DNA (3). Measurements of the amount of poly(A)-RNA in the fat bodies that could hydridize to the DNA from this clone, called λDm P1, showed about 50 times more hybridizeable poly(A)-RNA in the high ecdysone sample than in the low ecdysone sample (Table I), in agreement with the *in vitro* translation results.

TABLE I
HYBRIDIZATION OF POLY(A)-RNA FROM FAT BODIES OF LOW AND HIGH ECDYSONE SAMPLES TO CLONED *DROSOPHILA* DNA IMMOBILIZED ON NITROCELLULOSE MEMBRANES

	Hybridized poly(A)-RNA			
	cpm		% of total (above background)	
DNA source	Low ecdysone	High ecdysone	Low ecdysone	High ecdysone
No DNA	220	230		
λ(Charon 4)	230	240	0	0
λDm P1	590	18,300	0.017	0.87
λDm P2	970	8,520	0.036	0.40
λDm P3	35,000	16,000	1.68	0.76
λDm P4	1,275	5,640	0.050	0.26

The clones were isolated from a library of λ(Charon 4) phages carrying inserts of randomly sheared *Drosophila*

> *melanogaster* (Canton S) DNA (6). The DNA was extracted from purified preparations of the phages and immobilized on a nitrocellulose membrane (3). The poly(A)-RNA was isolated from late third instar larval fat bodies, dissected either from low ecdysone ecd^1 larvae shifted from 20° to 29° early in the third instar or from high ecdysone ecd^1 larvae grown continuously at 20°, and end-labeled by reacting with [γ-^{32}P]ATP and polynucleotide kinase (3). The hybridization, washing, and counting procedures are also described elsewhere (3).

Hybridization assays were also done with poly(A)-RNA isolated from the brain with attached ganglion and discs of high ecdysone third instar larvae, and from stage-14 oocytes, gastrulating embryos, 16-hour old embryos, and adult flies; in all cases, there was no measureable hybridization of poly(A)-RNA to the *λDm P1* DNA. These results provide further evidence that the ecdysone-induced expression of the *P1* gene occurs only in the fat bodies of late third instar larvae.

The location of the *P1* coding region in the *λDm P1* DNA was determined by cleaving the DNA with various restriction endonucleases, separating the fragments by gel electrophoresis and testing for hybridization with a high ecdysone sample of fat body poly(A)-RNA. The resulting restriction map (Fig. 1)

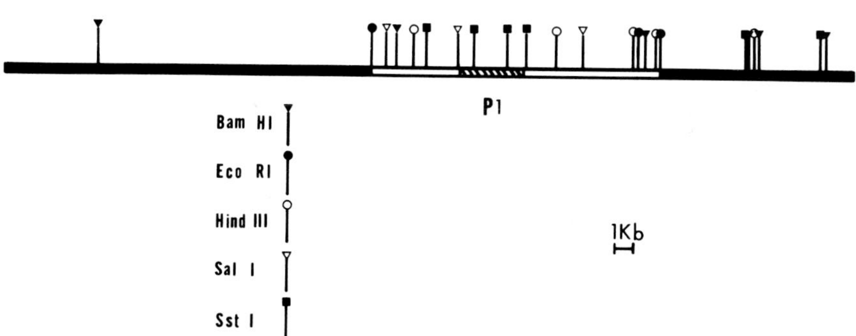

Figure 1. The *λDm P1* DNA was isolated from a purified phage preparation (3) and the products of endonuclease digestion were separated by electrophoresis in 0.7% agarose gels, transferred to nitrocellulose paper and hybridized with a high ecdysone sample of fat body [^{32}P]poly(A)-RNA (3). The solid black regions at the ends of the map are the *Charon-4* genome and the region between is the *Drosophila* insert. The *P1* region is the approximate length of DNA that hybridized with the probe after all the endonuclease cuts were made.

CELL LINEAGE IN INSECT DEVELOPMENT

Peter A. Lawrence,[1] and Gines Morata[2]

MRC Laboratory of Molecular Biology, Hills Road, Cambridge, CB2 2QH.

ABSTRACT Recent work with genetic cell markers has shown that development of the insect retina does not exhibit any rigid cell lineages. However, an unexpected type of lineage restriction operates within imaginal discs: while _individual_ embryonic cells have considerable flexibility, _groups_ of cells have a rigidly defined prospective fate. Each group of cells (or _polyclone_) constructs a specific _compartment_ in the adult.

INTRODUCTION

Traditionally, embryologists have classified animals as following one of two modes of development. In _mosaic development_ morphogenesis is rigidly programmed and unresponsive to experimental intervention. The most extreme model of mosaic development entails a rigid production of cell types by a defined tree of cell lineage. If an embryonic cell is killed that part of the adult that it would normally have generated is missing. In _regulative development_ organs develop flexibly, with cells easily replacing others. This view rather ignores the role of individual cells, and is instead concerned with fields, which are multicellular entities and are only loosely defined in the literature.

Whether particular groups of animals - and insects are a good example - exhibit mosaic or regulative development has long been controversial. Until recently there has been little precise information about cell lineage in insect development and, of course, prejudices grow and arguments tend to prosper in the absence of facts.

Here, we describe results of marking cells genetically which have produced detailed information on cell lineage, and suggest that insect development is neither strictly "mosaic" nor "regulative". We also show that studies on cell lineage may help us understand the way genes control development.

Some insect organs, with a small and defined number of

[2]Present address: C.S.I.C., Centro de Biologia Molecular, Universidad Autonoma de Madrid, Canto Blanco, Madrid.

ISBN 0-12-068350-4

cells (such as bristles and scales) are known to develop by rigid cell lineage (reviewed in, 1). The insect retina has been thought to develop similarly (2), no doubt because the cells are arranged in a perfect 'crystalline' lattice (3,4). Genetic mosaics have been used to analyse the cell lineage of the retina in two different ways:

(1) Clones of cells marked with white are made in red (white$^+$) Drosophila eyes. At the border between the clone and the surround, the cells of the two genotypes are apparently randomly distributed (4,5).

(2) Grafting experiments between mutant and wildtype eyes of Oncopeltus (Hemiptera) have also shown a random arrangement of cell genotypes at the host/graft border (6).

These two experiments eliminate the possibility that cell lineage of the retina is completely defined, but they do not rule out the suggestion of Campos-Ortega (e.g. 5,7) that there could be some lineage restriction in retina development. For example, there could be two or more classes of precursor cells, each of which would generate clones of cells of only one type. In particular, Campos-Ortega suggested that pigment and sensory cells came from separate precursor classes. Once formed, these two cell types would then assemble to form the lattice. Campos-Ortega's hypothesis has been recently tested by making red (white$^+$) clones in white eyes of Drosophila during the last few cell divisions. Analysis of a large number of clones has shown no lineage relationships or restrictions. For example, two-cell clones are frequent and often comprise one pigment and one sensory cell (8). We can conclude that the highly precise retinal lattice is made by mechanisms which do not depend on cell lineage.

The imaginal discs of Drosophila have been studied for many years by the traditional methods of experimental embryology - fate maps, extirpation and grafting (see 9 for review, 10). Ideas of fixed cell lineage, at least within the individual disc, have not featured largely in these studies. However, recent analysis of cell lineage in the development of imaginal discs has shown that, for example, the wing disc is subdivided precisely into two regions (compartments, 11,12), each made by all the descendents of a small group of embryonic cells (a polyclone, see 13 for review). Each cell in the polyclone has a flexible fate, but the group as a whole has a precisely defined role - to construct one compartment of the wing and notum. Even when endowed with a higher growth rate than the other cells in the disc (14), a clone is unable to cross the precisely defined frontier between anterior and posterior compartments (13,14). Thus a state of determination is acquired by each cell when the polyclone is established and, in the normal course of development, propagated to all its descendents. The polyclone is therefore the executive unit of development, and its job is to construct one particular compartment.

How might this state of determination be acquired? We know that some compartments precisely coincide with the realm of action of some homoeotic genes (15; see 16 for review). In particular, the engrailed gene operates only in the posterior polyclone (19,20). Further, this gene, when active, directs the development of each and every cell in the posterior polyclone so that the posterior, rather than the anterior, pattern is formed (17,18). Recent experiments on the antennal segment (which makes most of the head) (19,20), and the three thoracic segments (21,22), have shown that they are each subdivided into anterior and posterior compartments and that the engrailed gene has an equivalent role in all four segments (19,20,22). These experiments on engrailed strongly support the selector gene hypothesis (15,23), a basic tenet of which is that the combination of active selector genes in each polyclone is, in effect, a binary code which directs the development pathway undertaken by that polyclone.

Clearly, precise information about cell lineage has been valuable; it has shown that insect development cannot be comfortably described as either 'mosaic' or 'regulative' and has led to some new ideas about genetic strategy in insect development.

REFERENCES

1. Lawrence, P. A. (1966). J. Cell Sci. 1.
2. Bernard, F. (1937). Bull. Biol. France Belg. Suppl. 23, 1.
3. Trujillo-Cenóz, O., and Melamed, J. (1966). J. Ult. Res. 16, 395.
4. Ready, D. F., Hanson, T. E., and Benzer, S. (1976). Devl. Biol. 53, 217.
5. Campos-Ortega, J. A., and Hofbauer, A. (1977). Wilhelm Roux' Archiv. 181, 227.
6. Shelton, P. M. J., and Lawrence, P.A. (1974). J. Embryol. exp. Morph. 32, 337.
7. Campos-Ortega, J. A., Jurgens, G., and Hofbauer, A. (1978). Nature 274, 584.
8. Lawrence, P. A., and Green, S. M. (1979). Devl. Biol. In press.
9. Gehring, W. J., and Nöthiger, R. (1973). In "Developmental Systems: Insects" (S. J. Counce and C. H. Waddington, eds) pp. 211-290. Academic Press, London and New York.
10. Haynie, J. L., and Bryant, P. J. (1976). Nature 259, 659.

11. Garcia-Bellido, A., Ripoll, P., and Morata, G. (1973). Nature New Biol. 245, 251.
12. Garcia-Bellido, A., Ripoll, P., and Morata, G. (1976). Devl. Biol. 48, 132.
13. Crick, F. H. C., and Lawrence, P. A. (1975). Science 189, 340.
14. Morata, G., and Ripoll, P. (1975). Devl. Biol. 42, 211.
15. Garcia-Bellido, A. (1975). Ciba Found. Symp. 29, 161.
16. Morata, G., and Lawrence, P. A. (1977). Nature 265, 211.
17. Morata, G., and Lawrence, P. A. (1975). Nature 225, 614.
18. Lawrence, P. A. and Morata, G. (1976). Devl. Biol. 50, 321.
19. Morata, G., and Lawrence, P. A. (1978). Nature 274, 473.
20. Morata, G., and Lawrence, P. A. (1979). Devl. Biol. In press.
21. Steiner, E. (1976). Wilhelm Roux' Archiv. 180, 9.
22. Lawrence, P. A., Struhl, G., and Morata, G. (1979). J. Embryol. exp. Morph. In press.
23. Garcia-Bellido, A., Lawrence, P. A., and Morata, G. (1979). Sci. Amer. In press.

THE GENETIC ANALYSIS OF CHORION MORPHOGENESIS IN DROSOPHILA MELANOGASTER[1]

Mary Ellen Digan, Allan C. Spradling, Gail L. Waring[2] and Anthony P. Mahowald

Program in Molecular, Cellular and Developmental Biology, Department of Biology, Indiana University, Bloomington, Indiana 47401

ABSTRACT Electron micrographs of thin sections of the *Drosophila melanogaster* chorion show a complex pattern of structural organization. We have examined the effects of 2 female-sterile mutations, *ocelliless* and a new EMS-induced mutation, *chorion-36*, on chorion morphology and report on the effects of *chorion-36* on chorion protein production. Our results suggest that c36, a major protein present in preparations of purified chorions, is a structural component of the endochorion. Both female-sterile mutations map to a 16-band interval on the X-chromosome which has been shown to include the structural genes for c36 and c38, another major chorion protein. The dosage of these genes can be changed using a duplication or deficiencies for this chromosomal region. We show that the amount of these gene products which accumulates in stage 12-14 egg chambers increases with gene dosage. The *Drosophila* chorion appears to be a system in which a combination of genetic, biochemical and morphological approaches can be used to examine the regulation of structural assembly in a eucaryotic organism.

INTRODUCTION

The controlled expression of genes leads to the production of complex and highly ordered structures. The mechanisms which govern these processes can be examined through the production of mutations affecting specific steps. In favorable cases, the effect of alterations in the properties or amount

[1]This work was supported by a grant from the NSF (PCM-77-25427).

[2]Present address: Department of Biology, Marquette University, Milwaukee, Wisconsin 53233

ISBN 0-12-068350-4

of single components can lead to an understanding of their functional roles. As a prerequisite for such an analysis, a significant number of the genes involved in a specific morphogenetic event must be identified in an organism which is amenable to genetic manipulation. The process of chorion synthesis by the ovarian follicular cells of *Drosophila melanogaster* appears to satisfy these requirements. In this report, we describe the beginnings of a genetic study of the production of the chorion which has been based on this approach.

An elegant series of studies on the silkmoth chorion has demonstrated that the genes involved in chorion synthesis and their products are susceptible to detailed biochemical analysis (1,2,3). Recently, progress has been made in carrying out similar studies in the case of *Drosophila* (4,5,6). The *Drosophila* chorion contains at least 20 proteins, six of which are present in large amounts. Putative chorion messenger RNAs have been identified (5) and the genetic locus of at least two of the major proteins c36 and c38, has been determined to reside on the X chromosome in the region 7E10-8A4 (5,7). The chromosomal localization of chorion structural genes has made it possible to study the effects on chorion morphology of alterations in their dosage.

At least one mutation mapping in this interval, the female-sterile mutant *ocelliless* (oc), is known to result in the production of a morphologically abnormal chorion (8). The chorion defects in *ocelliless* have been correlated with an underproduction of four major chorion proteins, including c36 and c38 (7). In order to gain further insight into the role of c36 and c38 in chorion morphogenesis, new recessive female-sterile mutations mapping in the 7EF region were selected following EMS mutagenesis. A mutant, *chorion-36* (*cor-36*) was obtained which fails to produce detectable amounts of c36. We have compared the morphology of the abnormal chorions and the patterns of protein accumulation in females bearing various combinations of the *cor-36* and *ocelliless* mutations. Our results suggest the possibility of an interesting functional role for c36 in the organization of endochorion and exochorion structure.

MATERIALS AND METHODS

Stocks of Drosophila strains were maintained in half-pint bottles or shell vials at 21°C or 25°C. Standard medium containing cornmeal, dried yeast and dextrose was used. EMS mutagenesis was carried out according to the method of Lewis and Bacher (10). Prior to injection of flies with ^{14}C-amino acids for the analysis of chorion proteins, stocks were transferred for one to two days to bottles containing a thick paste of live yeast. Labeling of chorion proteins and analysis of proteins from egg chambers or purified chorions were as described previously (5,6). For electron microscopy, stage 14 egg chambers were fixed according to the method of Kalt and Tandler (9), embedded in DER 322 and thin sections were examined in a Philips EM 300.

RESULTS

The Effect of Chorion Gene Dosage on Production of c36 and c38. A genetic perturbation of the chorion genes in the 7EF region which can be readily produced involves alterations in the dosage of these genes. Females bearing from one to three doses of the 7EF region were constructed using chromosomes containing deficiencies or duplications for this interval and were shown to be fertile. Df(1)KA14 (Df(1)7E10±; 8C1±) and Df(1)RA2 (Df(1)7D10±; 8A4±) lack the structural genes for at least two major chorion proteins, c36 and c38. The insertional translocation FN107 (T(1;2) 7A8; 8A5, 58E + In(2LR) 32C; 58E) contains the region of the X-chromosome bearing these same chorion genes. These chromosomes were a gift of George Lefevre. Figure 1 illustrates the amount of c36 and c38 which are accumulated in egg chambers of females bearing from one to three doses of the structural genes for these chorion proteins. Although the staining intensity of numerous proteins is approximately equal in each case, it is clear that the amount of c36 and c38 increases with gene dosage. Consequently, the total amount of these proteins accumulated per egg chamber is not closely regulated.

Female-sterile Screen. Female-sterile mutations within the 7E10± to 8C8± region of the X-chromosome were obtained by feeding EMS to 3 day old y sn^3 mal males for 24 hours according to the method of Lewis and Bacher (10). After a day of feeding on normal media, males were mass-mated to C(1)DX/Y virgin females. Single y sn^3 mal males were mated in the F1 cross to females heterozygous for Df(1)KA14 and FM7c, a balancer chromosome marked with y w^a sn^{x2} v B. Two classes of females which can be distinguished on the basis of body color (y) and eye shape (B) result in the F2 generation: Df(1)KA14/ y sn^3 mal and FM7c/ y sn^3 mal. These females were allowed to mate with their FM7c sibs in fresh vials for 7 days. The absence of y^+ females in the F3 generation indicated that a female sterile mutation may have been induced on the y sn^3 mal chromosome within the region deleted on its

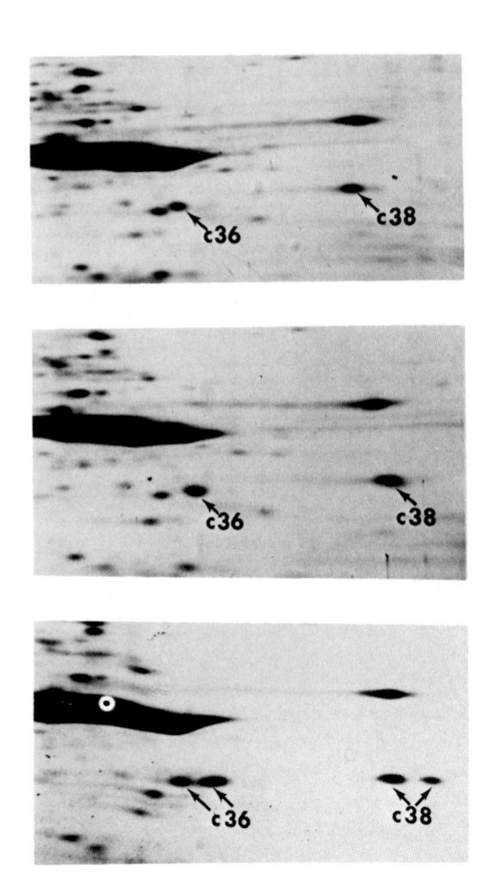

Figure 1: The effect of variation in dosage of structural genes for c36 and c38. Similar numbers of stage 12-14 egg chambers from females with one dose (top: Df(1)KA14/FM7c), two doses (middle: y cv v f car/y cv v f car) and three doses (bottom: C(1)Dx; T(1;2)FN107) of the chromosomal region containing the structural genes for c36 and c38 were prepared and run on the NEPHGE gel system as described in (6). The Coomassie Blue R staining pattern of the region of the gel containing c36 and c38 is enlarged in this figure. It should be noted that females with three doses of the c36 and c38 structural genes indicate that one of the structural genes for both c36 and c38 codes for a protein which differs slightly in charge.

homologue. Thus, vials containing only y progeny were kept; females homozygous for the mutagenized chromosome were tested for female sterility in the F4 generation. Descriptions of the compound chromosome and marker mutations used can be found in Lindsley and Grell (11).

Two female-sterile mutations were recovered in a screen of 1179 fertile pair-matings (0.17%). One of these mutations appears to be associated with semi-lethality which has prevented its analysis thus far. The other female-sterile mutant recovered has been named chorion-36 (cor-36) to indicate that its effect is on a specific gene product found in the chorion, c36. Females heterozygous for Df(1)RA2 (Df(1)7D10±; 8A4±) and cor-36 are sterile, placing the mutation in the overlap of Df(1)KA14 and Df(1)RA2 (7E10± to 8A4±).

cor-36 is fertile in heterozygous combination with oc, three alleles of goggle, gg^2, gg^3, gg^4 or with a representative of a series of lethal complementation groups in this region provided by Dr. G. Lefevre. Thus, it appears that cor-36 defines a new complementation group within this chromosomal region. Ovarian stage 14 oocytes appears shorter in general than normal oocytes and have very short chorionic appendages. The few eggs laid by cor-36/cor-36 females lack chorions. cor-36 males have normal fertility.

Synthesis of c36 and c38 in Mutant Egg Chambers. ^{14}C-labeled proteins were obtained from stage 12-14 egg chambers, electrophoresed on the NEPHGE gel system of O'Farrell, et al. (12) and subjected to fluorography as described in Waring and Mahowald (6). Enlargements of the region which contains c36 and c38 on the fluorographs from wild-type and cor-36/cor-36 females are shown in Figure 2. Although the labeling intensity of c38 in egg chambers from both genotypes is roughly equivalent, labeling of c36, by cor-36/cor-36 egg chambers is undetectable.

We have previously shown that the female sterility of ocelliless is associated with the underproduction of a number of chorion proteins, including c36 and c38 (7; see Fig. 3A) and an abnormal chorion (8; see Fig. 4C). Although both the oc and cor-36 mutation affect the production of c36, the heterozygous combination of these two mutations was fertile. Previous experiments have shown that the underproduction of c36 and c38 characteristic of the ocelliless chromosome is independent of the alleles of these genes carried on the homologous chromosome (7). The amount of c36 accumulated in this fertile heterozygote is not different from the amount found in homozygous ocelliless egg chambers when compared to reference proteins.

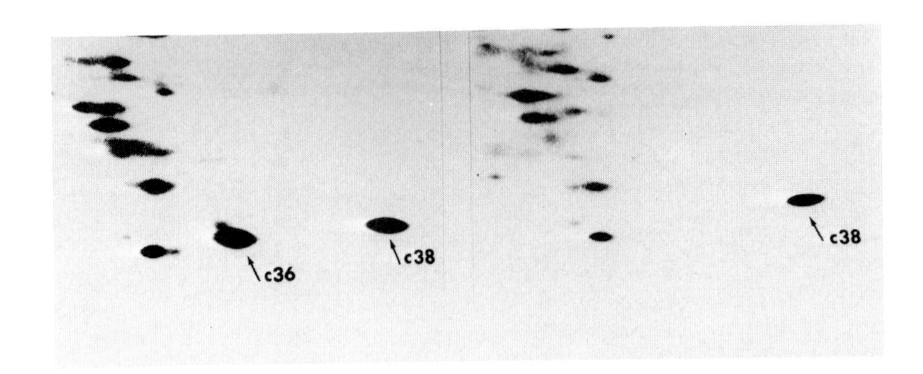

Figure 2: Production of c36 and c38 by $\underline{cor}$-$\underline{36}^{+}$/$\underline{cor}$-$\underline{36}^{+}$ (A) and cor-36/cor-36 egg chambers (B). Females homozygous for cor-36 and the wild type allele of this gene were injected with ^{14}C-amino acids and stage 12-14 egg chambers were dissected after approximately three hours. Samples were prepared, run on NEPHGE gels and fluorographs made as described in (6). Fluorographs were exposed to Kodak X-Omat R film for one month.

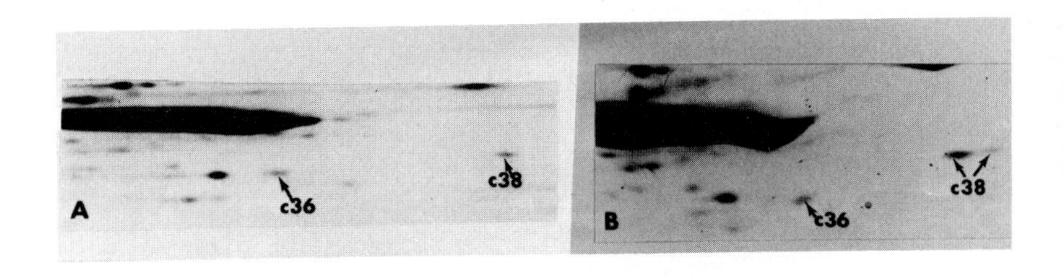

Figure 3: Accumulation of c36 and c38 in oc/oc (A) and cor-36/oc egg chambers. Stage 12-14 egg chambers from females of the appropriate genotype were dissected, prepared and run on the NEPHGE gel system as described in (6). The Coomassie Blue R staining pattern of the region of the gel containing c36 and c38 is enlarged in these pictures. Electrophoretic variants are available for these two proteins (7). The lightly staining c36 and c38 proteins in (B) are coded for by the oc chromosome and the heavily-staining c38 is coded for by cor-36.

Genetic Control of Chorion Morphology. The egg envelopes of the normal *Drosophila* chorion consist of four layers secreted by the follicle cells during oogenesis (Fig. 4A): (1) the vitelline membrane is closest to the oocyte and has separated from the other chorion layers during fixation in this figure; (2) the electron-dense intermediate chorionic layer (13); (3) a compartmental endochorion whose inner and outer layers are connected by pillars or trabeculae,; and (4) a fibrillar exochorion, the outermost layer. The compartmental structure of the endochorion is common to many insect species and serves a respiratory function (14). In Figure 4, electron micrographs of thin sections of chorions produced by females of different genotypes are shown. Chorions from females homozygous for either female-sterile mutation, *cor-36* (Fig. 4B) or *oc* (Fig. 4C) show different structural abnormalities, many of which are complemented in the fertile *cor-36/oc* heterozygote (Fig. 4D). *cor-36/cor-36* chorions show a total disorganization of the endochorion. In addition, components of endochorion extend into the exochorion layer. The electron-dense intermediate chorionic layer is normal in these chorions. In contrast, *oc/oc* chorions possess a perforated intermediate layer; the inner endochorion is missing and the pillars which normally connect it to the outer endochorion are greatly reduced in number. The outer endochorion and exochorion, however, are of normal structural organization in *oc/oc*. As mentioned previously, females heterozygous for *oc* and *cor-36* are fertile. The chorions produced by *cor-36/oc* females (Fig. 4D) approach wild type in morphology; the main distinguishing feature is the thinness of the outer endochorion.

DISCUSSION

Genetic analyses of mutations affecting oogenesis in *Drosophila melanogaster* have contributed a significant amount of information about this developmental process (reviewed in 15). Several laboratories have begun systematic collections and complementation analyses of female-sterile and maternal effect mutations on the X-chromosome (16,17) and on the autosomes (18,19). Mutations characterized by fragile chorions or malformed dorsal appendages have been described (see 15 for a review), but the biochemical basis for the mutant phenotypes have not been determined. A developmental approach to the identification of chorion-defective mutants has been pioneered by Wieschaus, et al. (20). These workers constructed ovarian mosaics by reciprocal transplants of pole cells between wild type and a mutant with abnormal chorionic appendages and showed that expression of the mutant phenotype in the progeny

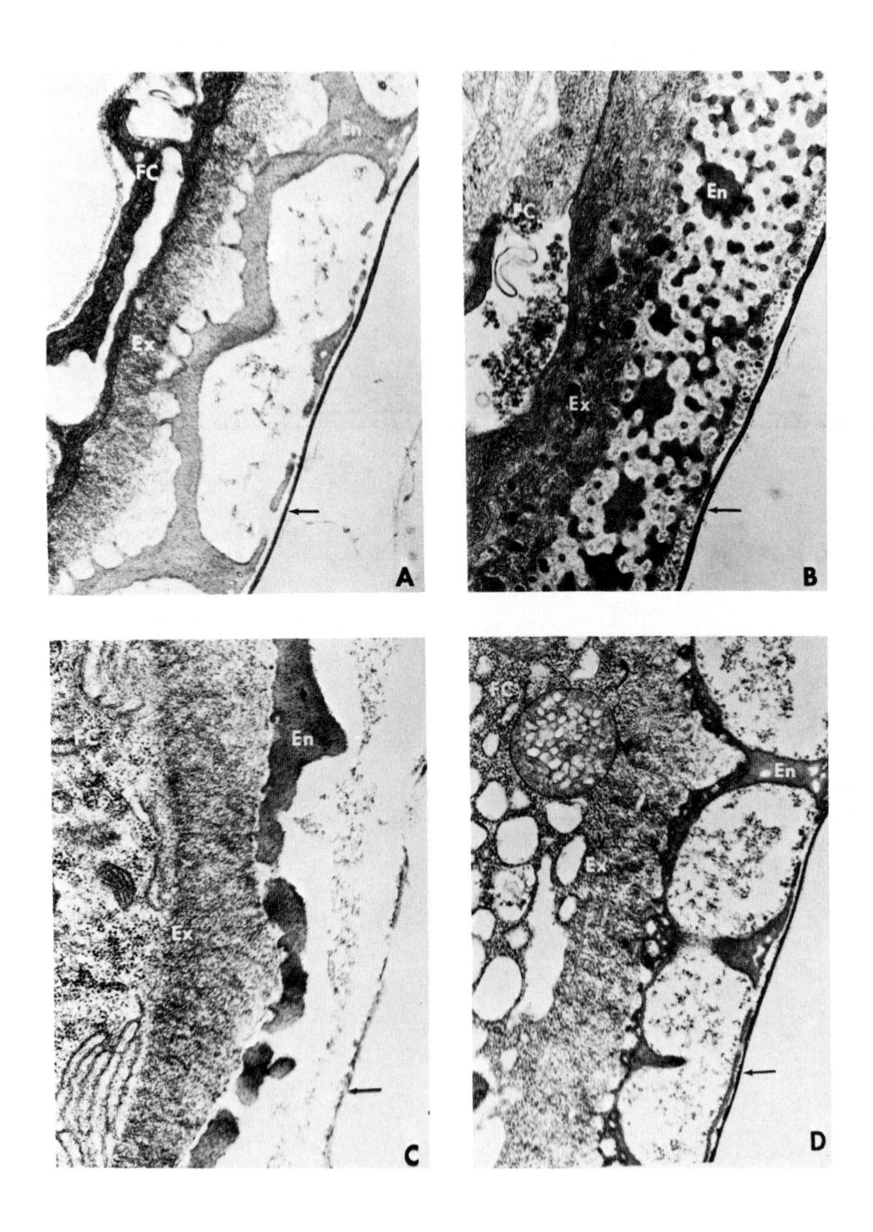

Figure 4: Genetic control of chorion morphology. The genotypes of the stage 14 egg chambers pictured are: (A) P2/P2 (wild type); (B) cor-36/cor-36; (C) oc/oc; and (D) cor-36/oc. En, endochorion; Ex, exochorion; FC, follicle cell. The arrow indicates the intermediate chorionic layer, which in the living egg chamber is opposed to the vitelline membrane.

depends on the genotype of the germ-line and not on the somatically-derived follicle cells (20). Using similar transplantation techniques, the effect of the _ocelliless_ mutation on chorion morphology has been shown to be follicle cell-specific (E. M. Underwood and A. P. Mahowald, ms. in preparation). Our search for female-sterile mutations which alter the electrophoretic migration or amount of proteins present in the chorion was facilitated by two fundamental advantages: we knew the region to screen for mutants from _in situ_ hybridization studies (5), and we could recognize altered electrophoretic patterns from knowledge of the pattern produced by normal chorion proteins (6).

Observations on the biochemical and morphological defects in chorions produced by _ocelliless_ females have led to insights concerning the possible chromosomal arrangement and regulation of the genes coding for four major chorion proteins (7). The _cor-36_ mutation has allowed us to distinguish the role of c36 in the chorion and has furthered our knowledge about the relationship of c36 to c38. The common reduction in the synthesis of c36 and c38 by the _oc_ mutation suggests the possibility that these two proteins differ only by secondary modifications. The likelihood of this possibility is reduced in light of the fact that c38 synthesis is normal in _cor-36/cor-36_ egg chambers, while synthesis of c36 is undetectable. In addition, experiments in progress indicate that c36 and c38 are synthesized by different mRNAs.

The ultrastructural abnormalities seen in _cor-36/cor-36_ chorions involve mainly the endochorion. The exochorion defects may be a secondary effect of the disorganization of the endochorion. The interpretation that c36 performs a necessary structural function in the endochorion is consistent with the facts that c36 is found in purified preparations of chorions and that c36 is predominantly synthesized during stages 12 and 13 (6), the stages during which the endochorion is synthesized (21). The restoration of fertility and the partially repaired morphology observed in _cor-36/oc_ heterozygotes is indicative that some c36 is necessary for fertility, but that wild type ratios of c36 to c38 are not needed. The ultrastructural defects present in _oc/oc_ chorions are somewhat harder to interpret due to the number of proteins affected and the unknown nature of their interactions.

The amount of c36 and c38 which accumulated in egg chambers of flies bearing from one to three copies of the structural genes for these proteins was found to be proportional to the gene dosage. Differences among these strains in the amount of c36 and c38 in the chorion itself is currently under investigation. Preliminary observations reveal that some parameters, such as chorionic filament length, do show a

several criteria be met. It should be shown that the element is located close to the structural gene, that the structural gene has not been altered and that the number of enzyme molecules produced has been modified.

Dopa decarboxylase (DDC) possesses a number of properties that make it suitable for the study of developmental regulation. DDC activity is a developmentally vital function which is under stringent tissue and temporal control. The requirement for DDC at well defined developmental stages, its tissue distribution and the availability of toxic analogue inhibitors provide several possible means of screening for putative regulator mutants. Of primary importance is the availability of a sensitive biochemical assay for the gene product which permits characterization of both normal and mutant enzymes *in vitro*, purification of the enzyme and production antisera. Our analysis is aided by the presence of an adjacent gene of apparently related function *l(2)amd*. In the following sections we describe the properties of the DDC system which make it suitable for the recovery of regulatory mutants and we describe the initial characterization of two putative control mutations which seem to lead to the increased expression of the adjacent genes *l(2)amd* and *Ddc*.

RESULTS

Dopa decarboxylase (DDC) catalyzes the conversion of dopa to dopamine. In Drosophila DDC activity has been found in three tissues: in the epidermis it acts in sclerotization (cuticular hardening), in the central nervous system it produces the neurotransmitter dopamine, and in the proventriculus it probably acts in the sclerotization of the cuticular peritrophic membrane. During development, peaks of DDC activity occur at the times of embryonic hatching, the first and second larval molts and pupariation. These four peaks of activity coincide with four peaks of ecdysone titer during development and are thought therefore to be hormone induced. There is experimental evidence that DDC is induced by ecdysone prior to pupariation in Calliphora (3,4) and Sarcophaga (5). Additional evidence that DDC activity in Drosophila is hormone induced comes from the observation that no DDC peak is found in larvae homozygous for the temperature sensitive ecdysoneless mutant, $\underline{ecd}^1$, at restrictive temperatures (6). Thus DDC activity increases sharply in conjunction with increased titers of ecdysone and decreases abruptly in response to decreasing ecdysone titers during these four periods of ecdysone production. However, a fifth and major peak of ecdysone during mid-pupal development is not accompanied by an increase in DDC activity although some 60 hours later, as the adult ecloses, a fifth peak of DDC is observed. Whether this DDC peak at eclosion arises as a delayed response to the mid-pupal ecdysone peak or whether this final increase in DDC activity is controlled by a completely different mechanism is unknown.

Genetic Localization. Hodgetts (7) using segmental aneuploids to vary gene dosage demonstrated that the only dose sensitive region for DDC in the genome is on the left arm of the second chromosome in the region between 36EF and 37D. Subsequent induction of overlapping deficiencies including in particular the small 8-10 band deficiency Df(2L)130, 37B9-10;D1-2, permitted the delineation of the DDC dosage sensitive region to nine bands, 37B10-C6 (8). Two mutagenesis schemes have been employed to search for point mutants affecting DDC.

DDC Structural Gene. The first scheme involved screening for recessive lethals by saturation mutagenesis in the Df130 region (10). A total of 132 recessive lethal mutations have been isolated in this Df130 region, and one hundred and four of them have been placed in 12 lethal complementation groups (genes). A recently acquired overlapping deficiency, Df(2L)TE42-1,37C1,2-5;39 makes it possible to assign eleven of these lethal genes and the visible mutation hook (2-53.9) to a maximum of seven bands, 37B10-C4, and one gene to a maximum of four bands, 37C3-6. No lethal mutations have been recovered that are assignable to band 37D1, deleted by Df130 but not by DfE71.

When representatives from each of the 12 genes were assayed for DDC activity as heterozygotes a single group in the distal Df130 region, 37B10-C4, was found which reduced DDC activity. All 17 mutant alleles of this gene lower DDC activity to 25 to 75% of normal when they are heterozygous over a wild type allele (6,10). In addition some pairs of alleles within this group exhibit intracistronic complementation (10), i.e. some heteroallelic heterozygous individuals are viable, an observation consistent with the contention that DDC is a homodimer (subunit MW=54,000) (9). Two complementary heteroallelic pairs show 5-10% and 0.5-3% DDC activity, and both produce DDC which is clearly thermolabile *in vitro* (6). In addition the DDC from a recently recovered temperature sensitive allele of this locus is also found to be thermolabile *in vitro* (6). Thus, on the basis of their effect on DDC activity and thermolability *in vitro*, this group of recessive lethals identify the structural gene, *Ddc*, for the enzyme, DDC.

Although Ddc^{n5}/Ddc^{n8} complementing heteroallelic heterozygotes with 5-10% of wild type DDC activity show no visible mutant effects and exhibit almost normal viability, other complementing heteroallelic heterozygotes with reduced viability and less than 3% DDC activity exhibit a striking escaper phenotype including prolonged development time, death within 24 hours of eclosion, and very thin, long and straw-colored bristles (10). In addition, the body remains light, not taking on the normal pigmentation after eclosion, the abdominal markings are apparent but do not darken, the wing axillae and leg joints are melanized in a manner similar to a wound reaction, and the flies walk on their tibia rather than on their tarsi. All these pleiotropic effects can reasonably be inferred to arise as a result of incompletely sclerotinized cuticle. This escaper phenotype is quite distinctive, being expressed in the very rarely (1 in 25,000) surviving Ddc^{n5}/Df130 heterozygote and in all Ddc^{ts1}/Df130 heterozygotes raised at 20° and 22° but not in those raised at 18°C (6).

The effective lethal phase of Ddc alleles as heterozygotes over Df130 has been determined to occur during embryogenesis. Complete, active larvae develop which are unable to hatch from the embryonic membranes. The only observable defect is that the larval mouthparts and setae do not take on their normal pigmentation (6). Whether this indicates that the mouth parts are insufficiently sclerotinized to be effective in breaking the embryonic membrane is unknown.

Dissected central nervous systems and proventriculi from white pre-pupae heterozygous for Ddc alleles, $Ddc^{n}/+$, exhibit DDC activities between 25% and 50% of wild type (6), indicating that the same structural gene codes for DDC in the epidermis, the central nervous system, and in the proventriculus. Furthermore, activity of a wild type Ddc allele appears to be required for female fertility. Females heterozygous for a temperature sensitive allele over Df130, $Ddc^{ts1}/Df130$, are completely sterile, laying reduced numbers of eggs which do not develop (6). Heterozygotes of Ddc^{ts1} with many, but not all the other Ddc deficient alleles, exhibit significantly reduced fertility and fecundity, although none are completely sterile. The presence of normal sperm in the spermathecae of $Ddc^{ts1}/Df130$ females indicates that the infertility does not arise as a result of an inability to mate. The effect of the Ddc mutants on female fertility is not unexpected since DDC activity has been determined to be necessary for female fertility in mosquitoes (11, 12), and we have observed within two or three days after being fed continuously on food containing high levels of α methyl dopa, that Drosophila females begin laying eggs which do not complete embryogenesis.

Thus the Ddc gene is expressed in three and probably four different tissues and at five well-defined developmental stages during development.

α Methyl Dopa Hypersensitive Mutations. The second scheme to recover mutations affecting DDC activity was the use of dietary α methyl dopa (α MD), an analogus inhibitor of DDC, as a possible discriminator of abnormally high and abnormally low levels of DDC *in vivo* (13, 14). When fed to adults, α MD has no effect except to sterilize females (see above), but growing larvae fed sufficient concentrations of α MD die at the next molt, and those puparia formed by larvae fed lower concentrations of α MD are quite flexible in contrast to the very rigid structures normally formed. These observations suggest that α MD interferes with normal sclerotinization in some way (14).

A screen was run for dominant α MD hypersensitive mutations which when heterozygous over a wild type allele, $amd^{H}/+$, effect lethality on levels of dietary α MD on which homozygous wild type, +/+, live. A screen of mutagenized second chromosomes produced seven such mutants, all of which were determined to be recessive lethals on standard food, and to be allelic mutations in the same gene, l(2)amd (14). Both phenotypes, dominant α MD hypersensitivity and recessive lethality, were mapped to a locus near 53.9 on 2L (14) which is in the middle of the DDC dosage sensitive region. It is now known that l(2)amd, like Ddc, is one of the eleven lethal genes located in the 37B10-C4 region encompassing a maximum of seven bands.

Neither as heterozygotes over a wild type allele, amd^{H}/+, nor as heteroallelic complementing heterozygotes, amd^{H1}/amd^{H89}, do mutants at this locus have any effect on DDC (14). This includes DDC activity at various stages in development, in the epidermis, CNS and proventriculus, the *in vitro* inhibition of DDC by α MD or by N-acetyldopamine, *in vitro* thermolability of DDC, the *in vitro* decarboxylation of α MD, soluble phenoloxidase activity in pupae and adults, and dopamine acetyl transferase activity in adults and molting first instar larvae. In short, no biochemical phenotype or protein product of this gene has been identified. That the l(2)amd gene product is in someway involved in sclerotization of cuticle is suggested by the phenotype of l(2)amd homozygotes which die at the embryonic/larval boundary. The cuticle of both unhatched and hatched larvae is exceptionally friable, and for all alleles examined both hatched and unhatched larvae exhibit abnormal, necrotic, melanized anal organs. These organs on each side of the anus are specialized epidermal cells involved in regulating the ionic balance of the larvae. Examination of electron micrographs of wild type anal organs suggest that their integrity is dependent on a heavy, cuticular suture between them and the surrounding normal epidermal cells. In EMs of homozygous l(2)amd embryos it is evident that the cuticular suture is inadequate or defective resulting in anal organ cell death and subsequent melanization (15). Further, it should be mentioned that the α MD hypersensitive alleles also effect dominant hypersensitivity to the dietary administration of another analogue inhibitor of dopa decarboxylase, N'(DL-seryl)-N^2-(2,3,4-trihydroxybenzyl) hydrazine (10) suggesting that the protein coded for by the amd locus may in some way recognize a molecule resembling dopa or dopamine.

Relationship of *Ddc* to *l(2)amd*. All 323 heterozygotes produced by crossing all 17 *Ddc* alleles to 19 of the 20 *l(2)amd* alleles (1 was lost) are completely viable at 25 and 30°C (6) indicating that two distinct functional units (genes) are represented by these two groups of mutations. This conclusion is supported by a comparison of the mutant phenes produced by the two groups of mutants. The *Ddc* mutants reduce DDC activity and alter the structures of the DDC proteins; the *amd* mutants do not effect DDC in any known way. The *amd* mutants effect dominant hypersensitivity to αMD; the *Ddc* mutants are not hypersensitive to αMD even as heteroallelic heterozygotes, Ddc^{n5}/Ddc^{n8}, with DDC activities as low as 5-10% (6). Although both groups of mutants have a similar effective lethal phase at the end of embryogenesis, all *amd* homozygotes exhibit the strikingly abnormal anal organ phenotype along with weak, easily torn cuticles; unhatched Ddc^{n7}/Df130 heterozygous embryos show neither of these mutant phenes but instead have unpigmented mouthparts and setae, a mutant phene not expressed by the *amd* homozygotes. Although the *Ddc* and *amd* loci are related through their interactions with the structurally related compounds dopa and αMD, there is no functional evidence to support the notion that both groups of mutations are lesions in the same gene.

This conclusion is of some importance when one considers the fact that lesions in the two genes have been mapped as close as .002 map units apart, (10, 6), a distance more suggestive of intracistronic rather than intercistronic map distances. If one uses the estimate of 5800 nucleotide pairs per .01 map units (16), lesions in the two genes are approximately 1200 nucleotide pairs apart, suggesting that *Ddc* and *amd* are adjacent genes.

Analysis of Two Putative Regulator Mutations. In screening for dominant resistance to αMD, Sherald and Wright (13), recovered 3 mutations which exhibited elevated resistance to αMD. Assays of DDC activity indicated that two of these exhibited the dual phenotype of elevated resistance and elevated DDC activity. One of these strains, R^E, was ethylmethane sulfonate induced, and the other R^S, presumably arose spontaneously being found in a laboratory stock.

In the original isolates, the increase in αMD resistance could be attributed to elements on both the left arm and right arm of chromosome 2 but the increase in DDC activity to the left arm only. In order to isolate the left arm component(s), which might be contiguous to the amd and Ddc genes, the region between rdo (2-53) and pr (2-54.5), 1.5 map units apart, was moved by crossing over into a lethal-free, control second chromosome. The rdo-pr regions from both the R^E and R^S strains continued to express the dual phenotype when moved into the control 2nd chromosome. To test whether the increased DDC activity is due to a structural gene alteration, the DDC from both strains was tested for in vitro thermostability and in vitro inhibition to αMD and no differences were found. Mixtures of crude extracts were additive, suggesting that the increased activity was not due to the presence of activators or absence of inhibitors.

Using an antiserum against Drosophila DDC which precipitates, but does not inactivate the enzyme, we found that the amount of anti-DDC cross reacting material (CRM) in the resistant, overproducer mutants was increased to the same extent as the DDC activity in homozygous R^S and R^E flies and larvae. Homozygous R^E elevates DDC activity 58% above the wild type control and elevates DDC - CRM 56% and homozygous R^S elevates DDC activity 41% and DDC - CRM 37%. The conclusion is evident that in these two strains increased DDC activities are the result of increased levels of DDC molecules.

Is the elevated level of resistance to dietary αMD in the R^E and R^S strains a result of the increased DDC activity? Probably not. It is clear that flies with reduced DDC levels are not hypersensitive to αMD, e.g. Ddc^{n2}/+ with 53% DDC activity, Ddc^{n4}/+ with 35%, and Ddc^{n5}/Ddc^{n8} with 5-10% DDC activity exhibit wild type levels of sensitivity to αMD. On the other hand, flies with 150% normal DDC activity carrying three doses of Ddc^+ and either 1 or 2 doses of amd^+ not only fail to show increased resistance, but are actually significantly more sensitive. Thus the increased resistance of the R^S and R^E strains probably is not due to the increase in DDC activity in these strains.

It is clear that sensitivity to the dietary αMD is controlled by the amd locus alone and since R^S and R^E map to this region of the genome, it is reasonable to conclude that the increased resistance of R^S and R^E is due to an alteration or increase in the amd gene product. However, at this point there is no test to demonstrate genetic allelism between R^E or R^S and amd nor do we have a biochemical assay for the amd gene product.

The two phenotypes, elevated resistance and elevated DDC activity, are most readily explained by mutations in a region responsible for the coordinate regulation of the Ddc and amd genes or by a small duplication for the two genes.

Two lines of evidence argue against a duplication for the region. First no duplication is visible in cytological preparations and therefore a duplication, if present, must be very small.

Secondly flies with 3 doses of Ddc have 150% of the activity of normal 2 dose flies. If the resistant strains were duplications for the Ddc locus we would expect to see 200% of normal activity in R^S and R^E homozygotes. In fact we see 140% and 160% respectively. Since the increase in DDC activity in R^S and R^E is not an integral multiple of the single dose activity as is expected in duplication homo- and heterozygotes and since no duplication is cytologically detectable, a duplication would have to be very small and only partially active to account for the observed behavior of the R^E and R^S strains. At this point it appears more likely that these two mutations define a contiguous regulatory site which increases the expression of the amd and Ddc genes coordinately.

DISCUSSION

In the DDC system one can envision recovering a number of different types of regulator mutations.

Overproducer Mutations. These include overproducer mutations recovered by screening for resistance to αMD or other specific inhibitors of DDC. Other overproducer mutations might be recovered by screening for suppressors of "leaky" Ddc alleles such as Ddc^{n5} which is only very rarely viable; exhibiting escaper phenotype when it does survive. Such suppressors can be isolated by screening for viable Ddc^{n5}/Df130 flies which do not exhibit the escaper phenotype. These should include mutations which increase the expression of the Ddc^{n5} allele. That DDC production is in fact increased must ultimately be demonstrated by showing an increase in the level of anti-DDC CRM.

Down Regulation or Underproducer Mutations. Since the intracistronic complementing heterozygotes Ddc^{n5}/Ddc^{n8} or Ddc^{n1}/Ddc^{n8}, which have 5-10% and 0.5-3% DDC activity respectively, experience a two to four day delay in development, one can enrich for mutations which reduce DDC activity by collecting late hatching flies, and then confirm the presence of an underproducer mutation by assaying the progeny for reduced DDC activity. Ultimately the distinction between a down regulator and a leaky *Ddc* allele must be made on the basis of the regulator mutation mapping outside the limits of the structural gene and the integrity of DDC protein.

Tissue Specific Regulator Mutations. Tissue specific regulators can be identified by comparing DDC activities in the epidermis, the central nervous system and the proventriculus of all lethal and non-lethal alleles of *Ddc*. A more direct approach is suggested by the observation that DDC is required for female fertility. A screen for non-lethal, female sterile alleles of *Ddc* should allow recovery of tissue specific control mutations that result in reduced *Ddc* gene activity in the ovary and unaltered levels of activity in the epidermis, central nervous system, and proventriculus.

Stage Specific Regulator Mutations. Mutations that affect DDC activity at one developmental stage and not another can be identified by assaying *Ddc* mutations as heterozygotes over a wild type allele at different times during development. Those that exhibit significantly different levels of activity at different times in development relative to homozygous wild type are putative stage specific regulator mutations.

Given the stringent developmental control of *Ddc* and our ability to screen for different types of putative regulator mutations at the *Ddc* locus, we should be able to test the hypothesis that mutable controlling elements exist contiguous to the Ddc structural gene. The nature of the mutations recovered will be indicative of the modes of control which are exerted on the Ddc gene. The recovery of the R^S and R^E overproducing variants suggests that at least one of the predicted types of regulation is operative. These genetic studies coupled with the physical characterization of the Ddc gene itself should help us understand the different types of control which are exerted on a developmentally vital gene during development.

ACKNOWLEDGEMENTS

The authors gratefully acknowledge the excellent assistance and efforts of E.Y. Wright, L.E. Hendrickson, B.J. Robertson, A.D. Tomsett, and K.W. Bentley on this project.

REFERENCES

1. Davidson, E. H., and Britten, R. J. (1973). Quart. Rev. Biol. 48, 565-613.
2. Beermann, W. (1972). In "Developmental Studies on Giant Chromosomes" (W. Beermann, Ed.), pp. 1-33. Springer-Verlag, New York - Heidelberg - Berlin.
2a. Wold, B. J., Klein, W. H., Hough-Evans, B. R., Britten, R. J. and Davidson, E. H. (1978). Cell 14, 941-950.
3. Fragoulis, E. G. and Sekeris, C. E. (1975a). Eur. J. Biochem. 146, 121-126.
4. Fragoulis, E. G., and Sekeris, C. E. (1975b). Biochem. J. 146, 121-126.
5. Chen, T. T., and Hodgetts, R. B. (1974). Devel. Biol. 38, 271-284.
6. Wright, T. R. F. Unpublished Data.
7. Hodgetts, R. B. (1975). Genetics 79, 45-54.
8. Wright, T. R. F., Hodgetts, R. B., and Sherald, A. F. (1976). Genetics 84, 267-285.
9. Clark, W. C., Pass, P. S., Venkataraman, B., Hodgetts, R. B. (1978). Molec. Gen. Genet., 162, 287-297.
10. Wright, T. R. F., Bewley, G. C., and Sherald, A. F. (1976). Genetics 84, 287-310.
11. Schlager, D. A. and Fuchs, M. S. (1974). Devel. Biol. 38, 209-219.

12. Schlager, D. A., and Fuchs, M. S. (1974). J. Insect. Physiol. 20, 349-357.
13. Sherald, A. F., and Wright, T. R. F. (1974). Molec. Gen. Genet. 133, 25-36.
14. Sparrow, J. C., and Wright, T. R. F. (1974). Molec. Gen. Genet. 130, 127-141.
15. Sparrow, J. C. Unpublished Data.
16. Gelbart, W., McCarron, M., and Chovnick, A. (1976). Genetics 84, 211-232.

FAMILIES OF INTERSPERSED REPEAT SEQUENCES AT THE 5' END OF DICTYOSTELIUM SINGLE-COPY GENES[1]

Alan R. Kimmel,[2] Cary Lai and Richard A. Firtel[3]

Department of Biology, B-022, University of California at San Diego, La Jolla, California 92093

ABSTRACT M4 and KH10 are recombinant plasmids which contain Dictyostelium nuclear DNA. Both plasmids contain sequences which are repeated in the genome and are interspersed between single-copy DNA. The repeat sequences hybridize ~1-5% of poly(A)$^+$ mRNA, but only 10-20% of the hybridization is resistant to low levels of RNase. The complementary mRNA is heterogeneous in size and 80-90% of the mass is complementary to single-copy DNA. In M4 the repeat sequence and an adjacent single-copy region are part of a single transcription unit. The repeat sequences are adjacent to sets of single-copy genes and are transcribed to produce different mRNA molecules carrying a common repeat sequence at the 5'-end.

INTRODUCTION

Dictyostelium is an ideal organism for the study of the eukaryotic genome. The genome possesses a pattern of interspersion of single-copy and short repeated sequences similar to that found in most other eukaryotes (1,2). In Dictyostelium short interspersed repeat sequences are present on ~60% of hnRNA and ~25% of mRNA (3). We are interested in the function of these repeat sequences and have isolated recombinant plasmids which contain sequences that are repeated in the Dictyostelium genome and interspersed with single-copy DNA (4). We shall present evidence which implies that families of repeat sequences are located at the 5' end of sets of single-copy genes. We suggest that initiation of transcription occurs within the repeat. The possible function of these short repeats in Dictyostelium will be discussed.

[1]This work was supported by grants from the National Institutes of Health and the American Cancer Society.
[2]American Cancer Society Postdoctoral Fellow
[3]Recipient of an American Cancer Society Faculty Research Award.

ISBN 0-12-068350-4

RESULTS

Selection and characterization of recombinant plasmids containing short interspersed repeat sequences. *Dictyostelium* DNA was randomly sheared and inserted into the Eco RI site of pMB9 by poly(dA)-(dT) homopolymer extension (4). Several recombinant plasmids (eg. M4, KH10) identified by the Grunstein-Hogness (5) colony hybridization procedure hybridize a large fraction (1-5%) of total poly(A)$^+$ mRNA. They each possess a short repeat sequence which is flanked by single-copy DNA (4).

The restriction maps of KH10 and M4 are shown in figure 1a. DNA excess renaturation kinetics have shown that ~10% of the insert of either M4 or KH10 is repetitive; 90% of the insert is complementary to single-copy DNA (4). M4 has been more thoroughly examined. Restriction fragments 1, 2 and 3 are present in only one copy in the genome as determined by Southern (6) DNA filter hybridization (fig. 1b) and hybridization kinetics. Fragment 4 hybridizes to more than 50 bands on *Dictyostelium* genomic Southern blots, and DNA excess renaturation kinetics show that a ~300 bp sequence is repeated 100 times in the genome. Similar results have been obtained with KH10 DNA (4). We conclude that M4 and KH10 contain short repeat sequences interspersed with single-copy DNA in the *Dictyostelium* genome.

Analysis of M4 and KH10 complementary mRNA. The mRNA complementary to these clones has been analyzed in considerable detail. The complementary mRNAs are heterogeneous in size (4). In addition, most of the hybridization is sensitive to low levels of RNase suggesting that a number of different mRNAs contain a relatively short homologous sequence.

Each of the major restriction fragments of M4 (labeled 1-4 in Fig. 1) has been isolated and subcloned into pBR322. M4 fragments were tailed with poly(dC) and inserted into the pBR322 Pst I site which had been tailed with poly(dG). This method reconstructs Pst I sites flanking the insert and permits the excision of the fragment (7,8). DNA was isolated from each of these subclones, immobilized on nitrocellulose filters and hybridized with *in vivo* [^{32}P]-labeled poly(A)$^+$ mRNA (fig. 2). Approximately 1-1.5% of mRNA hybridizes to total M4. This mRNA migrates heterogeneously in urea-SDS polyacrylamide gels. Subclones 1 and 2 hybridize to a single 0.9 kb mRNA. It is a middle abundancy mRNA, present at approximately 0.1% in poly(A)$^+$ mRNA as measured by hybridization of pulse-labeled mRNA to DNA filters and by mRNA excess hybridization studies. Subclone 3 hybridizes a low abundancy (0.01%) mRNA. This has been confirmed by mRNA excess

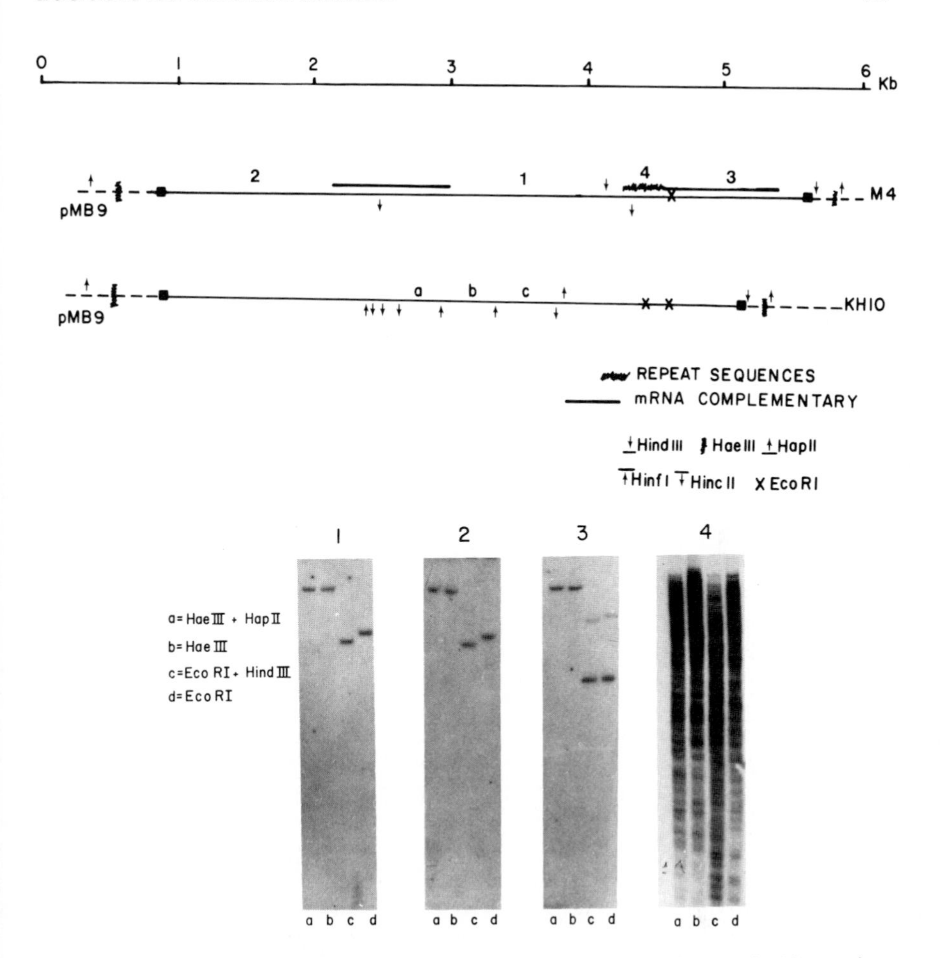

Figure 1a. Restriction Map of KH10 and M4. KH10 and M4 each contain a fragment of Dictyostelium DNA inserted into the Eco RI site of pMB9 by poly(dA)-(dT) homopolymer extension as described previously (4). There is a deletion in the pMB9 vehicle from the Eco RI site to near the Hind III site. The bars above the M4 restriction map indicate the regions transcribed into mRNA. The polarity of the band 4-3 mRNA is indicated. In plasmid KH10, fragments labeled a, b, and c have been purified and shown to be repeated in the genome and asymmetrically represented in mRNA. Regions to the right and left of these sequences are mostly single-copy.

b. Bands 1-4 were purified from their respective subclones. Probes were labeled in vitro by nick-translation with [α^{32}P]-dXTPs and hybridized to various restriction digests of Dictyostelium DNA which had been size fractionated on agarose gels and transferred to nitrocellulose filters. After appropriate washes the filters were autoradiographed.

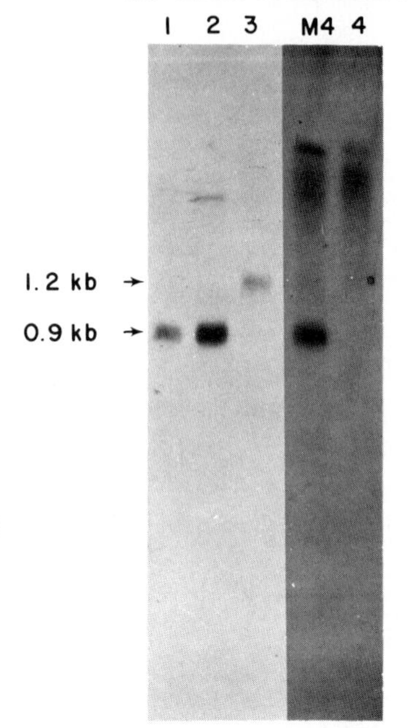

Figure 2. Electrophoresis of [^{32}P]-mRNA Complementary to M4 Clones on Polyacrylamide Gels. DNA was purified from total M4 and M4 subclones (see restriction map, Fig. 1) and immobilized on nitrocellulose filters. Filters were hybridized with in vivo labeled [^{32}P]-poly(A)$^{+}$ RNA (as described, 4). RNA was size fractionated on urea-SDS polyacrylamide gels and RNA bands identified by autoradiography. Lengths of mRNA were determined by comparison with the mobilities of [^{32}P]-poly(A)$^{-}$ RNA electrophoresed in adjacent slots.

hybridization kinetics at 100 fold sequence excess. This mRNA is present at approximately 5 molecules per cell. Pulse-labeled band 3 mRNA was isolated by hybridization to subclone 3 DNA immobilized on nitrocellulose filters. After elution, the mRNA migrated as a single band of approximately 1.2 kb. Subclone 4, a 0.45 kb fragment containing the repeat sequence, hybridizes ~1% of total mRNA. This mRNA is heterogeneous in size. DNA excess hybridization has shown that 90% of mRNA that hybridizes to subclone 4 is complementary to single-copy DNA. These data suggest that a large fraction of the repeat family and adjacent single-copy sequences are transcribed. The mRNA complementary to subclone 3 is one of these transcripts. Using a sandwich hybridization technique (9), we have shown that unlabeled mRNA that has hybridized to band 3 is a site for the hybridization of [^{32}P]-labeled band 4 (fig. 3). By the same criteria, the mRNA transcribed from bands 1 and 2 does not possess the repeat sequence; bands 1 and 2 comprise an independent single-copy transcription unit. Bands 4 and 3 represent one transcription unit composed of a short interspersed repeat sequence and an adjacent single-copy region. Experiments using exonuclease III or λ-exonuclease indicate that the repeat sequence is located at the 5'-end of the band 4-3 transcription unit.

A heterogeneous size population of mRNA hybridizes to repeat sequences on the KH10 insert. Linkage hybridization studies indicate that 10-20% of each KH10 mRNA molecule is transcribed from repeat sequences while the remainder is complementary to single-copy DNA (4). The presence of specific

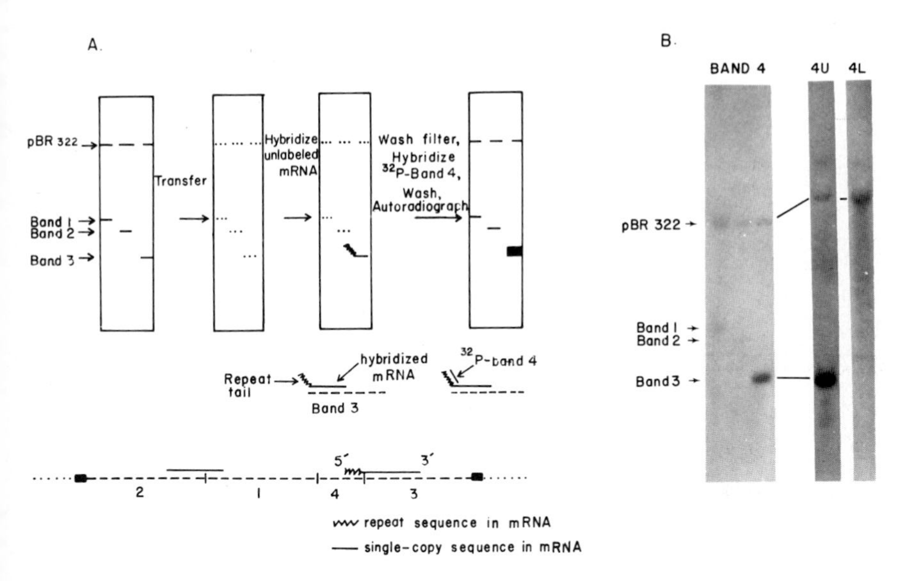

Figure 3. Sandwich Hybridization of [^{32}P]-band 4 probes to bands 1, 2 and 3 RNA-DNA Hybrids of M4.

A. Diagram of the sandwich hybridization technique (after 9): Bands 1, 2 and 3 were excised from their respective DNA subclones with Pst I and electrophoresed on agarose gels. The DNA was transferred to nitrocellulose filters by the Southern technique and hybridized with unlabeled poly(A)$^{+}$ RNA. The filters were washed and hybridized with [^{32}P]-band 4. Hybridizations were identified by autoradiography.

The mRNA hybridizing to band 3 possesses an unhybridized tail which is a site for [^{32}P]-band 4 annealing. The restriction map of M4 is presented along with the regions transcribed into mRNA.

B. Band 4 Sandwich Hybridization: Sandwich hybridization of [^{32}P]-band 4 to Pst 1 digested subclones 1, 2 and 3 DNA. Upper (4U) or lower (4L) strand separated [^{32}P]-band 4 sandwich hybridization to Pst I digested subclone 3 DNA.

---

mRNA coding regions adjacent to the repeat on KH10 is being examined. The M4 and KH10 data suggest that a large fraction of the members of the repeat family and adjacent single copy sequences are transcribed.

The M4 and KH10 repeats are asymmetrically represented in RNA. Restriction fragments containing the transcribed repeat sequences of M4 and Dd38 (a recombinant plasmid carrying the same repeat as KH10 adjacent to different single-copy sequences) were prepared, end labeled and strand separated on polyacrylamide gels. Separated strands were <3% contaminated with their complementary strands. Each strand was then hybridized with a vast excess of poly(A)$^+$ RNA (fig. 4). As can be seen for both M4 and KH10 only one strand is complementary to poly(A)$^+$ RNA. From the Rot$_{1/2}$ of each reaction (0.15 for M4; 0.015 for Dd38), and assuming that the full length mRNA hybridizing to the repeat sequence is ~10 times the length of the repeat fragment probe, we conclude that ~1% and ~10% of total poly(A)$^+$ RNA is complementary to the M4 and KH10 repeat sequences respectively. These results are similar to those obtained by the filter hybridization procedure. We do not detect any hybridization to the opposite strands of either KH10 or M4. The M4 repeat is also asymmetrically represented in total nuclear RNA, strongly suggesting that only one strand of interspersed repeats in *Dictyostelium* is transcribed.

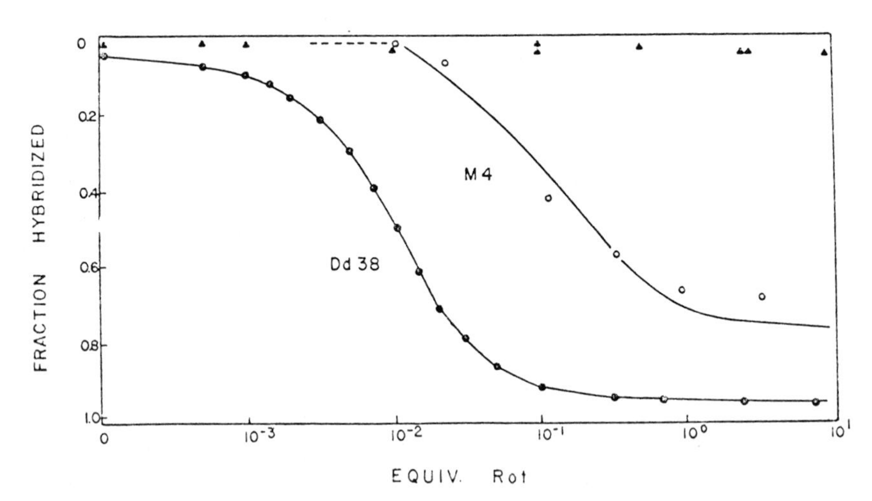

Figure 4. Hybridization of Strand Separated Repeat Sequence Fragments to Excess Poly(A)$^+$ RNA. Restriction fragments containing either M4 or Dd38 repeat sequences were end labeled with [P] using Klenow fragment, strand separated on polyacrylamide gels, and hybridized with 10^5 fold mass excess of poly(A)$^+$ RNA. Hybridizations were assayed on HAP.
O, ● - mRNA complementary strands
Δ - opposite strands

DISCUSSION

We have presented evidence that recombinant plasmids M4 and KH10 contain sequences which are repeated in the *Dictyostelium* genome and adjacent to single-copy DNA. These interspersed repeat sequences are transcribed asymmetrically onto mRNA and represent 10-20% of each mRNA molecule; the remaining 80-90% is transcribed from single-copy DNA. We suggest that a number of different mRNAs contain a relatively short sequence that hybridizes to these clones. One of these mRNAs is encoded completely by M4. In sea urchin, the interspersed repeat sequences are also transcribed. Unlike *Dictyostelium* these repeats are represented in hnRNA but are not present in mRNA. Also both strands of the repeat sequences are transcribed in sea urchin (10,11).

The M4 repeat and adjacent single-copy region are part of a single transcription unit. The short repeat is located at the 5'-end and the single copy sequence at the 3'-end. A model for the transcription of short, interspersed repeat sequences and their associated single-copy DNA is shown in figure 5. We propose that transcription initiates in these repeat sequences. We also suggest that most, if not all, of the repeat sequence in the mRNA is not translated but that translation may initiate in the single-copy complementary region. This view is supported by sequencing data of the M4 repeat. There are at least 3 stop codons in each of the three reading frames upstream from the Eco RI site suggesting that the repeat is not translated. There is an AUG site 34 bp from the Eco RI site near the junction of repeat and single-copy sequences. It is likely that this AUG is the translation start site. It is not followed by any in-frame stop codons and there are stop codons in the two other reading frames.

The function of the interspersed repeat sequences is unknown. The location of the repeat sequence at the 5'-end of a set of single-copy genes suggests that the repeat may be a regulatory recognition site. Britten and Davidson (12, 13) have proposed that repeated sequences may function as regulatory elements for the expression of adjacent single-copy structural genes. At present, there is no direct evidence for such a model, however, several facts have been taken as support for this theory. Most of the genes active in sea urchin development are single-copy and adjacent to repetitive sequences (14). In addition, there is differential expression of these short repeated sequences during sea urchin development (13). While we postulate a role for the repeat in transcriptional regulation it cannot be excluded that it serves a post-transcriptional function. It should also be noted that not all genes in *Dictyostelium* (eg.the gene for the 0.9 kb kb mRNA in M4) are adjacent to repeat sequences, and thus repeats are not essential to the expression of all genes.

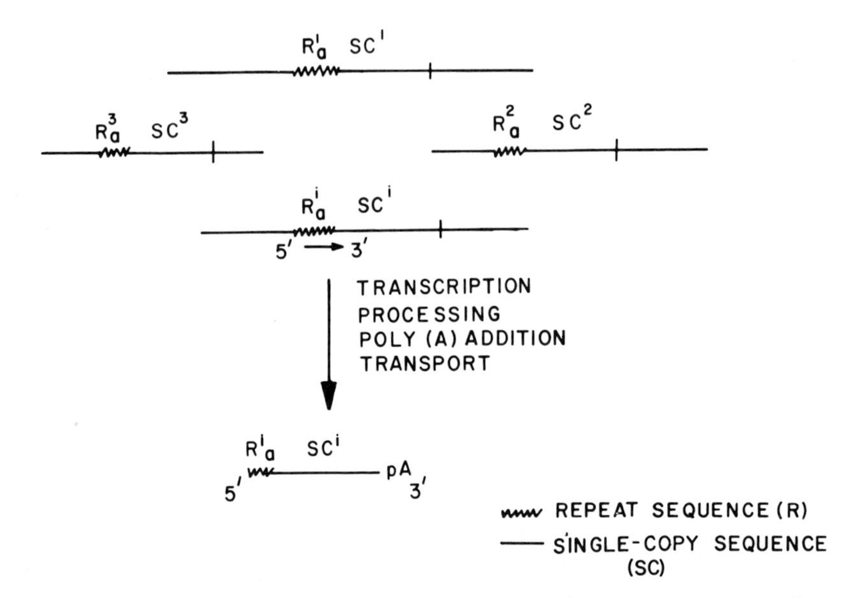

Figure 5. A Model for Transcription of Short Interspersed Repeat Sequences and their Associated Single-Copy DNA in Dictyostelium. Ra^1, Ra^2, Ra^3, Ra^i = different members of the short interspersed repeat family Ra. SC^1, SC^2, SC^3, SC^i = the corresponding set of adjacent single-copy DNA. Transcription initiates at the repeat and elongation proceeds into the single-copy DNA. Sequences complementary to the repeat are located at the 5'-end of the resulting transcript. The RNA is processed (including cap addition), poly(A) tailed and transported to the cytoplasm.

In order to focus on the function of interspersed repeat sequences in Dictyostelium we have isolated several genomic and [poly(A)$^+$mRNA] cDNA clones which possess members of the M4 or KH10 repeat family adjacent to different single-copy DNA sequences. Furthermore, we have identified additional recombinant clones containing transcribed interspersed repeat sequences which are members of other repeat families.

The evidence indicates that Dictyostelium repeat families code for non-overlapping sets of low abundance class mRNAs. We are now examining the transcription of the single-copy regions associated with these repeat families in relation to Dictyostelium development and, also, to identify the sites of initiation of transcription.

REFERENCES

1. Davidson, E.H., Galau, G.A., Angerer, R.C. and Britten, R.J. (1975). Chromosoma 51, 253-259.
2. Firtel, R.A. and Kindle, K. (1975). Cell 5, 401-411.
3. Firtel, R.A., Kindle, K. and Huxley, M.P. (1976). Fed. Proc. 35, 13-22.
4. Kindle, K.L. and Firtel, R.A. (1979). Submitted for publication.
5. Grunstein, M. and Hogness, D.S. (1975). Proc. Nat. Acad. Sci. USA 72, 3961-3965.
6. Southern, E.M. (1975). J. Mol. Biol. 98, 504-517.
7. Otsuka, A. (1979). Submitted for publication.
8. Rowekamp, W. and Firtel, R.A. (1979). Manuscript in preparation.
9. Dunn, A.R. and Hassell, J.A. (1977). Cell 12, 23-26.
10. Costantini, F.D., Scheller, R.H., Britten, R.J., and Davidson, E.H. (1978). Cell 15, 189-203.
11. Scheller, R.H., Costantini, F.D., Kozlowski, M.R., Britten, R.J., and Davidson, E.H. (1978). Cell 15, 189-203.
12. Britten, R.J. and Davidson, E.H. (1969). Science 163, 349-357.
13. Davidson, E.H. and Britten, R.J. (1973). Quart. Rev. Biol. 48, 565-613.
14. Davidson, E.H., Hough, B.R., Klein, W.H., and Britten, R.J. (1975). Cell 4, 217-238.

STABILITY OF THE C. ELEGANS GENOME DURING DEVELOPMENT AND EVOLUTION

David Hirsh, Scott W. Emmons, James G. Files
and Michael R. Klass

Department of Molecular, Cellular and Developmental Biology,
University of Colorado
Boulder, Colorado 80309

ABSTRACT Recombinant DNA methods have been used to characterize the genome of *Caenorhabditis elegans*. To determine if DNA rearrangements occur during somatic differentiation, fifteen randomly cloned *Bam* H1 fragments of somatic DNA were hybridized to *Bam* H1 digests of germ and somatic DNA's on Southern filters. In this way, 50 fragments representing 0.3% of the genome were compared and no size differences were detected. The DNA's of two interbreeding strains of *C. elegans* were also compared to determine the degree of evolutionary divergence. Fifteen percent of the fragments differed between the two strains. However, no differences could be found between the rDNA's. The DNA's of *C. elegans* and *C. briggsae* were compared and very little homology could be detected even though these species are morphologically very similar.

The fragments that differ in size between the two interbreeding strains are being genetically mapped. These experiments suggest that non-random segregation of chromosomes might be occurring in the interstrain hybrids.

INTRODUCTION

Recombinant DNA methods now make it possible to investigate whether rearrangements occur in the eukaryotic genome during somatic differentiation. By implication such rearrangements could be the cause of irreversible steps in differentiation. Alteration of the primary structure of DNA is known to occur during the differentiation of certain eukaryotic cell types. For example, some nematodes, insects, and crustaceans undergo chromosome diminution and fragmentation in the somatic cell lineages but not in the germ cells (1, 2). In protozoans, the vegetative nucleus

ISBN 0-12-068350-4

often contains only a subset of the sequences present in the germ nucleus (3). Rearrangements have been shown to occur in the differentiation of lymphocytes to produce specific antibodies (4). Somatic DNA rearrangements have been postulated to explain specialized differentiation events in the maturation of corn and in the establishment of mating types of yeast (5, 6).

We have looked for evidence of rearrangements in the genome during development of the nematode _Caenorhabditis elegans_. In addition, we have compared the DNA of our strain of _C. elegans_ to that of another strain and to a closely related species to explore the notion that rearrangements could be a mechanism of speciation (7, 8). From this DNA work and the availability of genetic markers in _C. elegans_, we propose a method for mapping genes represented in cloned DNA sequences.

C. elegans is a small, free-living nematode (9). The anatomy of _C. elegans_ has been described in great detail from both light and electron microscopic observations (10-14). The embryonic cell lineages are known from the fertilized egg to the 200 cell stage (15). The complete postembryonic lineages have been described from the 550 cells of the first stage larvae to the 808 somatic cells in the adult (10). The somatic lineages of the reproductive system have also been completely described (17). The genetics of _C. elegans_ have been studied by Brenner (18) and more recently by Herman (19). These genetic studies indicate from the frequency of lethal mutations that there are approximately 2,000 to 3,000 essential genes in _Caenorhabditis elegans_.

Sulston and Brenner (20) first characterized the DNA of _C. elegans_. They found that _C. elegans_ contains 8×10^7 base pairs per haploid genome, which is equal to about 20 times the _E. coli_ genome or one-half the Drosophila genome. About 15% of the DNA behaves as repetitive DNA according to its reassociation kinetics and approximately 85%, which is 6.7×10^7 base pairs of DNA, behaves as unique DNA. This amount of DNA is to be contrasted with the small number of essential genes mentioned above that are believed to exist on the basis of the frequency of lethal mutations. Chromosome diminution apparently does not occur in _C. elegans_ (20). Sulston and Brenner (20) also showed that there are about 50 copies of the ribosomal genes per haploid genome. The rDNA contains 50% G+C and remaining DNA, 36% G+C (20).

COMPARISON OF GERM LINE AND SOMATIC DNA'S

In order to investigate whether DNA rearrangements occur during somatic differentiation in _C. elegans_, preparations of somatic and germ-line DNA's were needed. Sperm served as a source of pure germ-line DNA. DNA was prepared by conventional methods of digestion with proteinase K in SDS, phenol extraction, chloroform-isoamyl alcohol extraction, ethanol precipitation and RNase digestion. The details of the methods will be published elsewhere. The mutant strain E879 with a mutation in the *him-1 III* gene was used for sperm isolation because there is a high incidence of males (21). The adult males were separated from the other worms by sieving on nylon screens. Sperm was extracted from the males by pressing them between two flat plates and the cells were further purified by sedimentation. Germ-line DNA was isolated from these preparations of sperm. It was known from the extensive studies on cell lineages that the first stage larva (L1) that hatches from the egg shell contains 550 somatic cells and only two germ-line precursor cells and, therefore, these animals served as the source of somatic DNA.

Recombinant DNA molecules consisting of the plasmid pBR313 with inserts of *Bam* H1 restriction fragments of *C. elegans* DNA were randomly cloned in the *E. coli* strain SF8 (22, 23). All of the cloned worm DNA fragments were from first stage (L1) larvae and therefore are somatic DNA. *Bam* H1 restriction digests of sperm and L1 DNA's were fractionated by agarose gel electrophoresis, and transferred to Millipore filters by the Southern method (24). Fifteen of the cloned larval *Bam* H1 restriction fragments were each radioactively labeled and hybridized to the Southern filters containing sperm or L1 DNA.

Several cloned fragments were hybridized together in these experiments because of the difficulty in preparing sperm DNA. The results are shown in Figure 1. No differences are seen in the molecular weights or the relative intensities of the hybridization bands when somatic and germ-line patterns are compared. There are more bands than cloned fragments applied as probes indicating that several of the cloned fragments contain sequences that are present more than once in the genome. This result has been corroborated in experiments where each probe was hybridized separately to a digest of adult DNA, which should be roughly an equal mixture of germ and somatic DNA's (25). A total of approximately 50 restriction fragments can be observed to hybridize, representing about 0.3% of the total *C. elegans* genome. We conclude that within the limitations of these experiments there are not rearrangements of the DNA during the germ and somatic differentiation in *C. elegans*.

This conclusion must be made with the following reservations: 1) very large rearrangements would not have been detected because only the terminal restriction fragments

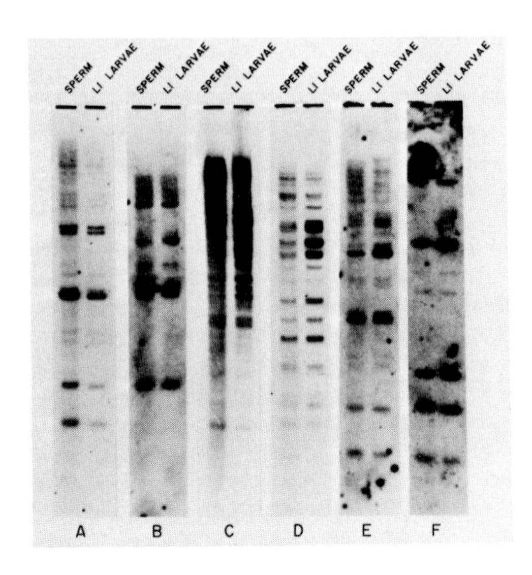

FIGURE 1. Comparison of germ-line (sperm) DNA and somatic (first stage larval) DNA. Cloned fragments of first stage (L1) larval DNA were hybridized to Bam H1 restriction digests of sperm DNA or larval DNA. Each lane contains 4 μg of DNA. The radioactive probes used in each pair of lanes were: A, pCe1, pCe4, pCe5; B, pCe2, pCe3, pCe5; C, pCe17; D, pCe18a, pCe18b; E, pCe1, pCe10, pCe14; and F, pCe5, pCe13, pCe19.

would be altered nor would very small rearrangements of less than 50 base pairs have been detected because the size differences would be below the resolution of the gel method; 2) fragments present in fewer than 10% of the cells of the organism would not have been detected judging from the intensities of the bands in the autoradiograms that correspond to fragments that are present once per genome as measured through reconstruction experiments; 3) chromosome diminution would not have been seen because cloned fragments were derived from somatic DNA. On the other hand, we would have found DNA rearrangements of intermediate sizes and we would have detected inversions if the Bam H1 restriction sites were asymmetrically placed. Rearrangements in the DNA that corresponded to the differentiation of the major embryonic tissue types, i.e., endoderm, ectoderm and mesoderm, or rearrangements that corresponded to the five basic embryonic stem cell lineages would have been detected because the numbers of cells that belong to each of these embryonic tissue groups or to each stem cell lineage contain more than 10% of the DNA of the organism (15, 26).

COMPARISON OF DNA'S OF RELATED STRAINS AND SPECIES

Similar hybridization experiments were carried out on the DNA's of another strain and a related species of *C. elegans* to examine the possibility that rearrangements are involved in the evolution of species. All of the experiments described above have been done with the English Bristol strain of *C. elegans* (17, 27). However, there is also the French Bergerac strain (28). The Bristol and Bergerac strains interbreed and produce fertile offspring (29). However, W.B. Wood (unpublished) observed three years ago that the individual F2 progeny of a cross between the Bristol and Bergerac strains vary markedly in their fecundity, suggesting that certain chromosomal segregation patterns were healthier than others. The significance of this casual observation will become apparent in a later section. Slight phenotypic differences exist between the two strains. Bergerac has a lower optimum growth temperature than Bristol, and Bergerac males are not as active as Bristol males.

The same 15 cloned larval (L1) DNA fragments from *C. elegans* var. Bristol that were used as hybridization probes in the experiments described above were also hybridized to *Bam* H1 restriction digests of DNA's from the Bergerac and Bristol strains, with representative results shown in Figure 2. Of the 13 cloned fragments that can be visualized after hybridization, two are of a different size in the Bergerac DNA, five differ in size from the corresponding Bristol fragments. This degree of difference of *Bam* H1 restriction sites, each of which contains a six base specificity, would imply that at least 1% of the nucleotides differ between the Bergerac and Bristol strains of *C. elegans* if the differences are due to single base changes in the restriction sites. This degree of difference is as great as that shown to exist between the DNA's of different species in other organisms, or for example between chimpanzee and man (30-32).

A similar experiment was done to compare the DNA of *C. elegans* var. Bristol with the DNA of the twin species *Caenorhabditis briggsae* (33). *C. briggsae* is very similar morphologically and developmentally to *C. elegans* but does not interbreed with it. The results showed that nine of the cloned *C. elegans* fragments have no homologous sequence in the *C. briggsae* DNA, four fragments hybridize weakly, and one hybridizes strongly. None of the fragments observed by hybridization is the same size in both *C. briggsae* and *C.*

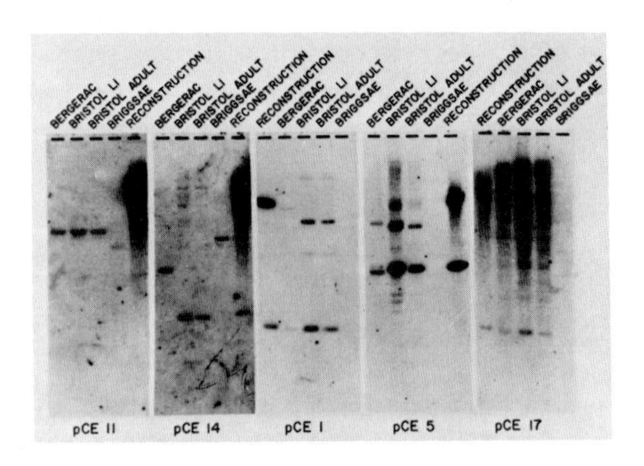

FIGURE 2. Comparison of C. elegans var. Bristol, var. Bergerac and C. briggsae DNA's by hybridization with five different recombinant plasmids. The Bergerac DNA was from L1 larvae, the C. briggsae DNA was from a mixed population of worms, and the Bristol DNA's were from L1 larvae or young adults. There was approximately 2 μg of Bam H1 restriction digest of each of these DNA's. The reconstruction consisted of 5 μg of sheared E. coli DNA plus 0.1 ng of plasmid to be used as probe, cleaved with Bam H1. The C. elegans genome consists of 8 x 10^7 base pairs (20). Therefore 2 μg of DNA will contain 0.1 ng of a 4,000 base pair fragment. The hybridization seen in the reconstruction is with the cloned nematode fragment, the pBR313 vector plasmid and the E. coli DNA that contaminates the preparation of the probe. The sizes of the cloned fragments are: pCe11, 5300 base pairs; pCe14, 2400 and 450 base pairs (two inserts); pCe1, 1800 base pairs; pCe5, 3750 base pairs; pCe17, 1650 base pairs. The probes were labeled by nick translation to greater than 10^7 dpm per μg with [α-^{32}P] dATP.

elegans. These results represent at least a 20% difference between the C. elegans and C. briggsae DNA's (34). This degree of divergence between these two DNA's is surprisingly great, especially since the two species are so similar morphologically. Perhaps, those DNA sequences that hybridize represent conserved genes of common functions.

In summary, the experiments thus far show that there are no differences between Bam H1 sites of germ and somatic DNA's, one out of seven fragments differs in size between the DNA's of interbreeding strains and there is relatively low homology between the DNA's of the two closely related species. Many of these results will be reported elsewhere in more detail (35).

The exact nucleotide nature of the differences observed in each case between the fragments of the Bergerac and Bristol strains is unknown. However, in the case of one clone, pCel4, the difference appears to be 1700 base pairs in the Bergerac DNA fragment that are not present in the corresponding Bristol DNA fragment because not only Bam Hl but also Eco RI and Hind III restriction enzymes produce fragments to which pCel4 hybridizes that differ by 1700 base pairs.

Whereas one out of seven DNA fragments differs between Bergerac and Bristol DNA's, no differences have been detected between the ribosomal DNA's of these two strains after digestion with 25 different restriction enzymes. Therefore, the restriction maps of the ribosomal DNA's from these two strains are the same. In contrast, the ribosomal DNA of C. briggsae differs from that of C. elegans (Fig. 3). A region of homology appears between the Sal I site and the Eco RI site and the other areas can be juxtaposed by simple deletions of the briggsae pattern or insertions into the elegans pattern.

We have also hybridized a cloned Dictyostelium actin cDNA to C. elegans DNA to search for the C. elegans actin genes

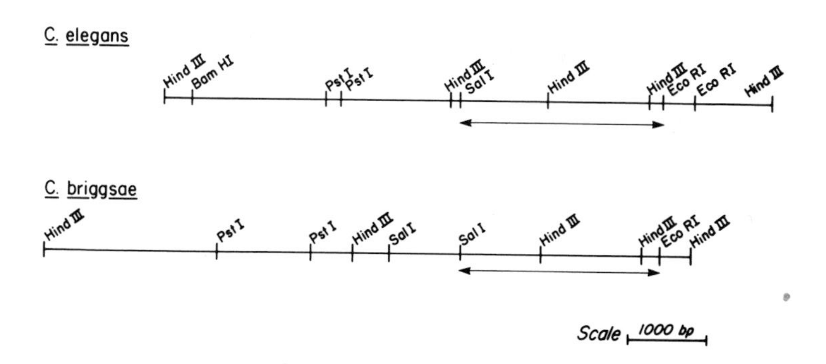

FIGURE 3. Restriction enzyme digestion maps of rDNA's from C. elegans and C. briggsae. The rDNA maps were constructed from hybridizations of the cloned C. elegans rDNA to Southern blots of restriction digests of total DNA's from mixed populations of C. elegans or C. briggsae.

and to explore further interstrain and interspecies relationships. The cDNA clone, pcDdBl-7c, was obtained from Dr. Richard Firtel (36). It hybridizes strongly and specifically with C. elegans DNA. If the hybridization patterns of the Dictyostelium actin probe with Hind III digested Bergerac and Bristol DNA's are compared, one of the fragments in the Bergerac DNA is 1700 base pairs larger than the corresponding fragment in the Bristol DNA. This 1700 base pair difference persists in three out of seven different restriction enzyme digests (Fig. 4). As in the case of pCel4, this implies that the size difference in the fragments between the two strains does not arise from a point mutation but that a segment of DNA is present in the Bergerac DNA fragment that is not present in the Bristol DNA fragment. The extra 1700 base pairs must be adjacent to the worm sequence homologous to the actin probe or they must be very near one end of the homologous sequence if they are within it because four restriction enzymes produced fragments that did not differ between the strains. No details are known about these extra 1700 base pairs that are in Bergerac but are not in Bristol;

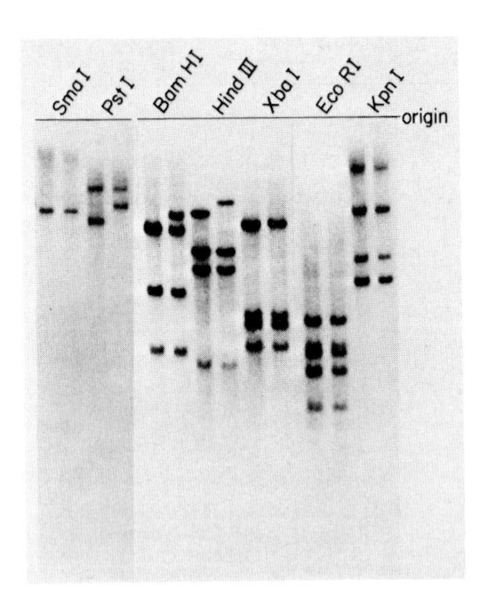

FIGURE 4. Hybridization of Dictyostelium actin probe, pcDdBl-7C, to restriction enzyme digests of DNA's from C. elegans var. Bristol and C. elegans var. Bergerac. The left lane of each pair contains the Bristol DNA. Each lane contains 5 μg of L1 larval DNA. The probe was nick translated to 3 x 10^7 dpm per μg (37).

it is apparently coincidence that the same 1700 base pair size difference exists between the Bergerac and Bristol DNA's for both the pCel4 and the Dictyostelium actin fragments.

GENETIC MAPPING OF CLONED FRAGMENTS

Caenorhabditis elegans is an extremely convenient organism for genetic studies (18, 19). It has a rapid generation time of 3 days at 20°, produces 280 progeny per parent, and is a self-fertilizing hermaphrodite. Because it is hermaphroditic, all progeny of a single parent can be obtained as identical genotypes and, after mutagenesis, recessive mutations appear as homozygotes in the F2 generation. True males exist that can be used for genetic crosses. Hermaphrodites are XX and have five pairs of autosomes and males are XO. Numerous mutant phenotypes, such as dumpy and uncoordinated, have been genetically mapped onto each of the six linkage groups of the Bristol strain of C. elegans. The differences in the DNA fragments in the Bergerac and Bristol strains that have been described above, such as for pCel4 and for the Dictyostelium actin probe, can also be used as phenotypes, and they can be genetically mapped because the Bristol and Bergerac strains interbreed. Therefore, experiments were first carried out to assign the DNA fragment size differences to linkage groups by the standard genetic methods. Basically, these involve measuring whether the fragment differences assort independently of standard phenotypic markers available on each of the six linkage groups in the Bristol strain or show linkage to such markers.

As an example, consider determination of linkage to chromosome I. The Bristol chromosome I was marked with mutations in two widely spaced genes, dpy-5 (e61) and unc-54 (e190). A strain homozygous for these markers is dumpy and uncoordinated. This homozygous doubly marked Bristol strain was crossed to the Bergerac strain. The F1 heterozygous cross progeny were allowed to undergo self-fertilization to produce the F2 generation. Among the F2 progeny 50 individual, uncoordinated dumpy worms were picked because these worms were homozygous for Bristol first chromosomes; the other five chromosomes presumably randomly assorted. Therefore, for each of the other unselected chromosomes, and in particular for the one on which the DNA fragment of interest resides, the F2 generation should have been homozygous Bristol, heterozygous Bristol/Bergerac and homozygous Bergerac in a 1:2:1 ratio. If the DNA fragment of interest is on chromosome I, then only the Bristol pattern for that fragment should be seen after hybridization, but if the fragment is on another chromosome that randomly assorted then both the Bristol and Bergerac patterns should be seen.

Such a protocol was carried out separately for each of six genetically marked strains representing the six Bristol chromosomes. About 50 marked F2 worms were selected and grown several generations to produce approximately 10^7 progeny; the DNA was isolated, digested with the appropriate restriction enzymes, fractionated on an agarose gel, transferred to a Millipore filter by the Southern technique and then hybridized to the pCel4 probe or the Dictyostelium actin probe. The results of the hybridization for pCel4 are shown in Figure 5. pCel4 hybridizes to a primary fragment in the Bristol DNA of the same size as the cloned Bristol fragment. This primary fragment is a different size in the Bergerac DNA. When chromosomes I, III, IV, V and X are kept as homozygous Bristol chromosomes, <u>both</u> the Bristol and

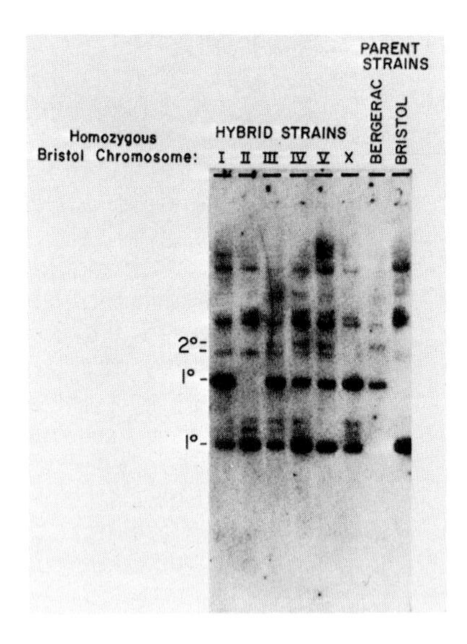

FIGURE 5. Hybridization of radioactive pCel4 probe to DNA's from the Bergerac strain, from the Bristol strain and from hybrid strains in which one chromosome was maintained as the Bristol type with genetic markers. The primary fragment, pCel4 1°, corresponds to the size of the cloned Bristol fragment. Among the other bands, the secondary one, pCel4 2°, is referred to in the text.

Bergerac DNA fragments were present. Therefore, the fragment does not reside on these chromosomes. However, when chromosome II was homozygous Bristol, only the Bristol hybridization pattern was present. Therefore, the primary fragment to which pCel4 hybridizes appears to come from chromosome II.

pCel4 also hybridizes to several other fragments. In the case of the secondary band denoted in Figure 5, the probe hybridizes to both Bergerac and Bristol fragments when chromosomes III, IV, V and X were homozygous Bristol, but not in the cases when Bristol chromosomes I and II were selected. The single fragment cannot be simultaneously on two chromosomes. Because the results with the pCel4 primary fragment indicate the pattern for a fragment on chromosome II, the pattern for the pCel4 secondary fragment indicates that this fragment is on chromosome I, and that there is apparently non-random assortment of chromosome I when Bristol chromosome II is homozygous. In other words, use of the pCel4 primary fragment as a marker for chromosome II shows that the Bergerac and Bristol second chromosomes assort randomly when chromosomes I, III, IV, V and X are homozygous Bristol. However, when chromosome II is maintained as homozygous Bristol, chromosome I assorts non-randomly and also remains homozygous Bristol; animals homozygous for Bergerac I and Bristol II presumably die or have a growth disadvantage. Therefore, only a Bristol fragment is seen in the hybridization with the pCel4 secondary fragment in the chromosome II lane in Figure 5, and this secondary fragment must be on chromosome I because only a Bristol band is seen in that lane.

In the case of the actin fragment, Bristol bands only are seen when chromosomes I, II or V are homozygous Bristol (Fig. 6). Because the preceding two examples define the patterns that result from linkages to chromosomes II and I, actin must be on chromosome V. The results also suggest that homozygosis for Bristol chromosome I or II leads to homozygosis of the unselected chromosome V. Because selection for Bristol II also selects for Bristol I, selection for II could select either directly or indirectly for Bristol V.

These results suggest, but do not prove, that there is pseudolinkage between certain chromosomes as they segregate from interstrain hybrids. This pseudolinkage does not appear to present an insurmountable problem for mapping the fragments for which size differences exist between the two strains. If one hybridization probe can be identified for each chromosome, as has presumably been done so far for I, II and V, then the behavior of each chromosome can be followed.

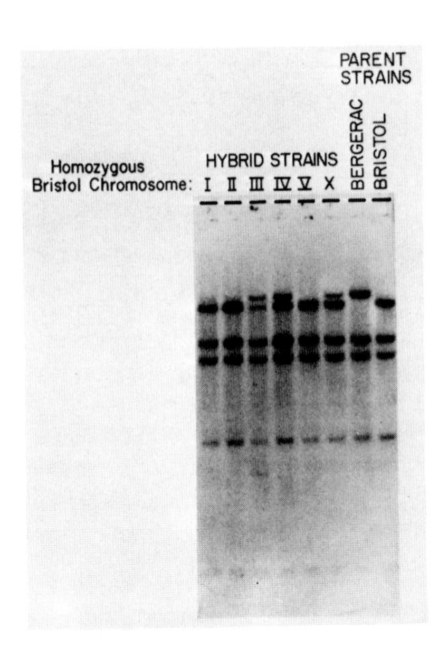

FIGURE 6. Hybridization of Dictyostelium actin probe to DNA's from the Bergerac strain, from the Bristol strain and from hybrid strains in which one chromosome was maintained as the Bristol type.

Genetic crosses of appropriately marked Bristol worms with Bergerac worms also can test the hypothesis of linkage. Such pseudolinkage might be due to Bergerac chromosomes containing essential genes that the homologous Bristol chromosomes do not contain, and vice versa. For example, a nonreciprocal translocation of an essential function from Bristol chromosome I to chromosome II during the evolution of the Bergerac strain would lead to linkage between Bristol chromosomes I and II during the segregation of the chromosomes from the hybrid strains, because a zygote containing homozygous Bristol II and Bergerac I would lack the essential function. Such nonreciprocal translocations might also be part of the explanation for the evolution of the physical DNA differences between the Bergerac and Bristol strains that are observed by hybridizations.

An important point is that these interstrain crosses assigned the positions of the fragments in the Bergerac genome and not the Bristol genome, because linkage assignments were made on the basis of exclusion of a Bergerac linkage group and correlative disappearance of a Bergerac hybridization band.

However, linkage assignments can be made for the Bristol genome by modifying the protocol to exclude a Bristol linkage group and looking for the correlative disappearance of a Bristol hybridization band. This modification involves picking homozygous wild type worms in the F2 generation instead of homozygous mutant worms. Once DNA fragments have been assigned to linkage groups, they can presumably be mapped by three factor crosses using hybridization to detect the different phenotypes.

ACKNOWLEDGMENTS

We thank Kimberly Johnson, Stephen Carr and Bradley Rosenzweig for help with the experiments. This work was supported by P.H.S. Research Grant GM 19851. S.W.E. was supported by P.H.S. Fellowship Award GM 05970. M.R.K. was supported by P.H.S. Grant AG 00310. J.G.F. was a Helen Hay Whitney Foundation Postdoctoral Fellow.

REFERENCES

1. Wilson, E.B. (1925). "The Cell in Development and Heredity." Macmillan, New York.
2. Moritz, K.B. and Roth, G.E. (1976). Nature 259, 55-57.
3. Prescott, D.M. and Murty, K.G. (1973). Cold Spring Harbor Symp. Quant. Biol. 38, 609-618.
4. Brack, C., Hirama, M., Lenhard-Schuller, R. and Tonegawa, S. (1978). Cell 15, 1-14.
5. McClintock, B. (1951). Cold Spring Harbor Symp. Quant. Biol. 16, 13-47.
6. Herskowitz, I., Strathern, J.N., Hicks, J.B. and Rine, J. (1977). In "Molecular Approaches to Eucaryotic Systems" (G. Wilcox, J. Abelson and C.F. Fox, eds.), pp. 193-202. Academic Press, New York.
7. White, M.J.D. (1977). "Modes of Speciation." W.H. Freeman and Co., San Francisco.
8. Wilson, A.C. (1976). In "Molecular Evolution" (F.S. Ayala, ed.) pp. 225-234. Sinauer Associates, Inc., Sunderland, Massachusetts.
9. Edgar, R.S. and Wood, W.B. (1977). Science 198, 1285-1286.
10. Albertson, D.G. and Thomson, J.N. (1976). Phil. Trans. R. Soc. Lond. B275, 300-325.
11. Sulston, J.E. (1976) Phil. Trans. R. Soc. Lond. B275, 287-297.
12. White, J.G., Southgate, E., Thomson, J.N. and Brenner, S. (1976). Phil. Trans. R. Soc. Lond. B275, 327-348.
13. Ward, S., Thomson, N., White, J.G. and Brenner, S. (1975). J. Comp. Neurol. 160, 313-338.

14. Ware, R.W., Clark, D., Crossland, K. and Russell, R.L. (1975). J. Comp. Neurol. 162, 71-110.
15. Deppe, U., Schierenberg, E., Cole, T., Krieg, C., Schmitt, D., Yoder, B. and von Ehrenstein, G. (1978). Proc. Natl. Acad. Sci. 75, 376-380.
16. Sulston, J.E. and Horvitz, H.R. (1977). Develop. Biol. 56, 110-156.
17. Kimble, J. and Hirsh, D. (1979). Develop. Biol., in press.
18. Brenner, S. (1974). Genetics 77, 71-94.
19. Herman, R.K. (1978). Genetics 88, 49-65.
20. Sulston, J.E. and Brenner, S. (1974). Genetics 77, 95-104.
21. Hodgkin, J., Horvitz, H.R. and Brenner, S. (1979). Genetics 91, 67-94.
22. Bolivar, F., Rodriguez, R.L., Betlach, M.C. and Boyer, H.W. (1977). Gene 2, 75-93.
23. Morrow, J.F., Cohen, S.N., Chang, A.C.Y., Boyer, H.W., Goodman, H.M. and Helling, R.B. (1974). Proc. Natl. Acad. Sci. 71, 1743-1747.
24. Southern, E.M. (1975). J. Mol. Biol. 98, 503-517.
25. Hirsh, D., Oppenheim, D., and Klass, M. (1976). Develop. Biol. 49, 200-219.
26. Chitwood, B.G. and Chitwood, M.B. (1950). "An Introduction to Nematology." Monumental Printing Company, Baltimore.
27. Nicholas, W.L., Dougherty, E.C. and Hansen, E.L. (1959). Ann. N.Y. Acad. Sci. 77, 218-235.
28. Nigon, V. (1949). Ann. Sci. Nat. Zool. 11, 1-132.
29. Fatt, H.V. and Dougherty, E.C. (1963). Science 141, 266-267.
30. Upholt, W.B. (1977). Nucleic Acids Res. 4, 1257-1265.
31. Angerer, R.C., Davidson, E.H. and Britten, R.J. (1976). Chromosoma 56, 213-226.
32. Kohn, D.E., Chiscon, J.A. and Hoyer, B.H. (1970). Carnegie Inst. Year. 69, 488.
33. Dougherty, E.C. (1955). J. Helminthol. 29, 105-152.
34. Bonner, T.I., Brenner, D.J., Neufeld, B.R. and Britten, R.J. (1973). J. Mol. Biol. 81, 123-135.
35. Emmons, S.W., Klass, M.R. and Hirsh, D. (1979). Proc. Natl. Acad. Sci., in press.
36. Bender, W., Davidson, N., Kindle, K.L., Taylor, W.C., Silverman, M. and Firtel, R.A. (1978). Cell 15, 779-788.
37. Maniatis, T., Jeffrey, A. and Kleid, D.G. (1975). Proc. Natl. Acad. Sci. 72, 1184-1188.

EXAMPLES OF EVOLUTION AND EXPRESSION IN THE SEA URCHIN GENOME[1]

Roy J. Britten,[2] Terrence J. Hall,[2], Ze'ev Lev, Gordon P. Moore, Amy S. Lee, Terry L. Thomas, and Eric H. Davidson

Division of Biology, California Institute of Technology, Pasadena, California 91125

In this communication we describe briefly three sets of measurements. First is a summary of single copy divergence and polymorphism for three sea urchin species of the genus *Strongylocentrotus*. Second is a measurement which indicates that the same set of repetitive sequence transcripts are stored at high concentration in the mature eggs of *S. purpuratus* and *S. franciscanus*, even though these two species have evolved separately for 15-30 million years, and large changes have occurred in the genomic frequency and actual sequences of most repeats. Finally, measurements with a cloned single copy DNA fragment are summarized which show an interesting quantitative modulation in the expression of this sequence in the RNAs of egg, embryo, and an adult tissue of the sea urchin.

Single Copy Sequence Evolution in Sea Urchins

Table 1 lists a set of measurements of interspecies and interindividual single copy DNA sequence differences, as they are reflected in the T_ms of reassociated duplexes. Data are from Hall *et al.* (1). The procedures are described in the footnote. One important technical addition to previously published measurements is the determination of the length of the duplex regions. Under our conditions of digestion S1 nuclease does not usually attack a single unpaired nucleotide. A small cluster of unpaired nucleotides will

[1]This work was supported by NIH grants HD-05753 and GM-20927. T.J.H. was supported by an NIH fellowship; Z.L. by a Weizmann fellowship; G.P.M. by an ACS fellowship; A.S.L. by an NIH fellowship and a senior fellowship from ACS, California Division; and T.L.T. by an NIH fellowship and a Lievre fellowship from ACS, California Division.

[2]Kerckhoff Marine Laboratory, California Institute of Technology, Corona del Mar, CA 92625.

ISBN 0-12-068350-4

TABLE 1

THERMAL STABILITY OF SINGLE COPY DNA DUPLEXES[a]

	S. purpuratus individual (A) vs:			
	S. purpuratus (A)	*S. purpuratus* (B)	*S. drobachiensis*	*S. franciscanus*
Extent of reaction (normalized) (%)	100[b]	97	85	75
S1 resistant duplex length[c] (NT)	450	430	290	215
T_m[d] (°C)	59	57	54	48
Interspecies T_m reduction[e] (°C)	2	4	7	13
Relative to *S. purpuratus* population[f] (°C)		0	3	9

[a]The procedure used was as follows: 1) Labeled single copy DNA was prepared by nick-translation of sperm DNA from individual A followed by reassociation to C_ot 100 and removal of repetitive DNA on hydroxyapatite; 2) The tracer was reassociated with an excess of 800 nucleotide long DNA of the same individual, another individual or another species in 1.5 M TEACL (tetraethyl ammonium chloride) to C_ot 5000; 3) The mixture was treated with S1 nuclease in 0.6 M TEACL, pH 4.5, 1 mM Zn; 4) The resistant fragments were isolated and transferred to 2.4 M TEACL, pH 7, and passed over a Sephadex column; 5) The samples were raised to a series of temperatures; 6) Each sample was diluted to 0.6 M TEACL, pH 4.5, 1 mM Zn; 7) The resulting denatured fragments were digested with nuclease and assayed on Sephadex or by TCA precipitation.

[b]Without heating about 66% of the tracer was resistant to the two S1 digestions.

(continued)

probably be digested but the minimum required degree of single strandedness is not known. It is possible that the length reductions shown in the table for interindividual and interspecies duplexes are due to local clusters of unpaired nucleotides. Further work will be required to determine if this simply reflects random base substitution. In any case, the length-corrected melting temperatures shown in Table 1 are the result of the average sequence divergence within the S1 resistant regions. As such they are likely to yield underestimates of the total amount of base substitution that has occurred, since some clusters of unpaired nucleotides are not included. At present the best estimate is that 1% unpaired nucleotides gives about 1°C reduction in melting temperature (2,3).

The interspecific T_m reductions (i.e., reductions from the T_m of native DNA) listed in Table 1 are larger than previously published values, even though they may be slight underestimates of total interspecific sequence divergence. The two reasons for this increase are the modification of the incubation and assay conditions ("lower criterion") to include lower melting duplexes and the recognition of single copy sequence polymorphism in *S. purpuratus* (4). The reduction in the thermal stability due to polymorphism in *S. purpuratus* as compared to perfect duplexes is about 4°C while the additional reduction in the interspecies comparison to *S. drobachiensis* is only 3°C. Thus previous estimates (5,6) of the divergence between the DNAs of *S. purpuratus* and *S. drobachiensis* must be increased by about a factor of two. This correction is, of course, more important for closely related species such as these, while the correction for reduced criterion is more important for more distantly related species such as *S. franciscanus* and *S. purpuratus*.

We have initiated measurements of the DNA sequence polymorphism of a few other species and the tentative results

---

TABLE 1 Footnotes (continued)

[c]After the first S1 digestion the length of the resistant duplexes was assayed by electrophoresis on alkaline agarose gels.

[d]The T_m has been corrected for duplex length by adding a correction (C=500/length) to the observed melting temperature. All temperatures have been rounded to the nearest degree.

[e]Below that of perfectly paired duplex or native DNA (61°C in 2.4 M TEACL).

[f]The interspecies T_m reduction calculated in the usual way by comparison with the melting temperature of interindividual duplexes of the reference species.

up to the present time can be summarized as follows: *S. drobachiensis* shows about half the polymorphism of *S. purpuratus* or a T_m reduction of about $2^{\circ}C$ for interindividual compared to precise duplexes. Within the limit of sensitivity of this technique, approximately $0.5^{\circ}C$, no polymorphism has been recognized for human DNA, *Mus musculus* DNA, or Eastern song sparrow DNA.

Conservation During Evolution of Patterns of Repetitive Sequence Expression

Costantini *et al.* (7) showed that transcripts of a subset of all of the repetitive sequence families of *S. purpuratus* were stored at high concentration in mature eggs. These RNA molecules are the product of a specific pattern of transcription during oogenesis. Using cloned repeats Costantini *et al.* (7) and Scheller *et al.* (8) studied the transcripts of nine families of repetitive sequence in mature sea urchin oocyte RNA, gastrula nuclear RNA, and adult intestine nuclear RNA. This work demonstrated a tissue specific pattern of repeat sequence transcription. The striking differences in the patterns of repeat sequence of expression in the three cases could indicate that these sequences carry out tissue specific functions.

Figure 1 shows a measurement of the representation of a repetitive DNA tracer in the egg RNAs of *S. franciscanus* and *S. purpuratus* (9). This tracer was selected on the basis of its rapid hybridization with *S. purpuratus* egg RNA (as described in ref. 7). The hybridization reaction goes to completion as well with *S. franciscanus* egg RNA as with *S. purpuratus* egg RNA. Clearly the same set of repetitive sequence families is highly represented in *S. franciscanus* egg RNA as in *S. purpuratus* egg RNA, though the concentration may be somewhat less in *S. franciscanus* RNA. Moore *et al.* (10) found that significant changes have occurred in the sequences and the frequencies of repetition of typical repeated sequence families since these two species diverged more than 15 million years ago (6,11). Nevertheless the distribution of repeat sequence transcripts in the egg RNAs of *S. franciscanus* and *S. purpuratus* is very similar. Presumably this similarity is due to selective pressure based on some significant function of these transcripts.

Expression of a Cloned Single Copy Sequence in Embryo mRNAs and Nuclear RNAs

The *S. purpuratus* egg is known to contain a large store of maternal mRNA, as well as repeat transcripts. Previous studies (reviewed in ref. 12) have resulted in a general

A direct measurement of increases in free calcium in response to specific stimulatory agents has previously been accomplished by utilizing the calcium-specific photoprotein aequorin. For example, aequorin-injected medaka eggs show a very low resting glow before they are fertilized but, upon activation by sperm, the calcium-mediated light emission increases by several orders of magnitude before returning slowly to the resting level (12). In the studies to be discussed below, we have used a similar approach to determine whether or not progesterone causes an increased calcium activity by monitoring changes in the light emission from aequorin-injected _Xenopus_ oocytes. We also report preliminary studies concerned with the isolation of a calcium dependent regulatory protein (CDR) from _Xenopus_ oocytes.

RESULTS AND DISCUSSION

Bellé _et. al_. (13) have reported that _Xenopus laevis_ oocytes injected with aequorin emitted no detectable resting glow nor any light output in response to progesterone. Our initial attempts to monitor changes in free calcium levels involved injecting 60-70 oocytes with aequorin and then maintaining them in a culture dish situated directly on top of a photomultiplier tube (PMT). Unfortunately, the results were uniformly negative. However, it should be pointed out that normal oocytes are heavily pigmented, at least in the animal hemisphere. Such pigmentation could mask the aequorin light emission.

Perhaps equally as important, the background noise level of the PMT averaged 100 photoelectrons per second at room temperature (23°C). This high level could further mask aequorin-luminescence, especially that produced by relatively small changes in Ca^{2+}. Suffice it to say that, while initial studies appeared to rule out the possibility of a large change in the level of free calcium, changes of the magnitude seen at fertilization or activation, the experiments were not designed to detect more subtle changes.

To circumvent the pigmentation problem, we utilized albino oocytes from the mutant albino frog described by Bluemink and Hoperskaya (14). A second modification involved utilization of a more efficient apparatus for the measurement of light output. Specifically, the photomultiplier tube was cooled and maintained at -70°C with dry ice. This reduced the background noise level (thermionic emission within the PMT) by 50 fold compared to room temperature operation, without a reduction of the aequorin signal. However, cooling the PMT required that it be physically separated from the oocytes, necessitating the use of a lens system to collect and focus the light emissions on the PMT. This in turn restricted measurements to one oocyte at a time.

Individual albino oocytes injected with aequorin were monitored continuously while maintained in a flow-through culture dish mounted on a microscope stage beneath the PMT. Control oocytes exposed only to the culture medium exhibited a resting glow of 1-3 photoelectrons above a background noise level of 1-2 photoelectrons per second. After an appropriate time interval, sufficient to establish that the resting glow was constant, a valve was actuated allowing progesterone (1-5 μg/ml) to enter the culture chamber. The data in Fig. 1 show the results obtained with one oocyte. Approximately 40 seconds after progesterone was introduced, the light output from the cell started to rise, reached a peak at about 80 seconds, and then slowly declined to the resting glow level about 6 minutes after progesterone exposure. The peak light output in this particular case was almost 15 fold greater than the resting glow. The magnitude of this response has been quite variable in other oocytes ranging from distinct spikes of the type shown to only a moderate (but more sustained) light emission.

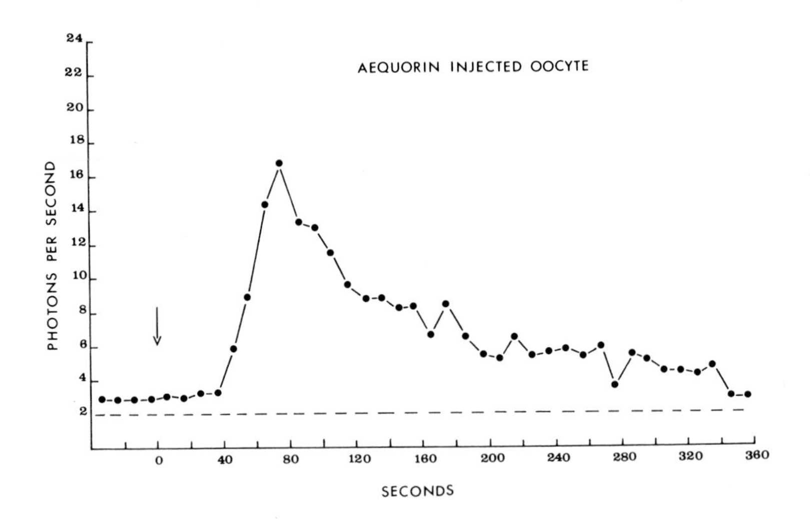

FIGURE 1. Aequorin luminescence from a single albino _Xenopus laevis_ oocyte following progesterone stimulation. The oocyte was injected with 40 nl of a 1% aequorin solution, allowed to equilibrate for 3 hours and then monitored with a PMT (EMI No. 9781A) operated at -70°C. Progesterone (1 μg/ml) was introduced at the time indicated by the arrow. The background noise level within the tube is represented by the dashed line.

Oocytes that did not mature in response to progesterone never gave an increased light output. However, only about 50% of the oocytes which did mature (from 5 different frogs) gave a positive result. We are uncertain as to the significance of the negative results. Since aequorin is, of necessity, injected at a single site, perhaps diffusion to the actual sites of calcium release varies. In this context, it is relevant to point out that the postulated release of bound calcium in response to steroids might be expected to occur largely, if not exclusively, within the oocyte cortex (9,10). It is also known that many (perhaps all) of the so-called early events, which occur in response to progesterone in oocytes from unstimulated females, either are bypassed or have occurred prior to progesterone exposure in the rapidly maturing oocytes from stimulated females (2). Stimulated females normally refers to those recently induced to ovulate with gonadotropins, but such animals can also result from changes (intentional or not) in several environmental parameters. In this sense, albino frogs have occasionally ovulated spontaneously under conditions in which wild-type frogs have never ovulated. Oocytes from such animals may already have undergone changes in free calcium levels to varying degrees prior to progesterone exposure *in vitro*. Considering these and other technical difficulties, we have chosen to emphasize the positive results. Thus, we suggest that the initial progesterone-oocyte interaction which leads to maturation involves the transient release of bound calcium located in or near the oocyte cortex.

If calcium does function in some way to carry the progesterone signal, then increased calcium activity must affect certain pathways within the oocyte that ultimately lead to the resumption of meiosis. An increased calcium activity could affect certain processes directly (15). Alternatively, calcium could act via the mediation of calcium binding proteins, such as the protein initially identified as an activator of cAMP phosphodiesterase in the mammalian brain and referred to as Calcium Dependent Regulator (CDR) or Calmodulin (16,17).

Recently, we have successfully isolated CDR from *Xenopus* ovarian oocytes utilizing essentially the procedures of either Dedman *et. al.* (18) or Watterson *et. al.* (19) for bovine brain CDR. The data summarized in Table 1, while not yet complete, indicate that the oocyte CDR is very similar, if not identical, to the CDR protein found in bovine brain (18) and sea urchin eggs (20).

TABLE I
PHYSICAL PROPERTIES OF XENOPUS OOCYTE CDR

	Parameter	Xenopus oocyte CDR	Bovine brain CDR[a]	Sea urchin egg CDR[b]
I	Molecular Weight Gel Filtration	30,000	28,200-31,000	—
	SDS-PAGE Electrophoresis	18,500	15,000-19,200	17,000
II	Isoelectric pH	4.3	3.9-4.3	—
III	Mol Ca^{2+} Bound per Mol Protein	2[c]	2-4	4
IV	Ca^{2+} Dissociation k_D	$1.2x10^{-5}M$[c]	$2.38x10^{-6}$ $3.5x10^{-6}M$	—

[a]Values for bovine brain CDR come from Dedman et. al. (18), Watterson et. al. (19) and Lin et. al. (30).
[b]From Head et. al. (20).
[c]Not determined with a Ca-EGTA Buffer.

Maller and Krebs (21) have proposed a model in which the initiation of meiosis (maturation) is regulated by a phosphoprotein subject to control by cAMP-dependent protein kinase. A key point in the model proposes that, after progesterone stimulation, an increase in free calcium occurs which activates the Ca^{2+}-dependent regulator of phosphodiesterase, resulting in a transient decrease in oocyte cAMP levels (see also 22).

Xenopus oocyte CDR stimulates cAMP phosphodiesterase from bovine brain by 4-6 fold (Fig. 2). This is comparable to the stimulatory effect of homologous CDR on bovine brain phosphodiesterase. In contrast, oocyte CDR did not appear to have any modulatory role in controlling oocyte cAMP phosphodiesterase activity (Fig. 2). This negative result was also reported for sea urchin egg CDR in its action on sea urchin egg phosphodiesterase (20). Thus, we suggest that the functional role of CDR in the process of oocyte maturation, if any, is not likely to be in the regulation of phosphodiesterase activity. On the other hand, CDR is now

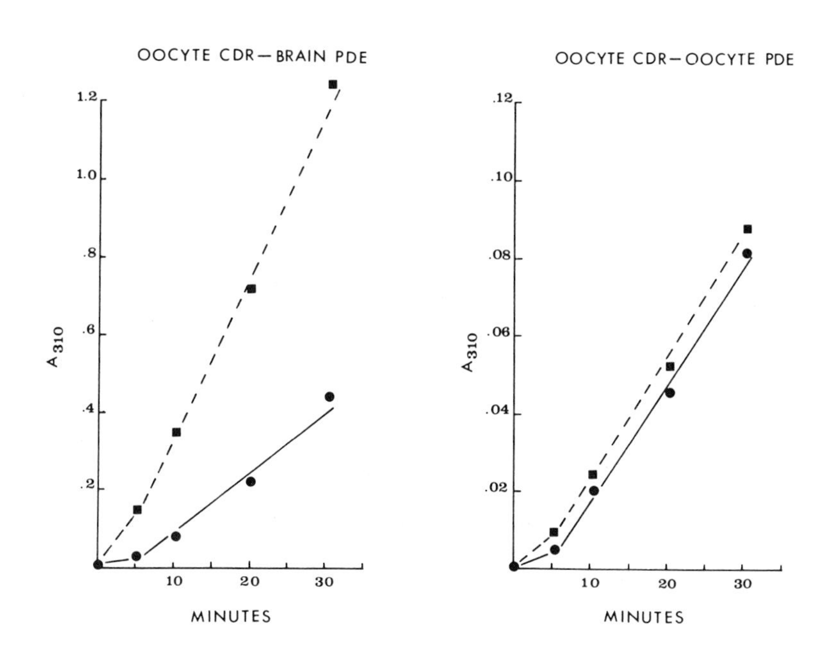

FIGURE 2. Effects of oocyte CDR on homologous and heterologous phosphodiesterase (PDE). 100 μg of DEAE-phosphodiesterase (19) was incubated in 1 ml, with (■- - - - ■) or without (●—●) oocyte CDR (10 μg), with 10^{-4}M Ca^{2+} at 30°C. PDE activity was monitored colorimetrically (19).

known to be involved in the regulation of a variety of cellular activities including adenylate cyclase, membrane ATPase and membrane-associated kinases (23) as well as tubulin depolymerization associated with mitosis (24). Perhaps oocyte CDR functions at one of these levels.

CONCLUDING COMMENTS

Recently, we have studied the activity of the maturation promoting factor (MPF) during the course of early cleavage in amphibian embryos (25). Oocytes which have completed maturation still contain high levels of MPF activity, but this is quickly lost after fertilization or artificial activation. MPF has previously been shown to be inactivated by low concentration of free calcium (11). Preliminary aequorin studies have indicated that a massive release of calcium occurs in Xenopus eggs at the time of activation (data not shown). Perhaps this increased calcium activity may account for the rapid loss of cytoplasmic MPF activity. Surprisingly, MPF activity later reappears and begins to cycle with time. The peak of MPF activity always coincides with the time embryonic nuclei are entering the G_2-M transition (mitosis), leading to the suggestion that MPF or an analogous factor plays a general role in regulating nuclear membrane breakdown during mitosis and meiosis (25). This view has been further supported by the report that cytoplasmic extracts from late G_2 or M phase HeLa cells (but not G_1 or S phase) induce germinal vesicle breakdown when injected into Xenopus oocytes (26). In the same sense, the induction of maturation by steroids represents more generally the stimulation of a cell, arrested in a particular phase of the cell cycle (late G_2), to complete the G_2-M transition.

Viewed in terms of the regulation of a cell cycle, the idea that a release of bound calcium is involved is not a novel suggestion. A number of studies have long implicated increased intracellular calcium in the initiation of cell division (27,28). The mechanism(s) by which calcium might be involved, directly or indirectly, in regulating oocyte maturation (or cell division in general) remains obscure. However, the presence of CDR within the oocyte does allow us to visualize a potential link between Ca^{2+} release and the eventual meiotic division, a link now under active investigation.

ACKNOWLEDGMENTS

The aequorin studies were begun in collaboration with Dr. Clare O'Connor and carried out largely in collaboration with Dr. Larry Pinto. The original research in this article was supported by NIH grant HD04229.

REFERENCES

1. Smith, L. D. (1975). In "Biochemistry of Animal Development" Vol. III (R. Weber ed.) pp. 1-46, Academic Press, San Francisco.
2. Wasserman, W. J. and Smith, L. D. (1978). In "The Vertebrate Ovary" (R. E. Jones ed.) pp. 443-468, Plenum Publishing Corporation, New York.

3. Shih, R. J., O'Connor, C. M., Keem, K. and Smith, L. D. (1978). Dev. Biol. 66, 172.
4. Maller, J., Wu, M., and Gerhart, J. C. (1977). Dev. Biol. 58, 295.
5. Masui, Y., and Markert, C. L. (1971). J. Exp. Zool. 177, 129.
6. Reynhout, J. K., and Smith, L. D. (1974). Dev. Biol. 38, 394.
7. Wasserman, W. J., and Masui, Y. (1975). Exp. Cell Res. 91, 381.
8. Baulieu, E., Godeau, F., Schorderet, M. and Schorderet-Slatkine, J. (1978). Nature 275, 593.
9. O'Connor, C. M., Robinson, K. R., and Smith, L. D. (1977). Dev. Biol. 61, 28.
10. Moreau, M., Dóee, M., and Guerrier, P. (1976). J. Exp. Zool. 197, 443.
11. Masui, Y., Meyerhof, P. G., Miller, M. A., and Wasserman, W. J. (1977). Differentiation 9, 49.
12. Ridgway, E. B., Gilkey, J. C., and Jaffe, L. F., (1977). Proc. Nat. Acad. Sci. USA 74, 623.
13. Bellé, R., Ozon, R., and Stinnakre, J. (1977). Mol. and Cell. Endocrinology 8, 65.
14. Bluemink, J. G., and Hoperskaya, O. A., (1975). Wilhelm Roux' Archiv, 177, 75.
15. Rasmussen, H. (1970). Science 170, 404.
16. Brostrom, C. O., Huang, Y. C., Breckenridge, B. M. and Wolff, O. J. (1975). Proc. Nat. Acad. Sci. USA 72, 64.
17. Cheung, W. Y., Lynch, T. J., and Wallace, R. W. (1978). Adv. Cyclic Nucleotide Res. 9, 233.
18. Dedman, J. R., Potter, J. D., Jackson, R. L., Johnson, D. and Means, A. (1977). J. Biol. Chem. 252, 8415.
19. Watterson, D. M., Harrelson, W. G., Keller, P. M., Sharief, F., and Vanaman, T. C. (1976). J. Biol. Chem. 251, 4501.
20. Head, J. F., Mader, J., and Kaminer, B. (1979). J. Cell Biol. 80, 211.
21. Maller, J. L., and Krebs, E. G. (1977). J. Biol. Chem. 252, 1712.
22. Mulner, O., Huchon, D., Thibier, C., and Ozon, R. (1979). Bioch. et Biophy. Acta 582, 179.
23. Schulman, H., and Greengard, P. (1978). Proc. Nat. Acad. Sci. USA 75, 5432.

24. Marcum, J. M., Dedman, J. R., Brinkley, B. R., and Means, A. (1978). Proc. Nat. Acad. Sci. USA 75, 3771.
25. Wasserman, W. J., and Smith, L. D. (1978). J. Cell Biol. 78, 15.
26. Sunkara, P. S., Wright, D. A., and Rao, P. N. (1979). J. Supramel. Struct. Supp. 3, abstr. No. 589.
27. Berridge, M. J. (1975). J. Cyclic Nucleotide Res. 1, 305.
28. Rebhun, L. I. (1977). Int. Rev. Cytol. 49, 1.
29. Sanui, H. (1974). Anal. Biochem. 60, 489.
30. Lin, Y. M., Liu, Y. P., and Cheung, W. Y. (1974). J. Biol. Chem. 249, 4943.

STAGE SPECIFIC CHANGES IN PROTEIN SYNTHESIS DURING *XENOPUS* OOGENESIS

Mary Lou Harsa-King, Andrea Bender, and Harvey F. Lodish

Department of Biology, Massachusetts Institute of Technology, Cambridge, MA 02139

ABSTRACT The pattern of proteins synthesized at different stages of *Xenopus* oogenesis was examined by two-dimensional polyacrylamide gel electrophoresis. Most of the approximately 230 proteins detected were synthesized throughout oogenesis. Approximately 30% of these proteins show stage specific changes in their relative rates of synthesis. The most dramatic changes in the pattern of protein synthesis coincide with the time the chromosomes are in the early and mid-lampbrush configuration - stages 2 and 3. At this time a total of 19 proteins are newly synthesized, and 58 increase in their relative rate of synthesis. A few new proteins are synthesized in response to hormonal (hCG) stimulation, although a dramatic increase in the number of new proteins made is not observed at any stage. Other specific effects of hCG on the pattern of proteins synthesized are a) an alteration in the time of synthesis for some proteins and b) the dramatic increase in the relative rate of synthesis of a few proteins.

INTRODUCTION

Amphibian oogenesis is a time of intense synthetic activ-ty and product accumulation. Some of these products are es-ential for later development [1]. In response to hormonal timulus mature oocytes are ovulated and immature oocytes are timulated to incorporate exogenous amino acids into proteins 2]. Because of the major morphological and biochemical vents which occur and of the potentially important role layed in these processes by the ovulation hormone, the regu-ation of gene expression during oogenesis is of considerable nterest. The first task is to identify genes whose expres-ion is regulated. To this end we have been investigating by

This work was supported by NIH grants #AI08814, AM15322 ıd NSF grant #PCM74-04869 A04.

ISBN 0-12-068350-4

two-dimensional gel electrophoresis the changes in the pattern of proteins synthesized during the six Dumont Stages of *Xenopus laevis* oogenesis before and after hormonal (human chorionic gonadotropin, hCG) stimulation.

RESULTS

We employed the two-dimensional gel system of O'Farrell [3] to analyze the synthesis of proteins during the six stages of *Xenopus* oogenesis described by Dumont [4]. Oocytes were manually separated according to stage and labeled *in vitro* for 6 hr with [^{35}S]-methionine (20μCi/ml; 985 Ci/mmol) added to the culture medium as detailed by Gurdon *et al.* [5]. Incorporation into protein was linear over this time period. Over 95% of the [^{35}S]-methionine incorporated was incorporated into oocytes, not follicle cells, since treatment with collagenase, which removes the follicle cells, released only 2-5% of the protein radioactivity. Furthermore, no significant differences in the protein pattern were observed in collagenase treated and control oocytes, indicating that the proteins observed are of oocyte not follicle cell origin. In all subsequent experiments oocytes were not treated with collagenase.

The autoradiograms were exposed uniformly to the equivalent of 1,000,000 cpm of [^{35}S]-methionine added to the gel for 24 hr. The amount of radioactivity in protein in each sample was used to calculate the time required for the desired exposure. Under these conditions, the absolute intensity of a given spot is an approximate measure of the relative rate of synthesis of that protein, assuming methionine residues are uniformly represented in the protein population. The method of analysis involve determining which spots on the fluorogram changed in intensity, relative to neighboring spots, from one stage to another. These changes were determined by visual inspection. Since not all polypeptides were optimally exposed for any one given exposure period, gels were exposed for different periods of time before final analyses were made.

The pattern of proteins synthesized by oocytes of any particular stage was very reproducible. The data presented here represents an analysis of 10 different frogs over a two year period. Approximately 230 different labeled polypeptides can be resolved on these gels. If gels are exposed for 5,000,000 cpm per 24 hrs some 400 different polypeptides can be distinguished. The majority of these proteins are made throughout oogenesis. Approximately 90 undergo quantitative or qualitative changes in their relative rates of synthesis at one or more oogenic stage, and these have been numbered in the following figures. For any given stage, approximately 30% of the proteins have changed in their relative rate of synthesis compared to the previous stage. To conserve space, only the gels from stage 2, 3, and 6 are shown (Figs. 1-3), although gels from all stages are summarized in Fig. 4, which represents a schematic summary of the most interesting proteins.

Stage 1 oocytes are previtellogenic. In stage 2 oocytes small yolk platelets and cortical granules first form, and chromosomes are in an early lampbrush configuration. Cytological and hybridization data support the idea that large amounts of RNA are transcribed from these chromosomes [6, 7]. The major change in the transition from stage 1 to stage 2 oocytes is a general increase in the relative rate of synthesis of 37 proteins (spots: 1, 3, 5, 7, 8, 13, 17, 18, 23-26, 30, 35-37, 39, 42, 43, 46, 47, 49-52, 56, 60, 61, 70-72, 76, 78, 82, 85, 88, 89)(Fig. 1). Eleven proteins are newly synthesized (spots: 4, 6, 11, 12, 31, 33, 38, 65, 77, 86, 87) while no proteins cease being made.

Stage 3 oocytes are characterized by the first signs of pigmentation, active accumulation of yolk, and maximumly extended lampbrush chromosomes. Eight new polypeptides are synthesized at this time (spots: 28, 48, 58, 59, 66, 67, 80, 81)(compare Figs. 1 and 2). The synthesis of three stage 2 proteins can no longer be seen (spots: 5, 27, 44). In addition, 36 proteins show an increase in their relative rates of synthesis (spots: 2-4, 7, 12, 14, 21, 23, 29-31, 33, 38, 20, 41, 42, 46, 47, 55, 57, 60, 65, 68, 70-79, 86-88), while six other proteins decline (spots: 1, 8, 13, 45, 50, 52). Of the 36 which show an increase, 14 are polypeptides which also increased in their relative rate of synthesis during stage 2.

During the rest of oogenesis the synthesis of only four other proteins will be initiated, making stage 2 and 3 together the time of major qualitative changes. Newly synthesized proteins during these two stage account for approximately 9% of the total polypeptides detected (19/214).

During stage 4, oocytes undergo the most rapid phase of vitellogenesis, increasing in diameter by some 400μ [4]. In succeeding stages vitellogenesis tapers off although oocytes increase another 200-300μ in diameter before maturation. From stage 4 to 6, the chromosomes progressively condense preparatory to the meiotic reduction divisions. During this time the animal and vegetal hemispheres become clearly delineated, the result of the pigment bearing organelles concentrating in the animal hemisphere. Stage 6 oocytes have reached maximum diameter and have completed yolk accumulation [4]. After hCG stimulation, these oocytes are ovulated and are fully competent to be fertilized.

Few changes in the pattern of polypeptide synthesis occur during the final three stages of oogenesis. Only two polypeptides are newly synthesized during stage 4 (spots: 22, 83)

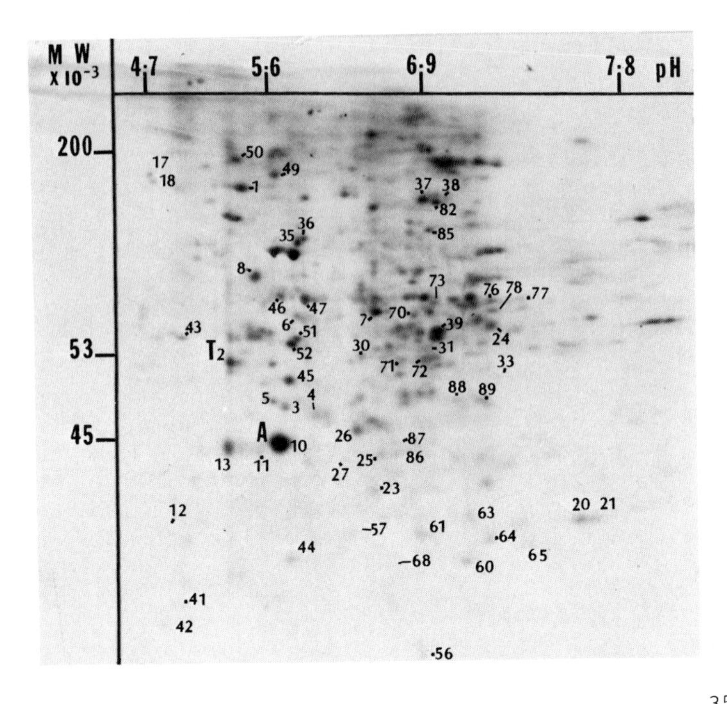

FIGURE 1. Fluorograph of two-dimensional gel of [^{35}S]-methionine labeled proteins synthesized in vitro by stage 2 Xenopus oocytes. Oocytes were removed surgically from a female Xenopus which had not been ovulated for 6 to 12 months. Stage 2 oocytes were manually isolated and labeled in vitro for 6 hr at 18°C with [^{35}S]-methionine (20μCi/ml) added to the culture medium as described by Gurdon et al. [5]. The oocytes were homogenized directly in isoelectric focusing buffer [8] containing 0.5% SDS, 9.5 M urea, 2% DTT, 2% NP-40, 5% ampholines (pH 3-10), 100mM NaCl and the samples centrifuged at 10,000g for 5 min to remove any insoluble material. The resulting supernatant was either stored at -20°C or run immediately on an isoelectric focusing gel [3]. The second dimension was run in 12% SDS-polyacrylamide gels as described by Storti et al. [8]. The fluorograph was prepared as previously described [9], and was exposed to the equivalent of 1,000,000 cpm for 24 hr. 86 proteins that clearly change in their relative rates of synthesis at sometime during oogenesis are numbered in the following figures. These changes are determined by visual inspection. The numbers at the top of the fluorograph refer to pH values which were measured directly. The pI of actin was empirically determined to be 5.6. Putative actin (A) and β-tubulin (T) protein spots are shown.

while three others cease being made (spots: 18, 21, 66). Polypeptides 3, 10, 44 show increases in their relative rates of synthesis. For the first time in oogenesis, more proteins decrease in their relative rate of synthesis than show increases (spots: 4, 33, 37-39, 41, 42, 55, 56).

The transition from stage 4 to 5 is accompanied by the synthesis of three new proteins (spots: 19, 84, 90) while the synthesis of one other protein ceases (spot: 1). In addition, the relative rate of synthesis of 15 polypeptides increases (spots: 4, 7, 12, 14, 18, 25, 32, 39, 41, 42, 51, 55, 56, 83, 85); six others decrease (spots: 31, 58, 59, 71, 86, 88).

During the last stage as oocytes reach their full size, the synthesis of no new proteins is initiated. The synthesis of two other proteins stops (spots: 8, 44). Twenty-six proteins show increases in their relative rates of synthesis at this time (spots: 2, 4, 5, 12, 17, 18, 21-23, 28-30, 50, 51, 55-57, 60, 63, 65, 66, 68, 71, 84, 86, 88). Only one protein shows a decrease in its relative rate of synthesis (spot: 3). Many of these changes can be seen by comparing stage 3 and stage 6 (see Figs. 2 and 3).

That few changes occur between stage 4 and 6 is supported by RNA-cDNA hybridization studies [10]. Approximately 20,000 different sequences were found in stage 4 and in stage 6 oocytes. These sequences were shown to be essentially the same in both stages. The numbers obtained by hybridization analysis emphasize that the level of dectection in the present study is only that 1-2% of the proteins representing the most prevalent translated messenger RNA species.

It is interesting to note that there is not one example of a protein being synthesized exclusively at one particular stage. All of the labeled proteins in this study were produced by at least two stages.

Actin is an abundant protein found in a variety of non-muscle cell types. It appears to be an important cytostructural protein playing a number of roles in the cell [for review see ref. 11]. Actin is one of the major proteins synthesized during oogenesis. After making allowances for the relatively high content of methionine in actin, incorporation into actin accounts for approximately 2% of the total amount of radioactivity in acid-insoluble material. Oocyte actin was isolated by DNAaseI affinity chromatography [12] and purified actin run in parallel and together with total oocyte protein. Actin identified in this manner (spot: 10) is synthesized at a fairly constant rate throughout oogenesis with only a minor increase in its relative rate of synthesis during the last three stages. Only one species of actin could be detected on our gel system.

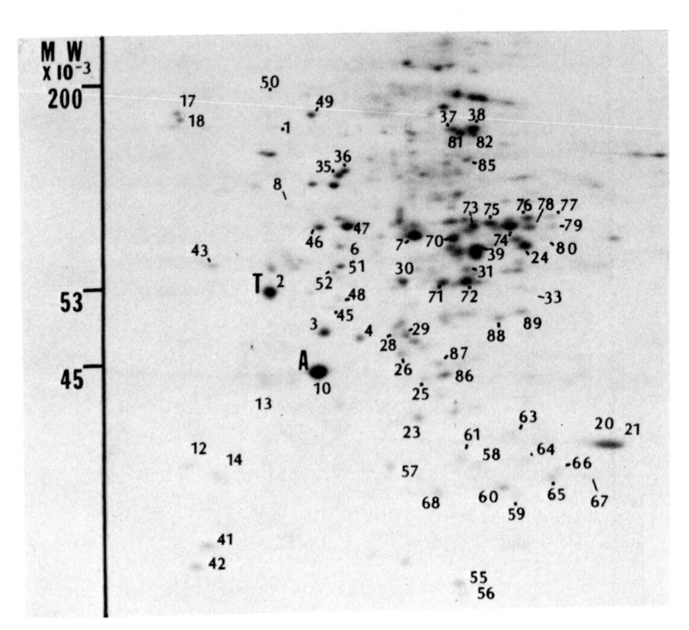

FIGURE 2. Fluorograph of two-dimensional gel of proteins synthesized by stage 3 Xenopus oocytes, otherwise the same as Figure 1.

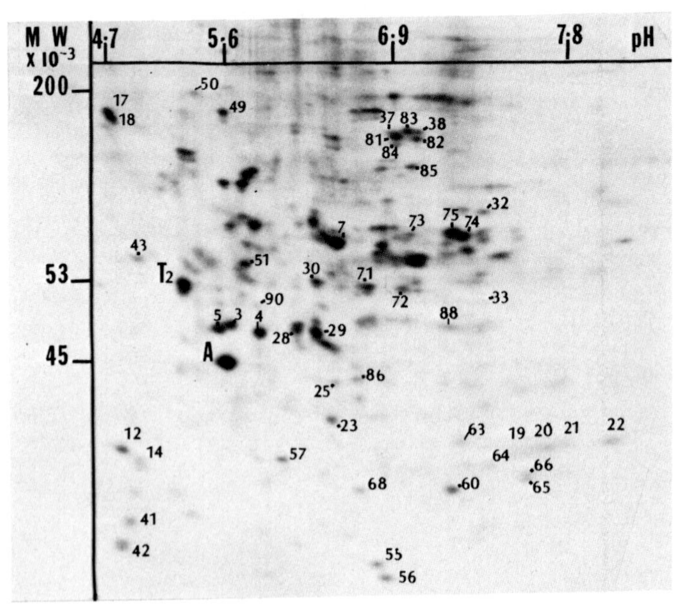

FIGURE 3. Fluorograph of two-dimensional gel of proteins synthesized by stage 6 Xenopus oocytes, otherwise the same as Figure 1.

Another important cytoskeletal protein is tubulin, and it too is synthesized as a major protein during oogenesis. The identification of this protein is based on comigration with purified sea urchin tubulin isolated from sperm tails. This is only a tentitive identification and will be confirmed by tryptic peptide analysis. α-tubulin could not be positively identified on our gels since two proteins comigrated with sea urchin α-tubulin. β-tubulin (spot: 2) is synthesized at a low relative rate during stages 1 and 2, increases during stages 3, 4, and 5, reaching a maximum relative rate of synthesis during stage 6.

The final steps in oocyte maturation are triggered by hormonal stimulation and include the breakdown of the geminal vesicle and the completion of meiosis up to second meiotic metaphase. At this stage the chromosomes are arrested until fertilization. A number of other effects of hCG on oocyte metabolism have been documented. These include the stimulation of rRNA synthesis in immature oocytes [13], the stimulation of vitellogenesis in stage 3-5 oocytes [14], an increase in the amino acid pool size [15], and the stimulation of incorporation of exogenous amino acids into proteins [2]. The net effect of hCG administration is that stage 6 oocytes are ovulated and all the earlier stages are stimulated to further growth and development. We asked whether the 2-3-fold stimulation of incorporation of [^{35}S]-methionine into proteins after hCG stimulation could be due to the synthesis of many new proteins or to the stimulation of synthesis of endogenous protens. To determine this, the same frogs that had been used in the above analysis and which had not been ovulated for 6 to 12 months, were injected with 500-800 units of hCG. Injection was intraperitoneally or into the dorsal lymph sac. 24 hrs later, after ovulation had occured, these frogs were anesthetized in ice water, a sample of their oocytes was removed surgically and stages 1 through 5 were separated manually. The results of the analysis are presented in Figs. 5 and 6 and summarized in Fig. 4.

The overall pattern of protein synthesized at any one stage is almost unchanged (compare Fig. 1 to Fig. 5 and Fig. 3 to Fig. 6). Approximately the same number of proteins detected before hCG stimulation (230) are detected afterwards. With the exception of three proteins, all of the approximately 90 proteins observed to change in their relative rates of synthesis before hCG stimulation also undergo changes after hCG stimulation. There are specific effects of hCG on the pattern of proteins synthesized and they fall into the following categories: 1) proteins synthesized *de novo* only after hCG stimulation, 2) proteins synthesized at greatly altered relative rates after hCG stimulation, and 3) proteins whose time of

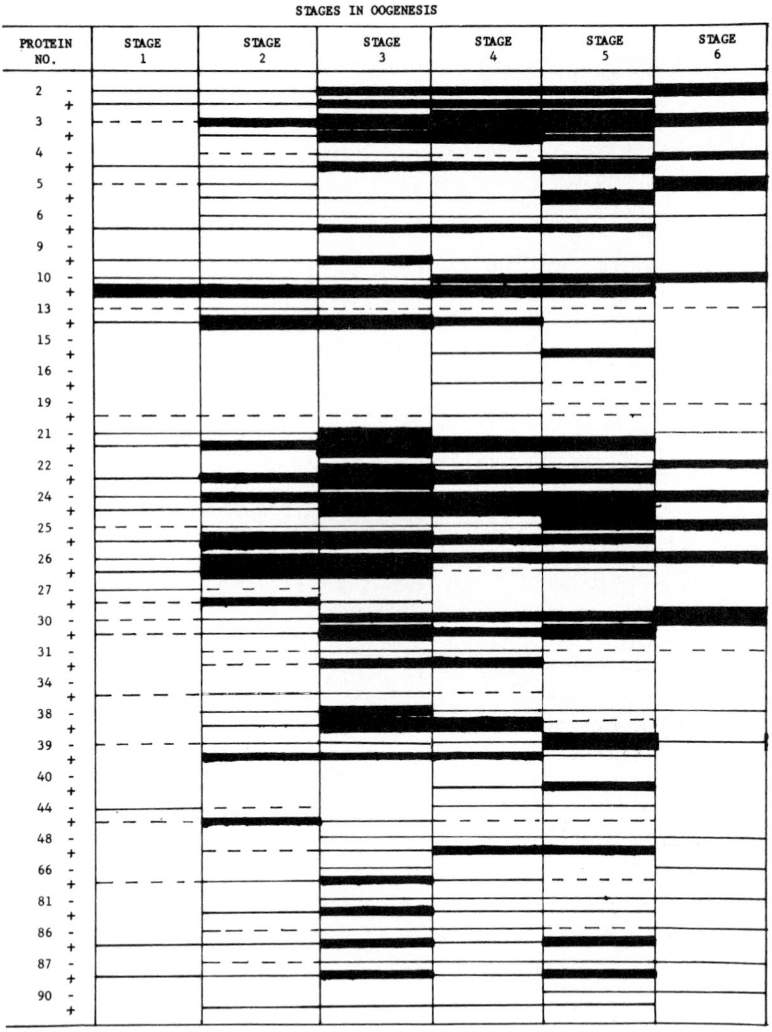

FIGURE 4. Schematic summary of the program of protein synthesis during Xenopus oogenesis. The numbers of the side correspond to the polypeptide spots labeled in Figs. 1-3, 5, 6. The + and - by these numbers indicate with or without hCG stimulation respectively. Dashed lines indicated barely detectable synthesis, solid lines indicate synthesis, heavy solid lines indicate increased relative rate of synthesis, and the absence of a line indicates that synthesis could not be detected.

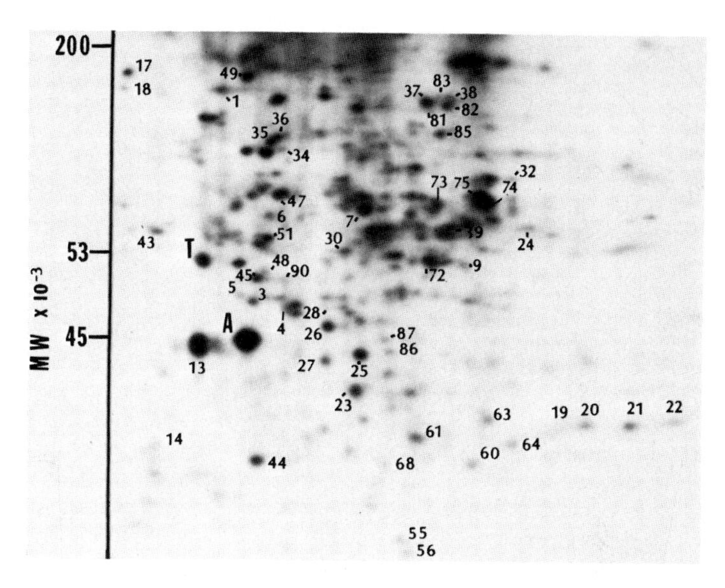

FIGURE 5. Fluorograph of two-dimensional gel of proteins synthesized by stage 2 oocytes isolated from a recently ovualted female _Xenopus_. Donor _Xenopus_ was injected with 500-800 units of hCG and oocytes removed 24 hr later for analysis. This fluorograph was exposed to the equivalent of 1,000,000 cpm for 24 hr.

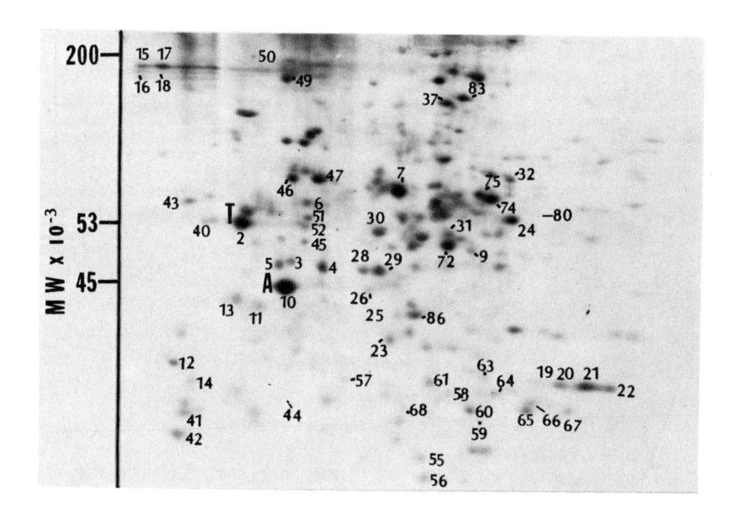

FIGURE 6. Fluorograph of two-dimensional gel of proteins synthesized by stage 5 oocytes isolated from a recently ovulated female _Xenopus_, otherwise the same as Fig. 5.

synthesis is altered after hCG stimulation.

After hCG stimulation 5 proteins (spots: 9, 15, 16, 34, 40) (Figs. 5 and 6) are synthesized which are never seen in unstimulated oocytes. The relative rate of synthesis of spot 9 is constant during stages 1 and 2, increases during stage 3, and falls off in stages 4 and 5, only after hCG stimulation. The relative rate of synthesis of spot 34 is low in stage 1, increases over stages 2 and 3, declines in stage 4, and ceases in stage 5 stimulated oocytes. Polypeptides 15, 16, and 40 are only synthesized during stages 4 and 5 and only after hCG stimulation.

Human CG stimulation alters the time at which some proteins are synthesized. Examples of proteins which are synthesized precociously after hCG treatment are spots 4, 6, 22, 28, 86, 90) (Figs. 4-6). Polypeptides 4, 6, and 86 are not synthesized in unstimulated stage 1 oocytes, but are in stimulated stage 1 oocytes. Spot 28 is not synthesized in unstimulated stage 2 oocytes, but is in stimulated stage 2 oocytes. Polypeptide 22 is synthesized in every stage after hCG stimulation but in only stages 4 and 5 before such treatment. Spot 90 is not synthesized until stage 5 in unstimulated oocytes, but is synthesized from stage 2 onwards after hCG stimulation. In the case of some proteins, synthesis is prolonged into later stages of oogenesis after hCG stimulation. For example, spot 21 is not synthesized in stages 4, 5, and 6 in unstimulated oocytes but is synthesized continuously in stimulated oocytes. Spot 5 is synthesized at stages 2 through 5 after hCG stimulation, but only during stages 1, 2, and 6 in unstimulated oocytes.

Dramatic increases in the relative rates of synthesis of some proteins after hCG stimulation are also observed (compare Fig. 1 and Fig. 5). Polypeptides 13, 25, 26, and 44 fall into this category. Spot 13 reaches its maximum rate of synthesis in stage 2 oocytes before or after hCG treatment. However, after hCG stimulation this protein appears to be synthesized at least at a 10-fold higher relative rate of synthesis than before. This phenomenon is observed in stage 3 oocytes (see spots: 21, 22, 24, 30, 31 and 38 in the summary in Fig. 4). Whether or not this increase is the result of increased levels of mRNA for these proteins remains to be determined.

DISCUSSION

The pattern of proteins synthesized at different stages of <u>Xenopus</u> oogenesis both before and after hCG stimulation was examined by two-dimensional polyacrylamide gel electrophoresis. We have identified several classes of developmentally regulated proteins based on this work: A) those proteins that show stage specific qualitative or quantitative changes in their relative rates of synthesis, B) those proteins synthesized only after hCG stimulation, C) those proteins whose relative rate of synthesis is dramatically increased after hCG stimulation.

In such an analysis, protein loss or modification during sample preparation and/or electrophoresis is important. Aproximately 35% of the [^{35}S]-methionine radioactivity incorporated into protein is lost as insoluble material during sample preparation. Roughly 2% of the radioactive protein loaded on the first dimension never entered the second dimension. Furthermore, polypeptides larger than 250,000 and smaller than 15,000 daltons and whose pI was less than 3.5 or greater than 8.2 would not be included in the analysis.

A variety of methods for sample preparation were tested and the one finally employed had the virtue of being the simplest and the one yielding very reproducible results. Samples were homogenized directly in isoelectric focusing buffer [8] as detailed in Fig. 1. Samples were never heated. Since the pattern of proteins synthesized was highly reproducible either no artifactual modifications were introduced or they were introduced uniformly. In either case, the results presented are not invalidated.

The majority of the proteins detected are made throughout oogenesis at the same relative rate of synthesis. Approximately 90 out of 230 proteins that were detected undergo quantitative or qualitative changes in their relative rates of synthesis. The most dramatic changes in the pattern of protein synthesis coincide with the time the chromosomes are in the early and mid-lampbrush configuration - stages 2 and 3. As stage 1 oocytes grow into stage 2 oocytes there is an increase in the relative rate of synthesis of many (37) proteins, while the synthesis of the largest number (11) of proteins is initiated at this time. Another 8 proteins are newly synthesized during stage 3 and 36 proteins increase in their relative rate of synthesis. There are also 6 proteins synthesized at decreased rates of synthesis at this time. Few new proteins (4) are synthesized in stages 4, 5, and 6 oocytes. The pattern of proteins synthesized is virtually unchanged from stage 4 to stage 5. No new proteins are made in the last stage of oogenesis, although 25 proteins increase in their relative rates of synthesis.

After hCG stimulation, 2-3 times more [^{35}S]-methionine is incorporated into protein than in unstimulated oocytes. By and large, this increased incorporation does not represent new proteins being synthesized, but represents increased incorporation into endogenous proteins. However, we identified 5 proteins which are made only after hCG stimulation and which are never made by unstimulated oocytes. In the case of some proteins hCG causes their synthesis to begin earlier and/or to continue to later stages.

Dramatic increases in the relative rates of synthesis are observed for some proteins after hCG stimulation. Whether or not this increase is the result of increased levels of mRNA for these proteins is an important point that could be determined by in vitro translation of polysomal RNA isolated from oocytes before and after hCG stimulation. Experiments of this nature are now underway.

REFERENCES

1. Davidson, E. H. (1976). In "Gene activity in early development" Second edition, Academic Press, New York, N.Y.
2. Hallberg, R. L. and Smith, O. (1976). Develop. Biol. 48, 308.
3. O'Farrell, P. N. (1975). J. Biol. Chem. 250, 4007.
4. Dumont, J. N.(1972). J. of Morphol. 136, 153.
5. Gurdon, J. B., Land, D. C., Woodland, H. R. and Marbaix, G. (1971). Nature 233, 177.
6. Callan, H. G. (1963). In "International Review of Cytology". (G. H. Bourne and J. F. Danielli, eds.) Vol. 15: 1-34. Academic Press, New York, N.Y.
7. Davidson, E. H., Crippa, M., Kramer, F. R. and Mirksy, A. E. (1966). Proc. Nat. Acad. Sci. USA, 56, 856.
8. Storti, R. V., Horovitch, S. J., Scott, M. P. Rich, A. and Pardue, M. L. (1978). Cell 13, 589.
9. Laskey, R. A., and Mills, A. O. (1975). Eur. J. Biochem. 56, 335.
10. Perlman, S. and Rosbash, M. (1978). Develop. Biol. 63, 197.
11. Pollard, T. D. and Weihing, R. R. (1973) In "Critical Reviews in Biochemistry" Vol. 2, 1-65.
12. Lindberg, U. and Eriksson, S. (1971) Eur. J. Biochem. 18, 474.
13. Brown, D. D. and Littna, E. M. (1964). J. Mol. Biol. 8, 669.
14. Wallace, R. A., Jared, D. W., and Nelson, B. L. (1970). J. Exp. Zool. 175, 259.
15. Eppig, J. J. and Dumont, H. N. (1972). Develop. Biol. 28, 531.

TERATOCARCINOMA CELL MUTATIONS AS PROBES OF MAMMALIAN DIFFERENTIATION AND DISEASE[1]

Beatrice Mintz

Institute for Cancer Research, Fox Chase Cancer Center, Philadelphia, Pennsylvania 19111

ABSTRACT The malignant stem cells of mouse teratocarcinomas have been found to lose their neoplastic properties and to undergo normal differentiation if placed in a normal early-embryo (blastocyst) environment. Under these conditions, the stem cells are developmentally totipotent: Tumor-derived cells can contribute to the formation of all somatic tissues and--if still close to euploidy--to germ cells, from which progeny are obtained. The teratocarcinoma cells thus provide a novel channel for deliberately introducing into mice specific, predetermined genetic markers. These can serve to dissect developmental processes; as probes for analyzing control of tissue-specific gene expression; and as agents for producing mouse models of human genetic diseases. The experimental possibilities rest upon the combined utilization of techniques of somatic cell genetics, molecular biology, and developmental biology. Genetic markers may be obtained by changes within the same species, or introduced from foreign species; they may be nuclear or cytoplasmic (mitochondrial). They may be first brought into the teratocarcinoma cells in culture following mutagenesis or DNA transfer and, after the application of appropriate selective or screening procedures, cells with the mutation of interest are placed in the *in vivo* setting, where gene expression may be optimally tested during differentiation.

The ultimate challenge, in the realm of eukaryotic gene regulation, is not only to understand how gene function is governed, but also how genes act selectively in specialized tissues of a complex organism. That challenge might be met if one could manipulate the genes--mutate them or redesign them, or their neighbors, to learn whether such changes

[1]This work was supported by U.S. Public Health Service Grants HD-01646, CA-06927, and RR-05539, and by an appropriation from the Commonwealth of Pennsylvania.

ISBN 0-12-068350-4

modify tissue-specific expression. This now seems feasible in mammals as a result of the broad developmental capacities and experimental options provided by malignant mouse teratocarcinoma cells.

Tumors may be regarded as developmental aberrations (1, 2). Normally, proliferating stem cells are the generative reservoir of a tissue; they yield mature, and ultimately non-dividing, cellular progeny; in some tissues, the stem cell population also continues to replenish itself throughout life. In malignancy, it is the initially normal stem cells that appear to be the targets in neoplastic conversion: The cells undergo an impairment of differentiation but retain (or even improve) their proliferative activities.

The earliest embryonic stem cells are developmentally versatile, in fact totipotent, each being capable of forming an entire organism. Teratocarcinomas are exceptional among tumors in that they contain a variety of tissues (3), a feature suggesting that their stem cells arose at a developmentally primitive stage. Thus, if these stem cells could be made to differentiate normally, each could in effect be regarded as a "potential mouse."

When teratocarcinoma stem cells were introduced into early embryos (at the blastocyst stage) by microinjection, the accompanying normal totipotent cells were in fact able to exert a permanent "normalizing" effect on their once-malignant companions (4-6). Cells from the tumor lineage now contributed to normal embryogenesis and to the formation of fully differentiated and functional somatic tissues in mosaic individuals. In the best cases, involving tumor lines that were still euploid, cells descended from the tumor also gave rise to functional germ cells yielding healthy offspring (7).

Thus, a basis was established for utilizing cultured teratocarcinoma stem cells as vehicles for introducing specific mutant genes into mice (8). Those genes might either be murine or foreign, nuclear or cytoplasmic. Some of the chief possibilities are briefly indicated in Figure 1.

The deliberate production of mice with specific heritable defects would obviously include constructing mouse models of human genetic diseases (10). In a large number of these diseases, the primary biochemical lesion is known but the complex _in vivo_ pathogenesis has remained obscure. Laboratory counterparts, which have hitherto rarely been available, are needed to clarify the syndromes. An example is human Lesch-Nyhan disease, a disorder (X-linked) of purine metabolism resulting from a severe deficiency of hypoxanthine phosphoribosyltransferase (HPRT) (11). Mutagenized mouse teratocarcinoma cultures were selected for resistance to the purine base analog 6-thioguanine and an HPRT-deficient clone

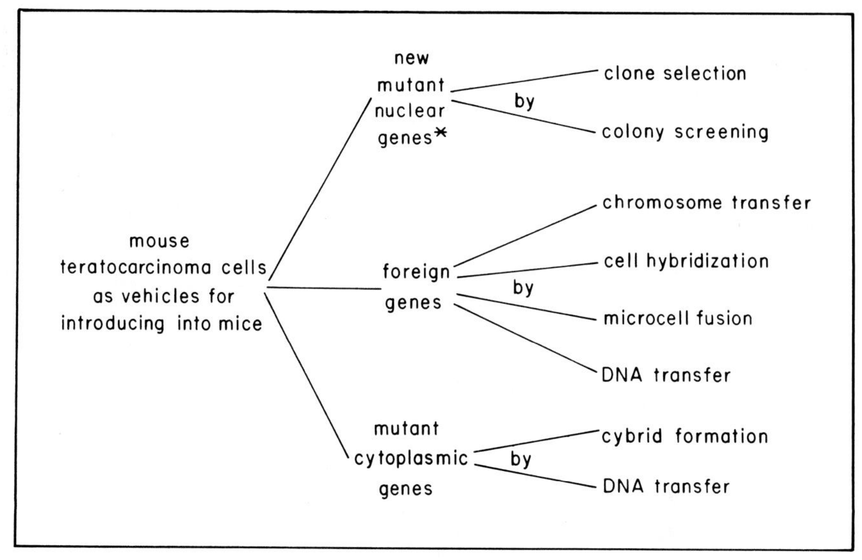

*e. g., mutations for production of mouse models of human genetic diseases

FIGURE 1. A summary of some of the chief methods by which developmentally totipotent mouse teratocarcinoma stem cells may serve as vehicles for the introduction of specific genes into mice (9). In each case, cells of a teratocarcinoma clone bearing the mutant gene (either murine or foreign, nuclear or mitochondrial) is microinjected into a blastocyst from another mouse strain. Mice with cellular genotypic mosaicism result. All-mutant progeny may be obtained in the F_1 generation, from mutant-strain germ cells. In some cases, the animals would be laboratory models of a human genetic disease.

was obtained (12). The cells were then introduced into blastocysts and mosaic animals resulted (Figure 2). In their tissues, tumor-lineage cells could be distinguished from embryo-lineage ones through various strain-specific independent markers. The tumor-strain cells had retained their HPRT deficiency throughout proliferation and differentiation *in vivo*. An interesting feature was the parallel to human heterozygous female carriers of the disease, in that a strong selection against the defective class of cells in blood was seen in the mosaic mice (12) as in the human heterozygotes (13).

Another example of a human disease with a known lesion but with a little-understood pathogenesis is familial hypercholesterolemia; here again, no animal model is available.

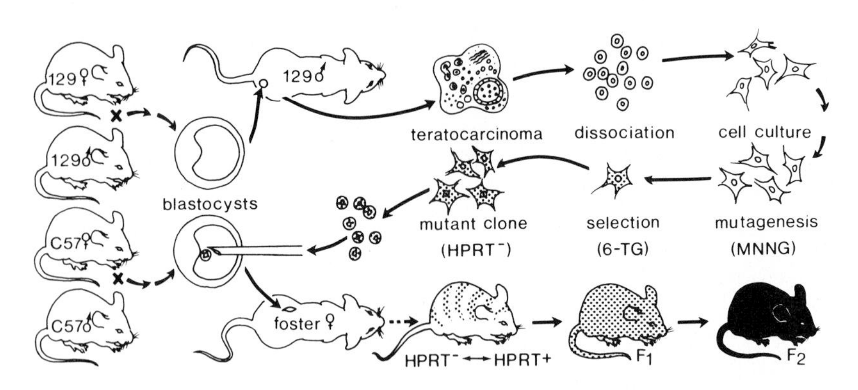

FIGURE 2. A diagrammatic summary of the experimental incorporation of HPRT-deficient ("Lesch-Nyhan") cells into mice (12). Teratocarcinoma cells in culture were mutagenized and selected for resistance to 6-thioguanine. Cells from a mutant clone were $HPRT^-$. They were then microinjected into blastocysts of an inbred strain different from the strain of origin of the tumor. After embryo transfer to uteri of foster mothers, mosaic animals were obtained. Their tissues included tumor-derived non-malignant cells still deficient in HPRT activity, along with embryo-derived normal cells. Progeny with all-mutant cells might occur in later generations from a fertile animal with tumor-lineage germ cells.

In some patients, the primary difficulty entails a deficiency in the specific cell surface receptor that binds low density lipoprotein (LDL) (14). In a recent study (15), we have found that teratocarcinoma stem cells in culture do uniformly express high-affinity LDL receptors similar to those of adult parenchymal cells. The LDL receptors on the stem cells were readily visualized by administering a fluorescent derivative of LDL. Cell colonies with depressed receptors, yielding little fluorescence, were distinguishable in the fluorescence microscope. Low-fluorescing cells also proved to be separable in the fluorescence-activated cell sorter. Thus, the necessary groundwork has been laid for the isolation of receptor-deficient mutant teratocarcinoma cells and, from them, the future synthesis of mice with the counterpart of this human disease.

Mitochondrial gene mutations as well as nuclear ones may also be introduced into animals via the teratocarcinoma route. Mitochondria are probably maternally transmitted and their genes surely play a significant role in development and disease. Yet specific _in vivo_ markers have not been available for investigation of these problems. By the indirect route of cybrid (cytoplasmic hybrid) formation (16), we have managed to introduce a mitochondrially encoded marker--chloramphenicol resistance (CAP^R)--into teratocarcinoma cells (17). Cytoplasts from a mutagenized melanoma line, first selected for resistance to the drug, were produced by enucleating those cells and the donor cytoplasts were then fused to teratocarcinoma cells. This afforded proof of the cytoplasmic origin of the mutant gene. The CAP^R teratocarcinoma cells were injected into mouse embryos and mosaic mice containing some CAP^R differentiated cells were later identified. The mutant trait remains stable *in vivo* even in the absence of the selective agent.

It was felt necessary to expand the possibilities for obtaining teratocarcinoma stem cells with specific mutations for which selective schemes are unavailable. We have recently succeeded in developing a replica plating method suitable for cultures of these stem cells (A.J.J. Reuser and B. Mintz, unpublished data). A photograph of a master plate and its replica, showing the very high replication efficiency, may be seen in Figure 4 of reference 9. This method is now receiving its first test in screening for a specific mutation.

Of the various technologies indicated in Figure 1, whereby predetermined genetic changes might be conveyed to mice from cultured teratocarcinoma cells, perhaps the most promising is the possibility of direct DNA or gene transfer. In the past, the only successful experimental introduction of foreign genes into a mammal was reported from our laboratory and resulted from microinjection of purified simian virus 40 (SV40) DNA into mouse blastocysts (18). From these injected embryos, healthy animals were born, yet by molecular hybridization tests of DNA extracted from their tissues, SV40-specific gene sequences were indeed present. This SV40 experiment was conducted before the advent of gene propagation by cloning of specific DNA segments in prokaryotic plasmids or vectors. Of course such cloning methods provide substantial amounts of pure genes which may now be introduced into mouse blastocysts (or into cells of earlier-stage embryos), thereby enabling numerous kinds of foreign genes to be brought into the cells of actual mice. However, the possibilities for selective pressures, so as to favor retention of the donated gene, are limited *in vivo*. Therefore, the use of teratocarcinoma-cell intermediaries, followed by injection of the stem cells into embryos, provides an arrangement whereby cells that have incorporated the gene can first be isolated by appropriate selection in culture.

Recent work in other laboratories has demonstrated that genes can in fact be taken up by mammalian cells in culture (19), either from high-molecular-weight DNA, from endonuclease restriction fragments, or from cloned segments. With mutant recipient cells, selective media can be used to obtain transformants; e.g., uptake of a foreign thymidine kinase gene can be monitored in thymidine kinase-deficient mouse cells. Moreover, an accompanying gene for which no selection is possible can be co-transferred with the selectable one even without being ligated to it (20).

The prospect of mice containing a "foreign" gene affords bold new possibilities for the analysis of tissue-specific control of gene expression. The fact of "foreignness" supplies the necessary discriminant between the donor gene and the corresponding native gene of the host. Particular tissues may then be assayed for presence and function of the donated gene. With already-existing methods (e.g., molecular hybridization, endonuclease restriction digestion, mRNA characterization, and specific protein-product identification), appropriate assays may be conducted for the presence, transcription, and translation of the gene in the "relevant" tissue (e.g., rabbit or human globin genes in mouse blood cells). By prior tailoring of the gene before its introduction (i.e., by rearrangements, deletions, or base substitutions), or by altering the flanking sequences of the gene of interest, the capacity for its expression to occur can be analyzed in relation to defined control mechanisms.

REFERENCES

1. Pierce, G. B. (1967). Curr. Top. Develop. Biol. 2, 223.
2. Mintz, B. (1978). Harvey Society Lectures, Ser. 71, 193. Academic Press, New York.
3. Stevens, L. C. (1967). Adv. Morphog. 6, 1.
4. Mintz, B., and Illmensee, K. (1975). Proc. Natl. Acad. Sci. USA 72, 3585.
5. Illmensee, K., and Mintz, B. (1976). Proc. Natl. Acad. Sci. USA 73, 549.
6. Dewey, M. J., and Mintz, B. (1978). Develop. Biol. 66, 550.
7. Cronmiller, C., and Mintz, B. (1978). Develop. Biol. 67, 465.
8. Mintz, B. (1977). In "Genetic Interaction and Gene Transfer." Brookhaven Symposia in Biology, 29 (C. W. Anderson, ed.) pp. 82-95. Brookhaven National Labs., Upton, N. Y.
9. Mintz, B. (1979). Differentiation 13, 25.
10. Mintz, B. (1979). In "Models for the Study of Inborn Errors of Metabolism." (F. A. Hommes, ed.) pp. 343-354. Elsevier/North-Holland Biomed. Press, Amsterdam.
11. Seegmiller, J. E., Rosenbloom, F. M., Kelley, W. N. (1967). Science 155, 1682.

12. Dewey, M. J., Martin, D. W., Jr., Martin, G. R., and Mintz, B. (1977). Proc. Natl. Acad. Sci. USA 74, 5564.
13. Nyhan, W. L., Bakay, B., Connor, J. D., Marks, J. F., and Keele, D. K. (1970). Proc. Natl. Acad. Sci. USA 65, 214.
14. Brown, M. S., and Goldstein, J. L. (1976). Science 191, 150.
15. Goldstein, J. L., Brown, M. S., Krieger, M., Anderson, R. G. W., and Mintz, B. (1979). Proc. Natl. Acad. Sci. USA 76, in press.
16. Bunn, C. L., Wallace, C., and Eisenstadt, J. M. Proc. Natl. Acad. Sci. USA 71, 1681.
17. Watanabe, T., Dewey, M. J., and Mintz, B. (1978). Proc. Natl. Acad. Sci. USA 75, 5113.
18. Jaenisch, R., and Mintz, B. (1974). Proc. Natl. Acad. Sci. USA 71, 1250.
19. Wigler, M., Silverstein, S., Lee, L.-S., Pellicer, A., Cheng, Y.-c., and Axel, R. (1977). Cell 11, 223.
20. Wigler, M., Sweet, R., Sim, G. K., Wold, B., Pellicer, A., Lacy, E., Maniatis, T., Silverstein, S., and Axel, R. (1979). Cell 16, 777.

STRUCTURE AND EXPRESSION OF OVALBUMIN AND CLOSELY RELATED CHICKEN GENES

P. Chambon, F. Perrin, K. O'Hare, J.L. Mandel, J.P. LePennec, M. LeMeur, A. Krust, R. Heilig, P. Gerlinger, F. Gannon, M. Cochet, R. Breathnach and C. Benoist

Laboratoire de Génétique Moléculaire des Eucaryotes du CNRS, Unité 184 de Biologie Moléculaire et de Génie Génétique de l'INSERM et Institut de Chimie Biologique, Faculté de Médecine
Université Louis Pasteur
Strasbourg - France

INTRODUCTION

Chicken ovalbumin accounts for 50 to 65% of total protein synthesis in laying hen oviduct tubular gland cells (1). As for other egg white proteins such as conalbumin, ovomucoïd and lysozyme, the rate of transcription of its mRNA (which represents 50% of the total mRNA population) is under hormonal control (for references, see 2-5). Therefore the genes corresponding to these proteins provide a very useful model system for the study of the transcriptional regulation of gene expression by hormones. Moreover, administration of oestrogen to immature chicks results in cytodifferentiation of tubular gland cells (1), which comprise up to 90% of the cells in the magnum portion of laying hen oviduct, thereby providing a very interesting model for investigating whether gene rearrangement could be one of the mechanisms involved in cell differentiation. With the advent of *in vitro* Recombinant DNA technology it is possible to undertake studies aimed at understanding, in molecular terms, how these hormonal regulations operate. This is mainly why we initiated a detailed analysis of the organization of the ovalbumin gene. Unexpectedly, we found that, as for many other eukaryotic genes (for references see 6-9), the ovalbumin gene is split (7). Very rapid progress has been made since this first report, and the organization

ISBN 0-12-068350-4

of the complete ovalbumin gene is now elucidated (8, 10-18). We will here first summarize our present knowledge of the ovalbumin gene structure, discuss the possible length of the ovalbumin transcription unit and provide evidence that the ovalbumin primary RNA transcript is colinear with the gene. We will next present some results which strongly suggest that gene triplication has occurred in the ovalbumin region during the course of evolution.

STRUCTURE OF THE SPLIT OVALBUMIN GENE

The 1872 nucleotides of ovalbumin mRNA (ov-mRNA) (see Fig. 1c, and Ref. 19, 20) are encoded for by a chicken genomic DNA region of about 7.7 kb which contains a leader-coding region (17) and 7 mRNA coding regions (exons 1 to 7) separated by 7 introns (intervening sequences) A to G. A schematic representation of this structure is given in Fig. 1b, which corresponds to the restriction enzyme map of the ovalbumin region inserted in a λ Charon 4A clone (λC4-ov5) (14). The split organization of the ovalbumin gene is visualized in Fig. 2 as a hybrid DNA-RNA molecule between ov-mRNA and the DNA of λC4-ov5 clone. The leader-coding region and the 7 exons (1 to 7) (thicker DNA-RNA hybrid regions) are separated by 7 single-stranded DNA loops (A to G) which correspond to the introns. The sizes of the leader-coding region (L) and of the exons 1 to 7 which are known from sequencing studies (14, 17, 20) are 47, 185, 51, 129, 118, 143, 156 and 1043 nucleotides, respectively (see Fig. 1c). It is unknown at present whether these exonic RNA domains could correspond to protein domains (for a discussion of this problem, see Ref. 8, 42). From electron microscopic measurements the sizes of the 7 introns A to G are 1560 ± 147, 238 ± 45, 601 ± 35, 411 ± 58, 1029 ± 71, 323 ± 36, 1614 ± 88 bp, respectively. For introns B, C and D which have been sequenced, these electron microscopic measurements are in very good agreement with the sequence results which give 251, 582 and 401 bp, respectively (C. Benoist, R. Breathnach and K. O'Hare, unpublished results). From all of these results, the length of the ovalbumin gene from the region coding for the 5' end of the ov-mRNA to the region coding for its 3' end is about 7.7 kb, approximately four times longer than the size of the mature mRNA.

In no case have we found a different arrangement of the ovalbumin split gene, when we compared the genomic DNA of cells in which the gene is expressed (for example, hen oviduct) with that of cells in which the gene is not transcribed (for instance, erythrocyte). This indicates that extensive gene rearrangement, such as found during lymphocyte differen-

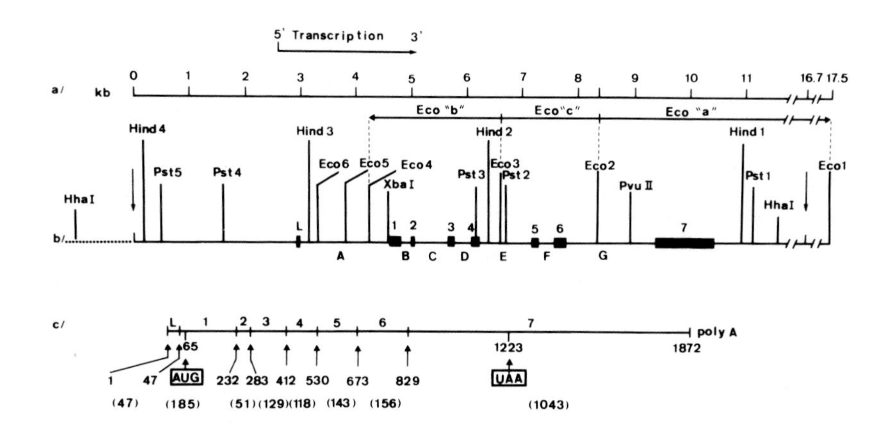

FIGURE 1. Organization of the ovalbumin gene.
a) Scale in kilobase pairs (kb) for Fig. 1b.
b) Localization of the exons (1 to 7, heavy lines) and introns (A to G) of the ovalbumin gene in λC4-ov5 (see Ref. 14) within and upstream from the Eco RI fragments Eco "a", Eco "b" and Eco "c" (8, 17, 18). The horizontal dotted line represents λ Charon 4A sequences. The arrows indicate the limits of the 16.7 kb genomic DNA fragments inserted in λC4-ov5. L corresponds to the location of the leader-coding sequences as determined by electron microscopy and DNA sequencing (see Ref. 14). The Eco, Pst, Hind and Hha I sites correspond to Eco RI, Pst I, Hind III and Hha I restriction enzyme sites (taken from references 8, 12 and 14).
c) Schematic representation of ov-mRNA. The total length, the position of the AUG and UAA codons are taken from references 17, 19 and 20. L refers to the leader sequence and 1 to 7 correspond to the 7 domains of ov-mRNA coded by exons 1 to 7. The arrows indicate the limits of these domains, assuming that the splicing events obey to the "GT-AG rule" (see Ref. 17). Numbers in parentheses indicate the length (in nucleotides) of the domains.

tiation (21), is not involved in the mechanisms leading to the expression of ovalbumin during differentiation. It is interesting that none of our studies (7, 8, 13, 14) have provided evidence that the ovalbumin intronic sequences are repeated in their entirety in the chicken genome. This may have some bearing on the mechanisms which have led to the split gene organization. In contrast, we have found that some amino-

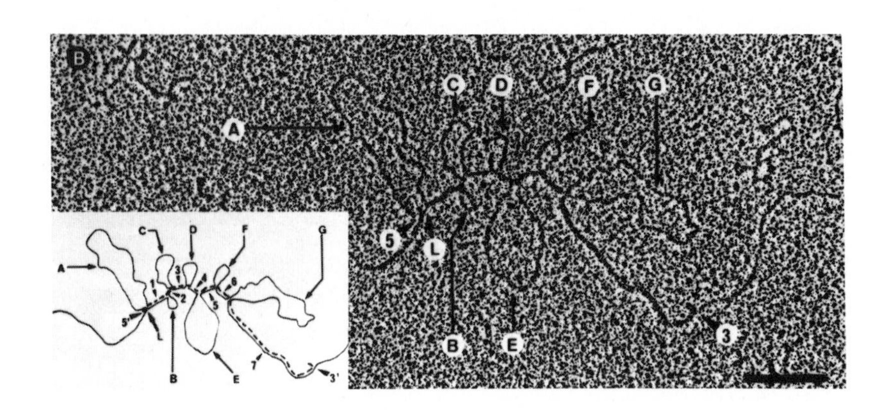

FIGURE 2. Electron microscopy of an RNA-DNA hybrid molecule between ovalbumin mRNA and the DNA of clone λC4-ov5 (see text and Ref. 14). In the line drawing, ov-mRNA is represented by a dashed line, whereas the solid line corresponds to single-stranded DNA. 5' and 3' arrow heads : 5' and 3' ends of ov-mRNA, respectively. 1 to 7 (the thicker DNA-RNA hybrid regions) correspond to the ovalbumin gene exons, whereas A to G correspond to the 7 introns. L is the hybrid region corresponding to the leader-coding sequences. The DNA-RNA hybrid region is 1893 ± 114 bp long. The lengths of intronic loops A to G and of the exonic mRNA-DNA hybrid segments 1 to 7 are (in nucleotides or base pairs) : 1560 ± 147 (A), 238 ± 45 (B), 601 ± 35 (C), 411 ± 58 (D), 1029 ± 71 (E), 323 ± 36 (F), 1614 ± 88 (G), 197 ± 34 (1), 85 ± 14 (2), 164 ± 25 (3), 143 ± 18 (4), 154 ± 25 (5), 201 ± 28 (6) and 1097 ± 51 (7), respectively. There is about 7.7 kb between the 5' end of loop A and the 3' end of exon 7. The bar represents 0.1 μm.

acid-coding regions of the ovalbumin gene are repeated elsewhere in the chicken genome (see Ref. 8). In fact our recent studies suggest that gene triplication could have occurred in the ovalbumin region during the course of evolution (see below).

THE OVALBUMIN TRANSCRIPTION UNIT : EVIDENCE FOR A COMPLETE COLINEAR RNA TRANSCRIPT AND PROCESSING INTERMEDIATES.

Our studies have established that about 7.7 kb separate the genome region coding for the 5' and 3' extremities of ov-mRNA. Recent *in vivo* (22) and *in vitro* (23) studies on primary transcripts from the major late transcription unit of Adeno-virus-2 strongly suggest that the cap site and the transcription initiation site are coincident and that the initiating residues of the primary transcript are conserved in mRNA as the 5' capped terminus. If the situation is identical in the case of ovalbumin, the minimal size of the ovalbumin transcription unit (Ov-TU) should be about 7.7 kb. A complete colinear transcript should then be in the range of 7700 nucleotides. The recent studies of Roop et al. (24) have provided evidence for ovalbumin RNA transcripts of at least 7800 nucleotides in length. The results shown in Fig. 3 confirm this result and indicate that the length of the largest RNA found in total hen oviduct RNA (lane 2) is in fact about 7900 nucleotides. This is in very good agreement with our estimate of the Ov-TU, when the existence of a poly A-tail is taken into account. This 7900 nucleotide-long molecule is indeed polyadenylated, since it is found in the RNA fraction which is retained on oligodT-cellulose (dt-RNA, lanes 1 and 5). Evidence that this 7900 nucleotide long RNA molecule contains transcripts of both exonic and intronic sequences is provided by its hybridization to both exonic (double-stranded cDNA, ds-cDNA, lanes 1, 2 and 5) and intronic (introns E or G, lanes 4 and 3, respectively) probes. Other hybridization results (not shown) indicate that this RNA molecule hybridizes also to a probe specific for intron A. In addition, electron microscopic studies have provided direct evidence that laying hen oviduct RNA contains RNA molecules which hybridize to the totality of the Ov-TU (F. Perrin, unpublished). There is therefore little doubt that the Ov-TU (as defined above) is transcribed in its totality yielding a complete colinear RNA transcript which can be polyadenylated.

Oviduct oligodT RNA contains many discrete RNA species which hybridize to the exonic ds-cDNA probe with molecular weights intermediate between that of the largest 7900 nucleotide RNA and of ovalbumin mRNA (the very heavy dark spot centred at a position corresponding to 1900 nucleotides - Fig. 3, lanes 1 and 5). As many as 9 individual bands can be clearly discerned on the original autoradiograms, with 3 main bands of 5050, 3250 and 2860 nucleotides in length, in addition to the 7900 nucleotide band. To demonstrate that these RNA molecules do in fact correspond to stepwise excisions of the intronic transcripts, we have hybridized them to

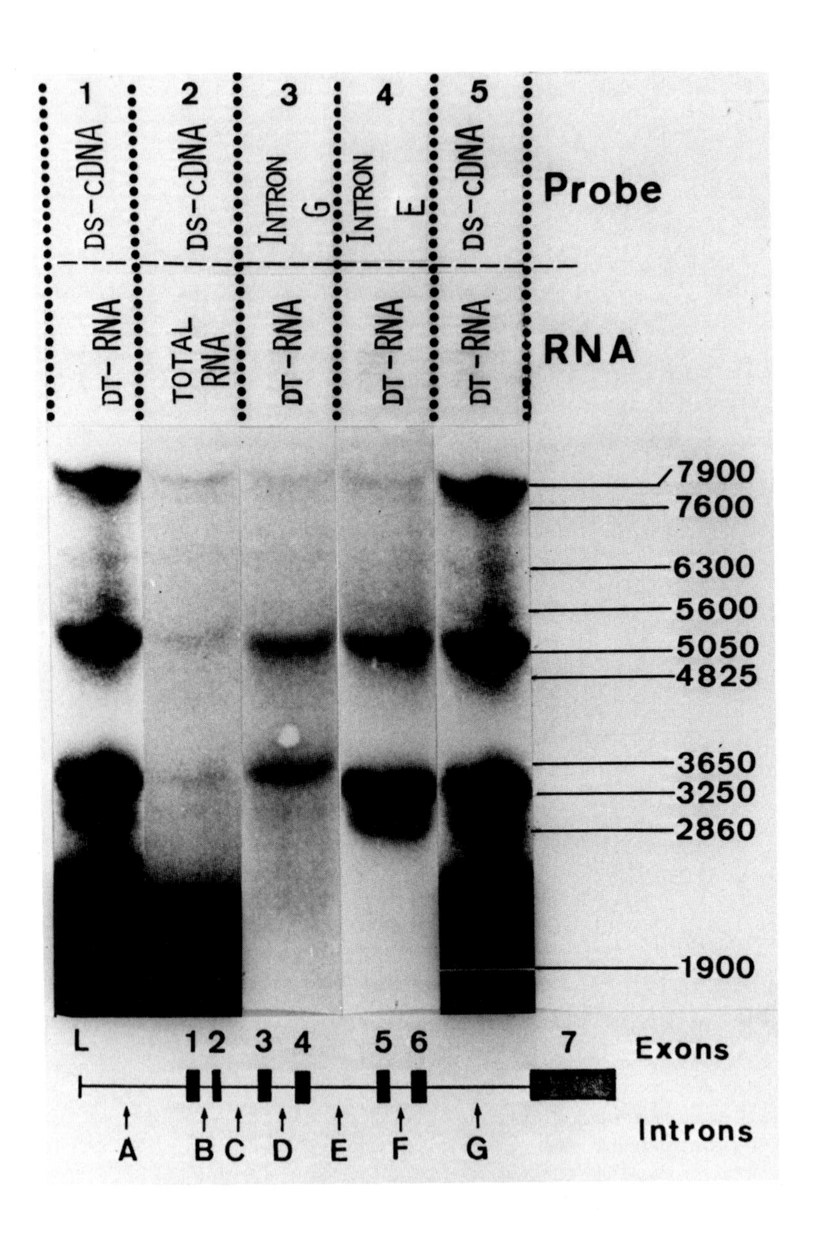

FIGURE 3. Identification of a complete colinear transcript of ovalbumin gene and of processing intermediate RNA species in total laying hen oviduct RNA (total RNA) and in oviduct RNA purified on oligodT-cellulose (dT-RNA). The purification of RNA, its fractionation by agarose gel electronphoresis in the presence of Methyl Mercuric Hydroxyde and its transfer to

specific intronic probes (Fig. 3, lanes 3 and 4). It is clear that some RNA molecules hybridize to both intron E and G probes (for instance the 5050 band), whereas other hybridize specifically to the intron E probe (3250 and 2860 bands) or to the intron G probe (the 3650 band). As expected ov-mRNA is not revealed by these intronic probes. These results show that one can find RNA molecules smaller than the complete colinear transcript, which contain both intron E and G transcripts, but also RNA molecules in which either the intron E or the intron G transcript has been "spliced out". In addition, the intronic transcripts do not appear to be necessarily removed in the same order as they were transcribed, since the intron E transcript is still present in the smaller 3250 and 2860 molecules, whereas the intron G transcript has already been removed. In any case, these results strongly suggest that the multiple hybridizing RNA bands which are seen on the autoradiogram correspond to intermediate species of RNA generated during the processing of the initial poly(A)$^+$ complete colinear RNA transcript.

To study in more detail these splicing steps, we have isolated the oviduct RNA molecules which contain specific intronic transcripts. Total laying hen oviduct RNA was hybridized to either an intron E or an intron G DNA filter, eluted, fractionated on agarose gel in the presence of Methyl Mercuric Hydroxide, transferred to DBM-paper and hybridized either to a probe representative of all of the ovalbumin gene sequences (Gene in Fig. 4) or to a probe specific for intron G. The results are shown in Fig. 4. In agreement with the result shown in Fig. 3, lane 4, 4 major RNA species (7900, 5050, 3250 and 2860 nucleotides in length) are specifically retained on the intron E filter (Fig. 4, lane 2). However, and in keeping with the results presented in Fig. 3 (lanes 3 and 4), only two of these RNA species (7900 and 5050) hybridize to the intron G probe (Fig. 4, lane 3), indicating that the intron G transcript has been exicised from the lower molecular weight species, whereas the intron E transcript is still present.

DBM-paper (31) will be described elsewhere. The immobilized RNA species were hybridized either to the [^{32}P]-labelled "Hhaov" ovalbumin double-stranded DNA (ds-cDNA, see Ref. 7) or to nick-translated [^{32}P]-labelled intron E and G DNAs (cloned in pBR322), as indicated in the figure. 20 μg of total laying hen oviduct RNA (lane 2) or of oligodT-cellulose purified RNA were electrophoresed (lanes 1, 3, 4 and 5) as indicated in the figure. Further experimental details will be published elsewhere. A schematic representation of the ovalbumin gene is shown below the autoradiogram.

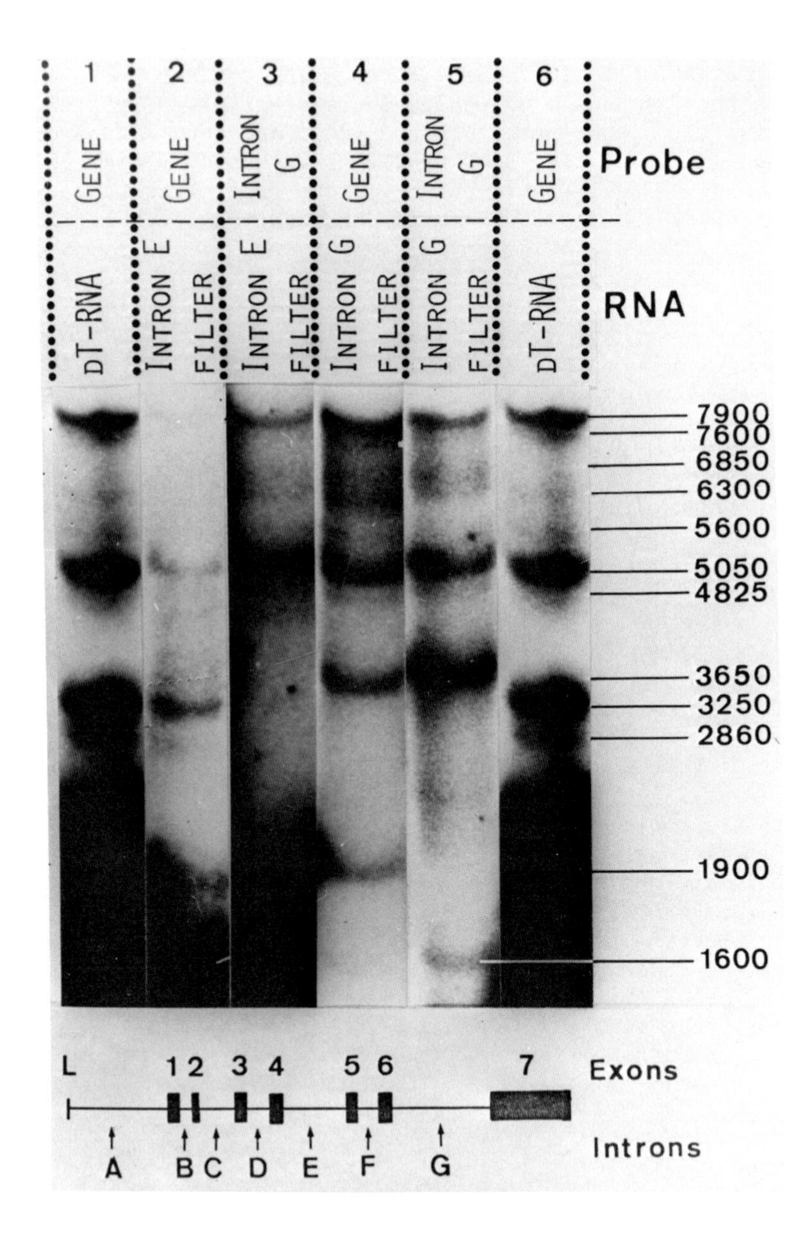

FIGURE 4. Identification of a complete colinear transcript of ovalbumin gene and of processing intermediate RNA species in total laying hen oviduct RNA purified by hybridization to specific intronic DNA. Total laying hen oviduct RNA was purified by hybridization either to intron E DNA (Intron E filter RNA) or to intron G DNA (Intron G filter RNA) immo-

When the laying hen oviduct RNA was fractionated by hybridization with an intron G filter and hybridized to a total ovalbumin gene probe (lane 4) at least 6 bands (7900, 6850, 6300, 5050, 4825, 3650) were revealed (the band at 1900 corresponds to a minor contamination by ov-mRNA). Not surprisingly the same RNA bands are found after hybridization to a probe specific for intron G (Fig. 4, lane 5), but a faint band at 1600 which was barely visible in lane 4 is now clearly seen : this RNA is an excellent candidate for the "spliced out" transcript of intron G, the length of which is precisely 1600 bp (see aobve).

To further support the conclusion that the multiple bands which are seen on the autoradiograms (below the 7900 complete colinear transcript) correspond to intermediate species in the processing of this initial transcript, we have analyzed by electron microscopy the DNA-RNA hybrid molecules which can be formed between the ovalbumin genomic DNA (clone λC4-ov5, see above) and the RNA species purified by hybridization to an intron DNA filter. The results are shown in Fig. 5. A control mature ov-mRNA-DNA hybrid molecule is presented in panel A. The single-stranded DNA loops B to G correspond to intronic regions whose transcripts are not anymore present in the mature ov-mRNA (loop A is not seen in this hybrid molecule as in other molecules presented in panel B and C, because the 47 bp DNA-RNA hybrid region between the mRNA and the leader-coding region (see above) is markedly unstable under the hybridization conditions used in this study). Panels B and C show two hybrid molecules between λC4-ov 5 DNA and RNA molecules containing sequences complementary to intron E. Examination and measurements of the hybrid molecule of panel B indicate that the transcripts of introns A, B, C, D and G have been removed, whereas the transcripts of introns E and F are still linked to the ov-mRNA sequences. A similar hybrid molecule, but from which the intron D transcript has not yet been excised, is shown in panel C. It is noteworthy that in both cases, and in agreement with the biochemical results presented in Fig. 3 and 4, one finds RNA molecules containing the intron E transcript, whereas the intron G transcript has already been "spliced out". The presence, at the 3' end of

---

bilized on DBM-paper (31). After elution the RNA was electrophoresed on agarose gels as described in legend to Fig. 3 and hybridized either to nick-translated [^{32}P]-labelled ovalbumin gene (Gene probe, lanes 1, 2, 4 and 6) or intron G DNA (Intron G probe, lanes 3 and 5) probes. Further experimental details will be published elsewhere. A schematic representation of the ovalbumin gene is shown below the autoradiogram.

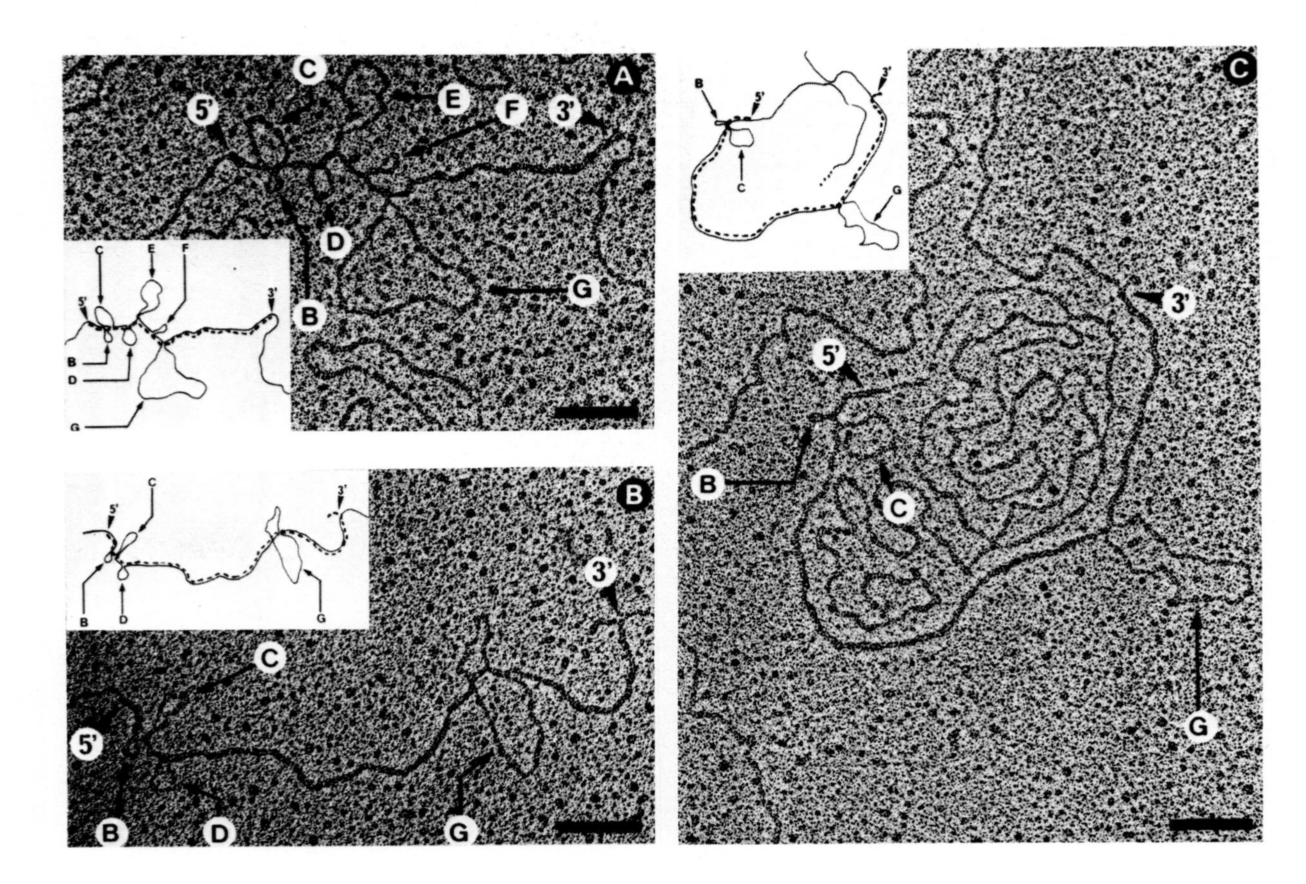
A
5'
C
E
F
3'
D
B
G
B
5'
C
3'
D
G
C
B
5'
C
3'
G

the molecule, of a poly A tail (100 to 200 nucleotides in length), much longer than that associated to mature ov-mRNA (panel A), is also remarkable. There is no doubt that one is here dealing with poly(A)$^+$ RNA species which are processing intermediates between the complete colinear ovalbumin gene transcript and the mature ov-mRNA.

From all of these results we conclude : 1) the ovalbumin gene is transcribed as a complete colinear RNA molecule which is very rapidly polyadenylated. A similar conclusion has been reached in the case of both Adenovirus-2 (25) and globin (26) initial RNA transcripts for which poly(A) addition precedes splicing. Whether, as it is the case for Adenovirus-2 late major RNA transcription (27), the RNA polymerase reads past the site where poly(A) is added and whether the 3' end which serves as the poly(A) addition site results from endonucleolytic cleavage of a longer primary transcript are at present unknown. It should however be pointed out that no discrete RNA transcripts longer than 7900 nucleotides were found among the RNA molecules which were selected by hybridizing total oviduct RNA to intron E or intron G filters (Fig. 4, lanes 2-5); 2) the processing of the complete colinear transcript is a stepwise mechanism involving the generation of a series of intermediates which differ in structure by an additional step of splicing. However, our data suggest that RNA splicing can occur through multiple pathways involving several combinations of intermediates, rather than being a direct RNA:RNA splicing of the ultimately conserved mRNA domains in a specific sequential temporal order in the 5' to 3' direction.

FIGURE 5. Electron micrographs of hybrid molecules between the ovalbumin gene integrated in λC4-ov5 DNA (see legend to Fig. 2 and text) and processing intermediate RNA species. (A): A control ov-mRNA-DNA hybrid molecule (see Fig. 2 for the symbols and line drawing); (B) and (C): RNA molecules were purified by hybridization to an intron E filter (see legend to Fig. 4). Length measurements on the hybrid molecule shown in panel B gave the following values : 230, 550, 351, 1579 for introns B, C, D and G, respectively; 204, 76, 128 and 1047 bp for exons 1, 2, 3 and 7, respectively (see line drawing on Fig. 2); 1850 bp for the hybrid region between intron D and G. The length measurements on the hybrid molecule shown in panel C were as follows : 244, 578 and 1488 bp for introns B, C and G, respectively; 220, 70 and 1100 bp for exons 1, 2 and 7, respectively (see line drawing on Fig. 2); 2400 bp for the hybrid region between introns C and G. The bar represents 0.1 μm. Further experimental details will be published elsewhere.

A similar situation may exist in the case of Adenovirus-2 (28), although evidence to the contrary has been reported (29); 3) the finding of a discrete complete colinear transcript and of discrete processing intermediates supports very strongly the concept that mature ov-mRNA, as other cellular mRNAs, like globin mRNA (30), is derived from an initial transcript which contains covalently linked intronic and exonic transcripts; 4) the excision of a given intron could be a single-step event, since at least in one case (intron G - see above and Fig. 4, lane 5), we have found evidence that a full-length intron transcript is "spliced out ".

GENE TRIPLICATION IN THE OVALBUMIN REGION.

Exploring the surrounding of the ovalbumin gene, Royal et al. (32) have recently isolated two genomic DNA fragments which include the ovalbumin gene and cover about 46 kb of the chicken genome. One of the cloned DNA fragments, about 30 kb long and present in a cosmid (pAR2), contains two other genes, X and Y which were identified by electron microscopic examination after hybridization of pAR2 DNA with total poly(A)$^+$ - RNA from laying hen oviduct (see Fig. 6). These genes are located upstream (with respect to the polarity of transcription) from the region coding for the 5' end of ov-mRNA and the gene order is 5'-X-Y-ovalbumin-3', the three genes being transcribed from the same strand. X and Y RNA are about 2400 and 2020 nucleotides long, respectively. Only the 3' three-fourth of X-RNA and the 5' third of ov-mRNA are coded by sequences present in pAR2, whereas Y-RNA is coded in its entirety in the genomic fragment cloned in pAR2. The 3' end of gene Y lies about 11.5 kb upstream from the ov-mRNA leader-coding sequence and there is about 5.5 kb between the 3' end of the X gene and the 5' end of Y gene. Although the 3 genes and their RNA transcripts are different, Royal et al. (32) found that they share partial sequence homologies. It was also noted that the split gene structures of Y and ovalbumin genes present similarities, suggesting that these genes could have arisen by at least partial duplications from a common ancestor. This possibility prompted us to further study the organization of X and Y genes, especially as the expression of these genes is under the same hormonal control as that of the ovalbumin gene (32), although X and Y RNAs are present at a lower concentration than ov-mRNA in laying hen oviduct cells.

The complete X gene which was not present in cosmid pAR2, was subsequently cloned from a λ Charon 4A chicken library (λC4-X1 clone, manuscript in preparation). Fig. 7 shows

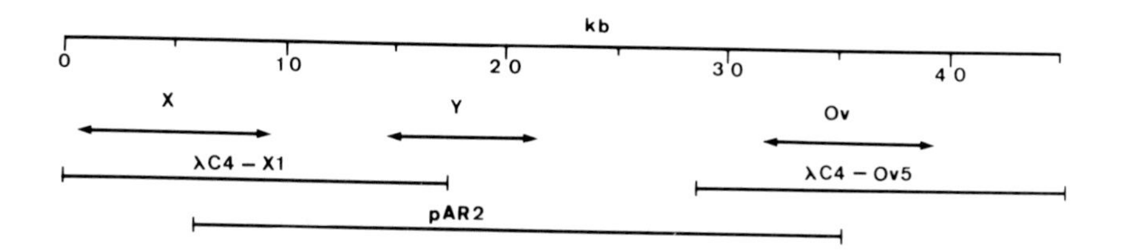

FIGURE 6. Map of the the chicken genome in the ovalbumin region. kb = kilo base pairs; X, Y, Ov : location of X, Y and ovalbumin genes, respectively; λC4-X1, λC4-ov5 : location of the λ Charon 4A clones containing the complete X and ovalbumin genes, respectively; pAR2 : location of the cosmid clone pAR2 (see text).

electron micrographs of DNA-RNA hybrid molecules between ov-mRNA (A), Y-RNA (B) and X-RNA (C) and the cloned ovalbumin, Y and X genomic DNAs, respectively. The split gene organization of X and Y genes is strikingly similar to that of the ovalbumin gene. In all three cases there are 7 introns (A to G) separating the RNA-coding sequences. In all three cases the first intron A separates a very short RNA-coding sequence, too small to be visualized by electron microscopy, from the neighbouring exon 1 (see line drawings in Fig. 7). Whether, in the case of X and Y genes, the short RNA coding sequences are equivalent to the ovalbumin mRNA leader-coding sequence (L, see above) is an open question which can be answered only by DNA sequencing studies. It is remarkable that not only the number of exons is identical for all three genes, but also that, within the accuracy of the electron microscopic measurements, the lengths of a given exon is the same for all three genes, with the exception of the last exon 7 which is the longest and, in the case of the ovalbumin gene, includes the sequences which code for the untranslated region of the mRNA (see above and below).

These striking similarities which strongly suggest that the three genes could have evolved from a common ancestor gene, have been confirmed by DNA sequencing studies. It has been shown (manuscript in preparation) that there are marked analogies between the DNA sequences of exons 5 and 6 of the ovalbumin gene and the corresponding exons of the X gene, as defined by the electron microscopic studies. For instance, comparison of the exon 6 sequences of ovalbumin and X genes indicates that both exons have the same lengths, the exon 6 -

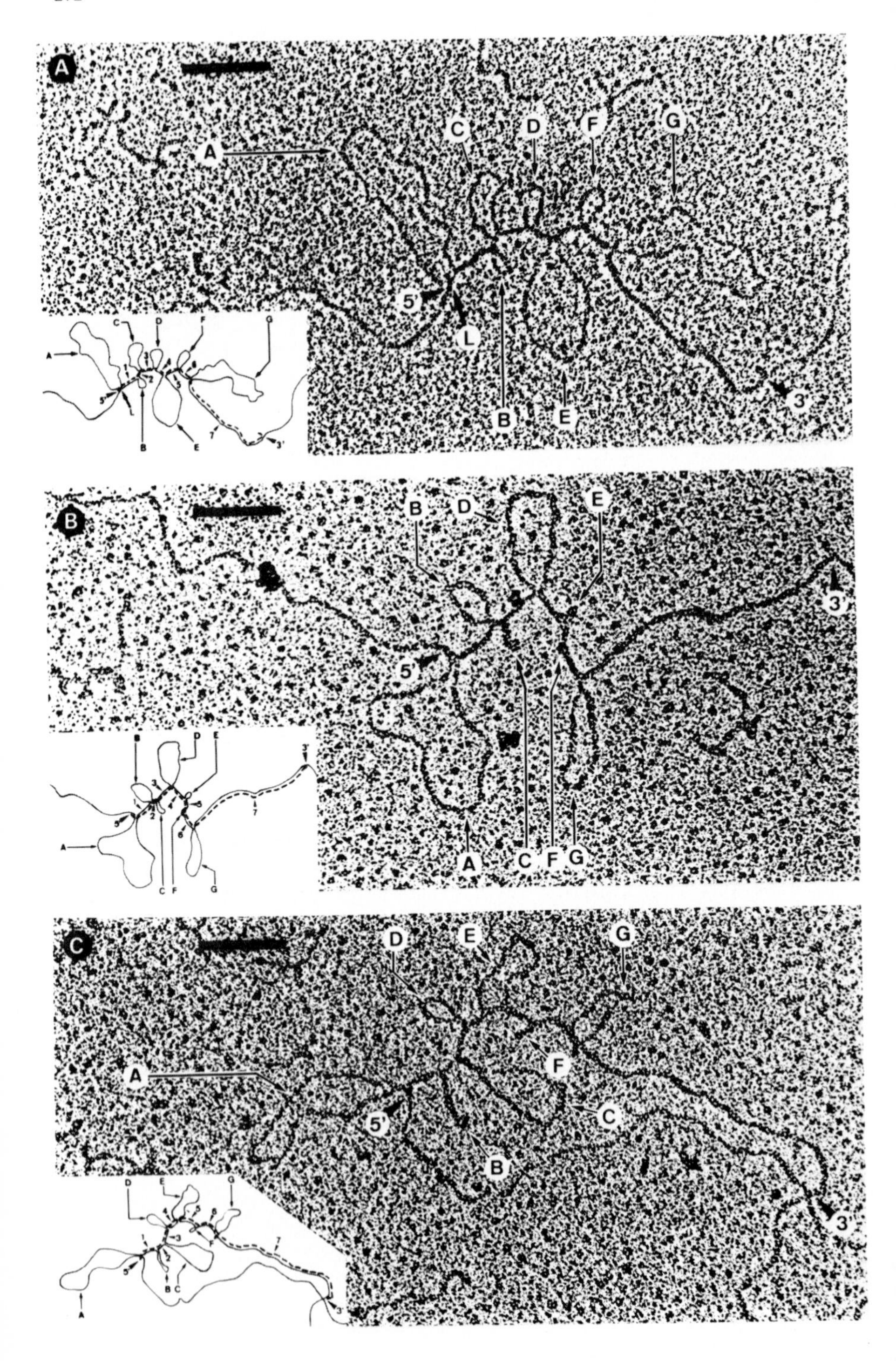
A
B
C
D
E
F
G
L
5'
3'

intron F and G junction sequences (17, 33) being highly conserved and similarly located. In addition, the amino-acid reading frame is the same for exon 6 of the two genes, but due to many base changes, 14 out of the 52 amino-acids which are encoded in the ovalbumin exon 6 are different in the Y gene. Fig. 8 shows a comparison of the DNA sequence of the ovalbumin and X genes within exon 7, in the region which codes for nucleotides 970 to 1280 of ov-mRNA and contains the sequence coding for the ov-mRNA UAA termination codon (boxed in Fig. 8). The two sequences present clear similarities. When the sequences are properly aligned, 184 out of the 252 base pairs of this amino-acid coding region of ovalbumin exon 7 are conserved in the exon 7 of gene X. In the two cases there is only one possible reading frame for coding amino-acids and possible deletions (see the 1120 to 1130 region) compensate each other to restore the same reading frame for the two genes. It is striking that the termination UAA codon is located in the same place in both exons. Other sequencing data (not shown) indicate that the boundary between intron G and exon 7

FIGURE 7. Electron microscopy of DNA-RNA hybrid molecules between (A) ovalbumin mRNA and the DNA of clone λC4-ov5 (see Fig. 2); (B) Y-RNA and the region of pAR2 cosmid (see text and Fig. 6) which contains the complete gene Y and was recloned in pBR322; (C) X-RNA and the DNA of clone λC4-X1 which contains the complete X gene (see text). In the line drawings, the RNA is represented by a dashed line, whereas the solid line corresponds to single-stranded DNA. 5' and 3' arrow heads, 5' and 3' ends of the RNAs, respectively. 1 to 7 (the thicker DNA-RNA hybrid regions) correspond to the exons, whereas A to G correspond to the introns. L (in A) is the hybrid region corresponding to the ovalbumin leader-coding sequences (see text and Fig. 2). Length measurements are as follows : (A) Ovalbumin gene : see Fig. 2; (B) Y gene, the lengths of intronic loops A to G and the exonic DNA-RNA hybrid segments 1 to 7 are (in nucleotides or base pairs) : 1729 ± 180 (A), 492 ± 130 (B), 303 ± 88 (C), 812 ± 135 (D), 206 ± 86 (E), 55 ± 20 (F), 890 ± 34 (G), 218 ± 71 (1), 93 ± 30 (2), 165 ± 40 (3), 153 ± 33 (4), 167 ± 35 (5), 183 ± 41 (6), 1320 ± 148 (7), respectively; (C) X gene, the lengths of intronic loops A to G and of the exonic DNA-RNA hybrid segments 1 to 7 are (in nucleotides or base pairs) : 1682 ± 341 (A), 624 ± 111 (B), 1275 ± 199 (C), 506 ± 173 (D), 887 ± 154 (E), 234 ± 73 (F), 808 ± 100 (G), 201 ± 70 (1), 106 ± 56 (2), 146 ± 33 (3), 130 ± 17 (4), 168 ± 24 (5), 188 ± 46 (6), 1530 ± 80 (7). The bar represent 0.1 μm. Further experimental details will be published elsewhere.

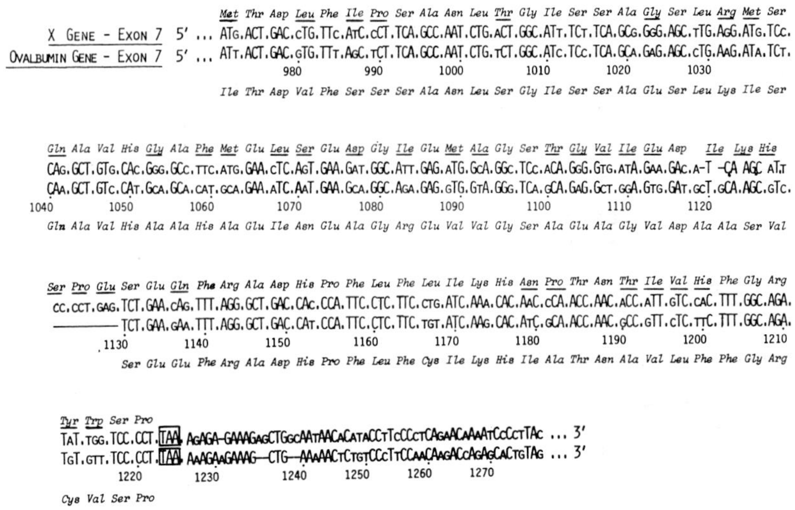

```
                               Met Thr Asp Leu Phe Ile Pro Ser Ala Asn Leu Thr Gly Ile Ser Ser Ala Gly Ser Leu Arg Met Ser
      X GENE - EXON 7  5' ... ATg.ACT.GAC.cTG.TTc.AtC.cCT.TCA.GCC.AAT.CTG.aCT.GGC.ATt.TCt.TCA.GCg.GgG.AGC.tTG.AgG.ATg.TCc.
OVALBUMIN GENE - EXON 7  5' ... ATt.ACT.GAC.gTG.TTt.AgC.tCT.TCA.GCC.AAT.CTG.tCT.GGC.ATc.TCc.TCA.GcA.GaG.AGC.cTG.AaG.ATa.TCt.
                                         980         990        1000        1010        1020        1030
                               Ile Thr Asp Val Phe Ser Ser Ser Ala Asn Leu Ser Gly Ile Ser Ser Ala Glu Ser Leu Lys Ile Ser

 Gln Ala Val His Gly Ala Phe Met Glu Leu Ser Glu Asp Gly Ile Glu Met Ala Gly Ser Thr Gly Val Ile Glu Asp  Ile Lys His
 CAg.GCT.GTg.CAc.Ggg.GCc.ttc.atg.GAA.cTC.AgT.GAA.Gat.GGC.Att.GAG.aTG.GcA.GGc.TCc.aCA.GgG.Gtg.atA.Gaa.GAc.a-T -CA AGC ATt
 CAa.GCT.GTc.CAt.Gca.GCa.cat.gca.GAA.aTC.AaT.GAA.Gca.GGC.Aga.GAG.gTG.GtA.GGg.TCA.gCA.GaG.Gct.ggA.Gtg.GAt.gcT.gCA.AGC.gTc.
1040      1050      1060      1070      1080      1090      1100      1110      1120
 Gln Ala Val His Ala Ala His Ala Glu Ile Asn Glu Ala Gly Arg Glu Val Val Gly Ser Ala Glu Ala Gly Val Asp Ala Ala Ser Val

 Ser Pro Glu Ser Glu Gln Phe Arg Ala Asp His Pro Phe Leu Phe Leu Ile Lys His Asn Pro Thr Asn Thr Ile Val His Phe Gly Arg
  cc.cct.gag.TCT.GAA.cAg.TTT.AGG.GCT.GAC.CAc.CCA.TTC.CTC.TTC.ctg.ATC.AAa.CAC.AaC.cCA.ACC.AAC.aCC.aTT.gTC.caC.TTT.GGC.AGA.
  ———————————TCT.GAA.gAa.TTT.AGG.GCT.GAC.CAt.CCA.TTC.CTC.TTC.tgt.ATC.AAg.CAC.AtC.gCA.ACC.AAC.gCC.gTT.cTC.ttC.TTT.GGC.AGA.
          1130      1140      1150      1160      1170      1180      1190      1200      1210
             Ser Glu Glu Phe Arg Ala Asp His Pro Phe Leu Phe Cys Ile Lys His Ile Ala Thr Asn Ala Val Leu Phe Phe Gly Arg

 Tyr Trp Ser Pro
 TaT.tgg.TCC.CCT.[TAA] AgAGA-GAAAGagCTGgcAAtAACaCataCCtTcCCctCAgaACaAAtCcCctTAc ... 3'
 TgT.gtt.TCC.CCT.[TAA] AaAGAaGAAAG—CTG—AAaAACtCtgtCCcTtCCaaCAagACcAgAgCaCtgTAg ... 3'
      1220      1230        1240      1250      1260      1270
 Cys Val Ser Pro
```

FIGURE 8. Comparison of DNA sequences (non sense strand) of part of the ovalbumin exon 7 and the corresponding region of exon 7 of gene X. The ovalbumin sequences which correspond to the ovalbumin mRNA nucleotides 971 to 1278 are taken from Ref. 20. The numbers correspond to the mRNA nucleotides starting from the 5' end. The ovalbumin amino-acid sequence is given below the DNA sequence. The gene X sequences are aligned above the ovalbumin sequences to show their similarities. The amino-acid sequence of the putative protein X is given above the DNA sequence. The amino-acids which are different from those of ovalbumin are underlined. Bases which are identical in ovalbumin and X genes are in higher cases, whereas those which are different are in lower cases. Further experimental details will be published elsewhere.

is found in the same place in the ovalbumin and X genes and that the intron-exon junction sequences are very similar for the two genes. Therefore, the length difference between ov-mRNA and X-RNA appears to be mainly due to the presence of a shorter 3'-untranslated region in ov-mRNA. In this respect, it is interesting to note that the ovalbumin and X sequences are very rapidly markedly different beyond the termination codon, which is confirmed by the lack of cross-hybridization between this region of the ovalbumin and X genes, whereas cross-hybridization is observed between the translated region of the ovalbumin exon 7 and the corresponding sequence in X gene (not shown, manuscript in preparation). (It should be pointed out that, although the X and Y RNAs are polyadenylated, we have not yet formally demonstrated that they are translated *in vivo* or translatable *in vitro*).

If the length (with the exception noted above) of a given exon is highly conserved in the three genes, the length of a given intron is highly variable from one gene to another. The most striking difference concerns introns E and F which are much smaller in the Y gene than in ovalbumin and X genes. This high degree of variation of the length of a given intron is correlated with a complete lack of cross-hybridization between the sequences of a given intron of the ovalbumin gene and the corresponding introns of X and Y genes.

From all of our present results it appears very likely that the ovalbumin, X and Y genes have evolved by triplication of an ancestor gene. In this respect these genes are similar to the globin genes (34-38). Although our results are still fragmentary, several important conclusions can be drawn from the structural organization of these genes. First, this ancestor gene was already split and very likely contained the same number of exons and introns as the present day ovalbumin, X and Y genes, which are all split in an identical manner. In this respect it would be interesting to investigate how the ovalbumin genes of other birds or even more distant species are organized to find out when, during evolution, the ancestor gene was triplicated. Second, the creation of new introns is an exceptional event, since, none has been added to the three genes since the duplication events. Third, the high degree of conservation of the exon lengths clearly reflects the existence of a selective pressure to conserve them. Alternatively, this could be due to difficulties in evolving new splicing signals. Fourth, it is clear that most of the intron sequences are not under selective pressure and are not required in order to splice out the intronic transcripts, since there are marked variations in the size of a given intron, although none has disappeared during the course of evolution. However, in agreement with the results of previous sequence studies of β globin genes of mouse and rabbit (39), the intronic sequences of ovalbumin and X genes which are close to a given exon-intron junction appear to be conserved, suggesting that they play a role in splicing. On the other hand, the strikingly reduced size of intron F in gene Y suggests that intron elimination accompanied by exon fusion is a process which could occur during evolution. Fifth, the position of the poly(A) addition-site has changed during evolution, since it is closer to the UAA termination codon in ovalbumin than in X gene. This observation suggests that one of the main purposes of a 3'-untranslated region is to link the end of the amino-acid-coding region to the poly(A) addition site. In some respect, such an untranslated region could be visualized as an unexcised intron transcript.

CONCLUDING REMARKS

With the elucidation of the organization of the ovalbumin gene and the characterization of its transcriptional unit, new possibilities are opened to investigate the molecular mechanisms involved in the hormonal control of its transcription. For example, direct binding experiments of steroïd hormone-receptor complexes to the region which contains putative promoter sequences (14, 33) are now possible. The finding of the two closely related genes X and Y, whose expressions are both under hormonal control should be very useful to identify the DNA sequences which could be involved in the control of transcription by steroïd hormones, and more generally in the control of RNA transcription and processing.

Besides the transcriptional and hormonal regulatory aspects, the finding of three highly split genes which have evolved from a common ancestor, allows one to draw some conclusions on the origin and evolution of split genes. It is clear from our findings that nature has chosen to create these new genes (whose proteins should differ widely in amino-acid sequence) by duplication of an already split gene without any modification of the splicing pattern. Our present studies eliminate the possibility that the ovalbumin, X and Y genes could have been created by new combination of preexisting exonic domains (40-42). How the ancestor split gene was created, is obviously unknown, but it should be possible to trace its original organization by analysing the structure of ovalbumin genes of distant species. Such a study should also reveal whether intron disapperance with concomitant exon fusion could occur in evolution as suggested by the size of introns E and F of gene Y. In the perspective that all genes were originately split (33, 40, 43, 44), such a possibility is very attractive, since it would explain the generation of the unsplit structure of prokayrotic genes as well as of some eukaryotic genes. Obviously, we are at the very beginning of a very long story.

ACKNOWLEGEMENTS

We are greatly indebted to Drs. J. Dodgson, R. Axel, D. Engel and T. Maniatis who provided the "chicken library" which was used to isolate complete ovalbumin and X genes. The technical assistance of C. Wasylyk, J.M. Garnier, M.C. Gesnel, G. Dretzen, E. Sittler, E. Taubert, A. Landmann and B. Boulay is greatly acknowledged. This work was supported by grants to

P. Chambon from the INSERM (CRT 76.5.462 and 76.5.468), the CNRS (ATP 2117) and the Fondation pour la Recherche Médicale Française. J.P. LePennec and K. O'Hare were supported by fellowships from the Université Louis Pasteur.

REFERENCES

1. Oka, T., and Schimke, R.T. (1969) J. Cell. Biol. 43, 123.
2. Palmiter, R.D. (1975) Cell 4, 189.
3. McKnight, G.S. (1978) Cell 14, 403.
4. Bellard, M., Gannon, F., and Chambon, P. (1977) Cold Spring Harbor Symp. Quant. Biol. XLII, 779.
5. Schütz, G., Nguyen-Huu, C., Giesecke, K., Hynes, N.Y., Groner, B., Wurtz, T., and Sippel, A.E. (1977) Cold Spring Harbor Symp. Quant. Biol. XLII, 617.
6. Chambon, P. (1977) Cold Spring Harbor Symp. Quant. Biol. XLII, 1209.
7. Breathnach, R., Mandel, J.L., and Chambon, P. (1977) Nature 270, 314.
8. Mandel, J.L., Breathnach, R., Gerlinger, P., LeMeur, M., Gannon, F., and Chambon, P. (1978) Cell 14, 641.
9. Dawid, I.B., and Wahli, W., Develop. Biol., in press.
10. Garapin, A.C., LePennec, J.P., Roskam, W., Perrin, F., Cami, B., Krust, A., Breathnach, R., Chambon, P., and Kourilsky, P. (1978) Nature 273, 349.
11. Perrin, F., Garapin, A.C., Cami, B., LePennec, J.P., Royal, A., Roskam, W., and Kourilsky, P. (1978) in Proceedings of the Internat. Symp. on Genetic Engineering, pp. 89-98. Elsevier North Holland.
12. Garapin, A.C., Cami, B., Roskam, W., Kourilsky, P., LePennec, J.P., Perrin, F., Gerlinger, P., Cochet, M., and Chambon, P. (1978) Cell 14, 629.
13. LePennec, J.P., Baldacci, P., Perrin, F., Cami, B., Gerlinger, P., Krust, A., Kourilsky, P., and Chambon, P. (1978) Nucleic Acids Res. 5, 4547.
14. Gannon, F., O'Hare, K., Perrin, F., LePennec, J.P., Benoist, C., Cochet, M., Breathnach, R., Royal, A., Garapin, A., Cami, B., and Chambon, P. (1979) Nature 278, 428.
15. Dugaiczyk, A., Woo, S.L.C., Lai, E.C., Mace, M.L. Jr., McReynolds, L., and O'Malley, B.W. (1978) Nature 274, 328.
16. Breathnach, R., Mandel, J.L., Gerlinger, P., Krust, A., LeMeur, M., Gannon, F., and Chambon, P. (1978) Cell 14, 641.
17. Breathnach, R., Benoist, C., O'Hare, K., Gannon, F., and Chambon, P. (1978) Proc. Natl. Acad. Sci. USA 75, 4853.
18. Kourilsky, P., and Chambon, P. (1978) Trends in Biochemical Sciences 3, 244.

19. McReynolds, L., O'Malley, B.W., Nisbet, A.D., Fothergill, J.E., Givol, D., Dields, S., Robertson, M., and Brownlee, G.G. (1978) 273, 723.
20. O'Hare, K. Breathnach, R., Benoist, C., and Chambon, P., in preparation.
21. Hozumi, N., and Tonegawa, S. (1976) Proc. Natl. Acad. Sci. USA 73, 3628.
22. Ziff, E.B., and Evans, R.M., Cell, in press.
23. Manley, J.L., Sharp, P.A., and Gafter, M.L., Proc. Natl. Acad. Sci. USA, in press.
24. Roop, D.R., Nordstrom, J.L., Tsai, S.Y., Tsai, M.J., and O'Malley, B.W. (1978) Cell 15, 671.
25. Nevins, J.R., and Darnell, J.E. (1978) Cell 15, 1477.
26. Ross, J. (1976) J. Mol. Biol. 106, 403.
27. Fraser, N.W., Nevins, J.R., Ziff, E., and Darnell, J.E., J. Mol. Biol., in press.
28. Chow, L.T., and Broker, T. (1978) Cell 15, 497.
29. Berget, S.M., and Sharp, P.A., J. Mol. Biol., in press.
30. Tilghman, S., Curtis, P., Tiemeier, D., Leder, P., and Weissmann, C. (1978) Proc. Natl. Acad. Sci. USA 75, 1309.
31. Alwine, J.C., Kemp, D.J., and Stark, G. (1977) Proc. Natl. Acad. Sci. USA. 74, 5330.
32. Royal, A., Garapin, A., Cami, B., Perrin, F., Mandel, J.L., LeMeur, M., Brégégère, F., Gannon, F., LePennec, J.P., Chambon, P.,and Kourilsky, P., Nature, in press.
33. Chambon, P., Benoist, C., Breathnach, R., Cochet, M., Gannon, F., Gerlinger, P., Krust, A., LeMeur, M., LePennec, J.P., Mandel, J.L., O'Hare, K., and Perrin, F. (1979) in Proceedings of the 11th Miami Winter Symp. "From Gene to Protein : Transfer in normal and abnormal cells", Academic Press, in press.
34. Lawn, R.M., Fritsche, E.F., Parker, R.C., Blake, G., and Maniatis, T. (1978) Cell 15, 1157.
35. Orkin, S.H. (1978) Proc. Natl. Acad. Sci. USA 75, 5950.
36. Tiemeier, D.C., Tilghman, S.M., Polsky, F.I., Seidman, J.G., Leder, A., Edgell, M.H.,and Leder, P. (1978) Cell 14, 237.
37. Leder, A., Miller, H.I., Hamer, D.H., Seidman, J.G., Norman, B., Sullivan, M., and Leder, P. (1978) Proc. Natl. Acad. Sci. USA 75, 6187.
38. Flavell, R.A., Kooter, J.M., DeBoer, E., Little, P.F.R., and Williamson, R. (1978) Cell 15, 25.
39. Van den Berg, J., Van Ooyen, A., Mantei, N., Schamböck, A., Grosveld, G., Flavell, R.A., and Weissmann, C. (1978) Nature 276, 37.
40. Crick, F., Sciene, in press.
41. Gilbert, W. (1978) Nature 271, 501.
42. Blake, C.C.F. (1978) Nature 273, 267.

43. Doolittle, W.F. (1978) Nature 272, 581.
44. Darnell, J.E. (1978) Science 202, 1257.

A COMPARISON OF THE SEQUENCE ORGANIZATION OF THE CHICKEN OVALBUMIN AND OVOMUCOID GENES

Bert W. O'Malley, Joseph P. Stein, Savio L.C. Woo, Achilles Dugaiczyk, James F. Catterall, and Eugene C. Lai

Department of Cell Biology
Baylor College of Medicine
Houston, Texas 77030

INTRODUCTION

In order to understand the mechanism of steroid hormone induction of specific mRNA's, we have for several years been investigating the molecular mechanism involved in the production of the major protein product of the chicken oviduct, ovalbumin (1-3). Even less understood than the regulation of a single gene, however, are the mechanisms by which steroid hormones coordinately control the expression of different genes in a single cell type. In order to gain a better understanding of this important problem, we have recently turned our attention to the isolation of a second estrogen-controlled oviduct gene, ovomucoid. We have succeeded in purifying ovomucoid mRNA, and have synthesized and cloned a double-stranded DNA from the message (4). Furthermore, the natural ovomucoid gene has been isolated from chicken DNA and cloned (5-6). We present here a summary of the structures of these two oviduct genes, and present a comparison of their salient features.

CLONING OF OVALBUMIN DNA FRAGMENTS

The structure of the natural ovalbumin gene has proven to be particularly complex due to the presence of multiple intervening DNA sequences within this gene. From restriction mapping of genomic DNA (7-10), it was originally thought that the entire ovalbumin gene was contained in three Eco RI DNA fragments of 2.4, 1.8 and 9.2 Kb in length. The two smaller Eco RI DNA fragments of the ovalbumin gene were partially purified by a combination of RPC-5 column chromatography (11-12) and preparative agarose gel electrophoresis and were cloned in the λgtWES vector (12-13) as previously described (14-15). These two fragments were subsequently recloned in *E. coli* X 1776 after ligation with Eco RI-digested pBR322 plasmid DNA and

ISBN 0-12-068350-4

designated pOV2.4 and pOV1.8. Both restriction enzyme mapping and electron microscopy demonstrated the complex structure of these ovalbumin gene fragments. The structural gene sequences within the 2.4 kilobase DNA were subdivided into four portions by four intervening sequences, while the structural gene sequences in the 1.8 Kb DNA were divided in two by three intervening sequences. Since the ovalbumin gene contained a structural sequence at the 5'-end, the structural gene appeared to be split into eight segments by seven intervening sequences.

However, the 5'- and 3'-ends of the ovalbumin gene were not represented in these fragments. In order to determine the detailed structures of the missing 5'- and 3'-ends of the gene, we have cloned additional overlapping restriction DNA fragments from genomic chick DNA, using the Charon 4A cloning system.

Ten μg of the 9.2 Kb DNA fraction recovered after preparative gel electrophoresis were ligated (17) with 30 μg of Charon 4A DNA arms in 500 μl and the entire mixture was employed for transfection of host cells (18). About 20,000 recombinant phage plaques were amplified according to a previously described procedure (19) and screened using a [^{32}P]OV_R probe (labeled by nick translation) which contains sequences corresponding to the 3'-terminus of ovalbumin mRNA with respect to the single Hae III site present in ovalbumin cDNA (10). Two positive phage plaques were obtained in this experiment.

According to our previous mapping of genomic DNA (10), a 3.2 Kb Hind III DNA fragment should contain 2.1 Kb of DNA sequence in common with the 2.4 Kb Eco RI DNA and an additional 1.1 Kb of DNA located toward the 5'-end of the gene. One μg of the enriched 3.2 Kb ovalbumin DNA was ligated with 1 μg of Hind III-digested Charon 21A DNA in a total volume of 20 μl. The phage DNA was packaged *in vitro* according to H. Faber, D. Kiefer and F. Blattner (personal communication). The ligation mixture was incubated for 15 min. at room temperature with 150 μl of buffer A (20 mM Tris, pH 8, 3 mM $MgCl_2$, 0.05% β-mercapto-ethanol, 1 mM EDTA), 20 μl buffer M1 (6 mM Tris, pH 7.4, 50 mM spermidine, 60 mM putresine, 18 mM $MgCl_2$, 15 mM ATP, 0.2% β-mercaptoethanol), 100 μl of sonic extract and 10 μl of protein A. After the addition of 750 μl of the freeze-thaw lysate this was further incubated for one hour at room temperature. Two to five μl of this material were employed to infect 0.2 ml of host cells and plated in soft agar. Plaques were screened using [^{32}P]-labeled OV2.4 DNA (19) as the hybridization probe. Three individual phage plaques yielded positive signals and were subcultured.

Mapping of the ovalbumin gene in total chick DNA has shown that there is a 4.5 Kb Pst I fragment which contains most of the 3.2 Kb Hind III fragment and an additional 1.3 Kb of DNA to the left of this fragment (10). Since no phage vectors are available to clone DNA fragments generated by Pst I digestion of DNA, ligation of this 4.5 Kb Pst I ovalbumin DNA (0.66 μg, isolated after preparative gel electrophoresis) with Eco RI-digested λgtWES DNA arms (2.2 μg) was performed in the presence of a natural 165 base pair Eco RI/Pst I DNA linker (46 ng). The DNA linker was obtained from digestion of the previously cloned 1.8 Kb ovalbumin DNA (16) with Eco RI and Pst I and isolated by preparative gel electrophoresis. Ligation was again carried out at 12° in a volume of 20 μl and the DNA in the ligation mixture was packaged *in vitro* as outlined above. An average of 400 plaques were obtained per plate. Without the linkers in the ligation mixture only 50-100 plaques were obtained which corresponds to background using the vector arms alone. Under identical conditions an Eco RI fragment of similar size yielded 2000 plaques per plate. Plaques were screened using the cloned [^{32}P]-labeled OV3.2 DNA as the hybridization probe. Four positive plaques were obtained. All of these ovalbumin gene fragments were later recloned in the plasmid BR322.

RESTRICTION MAPPING OF THE OVALBUMIN GENE

Characterization of the Cloned 9.2 Kb Eco RI Ovalbumin DNA (OV9.2). The cloned 9.2 Kb Eco RI DNA was isolated from the recombinant plasmid pOV9.2 and analyzed by restriction mapping and Southern hybridization. Bam HI digested this DNA into two fragments (4.7 and 4.5 Kb) but only the 4.7 Kb DNA fragment hybridized with the [^{32}P]OV$_R$ probe. The location of this Bam HI site within the 9.2 Kb Eco RI DNA fragment is in accordance with our previous mapping of the ovalbumin gene within genomic chick DNA (10). There are three Hind III cleavage sites within the 9.2 Kb DNA, and only the left-terminal 2.4 Kb DNA fragment was capable of hybridizing with [^{32}P]OV$_R$ as predicted from our previous restriction map of the gene (10). The remaining fragments did not hybridize because they are located outside of the DNA region that codes for the mRNA. Hae III cleaved the 9.2 Kb DNA into 11 fragments but only one (2.7 Kb) hybridized with the [^{32}P]OV$_R$ probe. Similarly, Pst I cleaved the 9.2 DNA into 6 fragments but only one (2.8 Kb) showed sequence homology to the OV$_R$ probe. These results identify the cluster of *Hind* III, *Hae* III, and *Pst* I sites present in genomic chick DNA about 2.4 Kb from the 5'-end of the 9.2 *Eco* RI DNA (10). This cluster of restriction sites is already beyond the DNA sequence coding for mRNA. The presence of the 3'-end of the ovalbumin gene within this 2.7 Kb fragment obtained from the 9.2 Kb DNA was verified subsequently by direct DNA sequencing (unpublished results).

Characterization of the Cloned 3.2 Kb Hind III Ovalbumin DNA (OV3.2). *Eco* RI digestion of this cloned DNA fragment gave rise to a 2.1 Kb fragment that hybridized with the [^{32}P]OV2.4 probe; the same fragment could be generated from the cloned 2.4 Kb *Eco* RI DNA by *Hind* III digestion. In addition, there were two *Eco* RI sites located 0.48 and 0.93 Kb to the left of the 2.4 Kb *Eco* RI DNA fragment. Both the 3.2 Kb *Hind* III and the 2.4 Kb *Eco* RI DNA fragments yielded a common 1.6 Kb *Sst* I fragment that hybridized with the probe. A *Hpa* I site unique to the entire ovalbumin gene was located within the 2.4 Kb DNA and this site was also present within the 3.2 *Hind* III DNA fragment. These and other restriction mapping analyses and cross-hybridization data (not shown) have confirmed the presence of a common 2.1 Kb DNA sequence between these two DNA fragments and that the 3.2 Kb *Hind* III DNA contains an additional 1.1 Kb of DNA located further toward the 5'-end of the ovalbumin gene. Since our previous data have demonstrated a *Taq* I site at position 41 of the structural gene (16,20), the DNA fragment containing the 5'-terminal structural gene sequences should be cleaved by this enzyme. Digestion of the 3.2 Kb *Hind* III DNA with *Taq* I, however, failed to reveal a cleavage site for this enzyme. Thus it appeared that the 5'-end of the gene was still not present within the 3.2 Kb *Hind* III ovalbumin DNA.

Characterization of the Cloned 4.5 Kb Pst I Ovalbumin DNA (OV4.5). The 4.5 Kb *Pst* I DNA fragment was isolated from a recombinant phage clone containing this DNA fragment and was analyzed by restriction mapping and Southern hybridization. The 4.5 Kb DNA fragment hybridized with [^{32}P]OV_L which is a probe that contains the left half of the structural ovalbumin gene (10). As expected, this 4.5 Kb *Pst* I fragment was not cleaved by *Bam* HI and was digested to a 3.2 Kb fragment by *Hind* III (data not shown). A partial digestion of this 4.5 Kb DNA by *Hpa* I generated a 1.8 Kb fragment that hybridized with OV_L, indicating that the unique *Hpa* I site common to both the 2.4 Kb *Eco* RI and 3.2 Kb *Hind* III ovalbumin DNA fragments was also present in the 4.5 Kb *Pst* I fragment. More importantly, the other 2.7 Kb *Hpa* I digestion product also showed a weak hybridization band, indicating the presence of additional structural gene sequence within this DNA fragment. Furthermore, this 4.5 Kb *Pst* I DNA contained at least one *Taq* I site and was cleaved by this enzyme to a 3.4 Kb DNA fragment that hybridized with the probe (not shown). Since this *Taq* I site in the 4.5 Kb *Pst* I DNA appeared to be the *Taq* I site in the structural gene (16,20) the structure of the entire ovalbumin gene could then be constructed based on the analyses of all the overlapping cloned fragments of the gene (Fig. 1). The overall length of the entire natural gene for ovalbumin was 7.55 Kb.

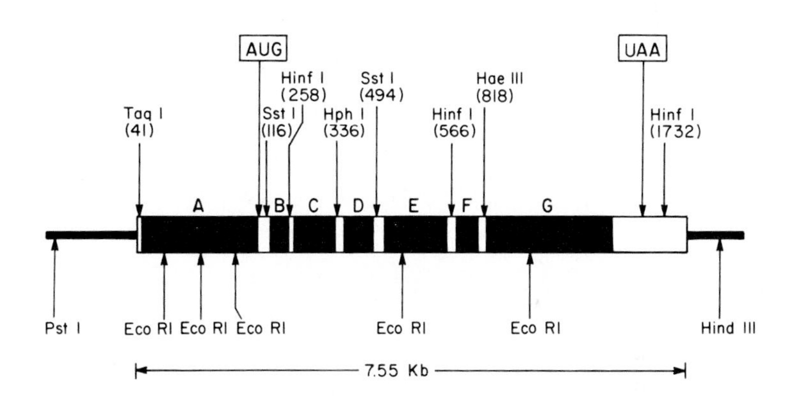

FIGURE 1. Physical map of the entire natural ovalbumin gene displaying some of the key restriction cleavage sites, the locations of the initiation and termination codons, and regions of interspersed structural and intervening sequences. This map is constructed from detailed analyses of the various gene fragments cloned in our laboratory, as described in this manuscript.

A CLONE OF THE CHICK GENOMIC DNA CONTAINING THE ENTIRE OVALBUMIN GENE

We have recently obtained from Drs. Axel, Engel and Dodgson a chick gene library generated by the method of Lawn *et al*. (21) using Charon 4A λDNA vector and chick DNA which was partially digested by Hae III and Alu I and ligated to synthetic Eco RI DNA linkers. Several million phage plaques from this chick gene library were screened for the ovalbumin gene, and twenty yielded a positive signal. One of the positive clones, designated 4A·OV, appeared to contain the entire ovalbumin gene. Upon Eco RI digestion, 4A·OV DNA yielded 8.0 and 5.5 Kb DNA fragments in addition to the 2.4, 1.8 and 0.5 Kb fragments (data not shown). Hind III and Pst I digestion of this DNA generated the expected 3.2 and 4.5 Kb fragments indicating the presence of additional chick DNA sequences to the left of the 2.4 and 0.5 Kb Eco RI DNA fragments of this ovalbumin gene clone.

4A·OV DNA was also analyzed by electronmicroscopic mapping in order to determine whether it contains the entire ovalbumin gene. A typical molecule as seen after hybridizing single-stranded λ4A·OV DNA with purified ovalbumin mRNA and its corresponding line drawing are shown in Fig. 2. A total of 7 single-stranded loops corresponding to the 7 intervening DNA sequences was evident. Loops B to D and loop F were identical to those observed previously with cloned 2.4 Kb and 1.8 Kb Eco RI DNA fragments respectively (16). While loop E contained the junction between OV2.4 and OV1.8 DNA fragments, loop G contained the junction between the OV1.8 and OV9.2 DNA fragments. Loop A was evidently to the left of OV2.4 DNA and could only be formed by the presence of additional structural gene sequences. Since loop A is approximately 1.6 Kb in length, the 5'-structural gene sequences must be located to the left of the 3.2 Kb Hind III DNA, but within the 4.5 Kb Pst I fragment. This observation agrees precisely with our analyses of the cloned 3.2 Kb Hind III and 4.5 Kb Pst I fragments. Furthermore, the sizes of loops B to G are in good agreement with the gene map generated by analyses of the various cloned gene fragments shown in Fig. 1.

IDENTIFICATION OF A COMPLETE OVOMUCOID GENE CLONED FROM THE CHICKEN GENE LIBRARY

We have previously reported the cloning of a 15 Kb chicken DNA fragment containing a portion of the ovomucoid gene (5). In order to determine the sequence organization of the complete ovomucoid gene and also to investigate the regulation of expression of this gene, a clone of the chicken genomic DNA containing the entire ovomucoid gene was desired. Therefore, approximately 750,000 phage plaques

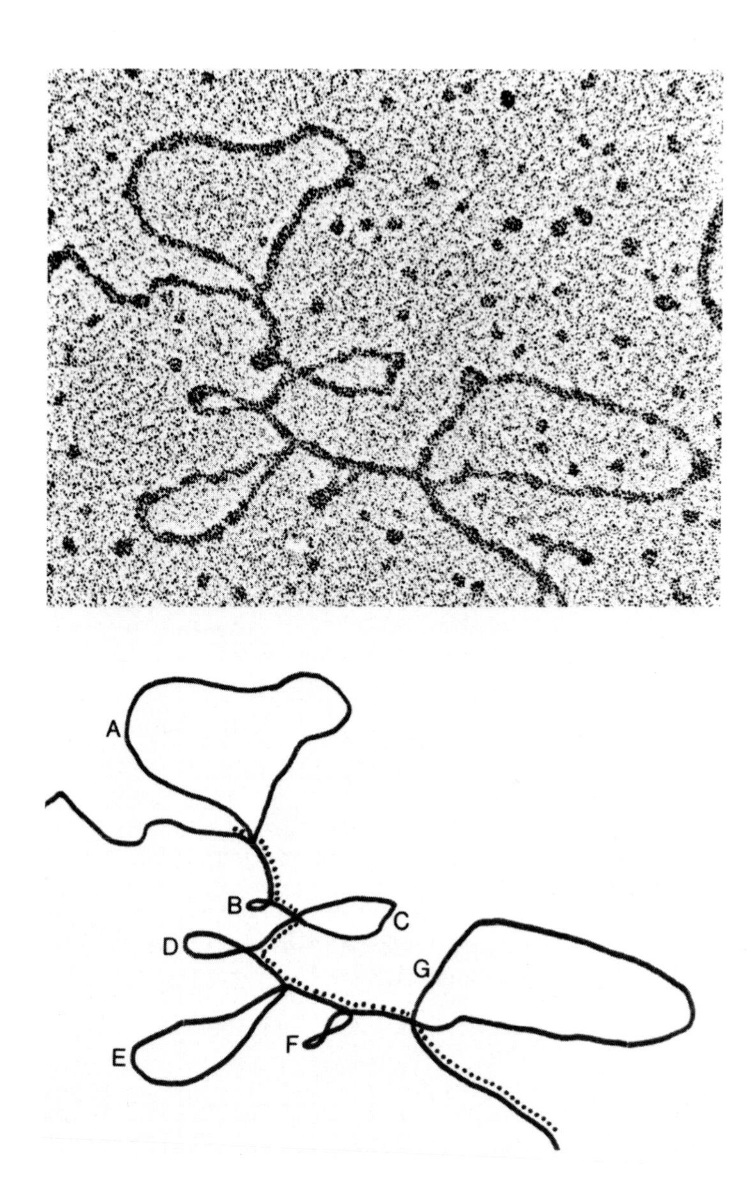

FIGURE 2. Electron micrograph and line drawing of a hybrid molecule formed between the ovalbumin gene (λ4A·OV) and ovalbumin mRNA. ——, 4A·OV DNA; ----- ovalbumin mRNA.

from the chick gene library were plated on petri dishes and screened for ovomucoid gene clones, using a [^{32}P] labeled ovomucoid cDNA clone as the hybridization probe. One positive clone, CL21, consisted of a chicken DNA insert of 15 Kb which was cleaved by Eco RI into a 9 Kb and a 6 Kb fragment. The 9 Kb fragment hybridized to both a probe specific for the left half of the ovomucoid structural gene (OM_L) and one specific for the right half (OM_R), while the 6 Kb fragment hybridized only to OM_R. This indicated that CL21 contains chicken DNA 9 Kb and 6 Kb to the left and right, respectively, of the unique Eco RI site in the 3'-noncoding region of the structural ovomucoid gene (4). So in order to define the sequence organization of the ovomucoid gene, we have focused our attention on CL21 since it appeared to contain the entire ovomucoid gene as well as DNA flanking both termini of the gene.

RESTRICTION MAPPING OF THE 5' PORTION OF THE CLONED OVOMUCOID GENE (OM9)

The 9 Kb DNA fragment (OM9) was isolated from CL21 DNA after Eco RI digestion and preparative agarose gel electrophoresis. Hind III was found to cleave this OM9 DNA into fragments of 3.6, 1.1, 0.8, and 3.2 Kb, respectively, from the 5' end to the 3' end (6). In order to characterize the sequence organization of the gene, these Hind III fragments of OM9 (i.e. OM 3.6, OM 1.1, OM 0.8 and OM 3.4) were subsequently cloned individually in pBR322. These fragments were then excised from the respective recombinant plasmids with restriction enzymes, separated from plasmid sequences by preparative agarose gel electrophoresis, and then mapped individually with various restriction endonucleases.

Structure of OM 3.6 DNA. The 3.6 Kb DNA (OM 3.6) was digested by Taq I into 3 fragments of 0.6, 1.7, and 1.3 Kb in size due to the presence of 2 Taq I sites located 0.6 and 1.3 Kb from the left and right hand termini of the DNA, respectively (data not shown). Only the 1.3 Kb fragment derived from the right hand portion of the DNA hybridized to the ovomucoid cDNA probe. Ava I cleaved the OM 3.6 into two fragments of sizes 3.4 and 0.2 Kb, but only the 0.2 Kb fragment yielded a hybridization signal. Thus there are structural sequences in this 0.2 Kb fragment but not the 3.4 Kb fragment. Since DNA sequences extending at least 12 Kb to the left of the 0.2 Kb fragment did not hybridize to the ovomucoid probe as we examined DNAs from the other genomic ovomucoid clones, the beginning of the ovomucoid gene coding for the 5' end of the mRNA is likely to reside within this 0.2 Kb fragment at the 3' end of OM 3.6.

Structure of OM 1.1 DNA. The 1.1 Kb Hind III fragment (OM 1.1) yielded a weaker hybridization signal than those of the other Hind III fragments indicating that the structural gene segment within OM 1.1 may be relatively short. In order to locate that structural sequence, OM 1.1 was digested with Hae III, Mbo II and Hinf I, separately. There are unique Hae III and Mbo II sites, 0.25 and 0.2 Kb from the left and right termini of OM 1.1 DNA, as well as 3 Hinf I sites, 0.05, 0.55, and 0.75 Kb from the left termini of the DNA, respectively (data not shown). Hybridization of the restricted DNA fragments with the [^{32}P] cDNA probe revealed that the single ovomucoid structural gene sequence in OM 1.1 DNA is located between the Mbo II and the third Hinf I sites.

Structure of OM 0.8 DNA. The 0.8 Kb Hind III fragment (OM 0.8) contains the unique Kpn I site in OM 9 DNA. This enzyme digested OM 0.8 DNA into a 0.63 and a 0.17 Kb fragment and only the 0.63 Kb fragment hybridized with [^{32}P] OM_L. Two fragments of 0.65 and 0.15 Kb in length were generated by Hinf I digestion; the 0.65 Kb fragment hybridized to the ovomucoid probe and the 0.15 fragment did not. This Hinf I site must be located 0.02 Kb to the right of the Kpn I site because a double digest of the OM 0.8 DNA by these two enzymes generated fragments of 0.63 and 0.15 Kb in size. There are two Pst I sites located at 0.12 and 0.14 Kb from the right hand end of the OM 0.8 DNA. This enzyme digested the DNA into 3 fragments of 0.54, 0.14 and 0.12 Kb in size but only the 0.14 Kb fragment hybridized to the probe. These experiments have indicated that the structural ovomucoid sequences are clustered between the Pst I and Kpn I sites in the OM 0.8 DNA, which are approximately 90 nucleotides apart. Since there is a Pst I site and a Hinf I site located at 82 and 15 nucleotides to the left and right of the unique Kpn I site in the structural ovomucoid gene, respectively, we conclude that the Kpn I site in the OM 0.8 DNA is the same one present in the structural gene. This segment of structural sequence is thus at least 97 nucleotides in length.

Structure of OM 3.2 DNA. Taq I has no cleavage sites in the ovomucoid structural sequence, but digested the OM 3.2 DNA into 3 fragments of 1.45, 1.25 and 0.7 Kb in size which all hybridized to the OM_T probe, indicating the presence of at least 2 intervening sequences within the OM 3.2 DNA. Each of these Taq I fragments were purified, individually digested with a variety of restriction endonucleases, and the resulting DNA fragments hybridized to OM_L and OM_R separately.

The 0.7 Kb Taq I fragment located at the right hand end of OM 3.2 DNA hybridized only to OM_R, indicating that it contained ovomucoid sequences to the right of the structural Hinc II site. This result was not unexpected because the Eco RI site at the right end of the OM 3.2 DNA has previously been identified to be the unique Eco RI site present in the ovomucoid structural sequence. There is a Bam HI cleavage site 0.2 Kb to the left of this Eco RI site and since both Bam HI DNA fragments hybridized only to OM_R, there must be at least two structural sequences within the 0.7 Kb Taq I fragment separated by an intervening sequence containing the Bam HI site. Further mapping with Hinf I located the two structural sequences at both ends of the 0.7 Kb Taq I DNA fragment separated by about 0.45 Kb of intervening sequences.

Pst I cleaved the 1.25 Kb Taq I fragment once, 0.45 Kb from the left terminus, to generate two small pieces of 0.8 Kb and 0.45 Kb in length. Only the latter hybridized to OM_T. Thus, there should be a third structural gene sequence located to the left of the Pst I site. There are three Hinc II sites within the OM 3.2 DNA. One of the sites is located at 0.3 Kb to the right of the left terminus of the 1.25 Kb Taq I fragment. This Taq I DNA fragment hybridized weakly to nick-translated OM_T, but interestingly not to OM_L or OM_R. Since the OM_L and OM_R probes were prepared by cleaving the ovomucoid structural sequence at its unique Hinc II site, this observation could be due to the presence of a short structural gene segment around the single Hinc II site present in the 1.25 Kb Taq I DNA fragment. Thus, when OM_T was cleaved into OM_L and OM_R, both probes did not contain sufficient sequence to form stable hybrids with the 1.25 Kb Taq I DNA fragment. This interpretation of the data requires that the Hinc II site within the 1.25 Kb Taq I fragment be the unique Hinc II site within the structural ovomucoid gene, which has subsequently been confirmed by direct DNA sequencing. Thus, a third structural gene segment must be present at that region of the OM 3.2 DNA.

The 1.45 Kb _Taq_ I fragment hybridized only to OM_L but not OM_R indicating that it contained structural sequence to the left of the _Hinc_ II site as expected. Further mapping with _Hinc_ II and _Pst_ I revealed the presence of two structural sequences separated by a 0.35 Kb intervening sequence near the middle of this 1.45 Kb _Taq_ I DNA fragment.

RESTRICTION ENDONUCLEASE MAPPING OF THE 3'-PORTION OF THE OVOMUCOID GENE (OM 6)

CL21 DNA was digested with _Eco_ RI, applied to a preparative agarose gel and the 6 Kb DNA was recovered from the gel. The DNA was digested by _Bam_ HI into two fragments of 5.0 Kb and 1.0 Kb in size and only the 1.0 Kb fragment hybridized to the ovomucoid cDNA probe. Thus, all of the 3' portion of the ovomucoid gene appeared to be contained within the 1 Kb fragment. There is a unique _Bam_ HI site in the ovomucoid structural sequence 14 nucleotides from the _Eco_ RI site towards the 3' end (4). This fragment, however, was too small to be retained in the 1% agarose gel and was not observed.

Hae III digested the OM 6 DNA into multiple fragments, but only a 0.4 Kb fragment derived from the left end of OM 6 DNA hybridized with the ovomucoid probe. Since only one hybridization band was observed and the _Eco_ RI site separating the 9 and 6 Kb DNA fragments in CL21 is the unique _Eco_ RI site in the structural gene, the 3' portion of the gene must be within 0.4 Kb from the _Eco_ RI site. Similarly, digestion of the OM 6 DNA by _Mbo_ II yielded multiple fragments but only one hybridization signal, at 0.13 Kb, derived from the fragment from the left terminus of OM 6 DNA. These results indicated that the 3' portion of the ovomucoid gene is contained within a small segment of the chicken DNA from the unique _Eco_ RI site. The _Eco_ RI site in the ovomucoid structural sequence is 119 nucleotides from the end of the mRNA sequence (6). Since our results from restriction mapping have shown that this entire sequence is contained within a 0.13 Kb fragment, it appears that there are no additional intervening sequence within this DNA region. This conclusion has since been confirmed by direct DNA sequencing.

This restriction mapping data from the OM 9 and OM 6 ovomucoid DNAs was combined to obtain a map of the organization of the entire ovomucoid gene (Fig. 3). The entire gene includes seven intervening sequences and is 5.6 Kb in length.

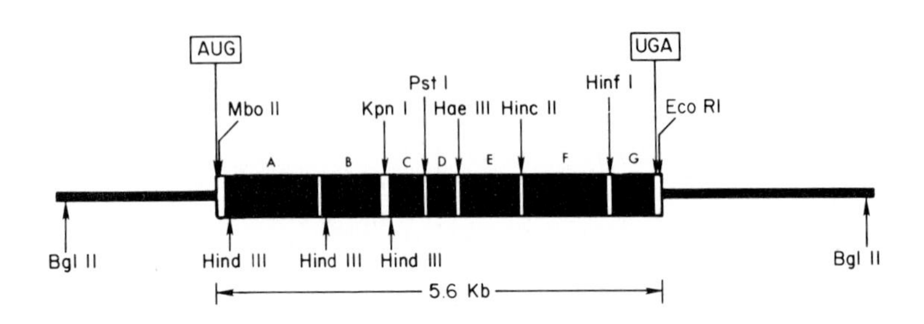

FIGURE 3. Physical map of the entire natural ovomucoid gene displaying some of the key restriction cleavage sites, the locations of the initiation and termination codons, and regions of interspersed structural and intervening sequences. This map is constructed from detailed analyses of the various gene fragments cloned in our laboratory, as described in this manuscript.

ELECTRON MICROSCOPIC ANALYSIS OF THE COMPLETE OVOMUCOID GENE

Based on electronmicroscopic data, we have recently reported the presence of at least six intervening sequences within the 15 Kb Eco RI fragment of the ovomucoid gene (5). Detailed restriction mapping however, has indicated the presence of 7 intervening sequences (Fig. 3). Since the structural gene segment in OM 1.1 DNA must be rather short, it was possible that this segment did not form a stable hybrid with ovomucoid mRNA, causing the appearance of only 6 loops. CL21 DNA, containing the entire natural ovomucoid gene, was thus thermally denatured and incubated with ovomucoid mRNA for 3 hrs under stringent conditions that permitted only RNA:DNA hybridization, but not DNA-DNA reassociation. The reaction mixture was then diluted 10-fold with hybridization buffer and further incubated at a lower temperature for 10 minutes to facilitate the formation of hybrids between short homologous regions.

Under these conditions, hybrid molecules with both 6 and 7 loops were observed. A molecule with 7 loops is shown in Fig. 4. Thirteen hybrid molecules with only 6 loops were measured, and the sizes and positions of loops C through G were similar to those present in molecules with 7 loops. However, there was a single large loop at one end and its size was that of loops A and B combined. Thus the entire chicken ovomucoid gene contains at least seven intervening sequences. The structural sequence between intervening sequences A and B appeared to be short and did not form a very stable hybrid with ovomucoid mRNA under the conditions employed previously.

DISCUSSION

The structure of these two hormone-regulated chicken oviduct genes show several similarities. Both genes contain seven intervening sequences dispersed among eight structural gene sequences. Liquid hybridization experiments with excess total chicken DNA as well as Southern blotting experiments have shown that both the ovalbumin and ovomucoid structural and intervening sequences are unique sequences in the chicken genome. They appear to be transcribed in their entireties during gene expression and are inducible by steroid hormones (22 and unpublished observations). Interestingly, the proportion of intervening sequences in both genes is quite high. The ovalbumin gene is 7.6 Kb in length, coding for a mRNA of 1859 nucleotides (20). Thus, only 24% of the ovalbumin gene is represented in mature mRNA. Similarly, the ovomucoid gene is 5.6 Kb in length, and codes for a mRNA of only 823 nucleotides (unpublished observations). So only 15% of the ovomucoid gene is represented in the mature mRNA, which represents the highest proportion of intervening sequences among all the genes studied to date (6,9,15,19,23-31). Furthermore, both genes contain intervening sequences which exhibit quite a diversity in size. The largest intervening sequence in the ovalbumin gene is 1.4 Kb in length, and the shortest .25 Kb. The largest intervening sequence in the ovomucoid gene is approximately 1 Kb in length, and the shortest .2 Kb. High molecular weight precursors containing these intervening sequences have been identified for both genes (22,32), and none of the precursor molecules appear to be common to both genes. Finally, the gene sequence coding for the 3' non-translated region of both genes is not interrupted by any intervening sequences.

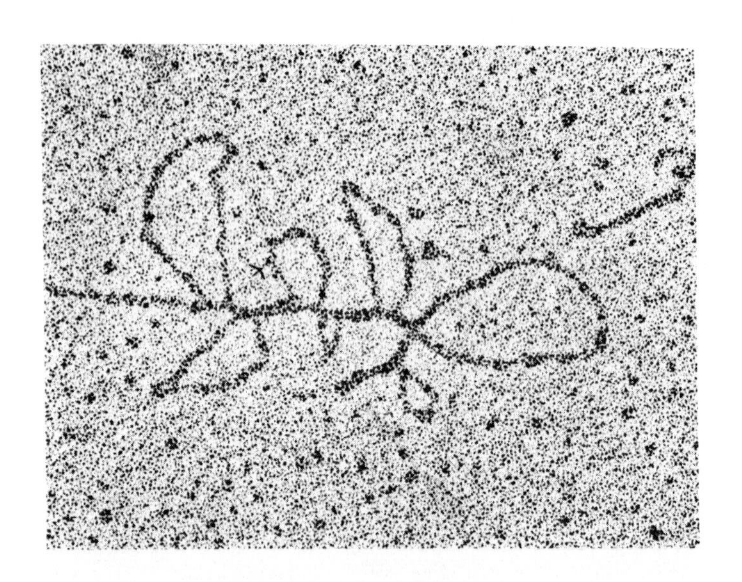

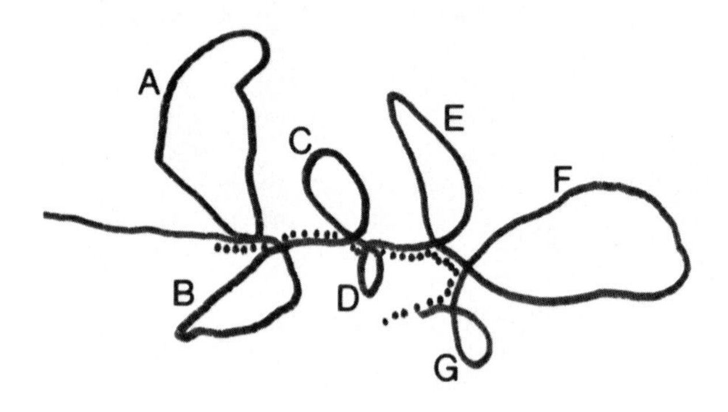

FIGURE 4. Electron micrograph and line drawing of a hybrid molecule formed between the ovomucoid gene (OM 15) and ovomucoid mRNA. ———, OM 15 DNA; ------, ovomucoid mRNA.

There are only two obvious gross structural differences in the two genes. Whereas the entire 5'-leader sequence of the ovomucoid gene is coded by a continuous segment of DNA, the ovalbumin gene contains an intervening sequence of about 1.4 Kb in length within this 5' leader sequence. Second, because the ovomucoid mRNA is less than half of the size of ovalbumin mRNA, the average size of the ovomucoid structural gene sequences is significantly smaller than the size of the ovalbumin structural gene sequences. Indeed, the second structural sequence from the 5' end of the ovomucoid gene is only 20 bp long!

All of the ovalbumin gene, and parts of the ovomucoid gene, have now been sequenced. Interestingly, there are short stretches of DNA sequences immediately flanking both genes that are remarkably similar. 19 nucleotides from the 3'-end of the ovomucoid mRNA is the hexanucleotide AATAAA (Fig. 5). This hexanucleotide is also found 19 nucleotides from the end of ovalbumin mRNA, and indeed appears near the end of all eucaryotic mRNAs examined thus far (33). It has been suggested that these preserved sequences may play some important functional role in the transcriptional termination of the gene and/or polyadenylation of the mRNA. Furthermore, seven nucleotides into the 3' flanking sequence of the ovomucoid gene is the sequence TTGT. Other eucaryotic genes that have been sequenced, such as chick ovalbumin, mouse β-globin and *Xenopus* 5S RNA genes, all contain the sequence TTTT at the same position in the 3' flanking sequence (Fig. 5). A second eukaryotic gene with TTGT flanking its 3'-terminus at that position is the one coding for rat insulin (40). It has been proposed that the T cluster flanking the 3'-end of eucaryotic genes may be a termination signal for transcription of these genes or may be responsible for insuring correct termination (34). It has been shown that following heat shock treatment of Drosophila, the primary transcript can terminate in a T cluster which is not normally the termination site (35). Furthermore, a mutation of the TTTT sequence to TTTC at the end of a *Xenopus* 5S RNA gene has allowed transcriptional read-through (Brown, personal communication). At present, it is not known whether the TTGT cluster at the end of the ovomucoid gene instead of TTTT as in other eucaryotic genes affects proper termination of transcription.

Up to 100 nucleotides flanking the 5' end of both the ovomucoid and ovalbumin genes have now been sequenced. At 27 nucleotides from the 5' end of the ovomucoid gene is an octanucleotide sequence TATATATT, which is identical to a sequence present in the ovalbumin gene at 32 nucleotides upstream from the 5'-end of the mRNA sequence (Fig. 6).

This AT-rich region is also present at approximately the same locations in the 5'-flanking sequences of other eukaryotic genes such as those coding for rabbit β-globin,

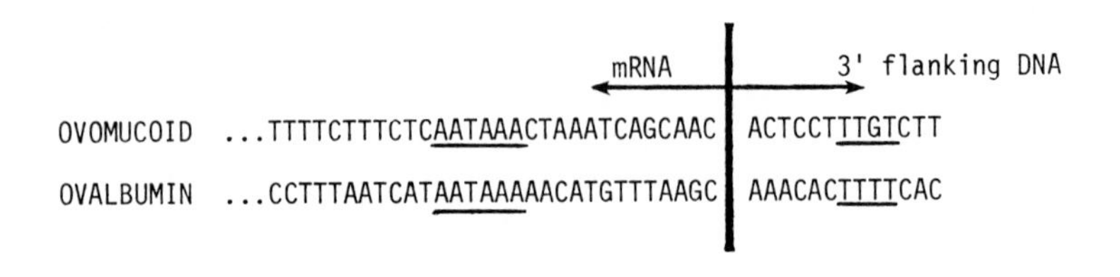

FIGURE 5. A similar DNA sequence flanking the 3'-termini of chicken ovomucoid and ovalbumin genes.

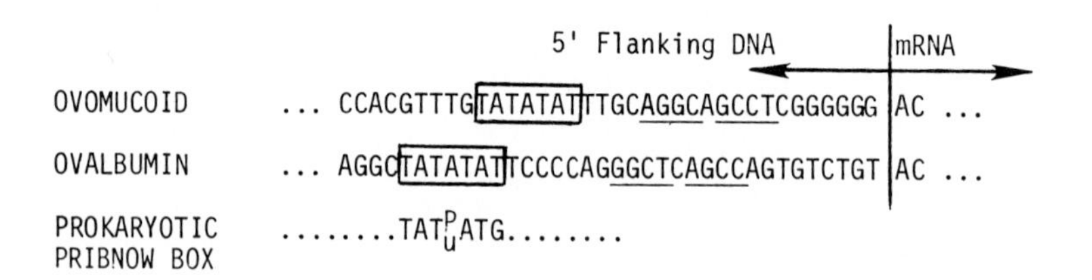

FIGURE 6. A similar DNA sequence flanking the 5'-termini of chicken ovomucoid and ovalbumin genes.

mouse β-globin, mouse immunoglobin λ light chain, adenovirus 2 major late mRNAs and the histones of sea urchin and Drosophila. These interesting sequences in eucaryotic genes bear a striking resemblance to that of the 'Pribnow box' sequence in prokaryotes (36). This box is a common site around promoters, with a sequence of TATPuATGT which precedes the transcription-initiation site by 5 to 7 nucleotides. Thus, procaryotes and eucaryotes may have similar signals for initiation of transcription.

Finally, a sequence of partial two-fold symmetry exists in both genes between the TATATATT sequence and the start of the structural genes. This sequence is GGCTCAGCC at -18 nucleotides in the ovalbumin gene, and AGGCAGCCT at -16 nucleotides in the ovomucoid gene. In both genes, the 2-fold symmetry is provided by a self-complementary 4-nucleotide sequence separated by a single nucleotide at the center. The presence of complete or partial dyad symmetry is common around procaryotic operators and at other sites controlling procaryotic gene transcription (37-39). However, whether these types of symmetrical sequences have any function(s) in the regulation of eucaryotic gene expression must await further analyses.

REFERENCES

1. O'Malley, B.W., McGuire, W.J., Kohler, P.O., and Korenman, S.G. (1969). Rec. Prog. Horm. Res. 25, 105-160.
2. Schrader, W., and O'Malley, B.W. (1978). In "Receptors and Hormone Action 2" (B.W. O'Malley and L. Birnbaumer, eds.), pp. 189-224. Academic Press, New York.
3. O'Malley, B.W., and Means, A.R. (1974). Science 183, 610-620.
4. Stein, J.P., Catterall, J.F., Woo, S.L.C., Means, A.R., and O'Malley, B.W. (1978). Biochemistry 17, 5763-5772.
5. Catterall, J.F., Stein, J.P., Lai, E.C., Woo, S.L.C., Dugaiczyk, A., Mace, M.L., Means, A.R., and O'Malley, B.W. (1979). Nature 278, 323-327.
6. Lai, E.C., Stein, J.P., Catterall, J.F., Woo, S.L.C., Means, A.R., and O'Malley, B.W. (1979, in press). Cell.
7. Doel, M.T., Houghton, M., Cook, E.A., and Carey, N.H. (1977). Nucl. Acids Research 4, 3701-3713.
8. Breathnach, R., Mandel, J.L., and Chambon, P. (1977). Nature 270, 314-319.
9. Weinstock, R., Sweet, R., Weiss, M., Cedar, H., and Axel, R. (1978). Proc. Natl. Acad. Sci. USA 75, 1299-1303.

10. Lai, E.C., Woo, S.L.C., Dugaiczyk, A., Catterall, J.F., and O'Malley, B.W. (1978). Proc. Natl. Acad. Sci. USA 75, 2205-2209.
11. Hardies, S.C., and Wells, R.D. (1976). Proc. Nat. Acad. Sci. USA 9, 3117-3121.
12. Leder, P., Tiemeier, D., and Enquist, L. (1977). Science 196, 175-177.
13. Tiemeier, D.C., Tilghman, S.M., and Leder, P. (1977). Gene 2, 173-191.
14. Benton, W.D., and Davies, R.W. (1977). Science 196, 180-182.
15. Woo, S.L.C. Methods in Enzymol. (in press).
16. Dugaiczyk, A., Woo, S.L.C., Lai, E.C., Mace, M.L., McReynolds, L., and O'Malley, B.W. (1978). Nature 274, 328-333.
17. Dugaiczyk, A., Boyer, H.W., and Goodman, H.M. (1975). J. Mol. Biol. 96, 171-184.
18. Mandel, M., and Higa, A. (1970). J. Mol. Biol. 53, 159-162.
19. Woo, S.L.C., Dugaiczyk, A., Tsai, M.-J., Lai, E.C., Catterall, J.F., and O'Malley, B.W. (1978). Proc. Natl. Acad. Sci. USA 75, 3688-3692.
20. McReynolds, L., O'Malley, B.W., Nisbet, A.D., Fothergill, J.E., Givol, D., Fields, S., Robertson, M., and Brownlee, G.G. (1978). Nature 273, 723-726.
21. Lawn, R.M., Fritsch, E.F., Parker, R.C., Blake, G., and Maniatis, T. (1978). Cell 15, 1157-1174.
22. Roop, D.R., Nordstrom, J.F., Tsai, S.Y., Tsai, M.-J., and O'Malley, B.W. (1978). Cell 15, 671-685.
23. Garapin, A.C., Cami, B., Roskam, W., Kourilsky, P., LePennec, J.P., Perrin, F., Gerlinger, P., Cochet, M., and Chambon, P. (1978). Cell 14, 629-639.
24. Tilghman, S.M., Tiemeier, D.C., Polsky, F., Edgell, M.H., Seidman, J.G., Leder, A., Enquist, L., Norman, B., and Leder, P. (1977). Proc. Natl. Acad. Sci. USA 74, 4406-4410.
25. Tiemeier, D., Tilghman, S., Polsky, F.I., Seidman, J.C., Leder, A., Edgell, M.H., and Leder, P. (1978). Cell 14, 237-245.
26. Maniatis, T., Hardison, R.C., Lacy, E., Lauer, J., O'Connell, C., Quon, D., Sim, G.K., and Efstratiatis, A. (1978). Cell 15, 687-701.
27. Tonegawa, S., Brack, C., Hozumi, N., and Schuller, R. (1977). Proc. Natl. Acad. Sci. USA 74, 3518-3522.
28. Brack, C., and Tonegawa, S. (1977). Proc. Natl. Acad. Sci. USA 74, 5652-5656.

29. Tonegawa, S., Maxam, A.M., Tizard, R., Bernard, O., and Gilbert, W. (1978). Proc. Natl. Acad. Sci. USA 75, 1485-1489.
30. Seidman, J.G., Leder, A., Edgell, M.H., Polsky, F., Tilghman, S.M., Tiemeier, D.C., and Leder, P. (1978). Proc. Natl. Acad. Sci. USA 75, 3881-3885.
31. Seidman, J.G., Edgell, M.H., and Leder, P. (1978). Nature 271, 582-585.
32. Nordstrom, J.L., Roop, D.R., Tsai, M.-J., and O'Malley, B.W. (1979). Nature 278, 328-331.
33. Proudfoot, N.J., and Brownlee, G.G. (1976). Nature 263, 211-214.
34. Korn, L.J., and Brown, D.D. (1978). Cell 15, 1145-1156.
35. Jacq, B., Jourdan, R., and Jordan, B.R. (1977). J. Mol. Biol. 117, 785-795.
36. Pribnow, D. (1975). Proc. Nat. Acad. Sci. USA 72, 784-788.
37. Bennett, G.N., and Yanofsky, C. (1978). J. Mol. Biol. 121, 179-192.
38. Goeddel, D., Yansuva, D.G., and Caruthers, M.H. (1978). Proc. Natl. Acad. Sci. USA 75, 3578-3582.
39. Otsuka, A., and Abelson, J. (1978). Nature 276, 689-694.
40. Efstratiatis, A. (1979). ICN-UCLA Symposia on Molecular and Cellular Biology, pp. 301-315, Academic Press, New York.

THE STRUCTURE AND TRANSCRIPTION OF RAT PREPROINSULIN GENES[1]

Argiris Efstratiadis*, Peter Lomedico**, Nadia Rosenthal*, Richard Kolodner†, Richard Tizard**, Francine Perler*, Lydia Villa-Komaroff‡, Stephen Naber‡, William Chick‡, Stephanie Broome**, and Walter Gilbert**

*Department of Biological Chemistry and ‡Joslin Research Laboratory, Harvard Medical School, Boston, MA 2115; **Biological Laboratories, Harvard University, Cambridge, MA 2138; †Sidney Farber Cancer Institute, Boston, MA 2115; ‡Department of Microbiology, University of Massachusetts, Worcester, MA 1605.

ABSTRACT. We inserted double-stranded cDNA copies of rat preproinsulin mRNA into the unique Pst site of plasmid pBR322, that lies in the region encoding penicillinase, and cloned the hybrid DNA in E.coli χ1776. One of the clones produces proinsulin in the form of a fused protein bearing both insulin and penicillinase antigenic determinants. The insulin material produced by the bacterium is biologically active. By using cloned insulin cDNA as probe we isolated from a rat DNA library chromosomal clones of the two non-allelic preproinsulin genes, I and II, that are equally expressed in the rat pancreas. Characterization of these clones by DNA sequencing and heteroduplex analysis shows that gene I lacks a large intron interrupting the coding region of gene II. This intron is transcribed and present in a preproinsulin II mRNA precursor.

INTRODUCTION

To understand the regulatory mechanisms involved in eukaryotic gene expression we must study the structure and organization of specific genes and their associated sequences in genomic DNA. Structural genes can be isolated from cloned libraries of eukaryotic DNA (e.g. ref.1) by the use of specific hybridization probes, such as cloned double-stranded cDNA (ds-cDNA) copies (2) of the corresponding mRNAs.

In this paper we summarize our studies along these lines, using as a model system the rat preproinsulin genes.

[1]This work was supported by Grant 1489-C-1 from the American Cancer Society, Massachusetts Division, Inc. to A.E. and by NIH Grants AM 21240 and G.M. 09541-17 to W.G. and AM 15398 to W.C.

ISBN 0-12-068350-4

To generate a pure probe, we converted the entire mRNA population of a rat insulinoma into ds-cDNA and isolated the cloned preproinsulin sequences (3). This was necessary since preproinsulin mRNA is a minor component of the RNA of the tumor. We used the cloned ds-cDNA as probe to isolate the two preproinsulin genes from a rat DNA library. Moreover, this ds-cDNA was employed in studies concerning the expression of insulin in bacteria.

The Phylogeny of Insulin (4,5). A separate islet organ appears for the first time in evolution in the most primitive vertebrates, the cyclostomes (lampreys and hagfish). However, cells of the gastrointestinal mucosa of invertebrates, like the mussel (a mollusc) and the starfish (an echinoderm) or of amphioxus (a cephalochordate), contain an insulin-like peptide detectable by immunofluorescence, using antisera against mammalian insulin. Insulin and proinsulin-like peptides were actually isolated from the oyster (a mollusc) (6). Biological assays indicate the presence of an insulin-like hormone in insects (Drosophila melanogaster, a dipteran, and Manduca sexta, a lepidopteran) (7,8). Unfortunately, none of the invertebrate "insulins" has been purified to homogeneity or sequenced; thus, sufficient data for the construction of a phylogenetic tree are not available. On the other hand, the endocrine pancreas of fish, amphibians, reptiles, birds and mammals has common cytological features and the amino acid sequence of several insulins is known (9,10). Bird and mammalian insulins (with the exception of the guinea pig insulin) are very similar, but different from fish insulins. It is noteworthy that in the rat, mouse, bonito, tuna and toadfish two different insulins, I and II, are found. At least for the rat, it is known that both of them occur in individual pancreata (11), and are synthesized in equal proportions (12). They are, therefore, the end-products of non-allelic genes.

The Chemistry and Biosynthesis of Insulin (10,13-16). Mature insulin consists of two polypeptide chains, A and B (21 and 30 amino acids, respectively), connected by two disulfide bridges, but it is initially synthesized as a single, longer chain (109 amino acids), termed preproinsulin (Fig.1). A hydrophobic leader sequence of 23 amino acids at the amino terminus of the nascent chain is cleaved off, producing proinsulin. The proinsulin chain folds up, the disulfide bridges are formed, and then a 31 amino acid peptide, the C-peptide, is cleaved out of the middle of the molecule by proteolytic cuts at pairs of basic amino acids.

Translation and processing of preproinsulin occurs in the β-cells of the islets of Langerhans (endocrine pancreas). Pre-

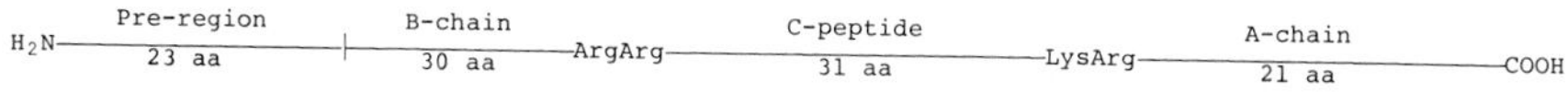

```
       Pre-region           B-chain               C-peptide               A-chain
H2N--------------------|--------------ArgArg-------------------LysArg------------------COOH
         23 aa               30 aa                   31 aa                   21 aa

II  AlaLeuTrpIleArgPheLeuProLeuLeuAlaLeuLeuIleLeuTrpGluProArgProAlaGlnAla                Pre
I    ?  ?     Met                            Val            Lys

II  PheValLysGlnHisLeuCysGlySerHisLeuValGluAlaLeuTyrLeuValCysGlyGluArgGlyPhePheTyrThrProMetSer   B
I                           Pro                                                         Lys

II  GluValGluAspProGlnValAlaGlnLeuGluLeuGlyGlyGlyProGlyAlaGlyAspLeuGlnThrLeuAlaLeuGluValAlaArgGln   C
I                        Pro                        Glu

II  GlyIleValAspGlnCysCysThrSerIleCysSerLeuTyrGlnLeuGluAsnTyrCysAsn                        A
I
```

FIGURE 1. The primary structure of rat preproinsulins. The data for the proinsulin sequences are from ref. 10 and 13. The amino acid sequence of the pre-region was deduced from our DNA sequencing data (this paper and ref.3).

proinsulin is synthesized on the rough endoplasmic reticulum; it is believed that the pre-region serves for the transfer of the nascent chain into the cisternal spaces through the microsomal membrane, before its rapid cleavage and subsequent degradation. Proinsulin is converted to insulin in the Golgi region by the action of trypsin- and carboxypeptidase B-like enzymes. It is then packaged in the form of secretory granules enclosed within limiting membranes. The mature granules appear to contain zinc and the C-peptide in addition to insulin. During the process of insulin secretion, which is physiologically stimulated by glucose, the granules move and their surface membrane fuses with the plasma membrane of the cell. Following rupture of the fused membranes, insulin is released into the pericapillary space.

Preproinsulin mRNA. A partial amino acid sequence of the pre-region was determined from the immunoprecipitable products of a cell-free translation directed by a crude RNA preparation from rat pancreatic islets (15). Partial purification of rat preproinsulin mRNA has also been reported (17).

Using islet RNA, Ullrich et al. (18) synthesized and cloned ds-cDNA copies of rat preproinsulin mRNA. From three clones they determined a total sequence of 354 nucleotides corresponding to preproinsulin I, and from one clone a sequence of 77 nucleotides corresponding to preproinsulin II.

RESULTS

Characterization of ds-cDNA Clones. We purified poly(A)-containing cytoplasmic RNA from an X-ray induced, transplantable rat insulinoma (19), and used it directly as template for ds-cDNA synthesis (3). The ds-cDNA was inserted into the unique Pst site of plasmid pBR322 by the oligo(dG)·oligo(dC) tailing procedure (in order to reconstruct this site). The Pst site lies in the plasmid region encoding penicillinase, about two thirds of the gene length away from the initiation codon (20). The hybrid plasmids were used to transform E.coli χ1776. A clone containing insulin DNA sequence (pI19) was identified by hybridization-arrested translation (21) and fully characterized by restriction endonuclease analysis and direct DNA sequencing (22). It corresponds to preproinsulin I. Pst-excised insert of pI19 was nick-translated and used as probe for further screening of the ds-cDNA library. We identified 48 positive clones. One of them (pI47), also corresponding to preproinsulin I, expresses a fused protein bearing both insulin and penicillinase antigenic determinants. Three more clones (pI7, pI41 and pI20) have been identified as corresponding to preproinsulin II.

Expression of Proinsulin in Bacteria. The proinsulin producing clone, pI47, was detected by the following solid-phase radioimmunoassay (23): Polyvinyl discs coated with an antibody (coating Ab) can bind speciffically antigens released from bacteria. For this purpose, cells from colonies are applied to an agarose/lysozyme/EDTA plate and the released antigens (if any) are absorbed to an antibody-coated polyvinyl disc applied on the plate. The immobilized antigen can then be detected by autoradiography, following incubation in a solution containing radioiodinated antibody (probing Ab). If both the coating and probing Ab are anti-insulin antibodies, a positive signal (spot) on the autoradiogram, corresponding to a colony, indicates the presence of insulin antigenic determinants. If, however, the coating Ab is anti-penicillinase and the probing Ab anti-insulin (or the coating Ab is anti-insulin and the probing Ab anti-penicillinase) a positive signal indicates the production of a fused protein. The positive responses of colonies of clone pI47 with all of the above combinations of antibodies indicated the production of a penicillinase-insulin hybrid polypeptide (Fig.2).

We then showed that this fused protein is secreted into the periplasmic space, as follows: When harvested E.coli cells are suspended in a hypertonic solution (20% sucrose, 30 mM Tris-HCl, pH 8.0, 1 mM EDTA) which does not penetrate them, they shrink. Following pelleting, the cells are rapidly

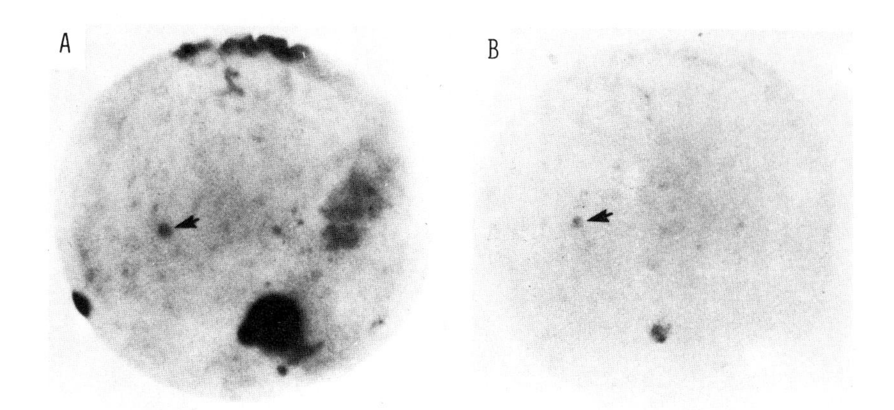

FIGURE 2. Cells from colonies of the 48 insulin cDNA clones and from control colonies, χ1776 and χ1776-pBR322, were applied to an agarose/lysozyme/EDTA plate. Positive controls (5 ng each of insulin and penicillinase) were also spotted on the plate. Antigen was absorbed to an IgG-coated polyvinyl disc. The autoradiograms are of discs with coating Ab anti-insulin (A) or anti-penicillinase (B). The probing Ab was radioiodinated anti-insulin IgG in both cases. The arrows indicate the signal generated by clone pI47. The large exposed area in the lower right of (A) is the positive control for insulin detection. (From ref.3).

dispersed into a hypotonic medium (cold water). Though the osmotic shock does not impair the viability of the bacterial cells, they swell and at the same time release into the water wash proteins, including various enzymes, located in the periplasmic space, between the inner membrane and the cell wall (24). Fig.3 shows that insulin antigen was present in the water wash of the osmotic shock procedure. Quantification by a standard liquid radioimmunoassay (25) indicated a recovery of about 100 molecules per cell.

By DNA sequencing of clone pI47 we deduced the structure of the fused protein: The ds-cDNA lies in the Pst site (which occurs in the region encoding amino acids 181-182 of penicillinase) in the correct orientation and in phase. An oligo(dG)·oligo(dC) bridge of 18 base pairs (encoding six Gly residues) connects the codon of amino acid 182 (Ala) of penicillinase to the codon of the fourth amino acid (Gln) of proinsulin. Since penicillinase is first synthesized as a preprotein with a 23 amino acid leader sequence, we conclude that the structure of the fused protein is penicillinase(24-182)-$(Gly)_6$-proinsulin(4-86) (Fig.4).

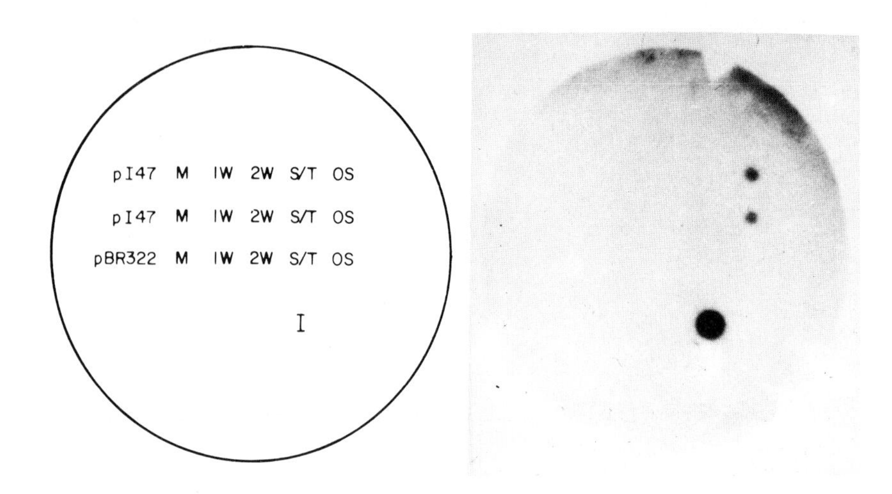

FIGURE 3. Release of insulin antigen from χ1776-pI47 by osmotic shock. Cells from one liter culture ($5x10^7$ cells/ml) were harvested and washed twice with Tris-NaCl. The cells were then osmotically shocked as described in the Text. Aliquots of various fractions were applied to the surface of an agar plate (left).M,medium;1W,first wash supernatant;2W, second wash supernatant;S/T(sucrose/Tris),hypertonic solution supernatant;OS(osmotic shock),water wash;I,insulin(control). Autoradiogram (right) of the solid-phase radioimmunoassay of these fractions. Anti-insulin IgG was used as coating and probing Ab. The labeled areas correspond to the water washes (duplicate) and the positive control (1 ng insulin). (From 3).

The detection of insulin antigenic determinants in the fused protein suggests that the proinsulin portion of the molecule has assumed its correct tertiary structure. However, immunoreactivity does not necessarily imply biological activity. To demonstrate whether or not the insulin material produced by the bacterium is biologically active we employed the standard rat epididymal fat pad bioassay (26). This assay measures the conversion of 1-^{14}C-glucose to $^{14}CO_2$, a process stimulated by insulin. Because proinsulin is only 3-5% as biologically active as insulin in this assay (27), we treated the material present in the water wash of the osmotic shock of χ1776-pI47 with trypsin, under conditions known to convert porcine proinsulin to dealanylated insulin (28). The exact structure of the trypsinization product of the fused protein is unknown. We assume that following the release of the C-peptide, the A-chain of the product is identical to the one of authentic rat insulin I, but its B-chain lacks the C-terminal Ser and has a minimum of twelve extra amino acids at the N-terminus (amino acids 177-182 of penicillinase and

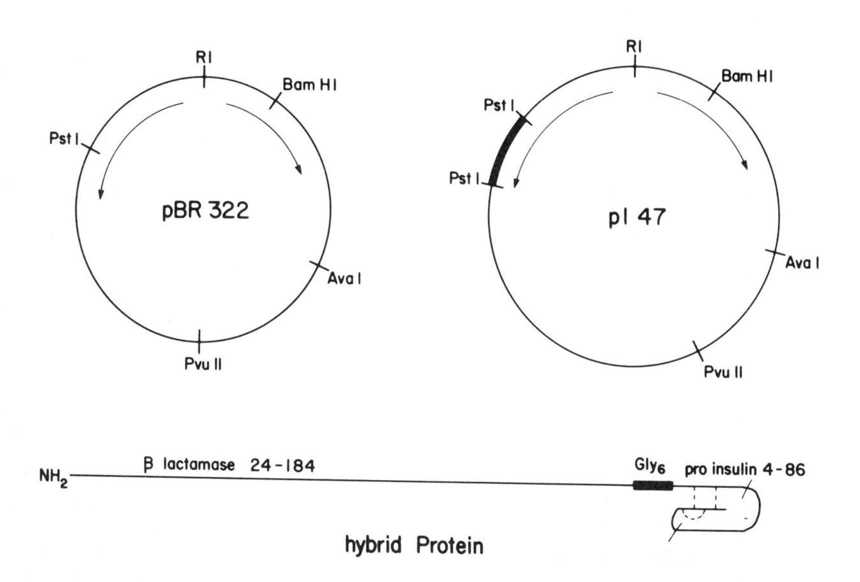

FIGURE 4. Schematic representation of a hybrid protein produced by insertion of ds-cDNA into the Pst site of pBR322. Upper left: pBR322. The position of several restriction sites are indicated for orientation. The arrows indicate the position and direction of transcription of the genes encoding penicillinase (β-lactamase) (left) and tetracycline resistance (right). Upper right: pI47. The inserted ds-cDNA lies between two regenerated Pst sites. The arrow to the left of the RI site indicates the region encoding the fused protein. Bottom: The structure of the fused protein, deduced from DNA sequencing data. The dotted lines indicate disulfide bridges. The boundaries of the C-peptide are also indicated.

six connecting Gly residues). This would be the case if the peptide bond between Arg176 and Asp177 of penicillinase is, under these conditions, susceptible to tryptic cleavage. Table 1 shows that a trypsinized osmotic extract from χ1776-pI47 stimulates the conversion of glucose to CO_2 to a greater extent than an equivalent extract from χ1776-pBR322. The stimulation by the extract from cells containing only the vector plasmid is probably due to the presence of proteolytic enzymes, known to cause nonspecific increase in glucose uptake in this assay (29). The addition of anti-insulin antibodies to the trypsinized extracts has no effect on the stimulation caused by the water wash of χ1776-pBR322, but abolishes the additional stimulation observed with χ1776-pI47 extracts. Though Table 1 presents only one representative experiment, these results were reproduced in six independent

TABLE 1
BIOLOGICAL ACTIVITY OF INSULIN SYNTHESIZED IN BACTERIAL CELLS*

	cpm/mg Tissue	Immunoreactive Insulin (μU/ml)	Biological Activity (μU/ml)
Porcine Insulin Standards			
25 μU/ml	18		
150 μU/ml	105		
Source of Tryptic Hydrolysate			
pBR322	86	0	
pBR322+GPAIS	83		
A. pI47	170	597	
B. pI47+GPAIS	92		
A-B	78		110 (18%)

*To maintain a ratio of substrate protein:trypsin:trypsin inhibitor of 100:1:100, protein concentrations in the water wash of the osmotic shock procedure less than 1 mg/ml were adjusted to this value by the addition of carrier protein (human serum albumin). The protein concentrations in this experiment ranged from 1 to 3 mg/ml. To 10 volumes of osmotic extract,prewarmed to 37^0, 0.25 volumes of trypsin (400 μg/ml) was added. Incubation was continued for 10 min at 37^0. The reaction was stopped by addition of 0.25 volumes of soybean trypsin inhibitor (40 mg/ml). Aliquots of the reaction mixtures were distributed to the flasks of the fat pad bioassay and lyophilized. They were redissolved in buffer containing 1-^{14}C-glucose, and,where appropriate,guinea pig anti-porcine insulin serum (GPAIS). A piece of epididymal fat pad was added to each flask. Incubation was for 2 hr at 37^0, with constant shaking. CO_2 was liberated by the addition of 0.1 volumes of 10 N H_2SO_4 to the medium and was trapped in 0.4 ml hyamine hydroxide contained in a cup suspended from the flask stopper. Following a 70 min incubation at room temperature, the cap containing the hyamine was placed into liquid scintillation fluid and counted.

trials. We conclude that at least 18% of the insulin detected immunologically can be converted to a biologically active form. Alternatively, the trypsinization product of the fused protein might be less active biologically than rat insulin because of the altered structure of its B-chain.

Identification and Characterization of Rat Preproinsulin Chromosomal DNA Clones. DNA sequencing of the insert of cDNA clone pI19 (preproinsulin I) revealed that the cloned DNA encodes the entire preproinsulin chain with the exception of the first two amino acid residues of the pre-region. It is missing, therefore, a copy of the entire 5' noncoding region of the mRNA. It is also missing the 3' noncoding region with the exception of 10 nucleotides immediately following the terminator TGA. Comparison of its restriction map with the maps of clones pI41, pI7 and pI20 showed that the latter correspond to preproinsulin II (Fig.5).

The inserts of clones pI41 and pI7 were completely sequenced. The DNA of clone pI41 encodes only the first 55 amino acids of preproinsulin, but contains 20 5' noncoding nucleotides upstream from the ATG initiator. These data establish the complete amino acid sequence of the preproinsulin II pre-region. Comparison with the pre-region sequence of preproinsulin I reveals that there are three amino acid replacements. These replacements, however, are very conservative, in the sense that in one case a basic amino acid replaces another basic residue and the other replacements are between hydrophobic amino acids.

The sequence of clone pI7 does not overlap with that of pI41. It begins with the codon of amino acid 58 of preproinsulin and ends immediately after the TAG terminator. Partial restriction mapping and DNA sequencing data showed that the insert of clone pI20 covers the sequence of both clones pI41 and pI7.

Though the restriction maps of preproinsulin I and II ds-cDNAs share common sites (Fig.5), there are also significant differences. For example, a Bam and a Sma site (both rare, since their recognition sequences are hexanucleotides) appear in the map of preproinsulin II cDNA. We used these sites to examine the chromosomal gene structures by the Southern DNA blotting technique (29). As a hybridization probe we used again nick-translated insert of clone pI19 and were able to derive preliminary restriction maps of the chromosomal genes. Digestion of rat DNA with Eco RI produced three hybridizing bands of 7.4, 3.5 and 0.8 kb, while Bam digestion produced only two bands of 5.8 and 1.25 kb. The 3.5 kb RI band and the 1.25 kb Bam band were cleaved by Sma. Therefore, they belong to gene II. (Another gene II-specific Bam fragment, predicted from the cDNA map, cannot be seen because the 5' noncoding sequence is missing from the probe). An Eco RI and Bam double digest showed that RI cleaves the 1.25 kb Bam fragment into two fragments, both of which hybridized to the probe. Therefore, since an RI site does not appear in the gene II cDNA sequence, we concluded that this

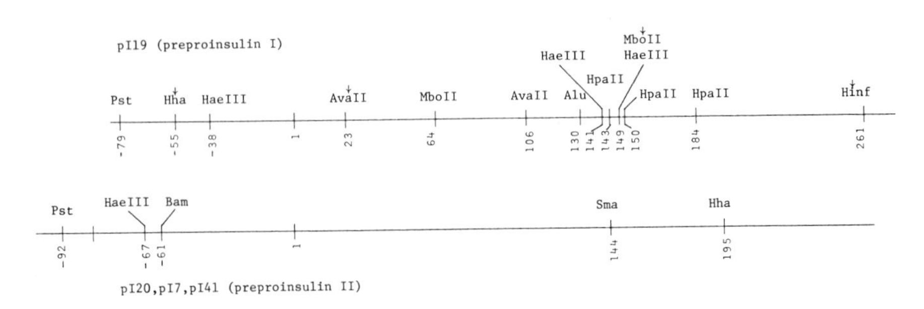

FIGURE 5. Restriction maps of preproinsulin I and II cDNA clones. Each restriction site is identified by a number indicating the 5' terminal nucleotide generated by cleavage at the message strand. Nucleotides are numbered beginning with the first base of the sequence encoding proinsulin. Nucleotides in the 5' direction from position 1 in the message strand are identified by negative numbers, beginning with -1. Arrows indicate sites of the preproinsulin I cDNA sequence that do not exist in the preproinsulin II sequence.

gene is interrupted by an intron.

The preliminary genomic restriction map proved important for the identification of insulin chromosomal DNA clones after their isolation from a rat DNA library. This library was constructed by Sargent, Wallace and Bonner, essentially as described by Maniatis et al. (1): Fragments produced by partial Eco RI digestion of the DNA purified from the liver of a single Sprague-Dawley rat were directly ligated to the purified RI arms of phage λ Charon 4A. Using nick-translated insert of clone pI19 we screened by the Benton and Davis method (30) 800,000 recombinant plaques and identified three preproinsulin gene clones. Blotting of recombinant phage DNA fragments produced by RI or Bam digestion revealed that one clone (20 kb insert) carries gene I, while the other two clones (16.7 kb inserts) carry gene II, but in opposite orientations.

A detailed restriction map of one of the gene II clones, designated λ Charon 4A-rI2 or simply rI2, was constructed and the region containing the gene plus 5' and 3' adjacent sequences was sequenced (Fig.6).

The DNA sequence verified the existence of an intron of about 490 bp interrupting the coding region between the codons of amino acids 61 and 62 of preproinsulin. To position the 5' noncoding region of the mRNA on this sequence we extended the sequence known from the preproinsulin II cDNA clones towards the 5' direction as follows: DNA of clone pI41 was digested with Bam and end-labeled. It was then digested with Hae III

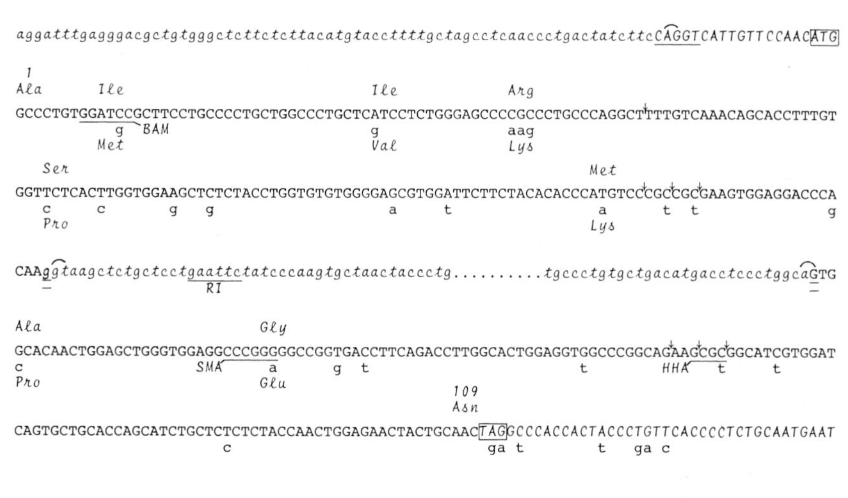

FIGURE 6. Nucleotide sequence of the rat preproinsulin II gene. The nucleotide sequence of the message strand is displayed 5' to 3'. The position of the first nucleotide following the cap structure in the mature mRNA is not known and probably lies upstream from the displayed sequence. The initiation and termination codons are boxed. Small letters indicate intronic and 3' adjacent sequence. The arrows indicate, in order, the boundaries of the pre-region and the peptides B,C and A. Small letters in a second line indicate nucleotide substitutions in the preproinsulin I cDNA sequence.

and a small DNA fragment (corresponding to amino acids 1-4 of the pre-region) was purified by gel electrophoresis. This fragment, already labeled at the 5' end of the antimessage strand, was denatured and used as primer in a reverse transcription reaction, employing as template oligo(dT)-cellulose bound RNA from the rat insulinoma. The cDNA product was sequenced directly. Thus the 5' noncoding sequence known from the cDNA clone pI41 (23 nucleotides including the initiator) was extended by 36 nucleotides. Though the full length of the 5' noncoding region is still unknown, this additional sequence revealed a leader sequence on the mRNA, because a small intron of 119 bp exists in the chromosomal DNA, 17 bp upstream from the ATG initiator. This small intron has a five nucleotide terminal redundancy at its boundaries with the exons, while the large intron has only a one nucleotide redundancy. By appropriate selection of the splicing point, it can be shown

(Fig.6) that both introns follow the general rule introduced by Chambon (31) (i.e. they begin with a GT and end with an AG dinucleotide).

The 3' noncoding region has a length of 56 nucleotides including the terminator and its end is position by comparison to the 3' noncoding region of preproinsulin I mRNA, reported by Ullrich et al. (18).

Restriction endonuclease analysis of the clone carrying gene I, designated rI1, did not reveal the existence of an intron in the coding region. This was supported by electron microscopic data. A heteroduplex formed between two large restriction fragments of the two clones (a 7.4 kb RI fragment of rI1 and a 4 kb BglII fragment of rI2) has a loop of about 0.5 kb interrupting the region of homology. The 5' and 3' direction of the heteroduplex can be deduced from the lengths of the nonhomologous tails of the restriction fragments, because the distance between their ends and the gene is known from the genomic DNA maps. The 3' regions following the looped-out intron are homologous, but not for a distance longer than the 3' noncoding region of the mRNAs. In contrast, the 5' region is homologous for a length of about 1 kb. This homology suggests that the small intron also exists in rI1. These data cannot exclude the possibility, however, that a very small intron (of the order of 5-10 bp) interrupts the coding region of rI1. For this reason, we sequenced the region of rI1 corresponding to the region of rI2 where the large intron occurs. The results show unambiguously that the coding region of rI1 is not interrupted by an intron.

Transcription of the Preproinsulin II Gene. The first indication for the presence of preproinsulin mRNA precursors in nuclear RNA purified from the rat insulinoma was derived from RNA blotting (32) experiments. The results, however, were not unambiguous, because of high hybridization background. We decided, therefore, to examine the RNA precursors by electron microscopy, using the R-looping technique (33). Nuclear RNA (contaminated with mature size preproinsulin mRNA, as shown by RNA blotting) was R-looped to slightly crosslinked clone rI2 DNA. Most of the R-loops observed were typical of mature mRNA: one DNA strand was displaced, while the middle of the hybridized strand (about 0.5 kb) was looped-out, presumably because the intronic sequence was spliced out from the mRNA. Looping out at the region of the 119 bp intron was not observed, probably because of the small size of this structure. A few larger R-loops (about 1100 bp) were also observed. They were smooth, without looping out of the hybridized strand. We conclude that as in the case of globin mRNA (34), preproinsulin mRNA is initially transcribed in the form of a precursor and goes through a maturation process. This result does not exclude the possibility that an even longer primary transcript might exist.

DISCUSSION

Except for the molecular characterization of yet another gene, these studies resulted in the following important observation: The two rat preproinsulin genes, presumably the products of gene duplication, have different structures, in the sense that one of them lacks the intronic sequence which interrupts the coding region of the other. Both these genes are equally expressed. This is an indication (at least in this system) that an intron per se is not directly involved in expression at the DNA or RNA level. (However, subtle regulatory mechanisms could be postulated. One could imagine, for example, two different β-cell populations, expressing separately each gene).

The insulin system provides now the exciting opportunity of studying the evolution of a duplicated gene by examining its structure before the duplication event, probably in an invertebrate species. It remains to be seen whether or not other mammals in which insulin II was never described have only one gene. The implications of the observed different structures of the two rat preproinsulin genes are discussed in more detail by Gilbert elsewhere in this volume.

REFERENCES

1. Maniatis,T., Hardison,R.C., Lacy,E., Lauer,J., O'Connell,C. Quon,D., Sim,G.K., and Efstratiadis,A. (1978). Cell 15,687.
2. Efstratiadis,A., and Villa-Komaroff,L. (1979). In "Genetic Engineering:Principles and Methods" (Setlow,J.K. and Hollaender,A., eds.), Vol.1, pp. 15-36. Plenum Press, New York.
3. Villa-Komaroff,L., Efstratiadis,A., Broome,S., Lomedico,P., Tizard,R., Naber,S., Chick,W., and Gilbert,W. (1978). Proc. Nat. Acad. Sci. USA 75, 3727.
4. Falkmer,S., Endin,S., Havu,N., Lundgren,G., Marques,M., Ostberg,Y., Steiner,D., and Thomas,N.W. (1973). Amer. Zool. 13, 625.
5. Falkmer,S., and Ostberg,Y. (1977). In "The Diabetic Pancreas" (Volk,B.W., and Wellmann,K.F., eds.) pp. 15-59. Plenum Press, New York.
6. De Martinez,N.R., Garcia,M.C., Salas,M., and Candela, J. (1973). Gen. Comp. Endocrinol. 20, 305.

7. Meneses,P., and Ortiz,M. (1975). Comp.Biochem.Physiol. 51A, 483.
8. Tager,H.S., Markese,J., Kramer,K.J., Speirs,R.D., and Childs,C.N. (1976). Biochem.J. 156, 515.
9. Dayhoff,M.O. (1972). Atlas of Protein Sequence and Structure, Vol. 5. Biomedical Research Foundation, Bethesda.
10. Humbel,R.E., Bosshard,H.R., and Zahn,H. (1972). In "Handbook of Physiology.Endocrinology" pp.111-132. Am.Physiol.Soc.,Washington,D.C.
11. Smith,L.F. (1966). Am.J.Med. 40, 662.
12. Clark,J.L., and Steiner,D.F. (1969). Proc.Nat.Acad. Sci. USA 62, 278.
13. Steiner,D.F., Kemmler,W., Clark,J.L., Oyer,P.E.,and Rubenstein,A.H. (1972). In "Handbook of Physiology. Endocrinology." pp. 175-198. Am.Physiol.Soc.,Washington.
14. Steiner,D.F., Kemmler,W., Tager,H.S., and Peterson,J.D. (1974). Fed. Proc. 33, 2105.
15. Chan,S.J., Keim,P., and Steiner,D.F. (1976). Proc. Nat.Acad.Sci.USA 73, 1964.
16. Chan,S.J., and Steiner,D.F.(1977). Trends Biochem.Sci. 2, 254.
17. Duguid,J.R., Steiner,D.F., and Chick,W.L. (1976). Proc. Nat. Acad.Sci. USA 73, 3539.
18. Ullrich,A., Shine,J., Chirgwin,J., Pictet,R., Tischer,E., Rutter,W.J., and Goodman,H.M. (1977). Science 196, 1313.
19. Chick,W.L., Warren,S., Chute,R.N., Like,A.A., Lauris,V., and Kitchen,K.C. (1977). Proc.Nat.Acad.Sci.USA 74, 628.
20. Sutcliffe,J.G. (1978). Proc.Nat.Acad.Sci.USA 75, 3737.
21. Paterson,B.M., Roberts,B.E., and Kuff,E.L. (1977). Proc.Nat.Acad.Sci.USA 74, 4370.
22. Maxam,A.M., and Gilbert,W. (1977).Proc.Nat.Acad.Sci.USA 74, 560.
23. Broome,S., and Gilbert,W. (1978). Proc.Nat.Acad.Sci.USA 75, 2746.
24. Neu,H.C., and Heppel,L.A. (1965). J.Biol.Chem. 240, 3685.
25. Makulu,D.R., Vichick,D., Wright,P.H., Sussman,K.E., and Yu,P.L. (1969). Diabetes 18, 660.
26. Renold,A.E., Martin,D.B., Dagenais,Y.M., Steinke,J., Nickerson,R.J., and Sheps,M.S. (1960). J.Clin.Invest. 39, 1487.
27. Kitabchi,A.E., Duckworth,W.C., Stentz,F.V., and Yu,S. (1972). CRC Crit.Rev.Biochem. 1, 59.
28. Chance,R.E., Ellis,R.M., and Broome,W.W. (1968). Science 161, 165.
29. Southern,E.M. (1975). J.Mol.Biol. 98, 503.
30. Benton,W.D., and Davis,R.W. (1977). Science 196, 180.

31. Breathnach,R., Benoist,C., O'Hare,K., Gannon,F., and Chambon,P. (1978). Proc.Nat.Acad.Sci.USA 75, 4853.
32. Alwine,J.C., Kemp,D.J., and Stark,G.R. (1977). Proc. Nat.Acad.Sci.USA 74, 5350.
33. Kaback,D.B., Angerer,L., and Davidson,N. (1979).(Submitted for publication).
34. Tilghman,S.M., Curtis,P., Tiemeier,D.C., Leder,P., and Weissmann,C. (1977). Proc.Nat.Acad.Sci.USA 74, 3184.

THE LINKAGE ARRANGEMENT OF MAMMALIAN β-LIKE GLOBIN GENES

T. Maniatis, E. T. Butler III, E. F. Fritsch, R. C. Hardison, E. Lacy, R. M. Lawn, R. C. Parker, and C.-K. J. Shen

Division of Biology, California Institute of Technology, Pasadena, California 91125

ABSTRACT The linkage arrangement of human and rabbit β-like globin genes has been determined by analyzing overlapping segments of cloned genomic DNA. In both systems, four different genes have been localized within a 40 kilobase region of chromosomal DNA. The two human fetal β-like genes designated $^{G}\gamma$ and $^{A}\gamma$ and the adult genes, δ and β are transcribed from the same DNA strand in the orientation 5' $^{G}\gamma$-$^{A}\gamma$-δ-β 3'. Similarly, two embryonic rabbit β-like genes which we have designated β4 and β3 and two adult genes (β2 and β1) are transcribed from the same DNA strand in the orientation 5' β4 β3 β2 β1 3'. At least one intervening sequence has been identified in each of the eight genes. The possible significance of the organization of globin genes in closely linked clusters is discussed.

INTRODUCTION

Until recently, studies of the mechanism of differential globin gene expression during development were limited to the analysis of globin polypeptides in erythroid cells obtained from embryos of increasing maturity. However, with the development of gene isolation procedures (1-3), it has become possible to study globin gene expression at the nucleic acid level. To this end, we have been isolating and characterizing rabbit and human globin genes and their flanking sequences. The rabbit provides a system in which we can examine the temporal expression of globin genes during early development, while the human globin gene system provides an opportunity for studying the molecular basis of mutations which alter the normal pattern of globin gene expression.

The approach we have taken to globin gene isolation is to construct libraries of random, high molecular weight genomic DNA and then to screen these libraries with gene-specific hybridization probes (2,4). With this procedure, sets of overlapping DNA segments can be obtained which makes

ISBN 0-12-068350-4

it possible to study the sequences extending many kilobases from the gene on either side. Analysis of DNA sequences adjacent to a given gene is particularly important in the study of mammalian globin genes since genetic linkage between different β-like genes has been demonstrated in man (5,6) and mice (7,8). The isolation of linked globin genes and the sequences which separate them will provide the opportunity for studying possible relationships between gene linkage and coordinate and/or differential gene expression, and may lead to the identification of linked, non-globin sequences which are expressed in erythroid cells. In this paper, we will describe our current understanding of the structure and linkage arrangement of the β-like globin genes in human and rabbit DNA. In both systems, a physical map of over 40 kb of chromosomal DNA containing four different β-like globin genes has been derived.

THE LINKAGE ARRANGEMENT OF HUMAN β-LIKE GLOBIN GENES

Five different human β-like globin genes have been identified. In the early embryo the ε-chain is produced in nucleated erythroid cells of the yolk sac (6). When the site of erythroid cell production changes from the yolk sac to the fetal liver, the ε-chains are replaced by the products of two non-allelic γ-globin genes. These genes code for polypeptide chains which have either glycine ($^{G}\gamma$) or alanine ($^{A}\gamma$) in position 136 of their amino acid sequence (9). Just prior to birth, the adult β- and δ-globin chains appear, while γ chain synthesis diminishes. By 6 months, hemoglobin A ($\alpha_2\beta_2$) and hemoglobin A_2 ($\alpha_2\delta_2$) represent greater than 98% of the hemoglobin in peripheral blood (6).

Analysis of two structural mutants, hemoglobin Lepore (10) and Kenya (11) suggested that the $^{A}\gamma$, δ- and β-globin genes are closely linked. In hemoglobin Lepore, the NH_2-terminal amino acid sequence of the δ-globin protein is joined to the C-terminal sequence of β-globin, while the $^{A}\gamma$ and β-globin polypeptide chains are fused in a similar fashion in hemoglobin Kenya. Both of these fusion proteins are thought to result from an unequal crossing-over event during meiosis between homologous sequences within linked β-like globin genes. These predictions were recently confirmed when the physical linkage between the δ- and β-globin genes was demonstrated using blot hybridization (12) and gene cloning (4) procedures. As shown in Figure 1, both the δ- and β-globin genes were isolated on a single 14.5 kilobase fragment of human DNA. The two genes which are separated by approximately 5.4 kilobases of DNA are arranged in the order 5' δ-β 3' with respect to the direction of transcription. Physical linkage between the two γ-globin genes in the arrangement 5' $^{G}\gamma$ - $^{A}\gamma$ 3' was demonstrated by blot hybridization experiments (13) and confirmed by the isolation of a cloned DNA fragment containing the entire $^{G}\gamma$ gene and the 5'-end of the $^{A}\gamma$ gene (unpublished results from our laboratory).

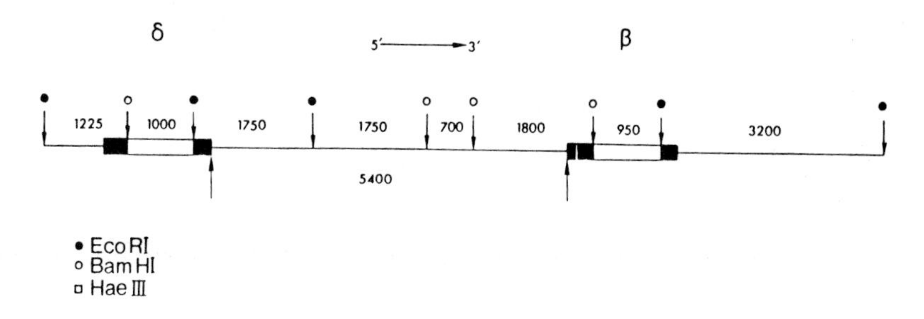

FIGURE 1. Location of δ- and β-globin genes in HβG2 DNA. The globin gene-containing region of HβG2 DNA is shown in this restriction endonuclease map. Arrows pointing down indicate restriction endonuclease sites of Eco RI (•) and Bam HI (o). Sizes of the fragments are given in base pairs. The boxed regions denote the δ- and β-globin genes. The filled boxes represent the mRNA sequence and the open boxes represent noncoding intervening sequences. The arrows pointing up delineate the distance between the linked genes. (Reprinted from Cell by permission.)

We have recently shown that the $^{A}\gamma$ gene is linked to the δ-globin gene in the orientation shown in Figure 2 (14). As shown in Figure 3, a 15.5 kb Bam HI fragment was identified in normal DNA by blot hybridization experiments using any of 3 different probes: 3'-specific γ-gene sequence; a DNA fragment which maps 4 kb to the 5'-side of the δ-globin gene (RIH); or a β-globin gene probe. This observation strongly suggested that a single 15.5 kb Bam HI fragment spans the region between the $^{A}\gamma$ and δ-globin genes. To test this prediction, we made use of a deletion in the DNA from an individual homozygous for hereditary persistence of fetal hemoglobin (HPFH). This deletion removes the δ- and β-globin genes and extends approximately 4 kilobases to the 5'-side of the -globin gene (see Figure 2; 14). If the linkage arrangement proposed above is correct, the size of the 15.5 kb Bam HI fragment observed in normal DNA should be altered by the HPFH deletion when either the RIH or γ-globin gene probes are used to detect the fragment. As predicted, the 15.5 kb Bam HI fragment of normal DNA is not seen in HPFH DNA. Instead, a 14 kb Bam HI fragment is observed when either the γ- or RIH probe is used to detect the fragment (Figure 3). A similar result was obtained when the enzyme Hpa I was used in an identical experiment (data not shown, 14). Since the RIH fragment was derived from a cloned DNA fragment containing both the δ- and the β-globin genes and since the experiment described above shows that the RIH fragment is also linked to the $^{A}\gamma$-globin gene, the $^{A}\gamma$ gene must therefore be physically linked to the δ-gene. The arrangement of the four linked β-like globin genes is therefore 5' $^{G}\gamma$-$^{A}\gamma$-δ-β 3' as shown in Figure 2.

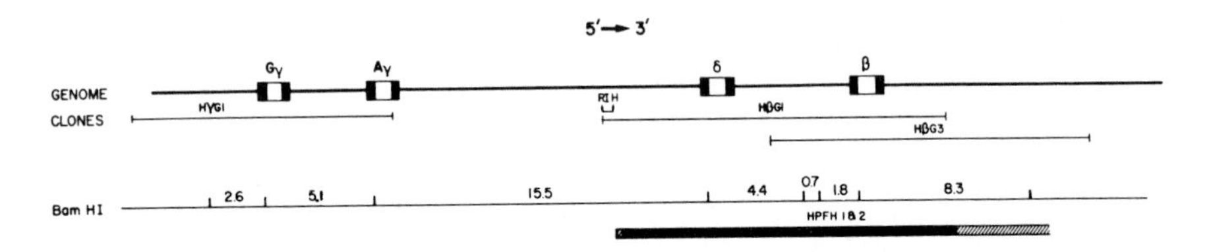

FIGURE 2. Linkage arrangement of the human γ-, δ-, and β-globin genes. The arrangement of fetal and adult β-like globin genes within a 43 kilobase (kb) segment of human DNA is presented. Physical linkage of the adult δ- and β-globin genes (4,12) and the fetal $^{G}\gamma$- and $^{A}\gamma$-globin genes (13) has been previously described. The physical linkage of the fetal and adult globin genes is described in this paper (see also 14). The direction of transcription of the four linked genes is left to right (5' → 3'). The solid boxes represent the locations of mRNA coding regions. The open boxes represent the large noncoding intervening sequence in each gene (4,12-15,17). The upward pointing arrows beneath the map delineate the position of the RIH fragment which is used as a hybridization probe (see Figure 3). The brackets below the gene map indicate the cloned DNA segments from this locus (4 and unpublished results). The locations of Bam HI cleavage sites are marked by a vertical line. The sizes for the restriction enzyme fragments are given in kb. The region of the β-like globin gene locus deleted in DNA from two different individuals with hereditary persistence of fetal hemoglobin (HPFH) DNA are indicated by solid boxes. The precise locations of the endpoints of the deletions are within the regions specified by the hatched boxes. The locations of the rightward ends of the deletions are not known.

In summary, a combination of genomic blotting (12,13,15) and gene cloning experiments (4) have provided a detailed physical map (14) of nearly 43 kilobases of human chromosomal DNA containing 4 different globin genes. This map and the well characterized cloned DNA fragments carrying each of the β-like globin genes within the cluster should prove to be important tools for studying the mechanism of the switch from fetal to adult globin gene expression at the level of RNA transcription and processing. Furthermore, with well characterized cloned DNA for comparison, it will be possible to identify the molecular basis of genetic diseases in hemoglobin expression. In fact, we have recently made use of this information to map deletions which alter the expression of fetal globin genes in adults (14).

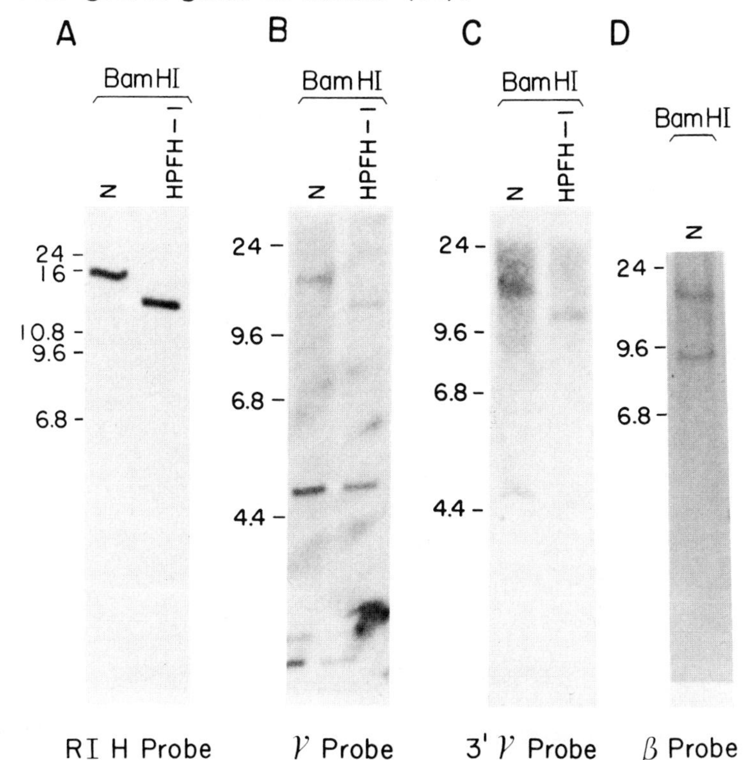

FIGURE 3. Hybridization of normal and HPFH DNA with cloned β- and γ-globin cDNA and the RIH fragment: linkage of fetal and adult β-like globin genes. Normal and HPFH DNAs were digested with Bam HI, electrophoresed in agarose gels, transferred to nitrocellulose filter paper, and hybridized with the specific hybridization probe indicated below each panel (see 14). The β- and γ-probes were the cDNA plasmids pJW102 and pJW151, respectively (30). The 3' γ probe was prepared from pJW151 DNA. The RIH probe is a 500 bp sequence isolated from HβG1 DNA and is located approximately 4 kb to the 5'-side of the δ-globin gene. The size markers are indicated in kb.

THE STRUCTURE OF HUMAN β-LIKE GLOBIN GENES

A large non-coding intervening sequence was first discovered within the rabbit (16) and mouse (1) β-globin genes. Structural analysis of the β- and γ-globin mRNAs and their corresponding cloned cDNAs made it possible to identify at least one intervening sequence in each of the 4 human β-like globin genes (4,12-15). Direct DNA sequence analysis of cloned δ- (4), β- (4) and $^{A}\gamma$- (17) globin genes indicates that each of these genes contains a large intervening sequence between codons 104 and 105. A second, smaller intervening sequence similar to that found in mouse (18) and rabbit (19, and Efstratiadis, Lacy and Maniatis, unpublished) globin genes has been identified in the β- (4) and $^{A}\gamma$- (17) genes. It is likely that a second intervening sequence will be found within the $^{G}\gamma$- and δ-globin genes. Figure 4 shows a comparison of the organization of intervening sequences in the rabbit and human β-globin genes. Although the sizes of the two intervening sequences in the two genes are different, their locations are identical. In fact, two intervening sequences within the mouse α-globin gene are also located in the same relative position (20). This is a remarkable observation considering that the β-globin genes probably arose by duplication of an α-like gene over 500 million years ago (20,21). Thus, the acquisition of the intervening sequences must have been a very early evolutionary event, possibly concomitant with the formation of a functional ancestral gene.

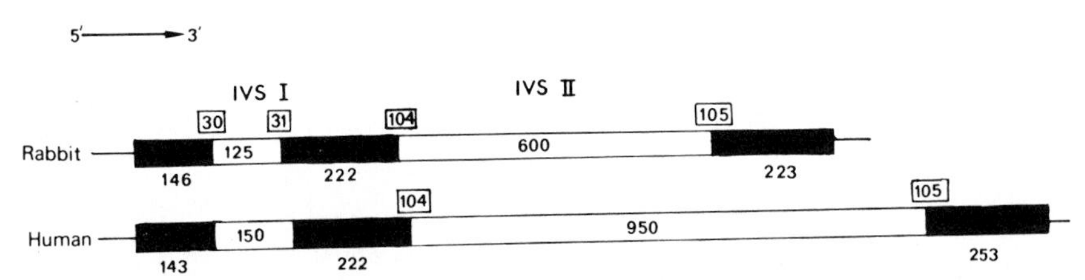

FIGURE 4. Intervening sequences in the rabbit and human β-globin genes. The location and sizes (in base pairs) of the intervening sequences in the rabbit and human adult β-globin genes are shown for comparison. In both genes, a small intervening sequence (IVS I) is located between codons 30 and 31 and a larger intervening sequence (IVS II) is found between codons 104 and 105 (14, 19, unpublished results). The black boxes represent coding plus untranslated mRNA sequences while the white boxes represent the non-coding intervening sequences. The direction of transcription is indicated by the arrow.

The sequence homology between the rabbit and human globin intervening sequences has not been studied. A comparison between the large intervening sequence of the rabbit and mouse β-globin genes (18) indicate that with the exception of 10 nucleotides at the junctions between coding and noncoding sequences, the two sequences are not highly conserved. Similarly, there appears to be only slight sequence homology between the intervening sequences in the mouse β-major and β-minor globin genes (22) and we (4) detect little (if any) cross-hybridization between the large intervening sequences in the human δ- and β-globin genes. In contrast, the coding sequences of mammalian β-like globin genes are highly conserved (19,23). This relatively rapid evolutionary divergence of intervening sequences suggests that only a small fraction of the noncoding sequences are required for accurate processing. Tiemeier et al. (22) have argued that rapid divergence of intervening and flanking sequences in duplicated globin genes would reduce the target size for unequal crossing-over events during meiosis and thus prevent the deletion or further amplification of genes which arise during evolution by duplication of an ancestral gene.

GENETIC POLYMORPHISM WITHIN INTERVENING SEQUENCES

Allelic polymorphism within the coding sequence of human globin genes has been described (6). We have detected sequence polymorphism within the large intervening sequence of the δ-globin genes from homologous chromosomes of a single individual (4). Two independently derived clones selected from a library prepared from the DNA of a single individual display different Pst I digestion patterns. As shown in Figure 5, digestion of a clone designated HβG1 (lane D) with Pst I yields two fragments (4.4 and 2.3 kb) which hybridize to the β-globin probe. When the DNA from another clone designated HβG2 (lane C) is similarly digested, three hybridizing fragments are produced (4.4, 1.35, and 0.95 kb), only one of which (4.4 kb) comigrates with a Pst I fragment from HβG1 DNA. The combined sizes of the smaller hybridizing fragments from HβG2 DNA (1.35 kb and 0.95 kb) are equal to the size of the smaller Pst I fragment (2.3 kb) of HβG1. This result indicates that either the δ- or β-globin gene in HβG2 DNA is cleaved by Pst I. Additional mapping and hybridization experiments revealed that this cleavage site is located within the intervening sequence of the δ-globin gene (4) (see Figure 6). Thus, it appears that the human DNA fragments in HβG1 and HβG2 were derived from homologous chromosomes which are heterozygous with respect to the Pst I site in the intervening sequence of the δ-globin gene. This heterozygosity is confirmed by the data presented in Figure 5

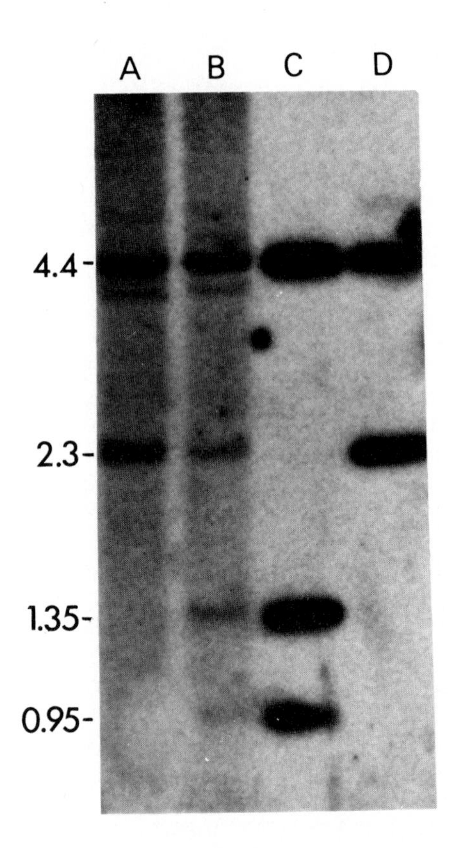

FIGURE 5. Blot hybridization analysis of polymorphism within the δ-globin intervening sequence. DNAs were digested with Pst I and analyzed by blot hybridization (see Figure 3 and ref. 14). The hybridization probe was a 2.3 kb fragment of HβG1 DNA containing the δ-globin gene and some flanking sequences (see Figure 6). Sizes are given in kb. The faint bands seen slightly above and below the strong band at 4.4 kb represent γ-globin gene containing Pst I fragments detected weakly by the δ-globin probe. (A) Normal human spleen DNA. (B) Fetal liver DNA from which the clones HβG1 and HβG2 were isolated. (C) HβG2 DNA. (D) HβG1 DNA.

(lanes A and B). The fetal liver DNA used to prepare the library from which HβG1 and HβG2 were derived was digested with Pst I, fractionated by agarose gel electrophoresis, transferred to nitrocellulose paper, and hybridized to a subclone of HβG1 containing the δ-globin gene. As can be seen in lane B of Figure 5, all of the hybridizing fragments which hybridize the δ-globin probe in HβG1 (lane D) and HβG2 (lane C) DNA are present in the fetal liver DNA. The fact that the large Pst fragment (2.3 kb) as well as the two smaller fragments (1.35 and 0.95 kb), which are generated by cleavage of the δ-globin gene intervening sequence, are present in this DNA clearly indicates that one of the two δ-globin gene copies in the diploid genome contains a Pst I site while the other does not. In contrast, a homozygous pattern was observed in the DNA from another individual. This pattern (Figure 5, lane A) is identical to the HβG1 pattern (Figure 5, lane D) indicating that neither chromosomal copy of the δ-globin gene contains an internal Pst I site. A similar polymorphism has been observed within one of the intervening sequences of the ovalbumin gene (24-26). In this case, the polymorphic alleles differ from each other by the presence or absence of an Eco RI cleavage site in one of the seven intervening sequences of the gene. Sequence analysis of cloned DNA suggests that the polymorphism results from a single base change in the Eco RI cleavage site (26).

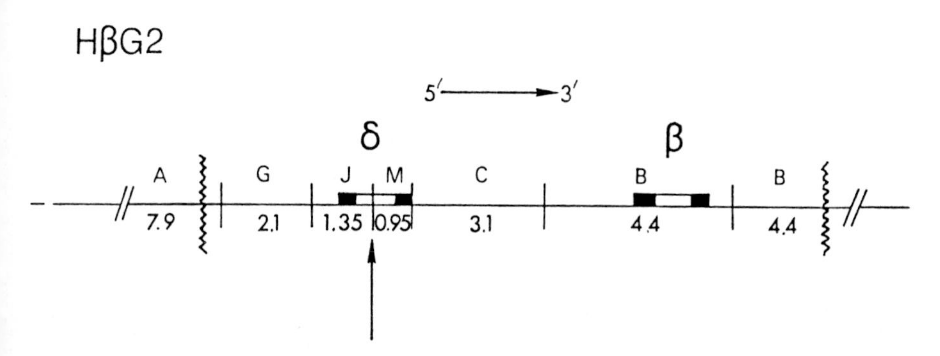

FIGURE 6. A map of Pst I sites in HβG2 DNA. A Pst I map of the region of HβG2 DNA containing the human DNA insert is shown. The Pst I site between fragments J and M of HβG2 (marked by an arrow) is absent in HβG1. A fragment of 2.3 kb in length replaces the Pst J and M fragments of HβG2. The wavy lines represent the junctions between the λ Ch4A DNA and inserted human DNA.

THE LINKAGE ARRANGEMENT OF RABBIT β-LIKE GLOBIN GENES

The rabbit globin gene system affords the opportunity for studying patterns of gene transcription during early development as well as during adult red cell maturation. For example, it is possible to obtain nucleated erythroblasts from the yolk sac blood islands of rabbit embryos in the amounts required for transcriptional mapping experiments. In addition, selective amplification of the red cell population in bone marrow can be accomplished by phenylhydrazine-induced anemia. Thus, relatively large numbers of adult immature nucleated red cells can be obtained. This makes it possible to identify globin mRNA precursors and study the mechanism by which they are processed (Flavell et al., this Symposium; our unpublished experiments).

The rabbit system is also interesting from a comparative point of view. As mentioned above, two switches in human β-like globin gene expression during development have been observed. In contrast, only one switch occurs in rabbits and many other mammals. For example, in rabbits two embryonic β-like polypeptide chains (ε_Y and ε_Z) are found in the nucleated erythroid cells derived from yolk sac blood islands and persist until the time that the fetal liver becomes the center for erythropoiesis (27,28). At this point the ε polypeptides are replaced by the adult β-globin protein. A similar pattern of mouse embryonic and adult globin gene expression has been reported (27,28). Although the possibility of genetic linkage between embryonic and adult β-related globin genes has not been studied in rabbits, such linkage has been demonstrated in the mouse. As will be described below, we have found that the rabbit genes are in fact closely linked and that their linkage arrangement is remarkably similar to that of the human β-like globin genes described above.

A library of 750,000 independent recombinant phage, containing large (15-19 kb) inserts of rabbit genomic DNA, was screened with labeled rabbit globin cDNA, and nine different hybrid phage were identified as containing globin sequences. Four of these clones hybridized strongly to an adult β-globin cDNA plasmid (pβG1, 29) while the remaining five hybridized more strongly to either embryonic rabbit globin cDNA or to a human γ-globin cDNA plasmid (pJW151, 30). Detailed restriction endonuclease cleavage analysis and hybridization experiments indicated that all nine clones could be arranged as overlapping segments of chromosomal DNA as indicated in Figure 7. The gene-containing regions of these clones were identified by Southern transfer and hybridization experiments and by the examination of R-loops between the cloned DNA and embryonic or adult globin mRNA. The locations of the mRNA hybridizing regions of the gene cluster are indicated in Figure 7.

IDENTIFICATION OF RABBIT β-LIKE GLOBIN GENES

The gene designated β1 in Figure 7 has been identified as the adult β-globin gene by restriction endonuclease cleavage analysis and by DNA sequencing studies (Efstratiadis,

Lacy and Maniatis, unpublished). As shown in Figure 4, the gene contains two intervening sequences, one of 125 bp located between codons 30 and 31 and another of 587 bp located between codons 104 and 105. Similar conclusions were recently reported by Van den Berg et al. (19). Thus, the structural organization of the rabbit adult β-globin gene is identical to that of the two adult mouse β-globin genes and to the human δ- and β-globin genes described above. The transcriptional orientation of gene β1 which is indicated in Figure 7 was determined by hybridization and restriction mapping experiments using hybridization probes specific for the 5' or 3' ends of the β-globin mRNA.

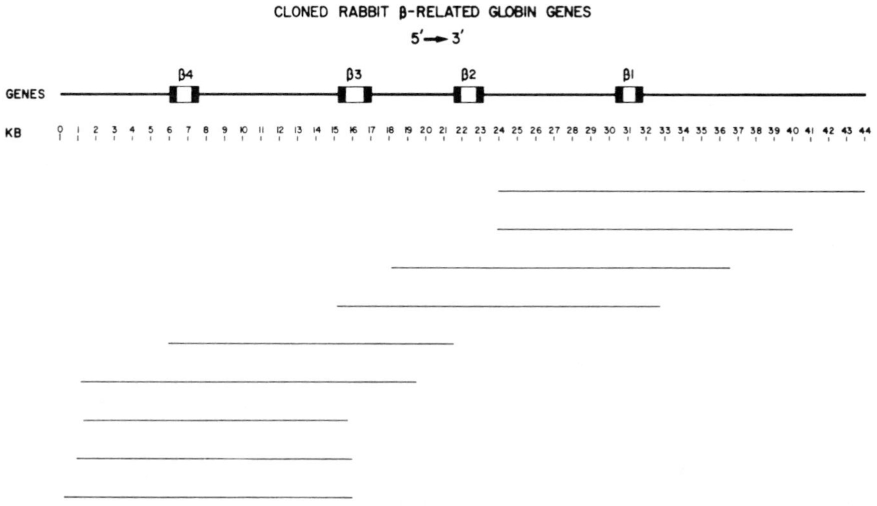

CLONED RABBIT β-RELATED GLOBIN GENES

5'→3'

GENES β4 β3 β2 β1

KB 0 1 2 3 4 5 6 7 8 9 10 11 12 13 14 15 16 17 18 19 20 21 22 23 24 25 26 27 28 29 30 31 32 33 34 35 36 37 38 39 40 41 42 43 44

FIGURE 7. The linkage arrangement of rabbit β-like globin genes. The relative positions of four different rabbit β-like globin genes are shown. The direction of transcription of all of the genes is indicated by the arrow. The series of horizontal lines below the map indicates the regions of rabbit chromosomal DNA carried within 9 independently derived recombinant phage. The scale below the map is in kilobase pairs.

The gene designated β2 is located to the 5' side of the adult β-globin gene (β1) and hybridizes to adult β-globin sequences. We have used hybridization probes corresponding to the 5' or 3' ends of the adult β-globin mRNA to localize the corresponding sequences on restriction fragments which map within the β2 region. These experiments revealed that genes β1 and β2 are transcribed from the same DNA strand and that β2 contains at least one intervening sequence. We have preliminary evidence for an RNA transcript in the bone marrow of anemic rabbits which specifically hybridizes to the β2 sequence. It is not known whether this transcript corresponds to a β-like mRNA sequence found in immature erythroblasts from anemic rabbits (31).

The genes designated β3 and β4 appear to encode embryonic ε-globins. Both genes hybridize efficiently to embryonic globin cDNA and they form stable R-loops with embryonic globin mRNA. Examination of these R-loops in the electron microscope reveals that both genes contain at least one intervening sequence. Since genes β3 and β4 efficiently hybridize to a cDNA clone of human γ-globin mRNA, hybridization probes specific for 5' or 3' γ-mRNA sequences were prepared from the cDNA clone to determine the transcriptional orientation of the β3 and β4 genes. As shown in Figure 7, the embryonic genes are transcribed from the same DNA strand as the adult β1 and β2 genes. Thus, the organization of the β-like rabbit globin genes is quite similar to that of the human globin genes.

Figure 8 shows a comparison of the linkage arrangement of the human and rabbit β-related globin genes. In both cases, all four genes are transcribed from the same DNA strand and the embryonic or fetal genes are located to the 5' side of the adult genes. The rabbit embryonic genes which function in both the early embryo and the fetus are approximately 6 kb away from the adult genes. In contrast, the human fetal genes are located approximately 15 kb away from the adult genes. Preliminary data indicate that an embryonic β-like gene (ε) is closely linked to the $^{G}\gamma$ globin gene.

REPEATED SEQUENCES WITHIN THE RABBIT β-GLOBIN GENE CLUSTER

We have recently begun to study the arrangement of nonglobin sequences which are repeated within the 40 kb region of rabbit DNA containing genes β1 and β4. The objective of this study is to determine whether the regions flanking the four different globin genes share sequence homology. If similar sequences are found in the same relative position with respect to two or more globin genes, it is possible that these common sequences are functionally significant. For example, transcriptional initiation sites, cis-acting regulatory sequences or RNA processing sites might be identified by this approach. As shown in Figure 9, an unexpectedly complex pattern of repeat sequences was found within the gene cluster. Approximately 20 different pairs of sequences which cross hybridize in filter hybridization experiments

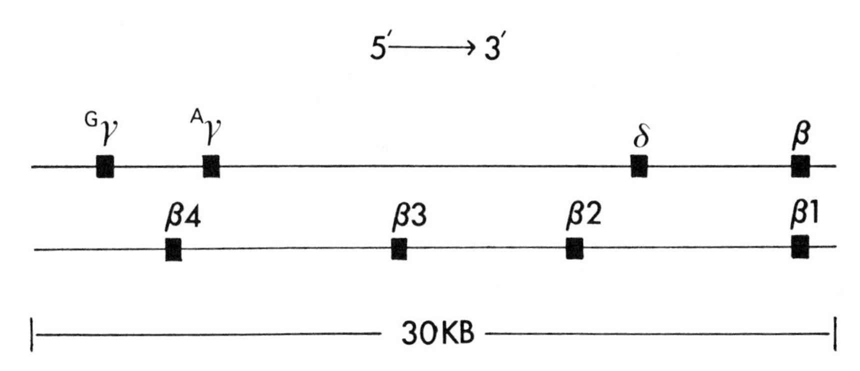

FIGURE 8. The linkage arrangement of human and rabbit β-like globin genes. The relative positions of the human (upper line) and rabbit (lower line) β-like globin genes are shown for comparison. The black boxes indicate the approximate location of the genes. The direction of transcription is indicated by the arrow.

have been identified and classified into five groups (A-E) on the basis of sequence homology. Repeated sequences in both tandem and inverted orientations are observed, and the size of the repeats varies from 140 bp to 1.3 kb. The 1.3 kb sequences in group D are inverted with respect to each other and are separated by approximately 38-40 kb of chromosomal DNA which includes all four globin genes. Interestingly, each of the four globin genes are flanked by inverted repeats. Although the functional significance (if any) of these repeated sequences is not known, it seems unlikely to us that such an intricate pattern of repeated sequences is a random occurrence. Analysis of these sequences with regard to their distribution in the genome and their possible occurrence in nuclear RNA is in progress.

DISCUSSION

We have reviewed evidence which indicates that four different β-like globin genes are arranged in a closely linked cluster in both human and rabbit DNA. As shown in Figure 8, the two sets of genes exhibit a number of similarities. In both cases, two genes are expressed during embryonic and/or fetal development and two genes are transcribed in adult erythroid cells. Since, in both organisms, all four genes are transcribed from the same DNA strand and since the embryonic (or fetal) genes are located on the 5' side of the adult genes, the two pairs of genes are arranged 5' → 3' in the order of their expression during development.

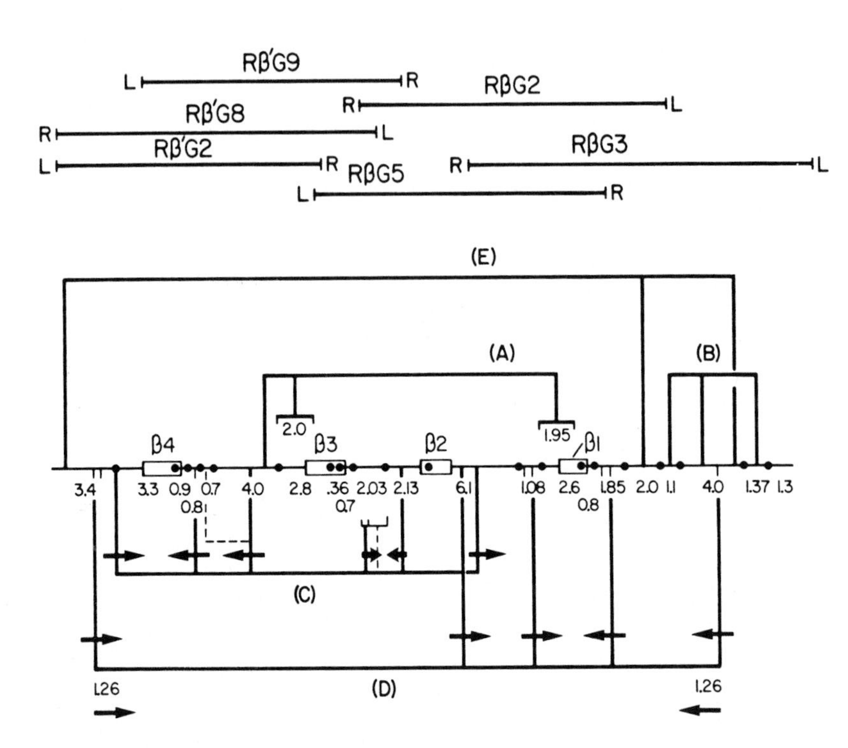

FIGURE 9. The arrangement of repeated sequences within the rabbit β-globin gene cluster. Each of the four genes (β1-β4) are represented by open boxes and the closed circles indicate the position of Eco RI cleavage sites. The horizontal lines flanked by the letters L or R (which designate the left or right arms, respectively, of the bacteriophage Charon 4A cloning vector) represent the rabbit DNA carried in each of six independent recombinant phage (RβGn). The lines which are drawn from one position on the restriction map to another indicate the approximate location of cross-hybridizing sequences. The arrows indicate the relative orientation of various pairs of repeated sequences. The letters A-D indicate five groups of repeated sequences which cross-hybridize.

Although the significance of the close linkage is not known, it is possible that the cluster acts as a functional unit in development. For example, it is possible that cis-acting regulatory sequences which affect the differential expression of each of the globin genes are located within the cluster and that the spatial relationships of these sequences to the genes are important. Furthermore, it is possible that genes which encode regulatory proteins or RNAs which are involved in the switch from fetal to adult globin gene expression are located within the gene cluster. A transcriptional map of globin as well as non-globin sequences which are expressed at different times during development may provide important information with regards to the latter possibility.

The existence of regulatory sequences within the human globin gene cluster is suggested by the fact that deletions which remove all or part of the human δ- and β-globin genes and their flanking sequences can apparently affect the normal switch from fetal to adult globin gene expression (32). For example, in one type of a genetic disorder designated hereditary persistence of fetal hemoglobin (HPFH) the entire δ-β globin region plus 4 kb of DNA on the 5' end of the globin gene are deleted (14), but this defect is compensated by continued expression of the $^{G}\gamma$ and $^{A}\gamma$ genes in cis. In contrast, individuals with one type of δβ thalassemia carry a deletion which removes most of the δβ-globin genes but leaves the region 5' to the δ-globin gene intact. In this case, the β-globin gene deficiency is not fully compensated for by continued fetal globin gene expression and the individual with this genotype suffers from a mild form of anemia. Thus, the only presently detectable genetic difference between these examples of HPFH and δβ-thalassemia is the presence of a region of the chromosome 5' to the δ-globin gene in δβ-thalassemia which is missing in HPFH DNA (14). It is therefore possible that sequences in this region play some as yet unknown role in the suppression of γ-globin genes in adults. It is interesting in this respect that in individuals with β^{o} thalassemia, a genetic disorder in which the β-globin gene is defective, but present, the level of expression of γ-globin genes in adult is inadequate and severe anemia results. The relative levels of continued γ-globin gene expression in adults with genetic disorders of β-globin gene expression are HPFH > δβ thalassemia > β^{o} thalassemia. This correlation suggests that sequences which are necessary for the suppression of γ-globin gene expression in adults are located within the δ-β-globin gene cluster and if they are removed by deletion, this control is relaxed.

The detailed restriction map derived for the human globin gene cluster and the availability of cloned hybridization probes corresponding to various regions within the linkage group will make it possible to precisely map other deletions which alter the normal pattern of globin gene expression. Hopefully, this analysis will lead to the ultimate identification of sequences which are involved in controlling the expression of these developmentally regulated genes.

ACKNOWLEDGEMENTS

We thank A. Efstratiadis for providing unpublished data and C. O'Connell and D. Quon for excellent technical assistance. This work was supported by grants from the National Institutes of Health and the National Science Foundation. E.F.F., R.M.L. and R.C.H. were supported by postdoctoral fellowships from the Damon Runyon-Walter Winchell Cancer fund, the American Cancer Society, and the Jane Coffin Childs Memorial Fund for Medical Research, respectively. T.M. is the recipient of a Rita Allen Foundation Career development award.

REFERENCES

1. Tilghman, S. M., Tiemeier, D. C., Polsky, F., Edgell, M. H., Seidman, J. G., Leder, A., Enquist, L. W., Norman, B., and Leder, P. (1977). Proc. Natl. Acad. Sci. USA 74, 4406.
2. Maniatis, T., Hardison, R. C., Lacy, E., Lauer, J., O'Connell, C., Quon, D., Sim, G. K., and Efstratiadis, A. (1978). Cell 15, 687.
3. Blattner, F. R., Blechl, A. E., Denniston-Thompson, K., Faber, H. E., Richards, J. E., Slightom, J. L., Tucker, P. W., and Smithies, O. (1978). Science 202, 1279.
4. Lawn, R. M., Fritsch, E. F., Parker, R. C., Blake, G., and Maniatis, T. (1978). Cell 15, 1157.
5. Weatherall, D. J. and Clegg, J. B. (1972). "The Thalassemia Syndromes" 2nd Edition, Blackwell Scientific Publications, Oxford.
6. Bunn, F. H., Forget, B. G., and Ranney, H. M. (1977). "Human Hemoglobins", W. B. Saunders Co., Philadelphia.
7. Russell, E. S. and McFarland, E. C. (1974). Ann. N.Y. Acad. Sci. 41, 25.
8. Gilman, J. G. and Smithies, O. (1968). Science 160, 885.
9. Huisman, T. H. J., Schroeder, W. A., Bannister, W. H., and Grech, J. L. (1972). Biochem. Gen. 7, 131.
10. Baglioni, C. (1962). Proc. Natl. Acad. Sci. USA 48, 1880.
11. Huisman, T. H. J., Wrightstone, R. N., Wilson, J. G., Schroeder, W. A., and Kendall, A. G. (1972). Arch. Biochem. Biophys. 153, 850.
12. Flavell, R. A., Kooter, J. M., DeBoer, E., Little, P.F.R., and Williamson, R. (1978). Cell 15, 25.
13. Little, P. F. R., Flavell, R. A., Kooter, J. M., Annison, G., and Williamson, R. (1979). Nature 278, 227.
14. Fritsch, E. F., Lawn, R. M., and Maniatis, T. (1979). Nature (in press).
15. Mears, J. G., Ramirez, F., Leibowitz, D., and Bank, A. (1978). Cell 15, 15.

16. Jeffreys, A. J. and Flavell, R. A. (1977). Cell, 12, 1077.
17. Smithies, O., Blechl, A. E., Denniston-Thompson, K., Newell, N., Richards, J. E., Slightom, J. L., Tucker, P. W., and Blattner, F. R. (1978). Science 202, 1284.
18. Konkel, D. A., Tilghman, S. M., and Leder, P. (1978). Cell 15, 1125.
19. Van den Berg, J., Van Ooyen, A., Mantei, N., Schamböck, A., Grosveld, G., Flavell, R. A., and Weissman, C. (1978). Nature 276, 37.
20. Leder, A., Miller, H. I., Hamer, D. H., Seidman, J. G., Norman, B., Sullivan, M., and Leder, P. (1978). Proc. Natl. Acad. Sci. USA 75, 6187.
21. Dayhoff, M. O. (1972). "Atlas of Protein Sequence and Structure", National Biomedical Research Foundation, Washington, D.C.
22. Tiemeier, D. C., Tilghman, S. M., Polsky, F. I., Seidman, J. G., Leder, A., Edgell, M. H., and Leder, P. (1978). Cell 14, 237.
23. Kafatos, F. C., Efstratiadis, A., Forget, B. G., and Weissman, S. M. (1977). Proc. Natl. Acad. Sci. USA 74, 5618.
24. Weinstock, R., Sweet, R., Weiss, M., Cedar, H., and Axel, R. (1978). Proc. Natl. Acad. Sci. USA 75, 1299.
25. Garapin, A. C., Cami, B., Roskam, W., Kounlsky, P., and Chambon, P. (1978). Cell 14, 629.
26. Lai, E. C., Wood, S. L. C., Dugaiczyk, A., and O'Malley, B. W. (1979). Cell 16, 201.
27. Kitchen, H. and Brott, I. (1974). Ann. N.Y. Acad. Sci. 241, 653.
28. Steinheider, G., Melderis, H., and Ostertag, W. (1974). In "Mammalian Embryonic Hemoglobins, International Symposium on the Synthesis, Structure and Function of Hemoglobin," Ed. by H. Martin and L. Novicke, p. 222. Lehmans, Munich.
29. Maniatis, T., Sim, G. K., Efstratiadis, A., and Kafatos, F. C. (1976). Cell 8, 163.
30. Wilson, J. T., Wilson, L. B., de Riel, J. K., Villa Komaroff, L., Efstratiadis, A., Forget, B. G., and Weissman, J. M. (1978). Nucl. Acids Res. 5, 563.
31. Clissold, P. M., Arnstein, H. R. V., Chesterton, C. J. (1977). Cell 11, 353.
32. Huisman, T. H. J., Shroeder, W. A. et al. (1974). Ann. N.Y. Acad. Sci. 232, 107.

THE STRUCTURE AND EXPRESSION OF GLOBIN GENES IN RABBIT AND MAN

R.A.Flavell, R.Bernards, G.C.Grosveld,

H.A.M.Hoeijmakers-Van Dommelen, J.M.Kooter and

E.De Boer

Section for Medical Enzymology and Molecular Biology, Laboratory of Biochemistry, University of Amsterdam, Jan Swammerdam Institute, P.O.Box 60.000, 1005 GA Amsterdam (The Netherlands)

and

P.F.R.Little

Department of Biochemistry, St. Mary's Hospital Medical School, University of London, London (Great Britain)

ABSTRACT The rabbit and human β-related globin genes have been analysed using genomic 'Southern blotting' and molecular cloning. The rabbit β-globin gene structure has been worked out in detail and its transcripts have been characterized by S_1 nuclease transcription mapping.

The arrangement of the human γδβ-globin gene locus has been largely elucidated. The gene order is (5' to 3') GγAγδβ and the intergene distances are Gγ-Aγ, 3.5 kb; Aγ-δ, 13.5 kb; δ-β, about 6 kb. All these β-related globin genes are transcribed from the same DNA strand. Several abnormal globin genes have been characterized by the same methods. Thus, δβ°-thalassaemia is the result of a deletion which begins approximately in the δ-globin gene large intron and extends beyond the β-globin gene. A form of β°-thalassaemia has been mapped where a 600 bp deletion, including the 3' exon of the β-globin gene, has occurred.

ISBN 0-12-068350-4

INTRODUCTION

In the past two years our ideas on gene structure have undergone a dramatic revision (see e.g. ref. 1). The structure of a large number of eukaryotic viral and cellular genes has been elucidated [2-11]. In this paper we shall discuss current knowledge on the structure of the rabbit and human non-α-globin genes, their transcription products and the structure of various abnormal globin genes.

THE STRUCTURE AND TRANSCRIPTION OF THE RABBIT β-GLOBIN GENE

Initial experiments, using Southern blotting and filter hybridization, showed that the rabbit β-globin gene consisted of at least two non-contiguous blocks of coding sequences separated by an intron [for a definition of intron, see ref. 12] of about 600 bp [6]. Analysis of a cloned rabbit β-globin gene has shown that in fact two introns are present, one of 126 bp in length present between the DNA sequences coding for amino acids 30 and 31, and the one already mentioned above of 573 bp present between the sequences coding for amino acids 104 and 105 [8; Van Ooyen, A. and Grosveld, G.C., unpublished].

The mouse β-globin gene has an essentially identical structure [7,9]. A mouse β-globin pre--mRNA of about 1200-1500 nucleotides has been described; this RNA forms a hybrid co-linear with the mouse β-globin coding regions and intervening sequences [13]. This RNA is presumably a precursor to β-globin mRNA; both intervening sequences must, therefore, be removed from this RNA and the mature mRNA obtained by splicing the exonic sequences.

We have investigated β-globin pre-mRNA in more detail in the rabbit β-globin gene system. Rabbit bone-marrow RNA is hybridized to segments of the cloned rabbit β-globin gene and then the hybrids are analysed by the S_1 nuclease transcription mapping of Berk and Sharp [14]. These results, which have been briefly reported in a recent Symposium volume [15] and will be described in detail elsewhere, can be summarized as follows:

1. The largest transcript detected up to now is 1250 nucleotides long. Its 5' and 3' termini map at the same position as the termini of the mature β-globin mRNA. This RNA is, therefore, a precise transcript of the β-globin exon-intron region; transcripts of the extragenic regions cannot be

detected on this RNA (resolution about 10-20 nucleotides).

2. A second major species is detected. From the mapping data we deduce that this RNA lacks the small intron but contains the large intron and the exons. This RNA is presumably a processing intermediate where the small intron has been removed by an excision-splicing mechanism. The alternative expected intermediate, lacking the large intron but containing the small one, cannot be detected in these experiments. This suggests that the major splicing pathway for the rabbit β-globin gene transcripts is first the removal of the small intron and subsequently the large intron. Whether this pathway is obligatory or not cannot be deduced from these experiments it is also possible that the rate of removal of the small intron is simply intrinsically greater than that for the large intron.

REPETITIVE DNA AROUND THE RABBIT β-GLOBIN GENE

We have previously localized a segment of repetitive DNA in a region between 640 and 2400 bp to the 3'-side of the rabbit β-globin gene [16,17]. This study also showed that less than about ten copies of the regions immediately flanking the gene at the 5'- and 3'-sides, as well as the intronic regions, were present in the rabbit genome.

The globin genes constitute a gene family in which simultaneous expression of the α- and e.g. β-globin genes must be coordinated; within a family such as the β-related globin genes, simultaneous expression of two globin genes occurs in man (δ and β) and probably also in rabbit [18]; alternative expression of β-related genes occurs during embryogenesis (e.g. in man $\varepsilon \longrightarrow \gamma \longrightarrow \delta+\beta$) of most organisms. One might envisage that such coordination of these and perhaps other genes expressed in erythroid cells, would be mediated via homologous DNA regions closely linked to these genes. Such homology would conceivably be only partial, but if these hypothetical sequences were long enough, they should be detectable by hybridization under non-stringent conditions. Our previous experiments excluded the presence of large numbers of such homologous sequences (∿10-20), but less than ten copies would be difficult to distinguish from a single copy in such experiments (see refs. 16, 17).

To test whether multiple copies of the flanking sequences exist in the rabbit genome we have hybridized Southern blots of rabbit DNA which had been cleaved with a variety of restriction endonucleases, with probes for the 5' extragenic regions (a ^{32}P-labelled HaeIII fragment of RβG1 DNA [8] extending from -1600 to -70 bp [O is defined as the 5' nucleotide of globin mRNA]) and the 3' extragenic regions (a ^{32}P-labelled HaeIII fragment extending from +1160, the coding site for amino acid 137 of β-globin, to +2200 bp). We also hybridized filter strips to a ^{32}P-labelled probe (a HaeIII fragment, extending from +2200 to +3400 bp), which contains the reiterated sequence previously characterized [16].

After hybridization the filters were subjected to increasingly stringent salt washes exactly as described in ref. 19. Washing of filters at low salt concentrations causes mismatched hybrids to melt and as a result, only well base-paired hybrids remain bound to the filter. On filters washed only in high salt, poorly base-paired, heterologous hybrids can also be detected (e.g. human β-γ [20]). As seen in Fig. 1, under stringent washing conditions, both the 5' or 3' extragenic probes only detect a single band. Both HaeIII probes only detect their corresponding genomic fragments in HaeIII-digested rabbit DNA; the 5' extragenic probe also detects the characteristic EcoRI (2.6 kb) and PstI x KpnI double digest fragments (1.3 kb and a faint band at 3.6 kb) predicted to be the 5' extragenic regions of the rabbit β-globin gene [19]. A similar conclusion may be reached for the fragments detected by the 3' extragenic probes. Essentially identical results are obtained for the filters washed at high salt (not shown) except that one or two extra components are detected faintly. These probably correspond to the rabbit β-related globin genes, since three components can be detected in Southern blots of rabbit DNA hybridized with cDNA probes and washed under non-stringent conditions [19]. We conclude that the 5' and 3' extragenic regions of the β-globin genes studied here are essentially single-copy DNA.

When the probe for the repeated sequence is used, this forms hybrids with DNA spanning the entire molecular weight range generated by PstI, KpnI or EcoRI (Fig. 1). This repetitive DNA, present in thousands of copies per cell, shows therefore the properties expected of interspersed repeated sequences rather than those of a 'satellite' DNA - in the latter case we would expect to find bands

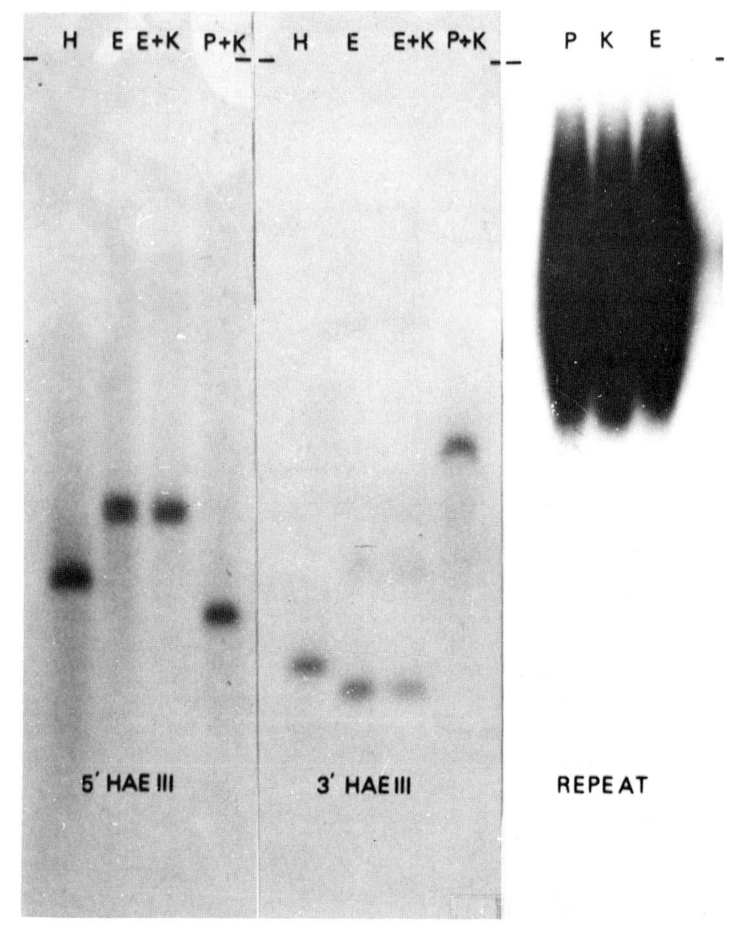

Fig. 1. Reiteration of the 5' and 3' extragenic regions of the β-globin gene in the rabbit genome. Rabbit DNA was cleaved with the restriction endonucleases indicated and Southern blots prepared of this DNA as described [19]. The filter strips were hybridized with ^{32}P-labelled, nick-translated probe (50-100 x 10^6 cpm/μg) as described [16,20]. The probes were prepared by HaeIII cleavage of 20 μg RβG1 DNA [8], followed by acrylamide gel electrophoresis to separate the fragments. The map positions of the probes used were 5' extragenic HaeIII, -1600 to -70; 3' extragenic HaeIII, +1160 to +2200; 'repeat' HaeIII, +2200 to +3400. The restriction endonucleases used were: HaeIII (H), EcoRI (E), PstI (P) or KpnI (K). The gel used to prepare the blot for the 'repeat' experiment was run for a shorter time than that used for the other two samples.

rather than a general smear along the filter.

THE HUMAN GLOBIN GENES

We have been interested in the human globin genes for two reasons:

1. Genetic evidence suggested close linkage of the non-α-globin genes in one of two arrangements, either GγAγδβ or AγδβGγ (the two non-allelic foetal γ-globin genes differ by a single amino acid at position 136; Gγ has glycine and Aγ alanine at this position; the adult δ- and β-globin genes are also closely related, but differ at 10 amino acid positions). We were interested in the intergene distance on a mammalian chromosome since the linkage of the β-related globin genes might well play a role in their coordinate expression.

2. Well-defined lesions in the globin genes have been described which fall into two class:

a) Abnormal globins are produced (e.g. as a result of amino acid substitutions).

b) The level of - otherwise normal - haemoglobin can be reduced, in some cases to zero, in the diseases known as the thalassaemias.

In this article we shall consider the β-thalassaemias which also fall into several classes. In some types of β°-thalassaemia no β-globin mRNA can be detected in the cell [21]; in other nuclear β-globin RNA is found, but no cytoplasmic globin mRNA [22]; in still other cases, cytoplasmic β-globin mRNA is found which is apparently inactive in translation. In β^{+}-thalassaemia reduced levels of β-globin mRNA are found with a parallel decrease in the levels of β-globin in the cell.

We have constructed physical maps of the normal human globin genes using the blotting procedure [23,6] and then compared these with the corresponding physical maps of the globin genes in DNA of patients with various thalassaemias. This makes it possible to detect large deletions in DNA regions within or around the globin genes.

THE STRUCTURE OF THE HUMAN γδβ-GLOBIN GENE LOCUS: THE USE OF GLOBIN-PROTEIN VARIANTS FOR GENE IDENTIFICATION

We have constructed maps of both the δβ region [20] and the γ genes [24], using cloned β-globin cDNA probes which detect the β- and δ-globin genes and a cloned Gγ-globin cDNA which detects both γ-globin genes. The fact that a single probe detects two genes poses a problem - How do we distinguish the two genes and identify them? To do this unequivocally we have made use of globin protein variants.

The β-globin gene was identified using DNA from a patient with Hb (O-Arab). This abnormal haemoglobin has lysine instead of glutamic acid at position 121. In the normal human β-globin gene amino acids 121-122 are coded for by GAATTC - the intragenic EcoRI site. In DNA from Hb (O-Arab) patients, however, this site is lost because of the G ⟶ A transition which gives AAATTC. Indeed, EcoRI digested DNA of an Hb (O-Arab) patient showed a novel globin-gene fragment of a size equal to the sum of two of the four EcoRI fragments. This established that these two fragments contained the β-globin gene [20].

The δ-globin gene was identified in DNA from patients with Hb Lepore [20]. This protein is a fusion product of the N terminal regions of δ-globin with the C terminal regions of β-globin. Likewise, the Hb Lepore gene is a fusion product of the δ- and β-globin genes. Fig. 2 shows the physical map of the normal δ+β locus and that present in the DNA of Hb Lepore patients.

In the case of the γ-globin gene, a similar problem arises: How do we distinguish the Gγ- and Aγ-globin genes? Fortunately, the mutation causing the only amino acid difference between these two proteins results in a PstI site being present in the Aγ-globin gene, but absent in the Gγ gene. This enabled us to identify these two genes. The γ--globin gene map [24] is shown in Fig. 3.

DETERMINATION OF THE γ-δ GENE DISTANCE

In our initial studies [20,24] we noted that a 15 kb BamHI fragment hybridized with probes for both β- and γ-globin genes under conditions where no cross hybridization between these two gene sequences occurs. This suggested the possibility that the 15 kb BamHI fragment of the Aγ globin gene in fact terminated at the BamHI site in the 5' region of the δ-globin gene. The alternative possibility would be that there are in fact two different 15 kb BamHI fragments, one containing the 3' regions of the γ-globin gene and one containing the 5' regions of the δ-globin gene. Several lines of evidence now show that the former model is the correct one.

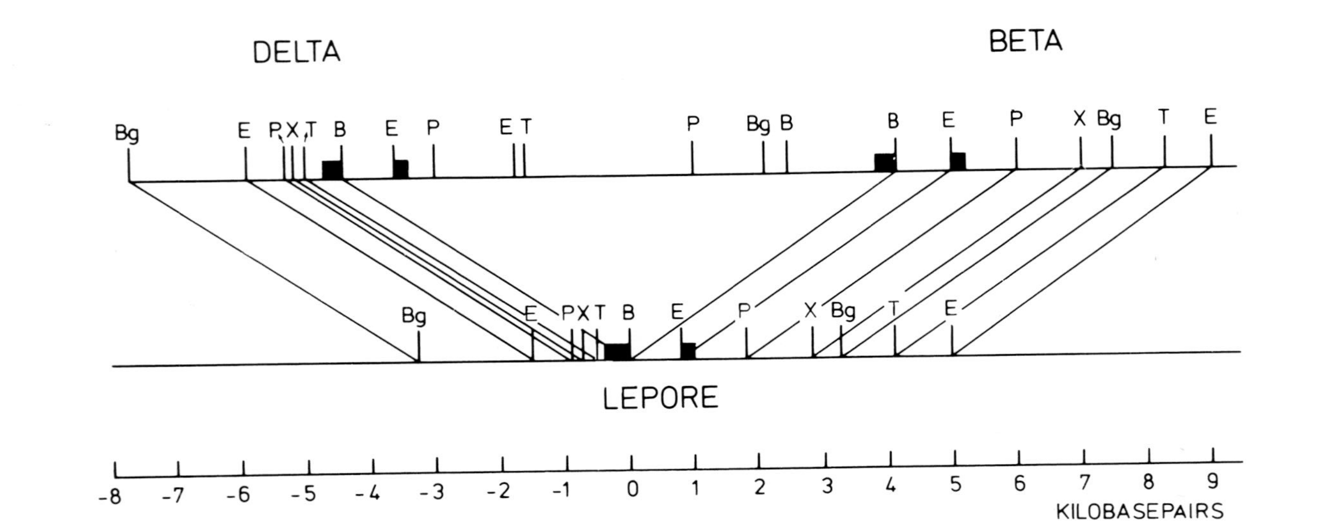

Fig. 2. A physical map of the β- and δ-globin genes in normal and Hb Lepore DNA. The probable positions of the coding regions of the two globin genes are shown as filled boxes. It should be stressed that the only extragenic cleavage sites which can be detected for a given enzyme by this analysis are those closest to the gene examined. Although the β- and δ-genes are presented as being composed of two coding segments in each case, the possibility that these segments are further split cannot be excluded from these data. BamHI, B; BglII, Bg; EcoRI, E; PstI, P; TaqYI, T; XbaI, X. Reproduced from ref. 20 with permission.

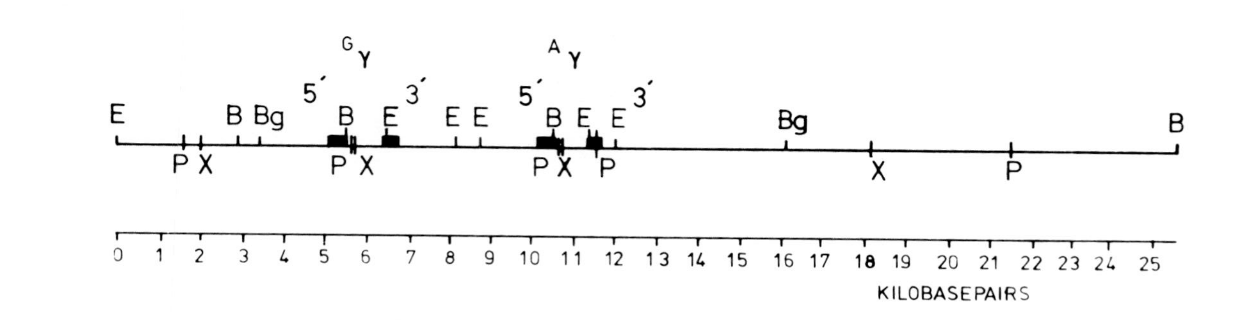

Fig. 3. A physical map of the γ-globin genes. The general comments in Fig. 2 apply here also. EcoRI, E; BamHI, B; BglII, Bg; PstI, P; XbaI, X.

1. BamHI-digested DNA has been fractionated by RPC-5 chromatography (which does not separate DNA primarily on the basis of molecular weight [25]) and then analysed by Southern blotting. The 15 kb 'bands' produced by hybridization with both β and γ probes are found in the same fraction, while all the other γ- and β BamHI fragments found partition differently over the column.

2. Both BclI and BglII cleave at the same position 5.4 kb to the 3'-side of the Aγ globin gene BamHI site. BclI and BglII also cut close to the δ-globin gene (2.1 kb and 3 kb upstream, respectively). If we perform a partial BclI or BglII digest or a limited BamHI digest, in both digests we detect a partial fragment of 9.2 kb which hybridizes with probes for the 5' regions of the δ-globin gene and which disappears in a limit BclI or BglII digest (BglII shows an additional partial of 4.9 kb, indicating a third BglII site between the Aγ and δ-globin genes). These two partials are of the size expected if the cleavage site closest to the 3'-side of γ has been cleaved, but that the respective sites to the 5'-side of the δ-gene are uncleaved. This approach of course selects these partials because a δ probe is used.

3. The 15 kb BamHI -globin gene fragment is cleaved by HpaI to give a 1.3 kb 5' δ-globin gene fragment. The corresponding 15 kb γ-globin gene fragment is trimmed approximately the same amount to give a 14 kb 3' Aγ-globin gene double digest fragment. No other HpaI sites are detected on these γ and δ BamHI 'fragments' in partial digests. In addition, neither KpnI nor BstEII cleave the 15 kb BamHI 'fragments' as measured with either γ or β hybridization probes. Thus either there is a single BamHI fragment linking the γ- and δ-globin genes, or alternatively - but less likely - there are two BamHI fragments, one containing the γ-gene region, the other containing the δ-gene, which contain BclI, BglII and HpaI sites at the same respective positions in both fragments.

4. KpnI cleaved human DNA contains a 46 kb fragment which hybridizes with probes for both the γ- and β-globin genes. The size of this fragment was estimated in 0.5% and 0.7% agarose gels using internal phage lambda DNA and EcoRI-cleaved DNA markers. A KpnI site close to the β-globin gene has been mapped by double digestion with BamHI, HpaI, HindIII or EcoRI to be 3.8 kb to the 3'-side of the end of the β-globin gene. In addition, we

have shown that KpnI does not cleave any of the XbaI, BamHI or EcoRI fragments in the γ, δ, β region except those fragments to the 3'-side of β described above. Assuming that the 15 kb BamHI fragment does link the γ- and δ-genes, the minimal length of this KpnI fragment possible on a hypothetical linkage map which is consistent with the lack of cleavage of the 5' EcoRI fragment of the Gγ--globin gene is 37 kb. The objection to this linkage model (see above) would, therefore, be that two tandem 15 kb BamHI fragments exist, one containing the 5' region of the δ-globin gene and one containing the 3' region of the γ-globin gene. In this case, however, this minimal distance becomed 53 kb (37 + 15 kb). This size is significantly larger than the size of the fragment (46 kb) estimated from the gel with the internal markers.

Taken together these results show that the best measure of the Aγ- to δ-globin gene distance is 13.5 kb although final proof of this point requires cloning of this region as a recombinant DNA. Fig. 4 shows the linkage map of the γδβ locus derived from our blotting data.

This map shows:

a) All four globin genes are transcribed from the same DNA strand.

b) The Gγ-globin gene is 3500 bp to the 5'-side of the Aγ-globin gene; the Aγ gene is 13500 bp to the 5'-side of the δ-globin gene; the δ-globin gene is about 6000 bp to the 5'-side of the β-globin gene.

c) All genes contain the large intron, previously found for the mouse and the rabbit β-globin genes, at the same position within the resolution of the Southern blotting analysis, i.e. between the codons for amino acids 101-120. The structure of the δ-β-gene locus has also been elucidated by Mears et al. [26] and Lawn et al. [27]. In the latter case [27], the linkage of the δ- and β-globin genes has also been demonstrated.

THE STRUCTURE OF β-THALASSAEMIC GLOBIN GENES: δβ°-THALASSAEMIA

δβ°-Thalassaemia is a rare condition in which the β- and δ-globin chains are completely absent. From cDNA titration-hybridization experiments it has been concluded that at least partial deletion of the β- or δ-globin genes has occurred, although the extent of the deletion could not be defined.

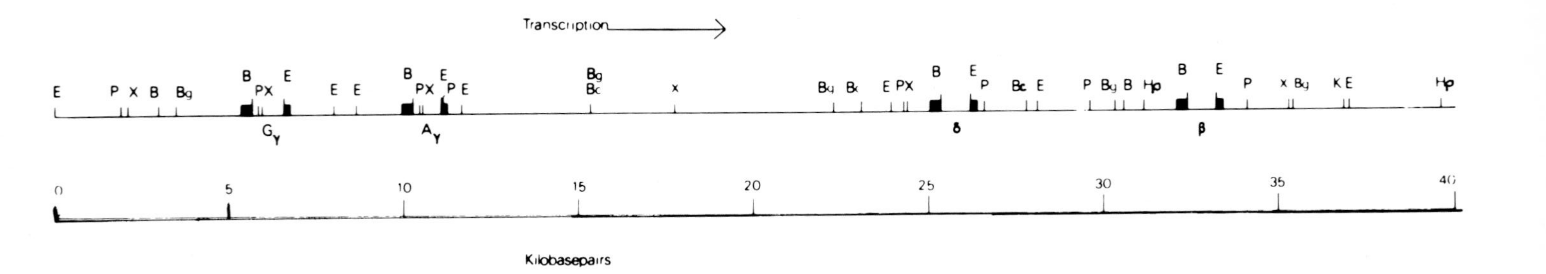

Fig. 4. A physical map of the normal human γδβ locus. See Fig. 2 for general comments. BamHI, B; BclI, Bc; BglII, Bg; EcoRI, E; PstI, P; XbaI, X; HpaI, Hp; KpnI, K.

Recently we have constructed a physical map of the '$\delta\beta^{\circ}$-thalassaemia' globin gene. Two (sibling) homozygous $\delta\beta^{\circ}$-thalassaemia patients (Italian) and two unrelated $\delta\beta^{\circ}$ heterozygotes (one Greek and one Italian) have been examined and shown to have the same deletion. This map (Fig. 5) shows that a large deletion of DNA has occurred to give the $\delta\beta^{\circ}$-thalassaemia genotype. The 5' break point of the deletion is in the δ-globin gene. The δ-globin gene up to the intragenic BamHI site in the second exon (coding for amino acids 31-104) is present. The deletion has occurred somewhere between this position and a HindII site which is present close to the 3'-end of the large intron of the δ-globin gene and which is absent in the corresponding gene in $\delta\beta^{\circ}$-thalassaemia patients. The β-globin gene seems to be entirely deleted in $\delta\beta^{\circ}$-thalassaemia: we have not yet identified the 3' break point of the deletion. Other very faint bands are seen in $\delta\beta^{\circ}$-thalassaemia DNA, but these probably derive from cross-hybridization to other globin genes such as ε. The regions containing the γ-globin genes show the normal γ-globin gene fragments already described [24] up to and including the 15 kb BamHI fragment which links the A- and δ-globin genes. This shows that no gross deletions have occurred in the foetal globin gene region in $\delta\beta^{\circ}$-thalassaemia. The same structure for the $\delta\beta^{\circ}$-thalssaemia gene has recently been deduced independently by T.Maniatis and his colleagues (see this Volume).

β-THALASSAEMIA: A SECOND CASE WHERE AN EXONIC DELETION HAS OCCURRED

Blotting hybridizations show that in the majority of cases of β°- or β^{+}-thalassaemia gross deletions of the DNA regions in or around the β-globin gene have not occurred. In PstI (Fig. 6), EcoRI or XbaI digests no differences in the fragment pattern can be seen between normal and β-thalassaemic DNA. In one exceptional case (No. 11 in Fig. 6) a deletion of about 600 bp in the β-globin gene region has occurred to give an abnormal PstI band. This clinically homozygous patient is apparently a compound heterozygote for two forms of β°-thalassaemia: the common type where no deletion can be seen and the deletion form described above.

We have mapped the deletion in this DNA by performing the relevant double digests and comparing

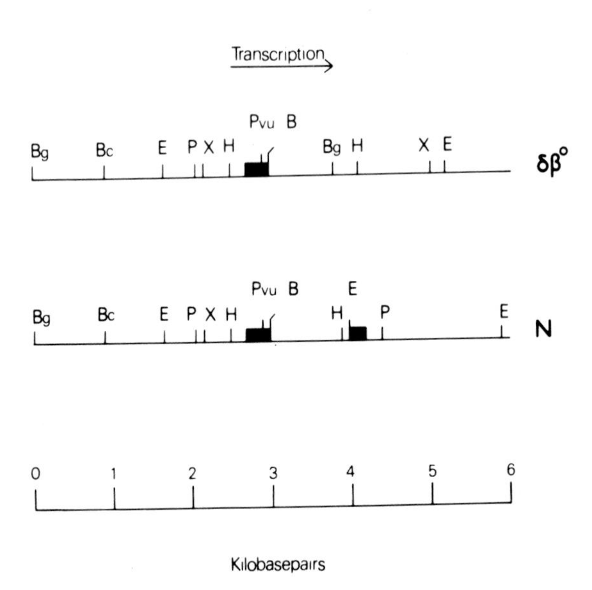

Fig. 5. A physical map of the δ-globin gene region in DNA from a patient homozygous for δβ°-thalassaemia. The DNA was isolated from a lymphocyte cell line, digested with the restriction enzymes indicated in single and in various double digests. The map is based upon the following fragments. BamHI: + BglII, 3.0 kb; + BclI, 2.1 kb; + EcoRI, 1.35 kb; + PstI, 0.95 kb; + XbaI, 0.85 kb; + HindII+III, 0.5 kb. BglII: + EcoRI, 2.1 kb; + PstI, 1.7 kb; + XbaI, 1.65 kb; + HindII+III, 1.3 kb. PvuII + XbaI, 0.8 kb. BglII, 3.8 kb. EcoRI, 3.5 kb. XbaI, 2.9 kb. HindII+III, 1.6 kb. In normal DNA, the δ-derived BamHI double digest fragments are as mentioned above. In addition, the following fragments were used to construct this map: EcoRI, 2.3 kb and 1.9 kb; HindII+III, 1.4 kb; EcoRI + HindII+III, 1.4 kb; PstI, 2.3 kb; PvuII + XbaI, 0.9 kb. BamHI, B; BclI, Bc; BglII, Bg; EcoRI, E; HindII+III, H; PstI, P; PvuII, Pvu; XbaI, X.

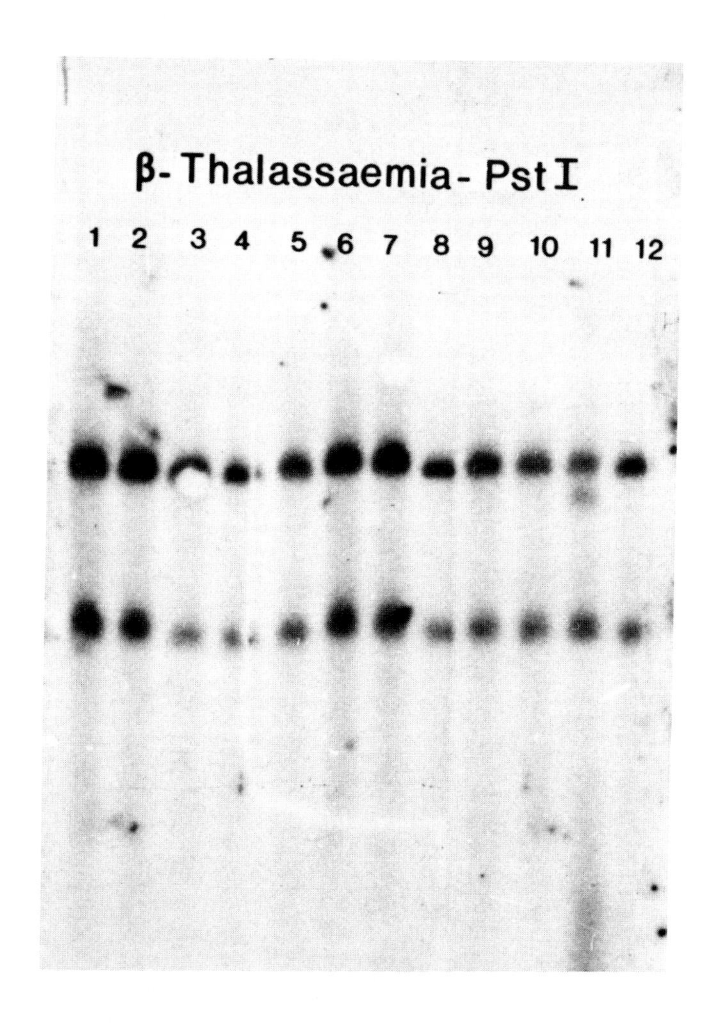

Fig. 6. β- and δ-globin gene fragments in digests of DNA from patients homogyzous for β-thalassaemia. DNA was isolated from β-thalassaemia as described [19] and digested to completion with PstI. The samples were analysed for the β-δ-globin genes as described [20]. Patient 5 is described by Comi et al. [22]: patient 11 in ref. 21. Patients 2, 3 and 10 have been diagnosed as β°-thalassaemics (Southern Italian; Ottolenghi, S., personal communication) and the remainder as homozygous β^+-thalassaemics. 1, 4, 8 and 12 show control normal DNA (derived from a placenta from a Dutch individual).

the fragments obtained with the pattern obtained in normal DNA (see Fig. 7).

These data show that the 600 bp deletion lies between the BamHI site in the second exon of the β-globin gene and the PstI site 1000 bp to the 3'-side of the end of the β-globin gene. Two convincing lines of evidence suggest that the third exon (containing the sequences coding for amino acids 105-146 plus the 3' non-translated region of the mRNA) has been deleted in this DNA.

1. The intragenic EcoRI site (amino acids 121-122) is absent in this β-thalassaemic gene; this results in the presence of an abnormal β-globin EcoRI fragment of 9 kb instead of the two normal fragments of 6 kb and 3.5 kb.

2. The 4.4 kb PstI fragment containing this abnormal globin gene only hybridizes with probes containing the regions to the 5'-side of the intragenic EcoRI site of the β-globin gene. Probes for the 3' regions do not detect this fragment.

The structure of this β-globin gene is, therefore, similar to the structure of the δ-globin gene in δβ°-thalassaemia. In both cases the 3' terminal exon, together with part of the intron and 3' extragenic DNA are missing; in both cases, transcripts of these genes cannot be detected in cDNA titration hybridizations. There are two possible explanations for this. First, it is possible that transcription of split genes requires the presence of the 3' (extragenic) regions. In this case these part-genes could not be transcribed. More likely, however, is the suggestion that the post-transcriptional processing of the transcripts of these abnormal genes is aberrant. For example, the 3' non-translated sequences on the RNA which are required for polyadenylation are deleted in these DNAs. Since polyadenylation seems to precede splicing in both globin [28] and adenovirus [29], it is possible that these hypothetical transcripts could not be spliced. In any case, it is unlikely that the large intron could be excised since the 3' intron-exon junction has certainly been deleted in δβ°-thalssaemia and probably also been deleted in this type of β°-thalassaemia. The net result of these processing defects might be the intranuclear degradation of the RNA. This would account for the low level of β-globin RNA sequences in the cell in this form of β°-thalassaemia and δ-globin RNA sequences in δβ°-thalassaemia.

One final word of caution is relevant. It is

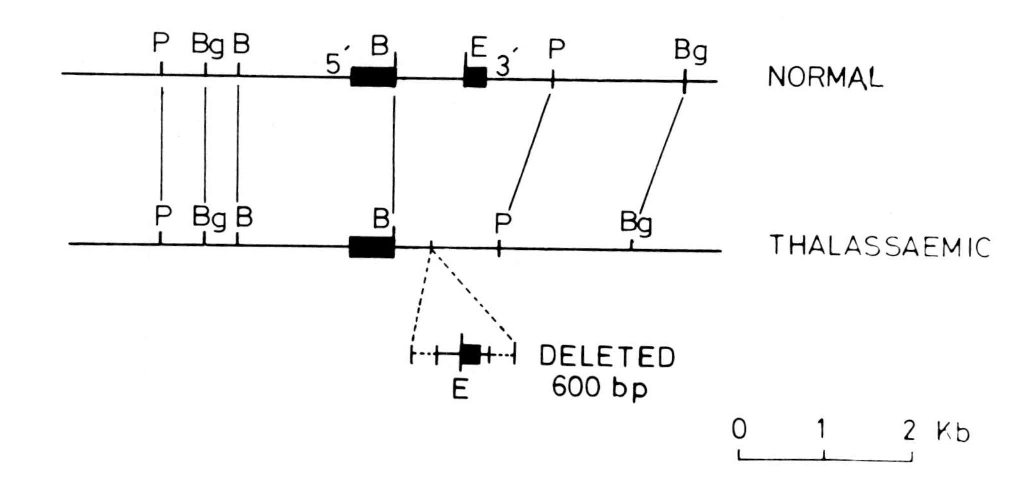

Fig. 7. A physical map of the β-globin gene region in a normal individual and a patient with β°-thalassaemia. The map shows a 600 bp deletion centered around the β-globin intragenic EcoRI site which is missing in this β°-thalassaemia gene. The 600 bp deletion is arbitrarily drawn centered on this EcoRI site. Also shown are the possible limites of the deletion which stretch 600 bp either side of this site. EcoRI, E; PstI, P; BamHI, B; BglII, Bg.

not established that the deletions which we have characterized here are the primary events which generated the two respective forms of thalassaemia. It is also possible that the initial lesion which produces the thalassaemic genotype was a point mutation or small deletion. A subsequent deletion could occur in an already defective gene with no further phenotypic consequences. Since the three cases of $\delta\beta^{o}$-thalssaemia described here all show identical deletions and since more recently others have found similar types of $\delta\beta^{o}$-thalssaemia with a deletion like that described above (Weatherall, D., personal communication), we consider this unlikely.

It is not yet clear what the lesion is in the other forms of β-thalassaemia where no deletions have been detected by Southern blotting. It is possible that different genetic defects have caused various forms of β-thalassaemia; alternatively, they may all turn out to be less extensive deletions of the type described above. Molecular cloning of these thalassaemic globin genes will provide the material with which to test this. In turn, the phenotypic analysis of these cloned thalassaemic genes can be performed in systems such as the TK^{-} mouse L-cell system to elucidate the nature of the defect at a molecular level.

ACKNOWLEDGEMENTS

We would like to thank Piet Borst, Frank Grosveld, Henrik Dahl and Bob Williamson for discussions. This work was supported in part by a grant to RAF from The Netherlands Foundation for Chemical Research (SON) with financial aid from The Netherlands Organization for the Advancement of Pure Research (ZWO) and a grant from the British Medical Research Council (for work by PFRL) to R.Williamson.

REFERENCES

1 Chambon, P. (1978) Cold Spring Harbor Symp. Quant.Biol. 42, 1209-1212.
2 Berget, S.M., Moore, C., and Sharp, P. (1977) Proc.Natl.Acad.Sci.U.S. 74, 3171-3175.
3 Chow, L.T., Gelinas, R.E., Broker, T.R., and Roberts, R.J. (1977) Cell 12, 1-8.
4 Klessig, D.F. (1977) Cell 12, 9-21.
5 Dunn, A.R., and Hassell, J.A. (1977) Cell 12, 23-36.

6 Jeffreys, A.J., and Flavell, R.A. (1977) Cell 12, 1097-1108.
7 Tilghman, S.M., Tiemeier, D.C., Seidman, J.G., Peterlin, B.M., Sullivan, M., Maizel, J.V., and Leder, P. (1978) Proc.Natl.Acad.Sci.U.S. 75, 725-729.
8 Van den Berg, J., Van Ooyen, A., Mantei, N., Schamböck, A., Grosveld, G., Flavell, R.A., and Weissmann, C. (1978) Nature 276, 37-44.
9 Konkel, D.A., Tilghman, S.M., and Leder, P. (1978) Cell 15, 1125-1132.
10 Breathnach, R., Mandel, J.L., and Chambon, P. (1977) Nature 270, 314-319.
11 Kourilsky, P., and Chambon, P. (1978) TIBS 3, 244-247.
12 Gilbert, W.(1978) Nature 271, 501-502.
13 Tilghman, S.M., Curtis, P.J., Tiemeier, D.C., Leder, P. and Weissmann, C. (1978) Proc.Natl. Acad.Sci.U.S. 75, 1309-1313.
14 Berk, A.J., and Sharp, P.A. (1978) Cell 12, 721-732.
15 Flavell, R.A., Grosveld, G.C., Grosveld, F.G., Bernards, R., Kooter, J.M., De Boer, E., and Little, P.F.R. (1979) in From Gene to Protein: Information Transfer in Normal and Abnormal Cells, Proceedings of the 11th Miami Winter Symposium, Academic Press, New York, in press.
16 Flavell, R.A., Jeffreys, A.J., and Grosveld, G.C. (1978) Cold Spring Harbor Symp.Quant. Biol. 42, 1003-1010.
17 Flavell, R.A., Waalwijk, C., and Jeffreys, A.J. (1978) Biochem.Soc.Trans. 6, 742-746.
18 Clissold, P.M., Arnstein, H.R.V., and Chesterton, C.J. (1977) Cell 11, 353-361.
19 Jeffreys, A.J., and Flavell, R.A. (1977) Cell 12, 429-439.
20 Flavell, R.A., Kooter, J.M., De Boer, E., Little, P.F.R., and Williamson, R. (1978) Cell 15, 25-41.
21 Tolstoshev, P., Mitchell, J., Lanyon, G., Williamson, R., Ottolenghi, S., Comi, P., Giglioni, B., Masera, G., Modell, B., Weatherall, D.J., and Clegg, J.B. (1976) Nature 259, 95-98.
22 Comi, P., Giglioni, B., Barbarano, L., Ottolenghi, S., Williamson, R., Novakova, M., and Masera, G. (1977) Europ.J.Biochem. 79, 617-622.
23 Southern, E.M. (1975) J.Mol.Biol. 98, 503-517.
24 Little, P.F.R., Flavell, R.A., Kooter, J.M., Annison, G., and Williamson, R. (1979) Nature

278, 227-231.

25 Leder, P., Tilghman, S.M., Tiemeier, D.C., Polsky, F.I., Seidman, J.G., Edgell, M.H., Enquist, L.W., Leder, A., and Norman, B. (1978) Cold Spring Harbor Symp.Quant.Biol. 42, 915-920.

26 Mears, J.G., Ramirez, F., Leibowitz, D., and Bank, A. (1978) Cell 15, 15-23.

27 Lawn, R.W., Fritsch, E.F., Parker, R.C., Blake, G., and Maniatis, T. (1978) Cell 15, 1157-1174.

28 Ross, J. (1976) J.Mol.Biol. 106, 403-420.

29 Nevins, J.R., and Darnell, J.E., Jr. (1978) Cell 15, 1477-1493.

ORGANIZATION OF HUMAN GLOBIN GENES IN NORMAL AND THALASSEMIC CELLS[1]

Arthur Bank, J. Gregory Mears, Francesco Ramirez, Alexander L. Burns, and John Feldenzer

Departments of Medicine, and Human Genetics and Development
Columbia University, N. Y., N. Y. 10032

ABSTRACT The linked human γ-δ-β globin gene complex provides a useful system for studying eukaryotic gene regulation. Several mutants in this system permit analysis of the effect of changes in gene structure on gene function. Restriction enzyme mapping has been used to study the globin gene organization in normal and abnormal human DNA. After Eco RI digestion and hybridization to plasmid-containing ^{32}P-labelled β globin cDNA, four β-like fragments are present in normal DNA. No β-like gene fragments can be detected in HPFH DNA and a unique 3.2 kb fragment is seen in homozygous δβ thalassemia DNA indicating less deletion of β-like genes in the latter disorder. Since δβ thalassemia has less γ globin gene expression than HPFH, these results support the hypothesis that sequences in the δ gene region or between the γ and δ genes may suppress γ synthesis in δβ thalassemia. The absence of these sequences in HPFH may permit full γ globin gene expression. By contrast, in β^+ and β^0 thalassemia there is no evidence for extensive deletion in the γ-δ-β gene complex, again consistent with the hypothesis that sequences persist in these disorders which suppress γ globin gene expression. The regulation of γ globin gene expression could also be associated with changes in the organization of the γ globin genes. Analyses to date using several restriction enzymes show no differences in the size or number of the γ globin gene-containing fragments in normal Lepore, HPFH, δβ, and β^+ and β^0 thalassemia DNA. Thus, it appears that extensive deletion of γ globin genes is not responsible for the variations in γ globin gene expression in these disorders.

[1] This work was supported by grants from the NIH, the NSF, the National Foundation-March of Dimes, and the Cooley's Anemia Foundation.

ISBN 0-12-068350-4

INTRODUCTION

The linked human γ-δ-β globin gene system serves as a useful model for studying eukaryotic gene regulation. In this system, there is a normal switch in late fetal life from the production of γ globin chains and hemoglobin F ($\alpha_2\gamma_2$), to β globin synthesis and hemoglobin A ($\alpha_2\beta_2$); hemoglobin A_2 ($\alpha_2\delta_2$) is a minor component of human adult red cells. Certain disease states, primarily the β thalassemia syndromes, provide a series of mutations resulting in abnormal regulation of β globin synthesis. In the β^+ thalassemias, there is decreased β globin synthesis and decreased β globin mRNA. When no β globin is produced, the disorder is designated as β^0 thalassemia. At least two general types of β^0 thalassemia have been described: Those associated with absent β mRNA, and those with abnormal β mRNA. In both β^+ and β^0 thalassemia, there is severe anemia due to the marked deficiency of hemoglobin A, and inadequate compensation by hemoglobin F production.[1] By contrast, in two related disorders, called $\delta\beta$ thalassemia and hereditary persistence of fetal hemoglobin (HPFH), there is absence of all β and δ globin synthesis, and more significant compensation by production of increased amounts of γ globin and hemoglobin F. In both of these disorders, physical deletion of β-like globin genes has been demonstrated by both hybridization in solution, and by restriction endonuclease mapping.[2-4] We have shown that there is greater deletion of β-like globin genes in HPFH than in $\delta\beta$ thalassemia, by these methods. These data support the hypothesis that there is a regulatory gene region between the structural γ and δ globin genes, which controls the expression of γ globin gene activity.

We now extend this analysis of restriction endonuclease studies to show that the deletion in HPFH involves the entire structural δ and β globin gene complex, while in $\delta\beta$ thalassemia there is deletion of the entire β globin structural gene with a portion of the δ globin structural gene and its 5' flanking sequences remaining intact. In addition, we can detect no significant deletion in either the δ or β globin genes in several β^+ and β^0 thalassemia patients by restriction endonuclease analysis. These data support the concept that deletion of the region between the δ and γ globin genes permits full expression of γ globin genes. On the other hand, the presence of gene material between the δ and γ globin genes prevents this full expression. In our analysis, we have also obtained evidence that the γ and δ globin structural genes are approximately 15 kilobases (kb) apart, and that there are two linked γ globin structural genes adjacent to each other and 5' to the δβ globin gene complex.

MATERIALS AND METHODS

Restriction Endonuclease Analysis of Human Cellular DNA. High molecular weight human cellular DNA was prepared as described previously,[5] with the use of the Blin[6] method to initially obtain DNA from spleen material. This modification is required to prevent degradation of spleen DNA samples. Restriction digestion was carried out as described previously.[3] DNA fragments are separated on 0.8 to 1.2% agarose gels under neutral conditions as described previously.[3] Transfer of DNA fragments to nitrocellulose filters and subsequent conditions of hybridization to globin cDNA were also as described. Globin cDNA probes were prepared as follows: The plasmids JW101 containing β globin double stranded cDNA and JW151 containing γ cDNA were grown in *E. Coli* strain HB101. The plasmid DNA was isolated as described.[7] The plasmids were cleaved with restriction enzyme Hha I and the 1.5 kb fragments separated by sucrose density gradient centrifugation. The β and γ cDNA probes were nick-translated as described[8] to specific activities of 1-5 X 10^8 cpm per μg. In this reaction, the labelled ^{32}P triphosphates were lyophilized and used at concentrations of 1.6 mM per ml at a specific activity of 300 C/mM of reaction mixture; C_ot values of 10^{-3} were attained by hybridization for 24 hours; the washing of filters following hybridization and subsequent radioautography were as described.[3]

RESULTS

Normal Globin Gene-Containing Fragments. With β globin cDNA prepared using β globin mRNA as probe, we previously identified four cellular DNA fragments generated by Eco RI which contains δ and β globin gene material.[3] Using the nick translated β globin cDNA probe prepared from plasmid DNA, four DNA fragments are seen: 5.5, 3.8, 2.3, and 1.8 kb in size (Figure 1); the 1.8 kb fragment is difficult to identify probably due to poor homology of the 3' untranslated region of the δ gene with β cDNA. Comparison with viral markers indicate that these fragments are all of proportionately smaller size than previously reported.[3,4] We have previously shown that the 2.3 kb fragment contains the 5' end of the δ globin gene, the 1.8 kb fragment the 3' end of the δ globin gene, the 5.5 kb fragment the region between the δ and β genes and the 5' end of the β globin structural gene, and 3.8 kb fragment the 3' end of the β globin structural gene.[4]

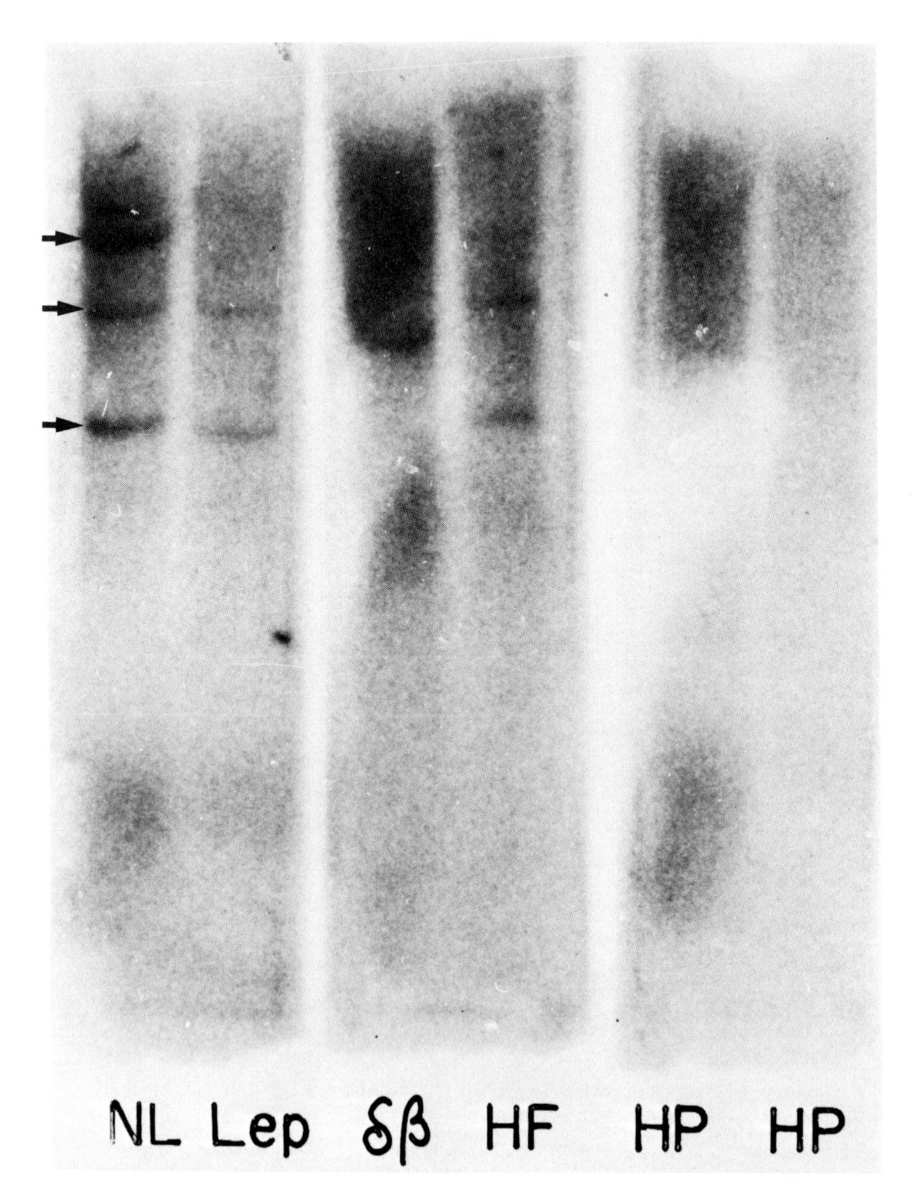

FIGURE 1. Hybridization of cellular DNAs to ^{32}P β globin plasmid cDNA. DNAs from normal (NL) subjects, and those homozygous for Lepore (Lep), δβ thalassemia, hydrops fetalis - α thalassemia (HF), and two different patients with hereditary persistence of fetal hemoglobin (HP) were restricted with Eco RI, hybridized and radioautographed as indicated.[3] The arrows refer to the 5.5, 3.8, and 2.3 kb bands.

of the δ gene and the 3' end of the γ^A gene, or two different 15 kb pieces each contain one of these gene regions. In either case, these data show that the γ^A and δ genes are at least 15 kb apart.

We have also determined, to a limited extent, the restriction sites within and surrounding the γ globin genes. We have found that the enzyme Bgl II results in a single γ cDNA fragment. These data are consistent with the map proposed by others,[11] and suggests that two γ globin genes are present on the 5' side of the δβ globin gene complex as suggested from genetic studies.[9]

Restriction Enzyme Analysis of β^+ and β^0 thalassemia DNA. Analysis of DNA from five individuals homozygous for β^+, and five homozygous for β^0 thalassemia with a variety of restriction enzymes has thus far not detected any significant deletion of β or δ fragments. Pst has been used to define the δ and β genes separated from each other, and no shortening of any Pst fragment has been seen (Figure 3). In addition, Bgl II digestion of β^+ and β^0 thalassemia DNA has detected no deletions (Figure 3). Eco RI digestion alone and in combination with Bam has also been indistinguishable from normal indicating no significant deletion of the region between the δ and β genes, or changes in size of the δ or β intervening sequence regions as analyzed by these enzymes.

Extent of δ and β Gene Deletion in δβ Thalassemia and HPFH. In HPFH DNA the lack of any β-like globin gene fragments (Figure 1) and the presence of normal γ globin gene-containing fragments indicates that there is deletion of the structural δ and β globin genes, and probable deletion of material between the γ and δ globin genes. The 15 kb Bam fragment in normal DNA is shorter in HPFH DNA. At the least, these data indicate that the 5' flanking region of the δ globin gene is deleted, and the structural γ globin genes are present and intact in HPFH DNA (Figure 4). The extent of deletion in the region between the δ and γ genes is uncertain.

In δβ thalassemia, there is persistence of a 5' δ structural gene fragment, and 5' δ flanking sequence. The presence of a single 3.2 kb Eco RI fragment indicates the persistence of β-like gene material which could represent either the δ gene or β gene. This 3.2 kb fragment does not hybridize to a 3' ended β cDNA probe, and thus, contains either 5' δ or 5' β structural gene sequence. If a single continuous gene deletion is present in δβ thalassemia, then it is most likely that the remaining β-like sequence is 5' δ (Figure 2). In addition, in δβ thalassemia, the 15 kb Bam fragment persists, and appears to be similar to that seen in normal DNA. This finding is consistent with a deletion which leaves the 5' δ region intact in $\delta\beta$ thalassemia DNA (Figure 4).

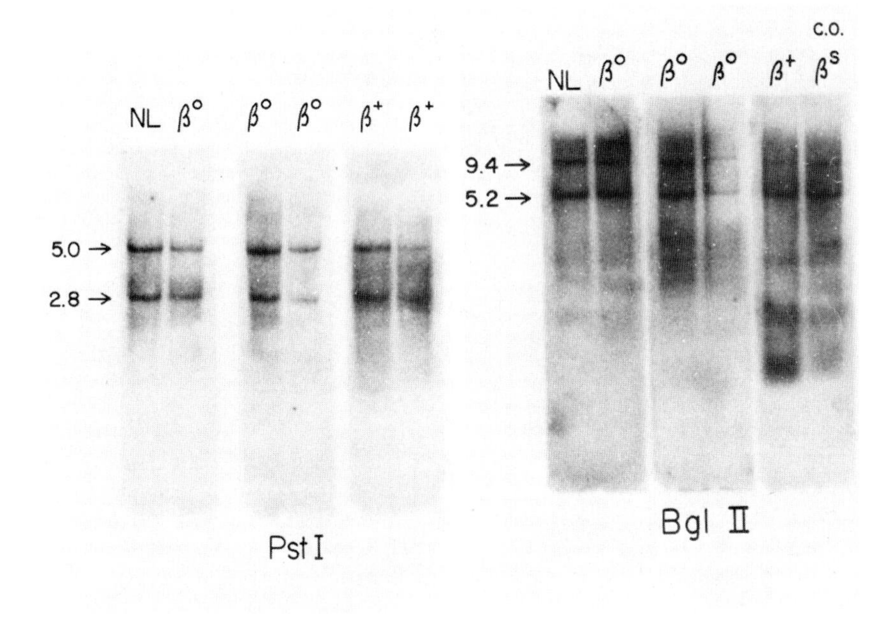

FIGURE 3. Pst I and Bgl II digests of normal, β^+, and β^0 DNAs hybridized to ^{32}P β globin cDNA from plasmid JW101.

DISCUSSION

The data presented in this paper indicate that specific β and γ globin cDNA probes can be used to distinguish cellular DNA fragments containing β and γ structural globin genes. From this analysis it has been shown that normal, β^+, β^0, HPFH, and δβ thalassemia DNA examined to date with one exception contain normal-appearing γ globin gene-containing fragments. By contrast, there is significant deletion of β-like gene-containing fragments in HPFH, δβ thalassemia, and Lepore DNA (Figure 5).

The use of specific β and γ cDNA probes has also led to the determination of the linkage of the δ and γ globin genes since a single Bam fragment 15 kb in size hybridizes to both β cDNA and γ cDNA. The γ cDNA probe has also been used to determine that the two γ genes are linked and 5' to the δ-β genes.

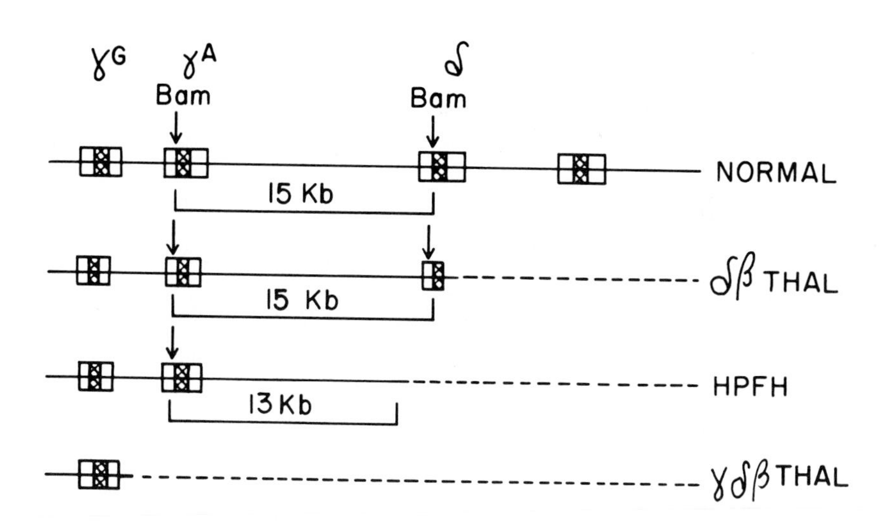

FIGURE 4. Linkage of γ and δ genes and extent of deletion in HPFH and δβ thalassemia.

Analysis of $\delta\beta$ thalassemia and HPFH DNA, using the purified γ and β cDNA probes, has confirmed and extended our previous conclusions that there is extensive deletion of β-like genes in both of these disorders, with more extensive deletion of β-like genes in HPFH (Figure 4). More specifically it appears that in HPFH there is complete deletion of the structural δ and β genes, while the γ globin genes are intact; the extent of the deletion 5' to the δ structural gene and 3' to the γ^A gene remains to be determined. By contrast, in $\delta\beta$ thalassemia, there is persistence of a part of the δ structural gene at the 5' end and its flanking sequence; the γ genes remain intact. Thus, the region 5' to δ structural gene or the 5' δ structural gene itself contains gene elements which may control the expression of the γ globin genes. In Lepore DNA and in most β^o and β^+ DNA, these sequences also persist and this is associated with a limitation in the amount of γ globin gene expression in these disorders (Figure 5). These results suggest that the sequences 3' to the γ globin genes may be active in the regulation of these genes. One possibility is that δ and β globin mRNA sequences in the nucleus are repressor-like and inhibit γ globin gene expression. This would explain the persistence of the δ structural gene

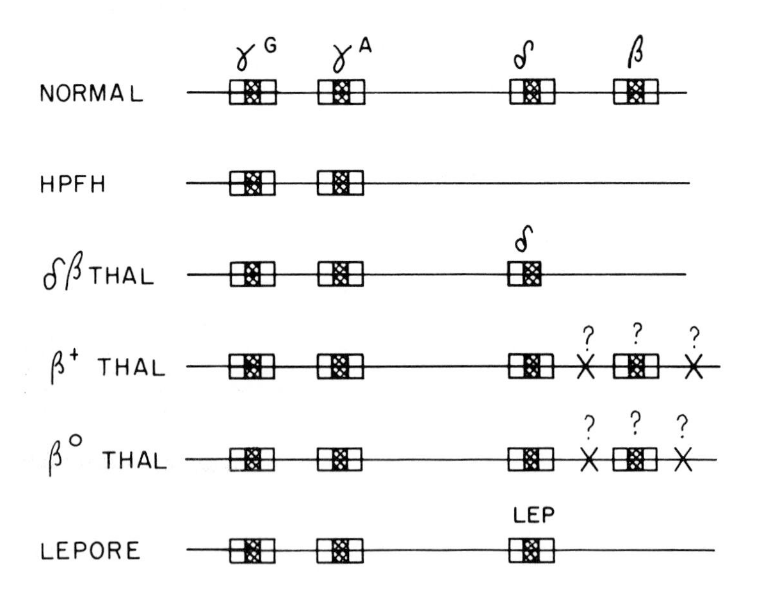

FIGURE 5. Gene defects in the β thalassemias and HPFH.

sequence so closely homologous to that of the β structural sequence despite non-homology of the intervening sequences of these genes, and despite the inability of δ globin mRNA to code for an adequate amount of δ globin gene product. It remains unknown whether the normal switch from γ to β globin gene expression in late fetal life involves the same or other regulatory gene sequences as those postulated above.

The use of cloned DNA fragments representing the intervening sequences, and flanking sequences of the γ, δ, and β globin gene complex may be useful in defining changes in the sequence of DNA in these disorders which may not have been detected by the use of cDNA probes which only recognize structural gene sequences. It is also possible that single base changes in either the flanking or intervening sequences of the β structural gene in β^+ and β^0 thalassemia may be responsible for the decreased or absent β globin synthesis in these disorders. It appears logical, therefore, to clone and sequence these genes, using methods recently described for normal globin genes.[12,13] The specific nucleotide changes in these disorders may provide clues to the sequences regulating gene expression in this linked gene system. These studies may also provide important insights into eukaryotic gene regulation which may be applicable to other gene systems. In addition, the finding of specific nucleotide changes in β^+ and β^0 thalassemia DNA may make it possible to obtain the prenatal diagnosis of these disorders, providing that the reproducibility of these defects can be confirmed. It is, of course, possible that there will be different types of changes responsible for different β^+ or β^0 phenotypes, and that it will be necessary to determine the specific nucleotide defect in different subpopulations in order to provide accurate prenatal diagnosis.

ACKNOWLEDGMENTS

We thank Unateresa Collins for preparing this manuscript.

REFERENCES

1. Bank, A. (1978). Blood 51, 369.
2. Ramirez, F., O'Donnell, J. V., Marks, P. A., Bank, A., Musumeci, F., Schiliro, G., Pizzarelli, G., Russo, G., Luppis, B., and Gambino, R. (1976). Nature 263, 471.
3. Mears, J. G., Ramirez, F., Leibowitz, D., Nakamura, F., Bloom, A., Konotey-Ahulu, F., and Bank, A. (1978). Proc. Natl. Acad. Sci. 75, 1222.

4. Mears, J. G., Ramirez, F., Leibowitz, D., and Bank, A. (1978). Cell 15, 15.
5. Gambino, R., Kacian, D., O'Donnell, J. V., Ramirez, F., Marks, P. A., and Bank, A. (1974). Proc. Natl. Acad. Sci. 71, 3966.
6. Blin, N., and Stafford, D. W. (1976). Nucl. Acid Res. 3, 2303.
7. Wilson, J. T., Wilson, L. B., de Riel, J. K., Villa-Komaroff, L., Efstratiadis, A., Forget, B. G., and Weissman, S. M. (1978). Nucl. Acid Res. 5, 563.
8. Maniatis, T., Jeffreys, A., and Kleid, D. G. (1975). Proc. Natl. Acad. Sci. 72, 1184.
9. Ramirez, F. Submitted for publication.
10. Orkin, S. Personal communication.
11. Little, P. F. R., Flavell, R. A., Kooter, J. M., Annison, G., and Williamson, R. Nature. In press.
12. Maniatis, T., Hardison, R. C., Lacy, E., Lauer, J., O'Connell, C., and Quon, D. (1978). Cell 15, 687.
13. Lawn, R. M., Fritsch, E. F., Parker, R. C., Blake, G., and Maniatis, T. (1978). Cell 15, 1157.

STRUCTURE OF THE HUMAN GLOBIN GENES[1]

B.G. Forget, C. Cavallesco, J.K. deRiel,
R.A. Spritz, P.V. Choudary, J.T. Wilson,[2]
L.B. Wilson,[2] V.B. Reddy, and S.M. Weissman

Departments of Medicine and Human Genetics,
Yale University School of Medicine,
New Haven, Connecticut 06510.

ABSTRACT By sequence analysis of human globin cDNA plasmids we have determined the nucleotide sequence of the coding portion of the human α and $^{G}\gamma$ globin mRNAs, and of the 3'-untranslated portion of $^{G}\gamma$ globin mRNA. In addition, nucleotide sequences adjacent to the extremities of the major intervening sequence or intron of a cloned human δ globin gene have also been determined. These studies have allowed us to establish detailed restriction endonuclease maps of the human α and $^{G}\gamma$ globin cDNAs and to examine sequence homology and divergence between the different human and rabbit globin mRNAs. The sequences involved in the splicing of human δ globin mRNA in the region of the major intervening sequence of the δ globin gene share considerable homology with the analogous sequences found in the rabbit and mouse β globin genes.

INTRODUCTION

The number and chromosomal arrangement of the human globin genes are illustrated diagrammatically in Figure 1. The evidence on which this model is based as well as the changes which occur in the expression of these different globin genes during development have been reviewed extensively elsewhere (1-3). The entire nucleotide sequence of the human α, β and γ globin mRNAs have now been determined and recombinant DNA technology has allowed the isolation of cloned DNA fragments containing the human globin genes from which partial nucleo-

[1]This work was supported in part by grants AM-19482, GM20124 and HL20922 of the USPHS, National Institute of Health.

[2]Present address: Department of Cell and Molecular Biology, Medical College of Georgia, Augusta, Georgia 30902.

ISBN 0-12-068350-4

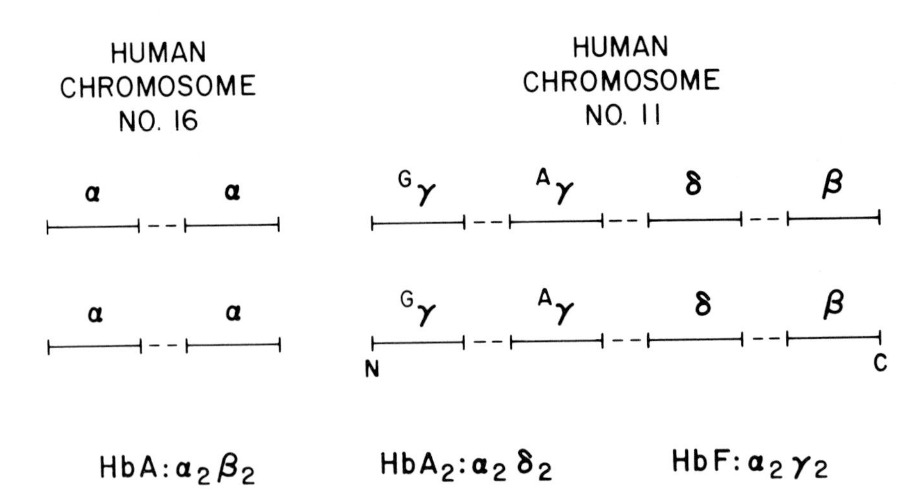

FIGURE 1. Number and chromosomal arrangement of the human globin genes (1). The chromosomal assignments were determined by Deisseroth et al. (2,3).

tide sequence information of the major intervening sequences or introns of the human β, δ and γ globin genes have been determined (4,5). In this manuscript we report the nucleotide sequences of the coding portions of the human α and $^{G}\gamma$ globin mRNAs determined by sequence analysis of cloned globin cDNA plasmids (6), as well as the nucleotide sequence of a portion of the human δ globin gene and its intervening sequence, determined from analysis of cloned human globin gene DNA (4). The three major human globin mRNA sequences (α, β and γ) are compared with respect to their base composition and codon usage, in parallel with those of the previously reported rabbit α and β globin mRNAs (7,8). Detailed restriction endonuclease maps of the human α and γ cDNAs are presented and the sequences involved in the splicing of the δ mRNA at the site of the major intervening sequence are discussed in comparison to those of rabbit and mouse β globin mRNAs (9,10).

MATERIALS AND METHODS

The isolation of human globin cDNA clones has been previously reported (6,11). The human α globin cDNA used in sequence analysis was that of plasmid JW101 (6) and the $^{G}\gamma$ cDNA analyzed was that of plasmid JW151 (6), but only for the coding sequence of $^{G}\gamma$ mRNA. Because the plasmid JW151 lacks the sequences complementary to the 3'-untranslated portion of $^{G}\gamma$ mRNA, these sequences were analyzed in plasmid pHγG1 (11) kindly provided by Dr. S. Malcolm. The human δ globin gene

DNA originated from the phage λ clone HβGl previously described by Lawn et al. (4); a Pst fragment, 2.5 kb in length, containing the entire δ globin gene with adjacent flanking sequences was subcloned by Dr. T. Maniatis and his colleagues in plasmid pBR322; the δ subclone was designated Hδ1 and was kindly provided to us by Dr. T. Maniatis for sequence analysis. Plasmids containing globin cDNA and gene DNA inserts were propagated in E. coli according to standard techniques and the plasmid DNA isolated (6), and digested with restriction endonucleases; the appropriate DNA fragments were then end-labeled using [γ-^{32}P] ATP and polynucleotide kinase, recut by another restriction endonuclease, and subjected to nucleotide sequence analysis by the chemical degradation method of Maxam and Gilbert (12).

RESULTS

Nucleotide Sequences of Human α and γ Globin cDNAs. The nucleotide sequence of the coding portion of the human α globin mRNA determined from sequence analysis of globin cDNA plasmid JW101 is shown in Figure 2. The nucleotide sequences

1				5					10					15					20
Val	Leu	Ser	Pro	Ala	Asp	Lys	Thr	Asn	Val	Lys	Ala'	Ala	Trp	Gly	Lys	Val	Gly	Ala	His
GUG	CUG	UCU	CCU	GCC	GAC	AAG	ACC	AAC	GUC	AAG	GCC	GCC	UGG	GGC	AAG	GUU	GGC	GCG	CAC
21				25					30					35					40
Ala	Gly	Glu	Tyr	Gly	Ala	Glu	Ala	Leu	Glu	Arg	Met	Phe	Leu	Ser	Phe	Pro	Thr	Thr	Lys
GCU	GGC	GAG	UAU	GGU	GCG	GAG	GCC	CUG	GAG	AGG	AUG	UUC	CUG	UCC	UUC	CCC	ACC	ACC	AAG
41				45					50					55					60
Thr	Tyr	Phe	Pro	His	Phe	Asp	Leu	Ser	His	Gly	Ser	Ala	Gln	Val	Lys	Gly	His	Gly	Lys
ACC	UAC	UUC	CCG	CAC	UUC	GAC	CUG	AGC	CAC	GGC	UCU	GCC	CAG	GUU	AAG	GGC	CAC	GGC	AAG
61				65					70					75					80
Lys	Val	Ala	Asp	Ala	Leu	Thr	Asn	Ala	Val	Ala	His	Val	Asp	Asp	Met	Pro	Asn	Ala	Leu
AAG	GUG	GCC	GAC	GCG	CUG	ACC	AAC	GCC	GUG	GCG	CAC	GUG	GAC	GAC	AUG	CCC	AAC	GCG	CUG
81				85					90					95					100
Ser	Ala	Leu	Ser	Asp	Leu	His	Ala	His	Lys	Leu	Arg	Val	Asp	Pro	Val	Asn	Phe	Lys	Leu
UCC	GCC	CUG	AGC	GAC	CUG	CAC	GCG	CAC	AAG	CUU	CGG	GUG	GAC	CCG	GUC	AAC	UUC	AAG	CUC
101				105					110					115					120
Leu	Ser	His	Cys	Leu	Leu	Val	Thr	Leu	Ala	Ala	His	Leu	Pro	Ala	Glu	Phe	Thr	Pro	Ala
CUA	AGC	CAC	UGC	CUG	CUG	GUG	ACC	CUG	GCC	GCC	CAC	CUC	CCC	GCC	GAG	UUC	ACC	CCU	GCG
121				125					130					135					140
Val	His	Ala	Ser	Leu	Asp	Lys	Phe	Leu	Ala	Ser	Val	Ser	Thr	Val	Leu	Thr	Ser	Lys	Tyr
GUG	CAC	GCC	UCC	CUG	GAC	AAG	UUC	CUG	GCU	UCU	GUG	AGC	ACC	GUG	CUG	ACC	UCC	AAA	UAC
141																			
Arg																			
CGU	UAA																		

FIGURE 2. Nucleotide sequence of the coding portion of human α globin mRNA. The sequences for codons no. 64 and no. 69 have been revised since the previous report of a preliminary sequence of the α mRNA (22).

--CU CCU AGU CCA GAC GCC AUG

1									10										20
Gly	His	Phe	Thr	Glu	Glu	Asp	Lys	Ala	Thr	Ileu	Thr	Ser	Leu	Trp	Gly	Lys	Val	Asn	Val
GGU	CAU	UUC	ACA	GAG	GAG	GAC	AAG	GCU	ACU	AUC	ACA	AGC	CUG	UGG	GGC	AAG	GUG	AAU	GUG
21									30										40
Glu	Asp	Ala	Gly	Gly	Glu	Thr	Leu	Gly	Arg	Leu	Leu	Val	Val	Tyr	Pro	Trp	Thr	Gln	Arg
GAA	GAU	GCU	GGA	GGA	GAA	ACC	CUG	GGA	AGG	CUC	CUA	GUU	GUC	UAC	CCA	UGG	ACC	CAG	AGG
41									50					55					60
Phe	Phe	Asp	Ser	Phe	Gly	Asn	Leu	Ser	Ser	Ala	Ser	Ala	Ile	Met	Gly	Asn	Pro	Lys	Val
UUC	UUU	GAC	AGC	UUU	GGC	AAC	CUG	UCC	UCU	GCC	UCU	GCC	AUC	AUG	GGC	AAC	CCC	AAA	GUC
61				65					70					75					80
Lys	Ala	His	Gly	Lys	Lys	Val	Leu	Thr	Ser	Leu	Gly	Asp	Ala	Ile	Lys	His	Leu	Asp	Asp
AAG	GCA	CAU	GGC	AAG	AAG	GUG	CUG	ACU	UCC	UUG	GGA	GAU	GCC	AUA	AAG	CAC	CUG	GAC	GAU
81				85					90					95					100
Leu	Lys	Gly	Thr	Phe	Ala	Gln	Leu	Ser	Glu	Leu	His	Cys	Asp	Lys	Leu	His	Val	Asp	Pro
CUC	AAG	GGC	ACC	UUU	GCC	CAG	CUG	AGU	GAA	CUG	CAC	UGU	GAC	AAG	CUG	CAU	GUG	GAU	CCU
101				105					110					115					120
Glu	Asn	Phe	Lys	Leu	Leu	Gly	Asn	Val	Leu	Val	Thr	Val	Leu	Ala	Ile	His	Phe	Gly	Lys
GAG	AAC	UUC	AAG	CUC	CUG	GGA	AAU	GUG	CUG	GUG	ACC	GUU	CUG	GCA	AUC	CAU	UUC	GGC	AAA
121				125					130					135					140
Glu	Phe	Thr	Pro	Glu	Val	Gln	Ala	Ser	Trp	Gln	Lys	Met	Val	Thr	Gly	Val	Ala	Ser	Ala
GAA	UUC	ACC	CCU	GAG	GUG	CAG	GCU	UCC	UGG	CAG	AAG	AUG	GUG	ACU	GGA	GUG	GCC	AGU	GCC
141				145															
Leu	Ser	Ser	Arg	Tyr	His														
CUG	UCC	UCC	AGA	UAC	CAC	UGA	↓ GCU	¦ CGC	UGC	CCA	UGA	UGC	AGA	↓ GCU	UUC	AAG	GAU	AGG	CUU
UAU	UCU	GCA	AGC	AAU	ACA	AAU	AAU	AAA	UCU	AUU	CUG	CUU	AGA	GAU	CA--				

FIGURE 3. Nucleotide sequence of human $^{G}\gamma$ globin mRNA, derived from analysis of cDNA plasmids JW151 (6) and pHγG1 (11). The underlined nucleotides and amino acids in the coding portion of the mRNA are those that differ from those of human β globin mRNA and β globin chains. The underlined nucleotides in the 3'-untranslated portion of the mRNA are those that differ from those of the human $^{A}\gamma$ mRNA (18). The arrows indicate cleavage sites for the restriction endonuclease Alu I and the dashed line shows the Sac I site.

of the 5'-untranslated and 3'-untranslated sequences of human α globin mRNA have been previously published (13-16). The nucleotide sequence of portions of the human $^{G}\gamma$ mRNA determined from sequence analysis of $^{G}\gamma$ cDNA plasmids JW151 and pHγG1 is shown in Figure 3. The 5'-untranslated sequence of mixed $^{A}\gamma$ + $^{G}\gamma$ globin cDNAs has been previously published (17) as well as the sequence of the 3'-untranslated sequence of $^{A}\gamma$ mRNA determined from sequence analysis of the $^{A}\gamma$ cDNA plasmid pRP10 (18). In Figure 3 the underlined amino acids and underlined nucleotides in the coding portion of γ mRNA are those which differ from those of human β globin mRNA, the sequence of which has been previously published (19-21). The underlined nucleotides in the 3'-untranslated portion of the $^{G}\gamma$ mRNA are those which differ from the sequence of the $^{A}\gamma$ mRNA. With respect to the sequence immediately adjacent to the poly(A) of the $^{G}\gamma$ mRNA we are thus far unable to determine whether the 3'-terminal A indicated in Figure 3 actually represents the beginning of the poly(A) sequence or whether it is followed by one additional nucleotide (C) as in the case of the $^{A}\gamma$ mRNA (18). The nucleotide substitution that occurs three nucleotides following the termination codon in $^{G}\gamma$ cDNA generates a new restriction endonuclease site for the enzyme Alu I which is not present in the corresponding position of the $^{A}\gamma$ cDNA. Both $^{G}\gamma$ and $^{A}\gamma$ cDNAs have an additional Alu I site 21 nucleotides further downstream. Another base change creates a Sac I site in $^{G}\gamma$ cDNA (Figure 3).

Base Composition of the Globin Genes. Table 1 lists the G+C base composition of the human α, β and γ globin cDNAs and of the rabbit α and β globin cDNAs. It can be noted that there is a striking difference in the G+C content of both α globin cDNAs compared to the non-α globin cDNAs: the human and rabbit α globin cDNAs both have a G+C content of 64%-65% compared to 50%-52% for the non-α globin cDNAs and 40%-44% for the respective total cellular DNAs (Table 1). In addition to the high G+C content of the α globin cDNAs, there is also a lack of avoidance of the dinucleotide CG which is present at a lower than the expected frequency in total human DNA and in most mRNAs including the non-α globin mRNAs (Table 1). The avoidance of CG dinucleotides is less striking in the case of human γ cDNA than in the two β cDNAs. The higher than expected G+C content of the α globin cDNAs is not dic-

TABLE 1
BASE COMPOSITION OF GLOBIN cDNAs

DNA	G+C Content	CG Dinucleotides no.	CG Dinucleotides % of expected
Human α	64.7%	39	65
Rabbit α	64.1%	28	49.6
Human β	51.2%	5	12
Rabbit β	50.6%	4	10.6
Human γ	51.6%	7	18
Total human DNA	40±1%		
Total rabbit DNA	44.2%		

tated by the requirements of the amino acid sequence of the α globin chains and in fact is a feature of the 3'-untranslated sequences of the α globin mRNAs as well as of the coding sequences.

Codon Usage in Globin mRNAs. Figure 4 lists the codon usage of the human α, β and γ globin mRNAs. A number of features can be noted. First of all there is a striking bias in the usage of certain codons for certain amino acids. This is especially striking in the case of leucine and valine, the amino acids that are present in the highest amounts in the globin chains. In the case of leucine, of the six possible codons, CUG is used 40 times out of 53 total possible opportunities in the three mRNAs, whereas in the case of valine, of four possible codons, GUG is utilized 31 times out of 44 total possible opportunities in the three mRNAs. In the case of a number of other amino acids as well there is a significant bias in codon usage. The same biases found in human α and β mRNAs are also found in the corresponding rabbit mRNAs (7,8,23). Overall the bias is similar between all of the globin mRNAs, but there are certain exceptions. Consistent with their high G+C content and lack of avoidance of the dinucleotide CG, the α globin mRNAs utilize CG containing co-

1 \ 2		U	α	β	γ		C	α	β	γ		A	α	β	γ		G	α	β	γ	2 / 3
U	Phe	UUU	0	5	3	Ser	UCU	3	1	2	Tyr	UAU	1	2	0	Cys	UGU	0	2	1	U
		UUC	7	3	5		UCC	4	2	5		UAC	2	1	2		UGC	1	0	0	C
	Leu	UUA	0	0	0		UCA	0	0	0	Term	UAA	1	1	0	Term	UGA	0	0	1	A
		UUG	0	0	1		UCG	0	0	0		UAG	0	0	0	Trp	UGG	1	2	3	G
C	Leu	CUU	1	0	1	Pro	CCU	2	5	2	His	CAU	0	2	4	Arg	CGU	1	0	0	U
		CUC	2	3	3		CCC	3	0	1		CAC	10	7	3		CGC	0	0	0	C
		CUA	1	0	1		CCA	0	2	1	Gln	CAA	0	0	0		CGA	0	0	0	A
		CUG	14	15	11		CCG	2	0	0		CAG	1	3	4		CGG	1	0	0	G
A	Ileu	AUU	0	0	0	Thr	ACU	0	3	3	Asn	AAU	0	1	2	Ser	AGU	0	2	2	U
		AUC	0	0	3		ACC	9	3	5		AAC	4	5	3		AGC	4	0	2	C
		AUA	0	0	1		ACA	0	1	2	Lys	AAA	1	3	2	Arg	AGA	0	0	1	A
	Met	AUG	2	1	2		ACG	0	0	0		AAG	10	8	8		AGG	1	3	2	G
G	Val	GUU	2	3	2	Ala	GCU	2	5	3	Asp	GAU	0	5	4	Gly	GGU	1	4	1	U
		GUC	2	2	2		GCC	12	9	6		GAC	8	2	4		GGC	6	8	6	C
		GUA	0	0	0		GCA	0	1	2	Glu	GAA	0	2	4		GGA	0	0	6	A
		GUG	9	13	9		GCG	7	0	0		GAG	4	6	4		GGG	0	1	0	G

FIGURE 4. Codon usage in human globin mRNAs.

dons that are not encountered in the β and γ mRNAs: the CGN codons for arginine, CCG for proline and GCG for alanine. Another striking example of exclusive use of a codon in one of the mRNAs but not the others is the case of the glycine codon GGA which is used six out of 13 possible opportunities in human γ mRNA but is not used in a total of 20 other possible opportunities in the human α and β mRNAs. Certain codons are not used at all in any of the three globin mRNAs and certain other codons are used exclusively in one globin mRNA but not the others.

Homology Between Different Globin mRNAs. Table 2 lists the degree of divergence between various globin mRNAs. Human α and β globin mRNA are compared to rabbit α and β globin mRNAs respectively and the human β and γ mRNAs which presumably evolved from a single ancestral gene are also compared. The degree of homology between the human and rabbit β globin mRNAs has been extensively reviewed elsewhere (23). In summary it has been found that there is a great deal of conservation or lack of divergence of the nucleotide sequences of the coding portions of these two mRNAs. As listed in Table 2, there is only an 11.2% difference in nucleotides between the two mRNAs associated with a 9.6% difference in amino acid sequence between the two globin chains. The most striking feature is a relative lack of substitutions in silent substitution sites of the mRNA where base substitutions would not necessarily lead to a change in amino acid sequence. Substitutions at such sites occurred at only 20.5% of the sites, an incidence that is significantly less than expected if there had been a random drift or neutral evolution between these two sequences. Using the fibrinopeptide model (23), a 55% substitution rate at silent substitution sites is the minimum neutrality standard and lower values are indicative of selec-

TABLE 2
HOMOLOGY BETWEEN CODING PORTIONS OF GLOBIN mRNAs

mRNAs compared	Amino acid differences	Nucleotide differences	Substitutions at silent sites
Human β vs. rabbit β	14/146 9.6%	49/438 11.2%	33/161 20.5%
Human α vs. rabbit α	25/141 17.7%	62/423 14.7%	34/155 21.9%
Human β vs. human γ	39/146 26.7%	103/438 23.5%	59/161 36.6%

tive pressure to maintain a sequence. When human and rabbit α globin mRNA sequences are compared one finds that the difference in nucleotides is 14.9% compared to an amino acid difference of 17.7%. The number of base substitutions at silent substitution sites of the mRNA is only slightly higher than in the case of the β mRNAs: 21.9%. This low figure is somewhat surprising because one might have expected more divergence and evolutionary drift between these two sequences at silent substitution sites since the human and rabbit α globin chains display almost twice as much divergence at the amino acid level as do the β globin chains. One must conclude again that there must have been selective pressure to maintain the sequences similar and that silent mutations in the coding region of the α and β mRNAs are not necessarily neutral and that some stabilizing selection operates on the mRNA itself. The constraints imposed by the nonrandom use of synonym codons (discussed in the prior section) may in fact constitute such a selective pressure.

When one compares human β and γ mRNAs on the other hand one observes a somewhat different pattern. The two globin chains differ by 39 out of 146 amino acids (26.7%) and the two mRNAs differ by 23.5% of their nucleotides. When one considers substitutions at silent substitution sites of the mRNAs the observed value of 36.6% is considerably higher than in the case of the human-rabbit comparisons just discussed. The value approaches that of the neutrality standard for neutral evolution and indicates that there appears to have been less selective pressure to maintain stability between the human β and γ mRNAs than there was to maintain stability of α and β mRNA sequences between the human and rabbit species. The silent substitution rate however is probably not as high as one might have expected in view of the almost threefold amino acid difference between human β and human γ globin chains when compared to the amino acid difference between human β and rabbit β globin chains. One might therefore have expected an even greater divergence at silent substitution sites. It was previously noted (23) that the substitutions between the human and rabbit β globin mRNAs were not randomly distributed but appeared to be clustered within the sequence leaving large stretches of the sequence without any substitutions. In the case of the human β and γ mRNAs the base substitutions appear to be widely distributed within the sequence, as shown by the underlined nucleotides in Figure 3, but there is one region of the sequence that is notably free of substitutions: in the sequence from codon no. 91 to no. 106 there are only two nucleotides that differ between the human β and γ mRNAs. There is similar strict conservation of this region of the nucleotide sequence between human β mRNA and rabbit and mouse β mRNAs.

Restriction Endonuclease Maps of Human α and γ Globin cDNAs. The detailed restriction endonuclease maps of the human α and γ globin cDNAs are shown in Figures 5 and 6. The map of the human β cDNA has been previously published(19,22). The following features are noteworthy. The human β and γ cDNAs share the following common single restriction endonuclease sites: Ava II (codon pos. 38-39), Bam HI (codon pos. 98-100), and Eco RI (codon pos. 121-122). The human α cDNA lacks these three restriction endonuclease sites although it does contain an Ava II site in its 5'-untranslated sequence near the 5'-terminus of the mRNA. The α cDNA however does contain two unique restriction endonuclease sites which are absent from β and γ cDNAs: Hind III (codon pos. 90-91) and Hpa II (codon pos. 95-96). Another striking difference between the α and non-α globin cDNAs is the presence of four Hha I sites in the α cDNA whereas no such sites are present in the β and γ cDNAs. A number of restriction endonucleases are helpful in differentiating β from γ cDNAs. The γ cDNA contains two Mbo II sites (codon pos. 21-22 and 131-133) and one Pvu II site (codon pos. 87-88) that are absent from the β cDNA. The three globin cDNAs also differ in the number of

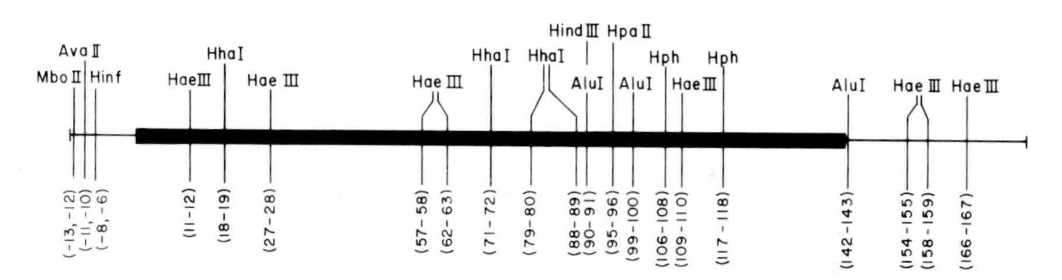

FIGURE 5. Restriction endonuclease map of human α globin cDNA.

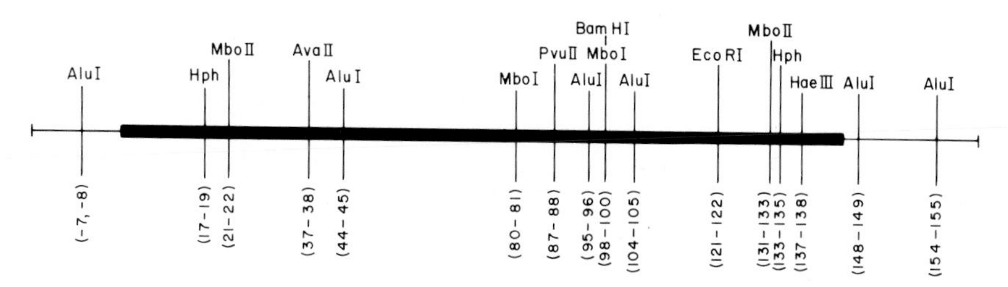

FIGURE 6. Restriction endonuclease map of human $^{G}\gamma$ globin cDNA. The Sac I site is not shown (see Figure 3).

Hae III sites that they contain: the α cDNA has 8, the β cDNA 5, but the γ cDNA only 1 (codon pos. 137-138). The nucleotide sequence of the $^{A}\gamma$ cDNA has been determined from codon 128 to the 3'-extremity of the mRNA. The nucleotide substitution that causes the amino acid replacement in the γ globin chain at position 136 from glycine ($^{G}\gamma$) to alanine ($^{A}\gamma$) also generates a cleavage site for the enzyme Pst I; Pst I therefore cleaves the $^{A}\gamma$ cDNA but not the $^{G}\gamma$ cDNA. As previously noted, two of the nucleotide differences between $^{G}\gamma$ and $^{A}\gamma$ cDNAs in their 3'-untranslated sequences generate an additional Alu I site and Sac I site in $^{G}\gamma$ cDNA that are not present in $^{A}\gamma$ cDNA (Figure 3).

Nucleotide Sequences of the Human δ Globin Gene. Only partial nucleotide sequence information of the human δ globin gene is available. Lawn et al. reported the nucleotide sequence of cloned δ gene DNA from the Eco RI site (codon pos. 121-122), going retrograde to codon pos. 105 and into the intron for a distance of 28 nucleotides (4); the corresponding portion of the β globin gene and its intron were also sequenced for comparison (4). We have obtained the same sequence information from our own studies of the δ globin gene DNA and have obtained additional δ gene sequence information from the Bam HI site (codon pos. 99-100) to codon pos. 104 and into the intron for a distance of approximately 135 nucleotides. The nucleotide sequences at the "joints" or junctions between coding sequences and intron sequences are shown in Figure 7. The sequences of the first six nucleotides at either extremity of the major intron and of the adjacent codons are the same in the δ globin gene as they are in the mouse and rabbit β globin genes (9,10), with the exception of one nucleotide (underlined in Figure 7) that also differs

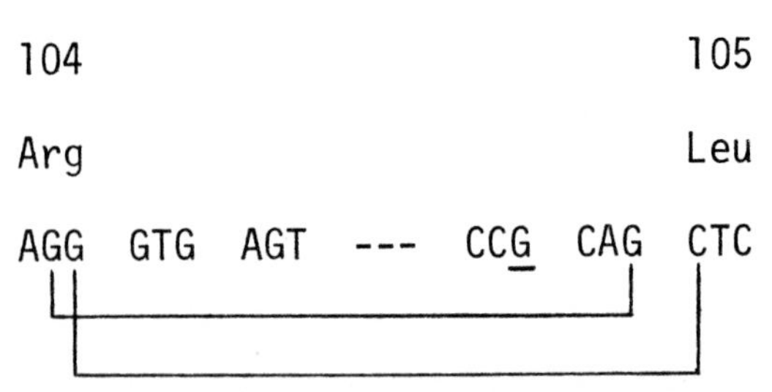

FIGURE 7. Nucleotide sequences of human δ globin gene DNA at the junctions between the major intron and adjacent codons. Possible splicing frames for excision of the intervening sequence are indicated by the brackets under the sequence. The underlined G is replaced by an A in the sequence of mouse, rabbit and human β globin gene DNA (4,9,10).

from the nucleotide in the same position of the human β globin gene intron (4). Nevertheless, possible splicing frames are the same in the human δ globin gene as in the rabbit and mouse β globin genes. As indicated in Figure 7 there are two possible splicing frames and these are identical to those of the rabbit and mouse β globin genes (9,10). Thus the human gene is not restricted in the number or position of splicing frames at the site of its major intron when compared to other β-like globin genes.

The sequence of the first 135 nucleotides of the 5'-extremity of the δ globin gene intron demonstrates some striking features. The A+T base composition of this portion of the sequence is extremely high: 69%. In addition there are numerous stretches of oligo T sequences: a total of 6 or 7 sequences consisting of runs of four or five consecutive T's. The available sequence information of the introns of the mouse and rabbit β globin genes(9,10) also indicates that the introns of these genes have a high A+T base composition but not as high as the δ gene. Furthermore, the 5'-extremities of the introns of the rabbit and mouse β globin genes do not contain multiple runs of T's as seen in the δ globin gene intron. It is conceivable that these runs of T's may have an attenuating effect on the transcription of the δ globin gene by signaling a slow-down or cessation of transcription of the gene at these sites and thereby be responsible, in part, for the low level of expression of the δ gene.

Finally one can make some observations concerning the partial nucleotide sequence information of the coding portion of the δ globin gene: the sequence from codons 100 to 104 does not differ from that of the human β gene (or the mouse and rabbit β genes). The sequence of the codons for amino acid residues no. 105 to 122 however demonstrates a number of differences between δ and β mRNAs as previously reported by Lawn et al. (4) and confirmed by our own sequence analysis. There are three nucleotide changes that resulted in the two amino acid replacements at positions 116 and 117 and in addition there are four other silent nucleotide substitutions at silent substitution sites. Two of the codons that resulted from these substitutions (UUG for leucine at position 106 and CGC for arginine at position 116) are used rarely or not at all in the other human globin mRNAs (Figure 4).

DISCUSSION

Knowledge of the total nucleotide sequences of the human α, β and γ globin mRNAs has allowed us to draw parallels and contrasts between these mRNAs that are expressed in the same type of very specialized cell but are the products of genes situated on different chromosomes or expressed at different stages of human development. The human α globin mRNA differs from the non-α mRNAs by its very high G+C content and by its lack of avoidance of CG dinucleotides. Codon usage in all three globin mRNAs demonstrates a marked bias in the use of synonym codons for certain amino acids: this is especially striking in the case of leucine and valine codons. The bias in codon usage is rather similar between the three human globin mRNAs with nevertheless some notable exceptions. The α globin mRNA utilizes a number of codons containing CG dinucleotides that are not utilized in the β and γ mRNAs. The γ mRNA on the other hand does not demonstrate as striking a selection against use of codons ending in A versus G as do the α and β globin mRNAs; as a result certain codons ending in A are more frequently used in γ than in α or β mRNAs. This is especially true of the glycine codon GGA.

The sequence homology of coding sequences has been compared between the human α and β globin mRNAs and their corresponding rabbit globin mRNAs as well as between the human β and γ mRNAs which presumably evolved from a single ancestral gene. There is more homology between the human and rabbit β globin genes than between their α globin genes, both from the point of view of total amino acid differences in their respective gene products and total nucleotide differences in the coding sequences of their respective mRNAs. However, the number of base substitutions at silent substitution sites is relatively low and fairly similar between the α and β mRNAs of both species and this finding suggests that there has been selective pressure for maintaining these sequences similar and that neutral evolutionary drift has not occurred between the human and rabbit globin genes. More divergence is seen between the human β and γ mRNAs than is observed between the human and rabbit α and β globin mRNAs respectively. There are a greater number of total nucleotide differences between the human β and γ mRNAs as well as a higher proportion of base substitutions at silent substitution sites between the two mRNAs. The number of the latter type of base substitutions is almost double the number that is observed between human and rabbit α and β mRNA species respectively but the number does not quite reach the value expected if there had been totally neutral evolutionary drift between the β and γ globin gene coding sequences. Of particular interest is the fact that one region of the nucleotide sequence of the γ mRNA has diverged very little from that of the β mRNA: the sequence between codon no. 91 and codon no. 106. This is precisely the region of the nucleotide sequence(between the sequences for codons no. 104 and no. 105) where the human β and γ globin mRNAs must be spliced from higher molecular weight precursors that contain transcripts of the major intervening sequences of the β and γ globin genes (4,5). The conservation of sequences in this region of the β and γ globin genes may be important for the processing of their globin mRNA precursors. Of note however is the fact that one of the base substitutions in the γ mRNA involves the codon immediately preceding the intervening sequence, i.e., codon no. 104 where the codon AGG for arginine has been changed to AAG for lysine. All of the other junctions between coding and intervening sequences in human, rabbit and mouse β mRNAs occur between arginine and leucine codons. The human γ globin gene is therefore the only β-like globin gene that has a lysine-leucine splicing site. It should be noted however that in the case of the mouse α globin gene, the junction between the coding sequence and the major intervening sequence is situated between a lysine and leucine codon at positions 99 and 100 (24).

Finally, partial nucleotide sequence analysis of human δ globin gene DNA has provided some information on possible mechanisms for the low level of expression of this gene in human erythroid cells. The sequences at the junctions of the coding sequence and the major intervening sequence of the δ gene show no significant differences from those of other β-like globin genes thus far sequenced and there is no restriction in the number of splicing frames at this site of the δ gene, compared to other β globin genes. However, the nucleotide sequence of the first 135 nucleotides at the 5'-extremity of the δ gene intron has an extraordinarily high A+T base composition including numerous runs of four and five consecutive T's which could conceivably attenuate transcription of the gene by acting as weak transcription termination signals and thus result in an overall decrease in the level of δ globin gene expression. Limited sequence information thus far available on the coding portion of the δ globin gene reveals the use in δ mRNA of two codons, UUG and CGC, that are underutilized in the three other human globin mRNAs. Theoretically, use of these codons might affect the rate of translation of δ mRNA, and constitute another possible mechanism for the low level of δ globin gene expression. However, at least in the case of the leucine codon UUG, this codon is utilized in at least two mutant human globin mRNAs (for Hb Tak and Hb Cranston) without a significant effect on the level of synthesis of the mutant β globin chains (25), and there is no direct evidence for an abnormally long translation time of δ chains in human erythroid cells (26). The codon UUG is also used once in human γ mRNA (Figure 4).

ACKNOWLEDGEMENTS

We thank V. Vellucci and E. Coupal for skilled technical assistance, and thank Dr. T. Maniatis and Dr. S. Malcolm for providing us with the plasmids Hδ1 and pHγG1 respectively.

REFERENCES

1. Bunn, H.F., Forget, B.G., and Ranney, H.M. (1977). In "Human Hemoglobins" pp. 101-112. W.B. Saunders Co., Philadelphia.
2. Deisseroth, A., Nienhuis, A.W., Turner, P., Velez, R., Anderson, W.F., Ruddle, F., Lawrence, J., Creagan, R., and Kucherlapati, R. (1977). Cell 12, 205.
3. Deisseroth, A., Nienhuis, A., Lawrence, J., Giles, R., Turner, P., and Ruddle, F.H. (1978). Proc. Natl. Acad. Sci. (USA) 75, 1456.
4. Lawn, R.M., Fritsch, E.F., Parker, R.C., Blake, G., and Maniatis, T. (1978). Cell 15, 1157.
5. Smithies, O., Blechl, A.E., Denniston-Thompson, K., Newell, N., Richards, J.E., Slighton, J.L., Tucker, P.W., and Blattner, F.R. (1978). Science 202, 1284.
6. Wilson, J.T., Wilson, L.B., deRiel, J.K., Villa-Komaroff, L., Efstratiadis, A., Forget, B.G., and Weissman, S.M. (1978). Nucleic Acids Res. 5, 563.
7. Efstratiadis, A., Kafatos, F.C., and Maniatis, T.(1977). Cell 10, 571.
8. Heindell, H.C., Liu, A., Paddock, G.V., Studnicka, G.M., and Salser, W. (1978). Cell 15, 43.
9. van den Berg, J., van Ooyen, A., Mantei, N., Schambock, A., Grosveld, G., Flavell, R.A., and Weissmann,C.(1978). Nature 276, 37.
10. Konkel, D.A., Tilghman, S.M., and Leder, P. (1978). Cell 15, 1125.
11. Little, P., Curtis, P., Coutelle, C., van den Berg, J., Malcolm, S., Courtney, M., Westaway, D., and Williamson, R. (1978). Nature 273, 640.
12. Maxam, A.M., and Gilbert, W. (1977). Proc. Natl. Acad. Sci. (USA) 74, 560.
13. Baralle, F.E. (1977). Cell 12, 1085.
14. Chang, J.C., Temple, G.F., Poon, R., Neumann, K.H., and Kan, Y.W. (1977). Proc. Natl. Acad. Sci. (USA) 74, 5145.
15. Wilson, J.T., deRiel, J.K., Forget, B.G., Marotta, C.A., and Weissman, S.M. (1977). Nucleic Acids Res. 4, 2353.
16. Proudfoot, N.J., Gillam, S., Smith, M., and Longley, J. I. (1977). Cell 11, 807.
17. Chang, J.C., Poon, R., Neumann, K., and Kan, Y.W.(1978).

Nucleic Acids Res. 5, 3515.
18. Poon, R., Kan, Y.W., and Boyer, H.W. (1978). Nucleic Acids Res. 5, 4625.
19. Marotta, C.A., Wilson, J.T., Forget, B.G., and Weissman, S.M. (1977). J. Biol. Chem. 252, 5040.
20. Proudfoot, N.J. (1977). Cell 10, 559.
21. Forget, B.G. (1978). Hemoglobin 1, 879.
22. Forget, B.G., Wilson, J.T., Wilson, L.B., Cavallesco, C., Reddy, V.B., deRiel, J.K., Biro, A.P., Ghosh, P.K., and Weissman, S.M. (1979). In "Cellular and Molecular Regulation of Hemoglobin Switching" (G. Stamatoyannopoulos and A.W. Nienhuis, eds.), pp. 569-591. Grune and Stratton, New York.
23. Kafatos, F.C., Efstratiadis, A., Forget, B.G., and Weissman, S.M. (1977). Proc. Natl. Acad. Sci. (USA) 74, 5618.
24. Leder, A., Miller, H.I., Hamer, D.H., Seidman, J.G., Norman, B., Sullivan, M., and Leder, P. (1978). Proc. Natl. Acad. Sci. (USA) 75, 6187.
25. Forget, B.G., Marotta, C.A., Weissman, S.M., and Cohen-Solal, M. (1975). Proc. Natl. Acad. Sci. (USA) 72, 3614.
26. Wood, W.G., Old, J.M., Roberts, A.V.S., Clegg, J.B., Weatherall, D.J., and Quattrin, N. (1978). Cell 15, 437.

THE ORGANIZATION OF CHICKEN GLOBIN GENES[1]

Jerry Dodgson[Θ], Judith Strommer[Φ] and James Douglas Engel[§]

[Θ]Department of Chemistry, California Institute of Technology, Pasadena CA 91125

[Φ]Molecular Biology Institute and Department of Microbiology and Immunology, University of California, Los Angeles, California 90024

[§]Department of Biochemistry and Molecular Biology, Northwestern University, Evanston, Illinois 60201

ABSTRACT A library of chromosomal chicken DNA fragments in the vector λ Charon 4A has been prepared. The library was screened with adult and embryonic globin cDNA probes. Seven independent recombinant phage have so far been isolated containing the adult and several embryonic globin genes. One of the recombinants contains the adult chicken β-globin gene and a closely linked embryonic β-like globin gene. Another recombinant contains two closely linked α-globin genes, probably the αA and αD globin genes expressed in both adult and embryonic erythrocytes.

INTRODUCTION

The genes that code for the various chicken globin polypeptides comprise a developmentally regulated gene family that is particularly suited to studies at the molecular level. Several distinct globin polypeptides are differentially expressed during embryonic erythropoiesis (1), and chicken embryonic development is readily accessible to biochemical experimentation (2). Furthermore, chicken chromatin structure has been extensively investigated (3), and studies on the regulation of a variety of other chicken genes should offer useful comparisons to those of chicken globin gene expression. We have isolated recombinant chicken DNA-λ Charon 4A phage containing several of the chicken globin genes. Our

[1]Supported by N.I.H. Research Fellowship No. GM06921 (JD), USPHS Training Grant GM7185 (JS) and Grant No. 78-77, American Cancer Society, Illinois Division (JDE).

ISBN 0-12-068350-4

preliminary studies on the structure and organization of these genes are summarized below.

RESULTS

Preparation of a Recombinant Chicken DNA-λ Charon 4A Library. The chicken recombinant library was prepared by the method of Maniatis et al. (4). Details of the preparation and characterization of the library will be published elsewhere (5). The library consists of about 10^{12} phage resulting from a single step amplification of about 4.5×10^5 independent recombinants. Chicken DNA sequences comprise 15 to 20 kilobase pairs (kb) of each of the recombinant phage genomes. The library should contain any single copy chicken gene at the 99% confidence level (6), and several genes other than globin genes have also been isolated from the library (P. Chambon, S. Woo and G. Schütz, personal communication).

Isolation of Chicken Globin Gene Recombinants. The chicken library was screened essentially by the method of Benton and Davis (7). Seven independent recombinants that hybridize globin cDNA have been isolated. The restriction maps of the chicken DNA insertions of these phage are shown in Figure 1. Three recombinants hybridize to β-globin specific probes (rabbit cDNA plasmid PβG1, (8); chicken cDNA plasmid pHb1001, (9)) and have been designated λCβG1, λCβG2 and λCβG3. Note that the chicken DNA sequences in λCβG2 and λCβG3 overlap. No overlap has been detected between either λCβG2 or λCβG3 and λCβG1. λCβG1 has been shown to contain both the adult chicken β-globin gene and an as yet unidentified embryonic β-like gene (described in greater detail below).

The other four recombinants do not hybridize to β-globin specific probes and have tentatively been designated λCαG1, λCαG2, λCαG3 and λCαG4. However, we cannot be sure at this time that all the hybridizing regions on these four recombinants code for globin polypeptides. Indeed, the hybridizing regions on λCαG4 hybridize to embryonic cDNA more slowly (and generally give rise to lower intensity bands) than, for example, those regions on λCβG1. Thus the coding regions on this recombinant may represent non-globin genes transcribed in erythroid cells at low levels relative to globin gene transcription. The recombinants λCαG1 and λCαG2 hybridize to both adult and 7 day embryonic globin cDNAs while λCαG3 and λCαG4 hybridize only to embryonic probes. The linked adult α-globin genes on λCαG2 will be discussed in more detail below.

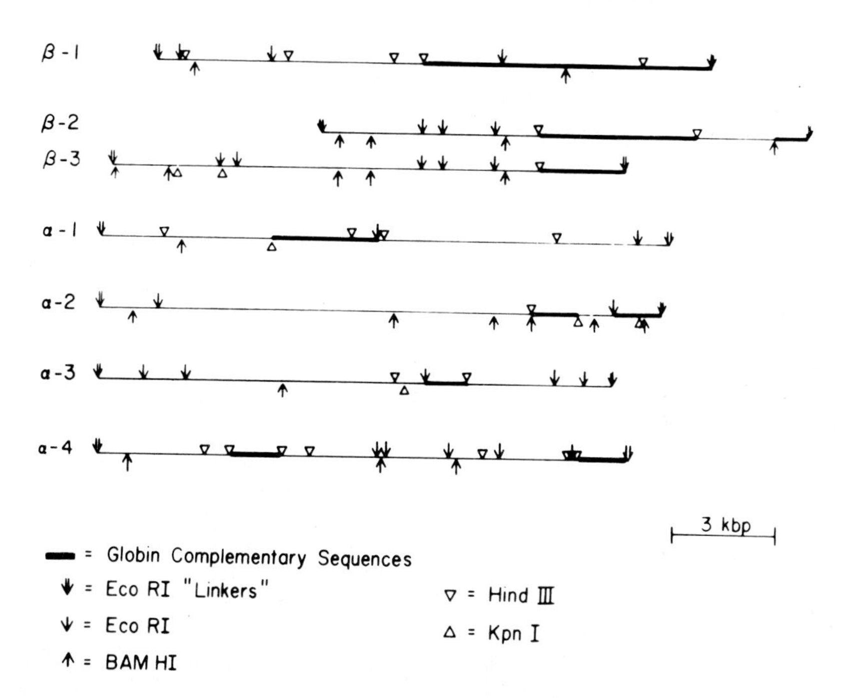

FIGURE 1. The restriction enzyme cleavage maps of the chicken DNA insertions of the seven independent recombinants isolated by their hybridization to globin cDNA. Restriction fragments that hybridize globin cDNA are shown in heavy lines.

A Recombinant Containing the Adult Chicken β-Globin Gene. The phage λCβG1 was initially selected for more detailed study because of its strong hybridization to adult globin cDNA and because the hybridizing region spanned several kb. One of the initial studies of λCβG1 employed the technique of electron microscopic R-loop visualization. The method we used was that of Kaback, Angerer and Davidson (10). This method involves the hybridization of globin mRNA to cross-linked phage DNA at temperatures and formamide concentrations where RNA:DNA hybrids can form and DNA:DNA duplexes are predominantly unstable. When the hybridization mixture is cooled, DNA:DNA hybrids reform but do not displace the RNA:DNA hybrids, and a characteristic displaced single-stranded DNA loop is formed. Furthermore, if there are internal sequences in the genomic DNA that are not present in

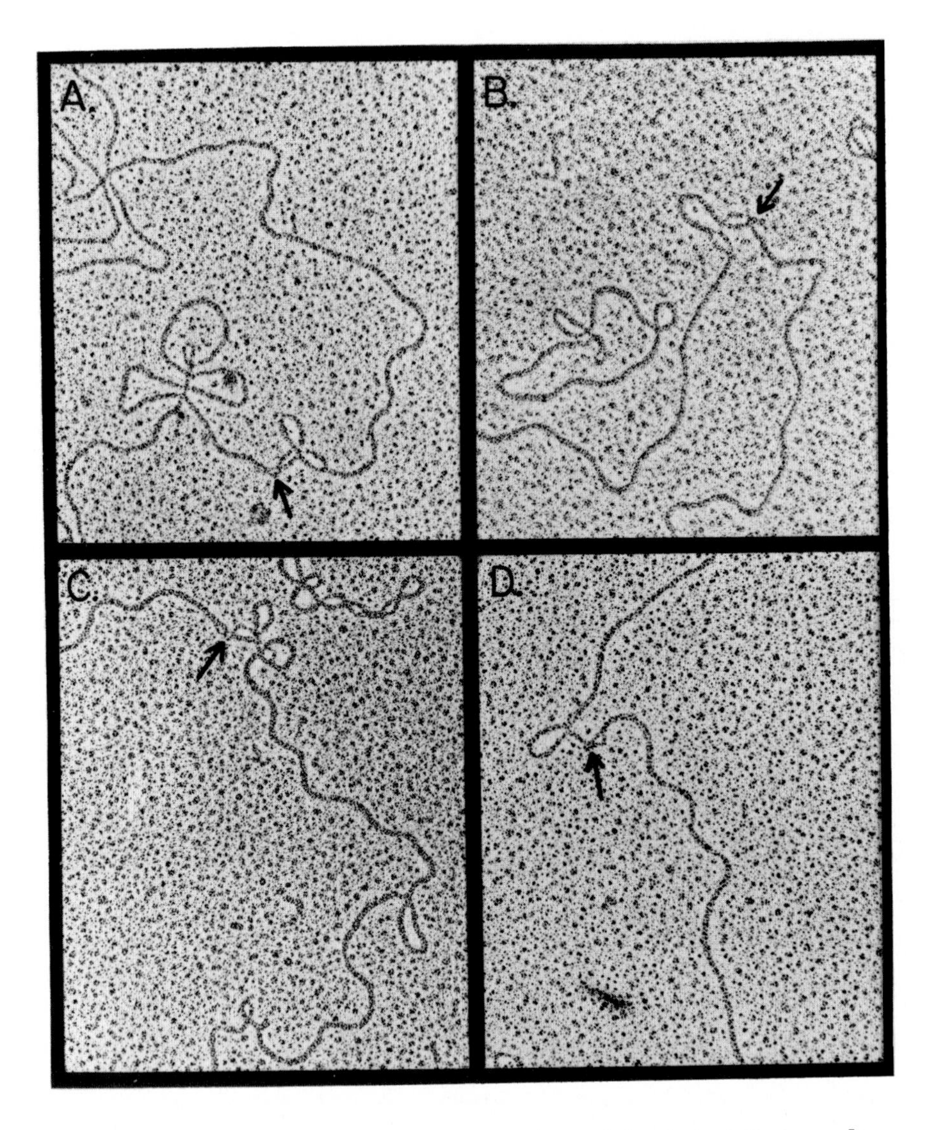

FIGURE 2. R-loop structures formed by hybridizing globin RNA to the adult β-globin gene on λCβG1. Arrows point to the loop or kink structures corresponding to the minor intervening sequence of this gene.

the mature mRNA, a DNA:DNA duplex loop is generally formed between the R-loop regions. (Small intervening sequences often do not reform a full DNA duplex structure, but rather are seen as a kink in the displaced DNA strand or as a small looped-out DNA on the RNA:DNA hybrid side of the R-loop.)

When λCβG1 phage DNA (cross-linked) was R-looped to adult globin mRNA, a characteristic R-loop structure was observed on most of the DNA molecules (Figure 2). This structure is similar to that observed in R-loop experiments with the cloned mouse β-globin gene (11), and consists of two R-loops, each approximately 250 base pairs (bp) in length, flanking a 750 bp intervening sequence (IVS) duplex DNA loop. The R-loop bubble proximal to the short λ arm was often seen to be divided by another small loop or kink (Figure 2). This loop corresponds to a ∿100 bp IVS analogous to that observed in the mammalian β-globin genes (5, 11-14).

When λCβG1 DNA is hybridized to embryonic globin mRNA (5 or 7 day), two R-loop structures are formed (5), both of which are similar to the one shown in Figure 2. One of the R-loop structures (i.e. two R-loops flanking a large IVS loop) is identical in form and location to that observed in the adult globin mRNA R-loop experiments. The other R-loop structure results from the hybridization of mRNA to an embryonic β-like globin gene. This structure is also comprised of two approximately 250 bp R-loops flanking a large IVS loop (about 800 bp long). We have not as yet definitively located a small IVS in the embryonic gene analogous to the 100 bp minor IVS in the adult β-globin gene, but such an embryonic minor IVS may be present but not form a readily observable loop under the hybridization conditions used.

The locations of the globin genes of λCβG1 detected by the R-loop experiments are such that, in the map of λCβG1 in Figure 1, the adult β-globin gene is cut in its major IVS by the right hand Hind III restriction site. The embryonic β-like globin gene is about 3.5 kb to the left of the adult gene (see Figure 1), and its major IVS appears to contain the right hand non-linker Eco RI restriction site of the chicken DNA insertion. Subsequent experiments (5) have confirmed the assignment of the right hand globin gene (with respect to Fig. 1) to the adult β-globin gene and the left hand gene to an embryonic β-like globin. We do not yet know which of the embryonic β-like globins is contained in λCβG1, but it may be the ρ- or ε-globin gene (1). Further experiments have shown that both genes on λCβG1 are transcribed in the right to left direction (relative to Fig. 1) and thus off of the same strand of DNA (5). Furthermore, fine structure restriction mapping of the adult chromosomal β-globin gene on λCβG1 (5) is completely consistent with the exact conservation of IVS locations with respect to amino acid sequence between the adult chicken and the major adult mammalian β-globin genes (12-14).

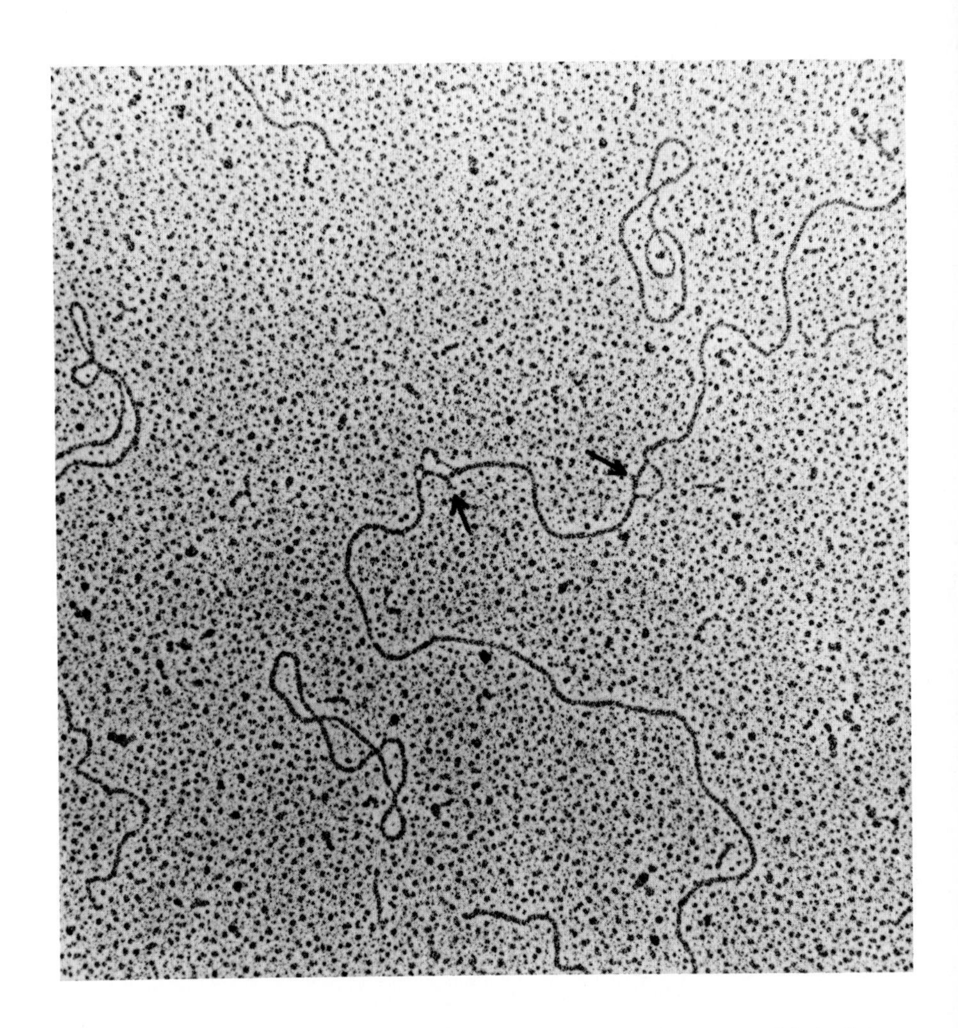

FIGURE 3. R-loops formed between λCαG2 cross-linked DNA and adult globin mRNA. Arrows point to looped-back DNA from the RNA:DNA hybrid regions suggesting small intervening sequence locations. The R-loop to the left of the figure represents the presumed αD globin gene and the one to the right represents the presumed αA gene.

A Recombinant Containing Two Linked α-Globin Genes. Restriction mapping of λCαG2 (Fig. 1) suggested that it contained two closely linked non-β globin genes. One of these genes is present on λCαG2 on a 13 kb Eco RI restriction fragment bordered by natural Eco RI sites (i.e. not the synthetic linker Eco RI sites used to make the recombinant phage (4)). This suggests that this corresponds to the αA globin gene detected at 12.5 kb in Eco RI-cleaved total chromosomal DNA Southern blots (15). The other gene on λCαG2 is on a synthetic linker-bounded Eco RI fragment, and therefore an analogous argument for its identification is not available. However, from its strong hybridization to adult globin cDNA, its close linkage to the proposed αA adult globin gene and the fact that it does not hybridize to rabbit globin cDNA or to the unusual chicken α-globin cDNA plasmid probes (pHb1003 and pHb1008, (9)) which are more closely related to the αA gene, it seems likely that this gene codes for the other α-globin expressed in adult chickens, the highly divergent αD globin.

When λCαG2 was R-looped to adult chicken globin mRNA, the phage DNA molecules generally contained two R-loop structures (Figure 3). The form of these R-loop structures varied considerably, but in general both R-loop structures were consistent with the α-globin gene structure observed in the cloned mouse α-globin gene (16). Both chicken α-globin genes appear to contain two IVS regions of roughly similar size (50 to 200 bp). We do not yet have data regarding the direction of transcription of the two α-globin genes or more accurately defining the location of the IVS regions. From the conservation of the β-globin gene structure between avians and mammals (above), it is likely that IVS locations in the chicken α-globin genes will be similar if not identical to those of the mouse α-globin gene.

DISCUSSION

While the comparison of chicken globin gene organization to that of mammalian globin genes is not the major reason for our interest in this system (5, 15), such comparisons may shed some light on what facets of gene organization may be essential for proper globin gene expression. Presumably mutational alterations in these essential features would be highly disfavored evolutionarily, and thus these aspects of gene organization would be observed in both avians and mammals.

It has already been noted that the IVS locations with respect to amino acid sequence are probably exactly the same in the adult chicken and mammalian β-globin genes (5, 12-14). This is not surprising since the amino acid codons bordering the IVS sites (Arg 30-Leu 31 and Arg 104-Leu 105) are identical in the chicken adult β-globin to those of mammalian β-globin genes (17). Furthermore, Leder et al. (16) have proposed that IVS locations are exactly conserved even between the much more divergent α- and β-type globin gene families. The results described above suggest that, in general, the structures of both α- and β-globin genes are very similar in chicken and mouse. While further studies of the chicken globin genes will be necessary to exactly pinpoint all of the IVS locations, it appears likely that these locations will be exactly conserved in all cases. Also, the sizes of the various IVS regions seem to be roughly the same in both species. This conservation of their size and location suggests to us that the IVS regions are of functional importance in the regulation of globin gene expression, but further work will be necessary to prove this.

The organization of globin genes into α- and β-type linkage groups also appears to be a feature common to both avians and mammals (13, 18). We do not yet know if the two β-type globin genes on λCβG1 are linked to the two on λCβG2, but this is likely to be the case, since mammalian β-globin genes seem to occur as linkage groups of at least four genes (13, 18, 19). While the two linked α-globin genes on λCαG2 are the first linked α-globin genes isolated in a single recombinant, it is known from Southern blotting studies that the two non-allelic α-globin genes in man are closely linked (20). In all cases studied to date, linked globin genes are transcribed off of the same DNA strand. In mammalian β-globin gene linkage groups, the adult β-globin gene always is located at the 3' terminus of the group (13, 19). This is not the case in chickens, however, since the embryonic β-like gene on λCβG1 is 3' to the adult β-globin gene.

The function of these linkage relationships is still unclear. While their evolutionary stability argues that they do have functional importance, linkage is not essential to coordinate globin gene regulation, since it is known that the α-type globin gene family is not linked to the β-type family (21-23). Again, further studies of globin gene expression will be necessary to understand the relationship of globin gene organization to the regulation of that expression.

ACKNOWLEDGMENTS

The continuing support and advice of Norman Davidson throughout all of this work is gratefully acknowledged. We also thank Richard Axel and Tom Maniatis for materials, advice and encouragement. We are also grateful to Winston Salser for cDNA plasmid samples and to David Kaback for advice and assistance regarding the electron microscopy.

REFERENCES

1. Brown, J.L., and Ingram, V.M. (1974). J. Biol. Chem. 249, 3960.
2. Romanoff, A.L. (1960). "The Avian Embryo, Structural and Functional Development." MacMillan, New York.
3. Weintraub, H., and Groudine, M. (1976). Science 193, 848.
4. Maniatis, T., Hardison, R.C., Lacy, E., Lauer, J., O'Connell, C., Quon, D., Sim, G.K., and Efstratiadis, A. (1978). Cell 15, 687.
5. Dodgson, J.B., Strommer, J., and Engel, J.D., submitted for publication.
6. Clarke, L., and Carbon, J. (1976). Cell 9, 91.
7. Benton, W.D., and Davis, R.W. (1977). Science 196, 180.
8. Maniatis, T., Sim, G.K., Efstratiadis, A., and Kafatos, F.C. (1976). Cell 8, 163.
9. Salser, W.A., Cummings, I., Liu, A., Strommer, J., Padayatty, J., and Clarke, P. (1979). In "Cellular and Molecular Regulation of Hemoglobin Switching." (G. Stamatoyannopoulos and A. Nienhuis, eds.) pp. 621-643. Grune and Stratton, New York.
10. Kaback, D.B., Angerer, L.M., and Davidson, N. (1979). Nucleic Acids Res., in press.
11. Tilghman, S.M., Tiemeier, D.C., Seidman, J.G., Peterlin, B.M., Sullivan, M., Maizel, J.V., and Leder, P. (1978). Proc. Nat. Acad. Sci. USA 75, 725.
12. Konkel, D.A., Tilghman, S.M., and Leder, P. (1978). Cell 15, 1125.
13. Lawn, R.M., Fritsch, E.F., Parker, R.C., Blake, G., and Maniatis, T. (1978). Cell 15, 1157.
14. van den Berg, J., van Ooyen, A., Mantei, N., Schamböck, A., Grosveld, G., Flavell, R.A., and Weissman, C. (1978). Nature 276, 37.
15. Engel, J.D., and Dodgson, J.B. (1978). J. Biol. Chem. 253, 8239.
16. Leder, A., Miller, H.I., Hamer, D.H., Seidman, J.G., Norman, B., Sullivan, M., and Leder, P. (1978). Proc. Nat. Acad. Sci. USA 75, 6187.
17. Matsuda, G., Maita, T., Mizuno, K., and Ota, H. (1973). Nature New Biol. 244, 244.
18. Little, P.F.R., Flavell, R.A., Kooter, J.M., Annison, G., and Williamson, R. (1979). Nature 278, 227.
19. Shen, C.K.J., and Maniatis, T. (1979). J. Supramol. Struc., Supplement 3, 66.
20. Orkin, S.H. (1978). Proc. Nat. Acad. Sci. USA 75, 5950.
21. Diesseroth, A., Nienhuis, A., Turner, P., Velez, R., Anderson, W.F., Ruddle, F., Lawrence, J., Creagan, R.,

and Kucherlapati, R. (1977). Cell 12, 205.
22. Diesseroth, A., Nienhuis, A., Lawrence, J., Giles, R., Turner, P., and Ruddle, F.H. (1978). Proc. Nat. Acad. Sci. USA 75, 1456.
23. Hughes, S.H., Stubblefield, E., Payvar, F., Engel, J.D., Dodgson, J.B., Spector, D., Cordell, B., Schimke, R.T., and Varmus, H.E. (1979). Proc. Nat. Acad. Sci. USA 76, 1348.

THE ORGANIZATION AND REARRANGEMENT OF HEAVY CHAIN IMMUNOGLOBULIN GENES IN MICE[1]

M. Davis, P. Early, K. Calame, D. Livant, and L. Hood

Division of Biology, California Institute of Technology, Pasadena, California 91125

ABSTRACT A preliminary analysis of several heavy chain variable (V) and constant region (C) gene segments from sperm (undifferentiated) and myeloma (differentiated) DNA has revealed the following: 1) the V_H and C_α genes are separate in the germ line; 2) the V_H and C_α genes are rearranged during the differentiation of the antibody-producing cell; 3) multiple rearranged C_α genes are present in the DNA of a single myeloma tumor; 4) small intervening sequences may separate the domains of the α and μ constant region genes; and 5) at least 8-9 germ line V_H genes exist for antibodies binding phosphorylcholine.

INTRODUCTION

The antibody gene families have several interesting organizational features. There are three distinct gene families - two code for light (L) chains, λ and κ, and the third codes for heavy (H) chains. They are composed of three distinct coding segments which are separated from one another by intervening DNA sequences - V (variable), J (joining) and C (constant). The V and J segments together comprise the V region of the antibody polypeptide which encodes the immunoglobulin domain concerned with antigen recognition. Moreover, each antibody gene family appears to contain multiple V and J segments.

The antibody gene families present two fascinating biological problems. First, it has been estimated that mammals can synthesize 10^5 to 10^8 different antibody molecules. What genetic mechanisms are responsible for this diversity of antibody molecules? We hope to assess the relative contributions of three genetic mechanisms: multiple germ line V genes (1), somatic mutation (2), and the joining in a combinatorial fashion of multiple V and J segments (3). Second,

[1]This work was supported by NSF grant PCM 76-81546.

ISBN 0-12-068350-4

how are antibody gene segments rearranged during the differentiation of antibody-producing cells? These DNA rearrangements presumably are fundamental components of the molecular events that commit the antibody-producing cell to the synthesis of a single type of antibody molecule as well as contributing to antibody diversity in the combinatorial joining of V and J segments (3,4).

We have focused on the analysis of the heavy chain gene family because, in addition to being an excellent system for studying the phenomena mentioned above, it has intricacies not exhibited in light chains. The heavy chain gene family of the mouse is comprised of an unknown number of variable (V_H) gene segments and at least eight different constant (C_H) gene segments (5) (Figure 1).

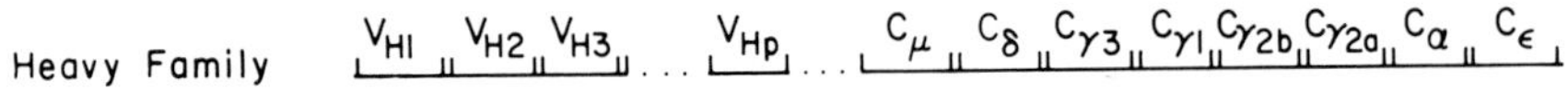

Heavy Family V_{H1} V_{H2} V_{H3} ... V_{Hp} ... C_μ C_δ $C_{\gamma 3}$ $C_{\gamma 1}$ $C_{\gamma 2b}$ $C_{\gamma 2a}$ C_α C_ϵ

FIGURE 1. Heavy chain antibody gene family in mice. The order of C_H gene segments is uncertain, although indirect evidence supports the following alignment: $C_{\gamma 3}C_{\gamma 1}C_{\gamma 2b}C_{\gamma 2a}C_\alpha$ (20). The number of V_H gene segments is still a matter of controversy. The heavy chain gene family also has multiple J segments that are not depicted in this figure (see text).

The various classes and subclasses of immunoglobulins are determined by the C_H gene segments (e.g., C_μ-IgM, C_γ-IgG, C_α-IgA, etc.). Moreover, during the differentiation of the antibody-producing cell, distinct classes of immunoglobulins are expressed in a reproducible order (Figure 2). First IgM is expressed; later IgD and IgM are expressed; and eventually the other classes of immunoglobulins are expressed (6). In the lineage of a particular antibody-producing cell, it appears that these developmental shifts in immunoglobulin class expression occur by associating a particular V_H gene segment with different C_H gene segments while maintaining the expression of the same light chain gene segments. Therefore, a question of particular interest is the nature of the DNA rearrangements which lead to sequential and at times, simultaneous, expression of different heavy chain classes. Fortunately, tumors of antibody-producing cells exist which "freeze" this developmental pathway at many different points. Thus in time we will understand how the antibody gene organization for sperm cells (undifferentiated DNA) differs from that of tumor cell lines producing IgM, IgM + IgD and IgA (i.e., various stages of differentiation). Accordingly, our our initial efforts are focused on understanding the gene organization in DNA at the beginning (sperm or embryo) and the end (IgA-producing myeloma) of a heavy chain differentiation pathway.

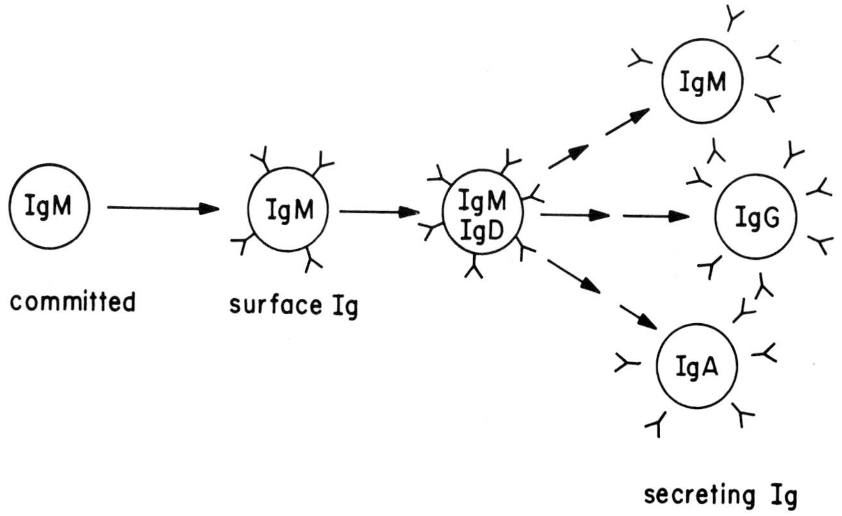

FIGURE 2. The differentiation of B cells. A B cell first becomes committed to the expression of a particular V domain (one V_L region and one V_H region) which is associated with cytoplasmic IgM molecules. Subsequently the IgM molecule is expressed on the cell surface. Later, cell-surface IgD molecules appear. Subsequent differentiation events lead to a terminally differentiated cell which specializes in the synthesis of soluble antibodies of one of a variety of immunoglobulin classes. For an individual B cell, the same V domain is associated with the various classes of immunoglobulins throughout the differentiation pathway.

THE PHOSPHORYLCHOLINE ANTIBODY SYSTEM

We have chosen to examine some of the questions posed above for a series of antibody-producing cells which synthesize immunoglobulin binding phosphorylcholine because this system allows us to analyze directly the biology of the immune response to phosphorylcholine (PC). Let us summarize the salient features of this system. First, several thousand myeloma tumors have been screened and twelve appear to synthesize immunoglobulins binding phosphorylcholine (7). Our laboratory has determined the amino acid sequences of the V_H regions for seven of these tumors (8,9) and other laboratories have analyzed several additional sequences (10) (Figure 3). The V_H sequences from myeloma proteins binding phosphorylcholine illustrate several features of V diversity. 1) Four V_H sequences are identical. Since these identical V_H sequences were expressed independently in different mice, it appears that they are encoded by a germ line V_H gene segment designated T15. This reasoning argues that it is unlikely that four somatic variants would be identical in amino acid sequence. 2) The variant sequences differ by one to eleven amino acid substitutions and also exhibit sequence gaps. Accordingly, one can hope to determine the nature and extent of diversity generated from somatic genetic mechanisms by sequencing germ line PC V_H gene segments and comparing them with the protein diversity patterns reflected in their myeloma counterparts. Second, antisera have been raised which are specific for the V domains of several myeloma proteins binding phosphorylcholine. These antisera are termed anti-idiotypic antisera. Anti-idiotypic antisera to T15 can be used to map genetic elements which control the expression of this V_H domain. The T15 idiotype maps about 0.4 centiMorgans (cM) from the C_H gene cluster (11) and simplistic genetic calculations suggest the PC V_H and C_H gene segments are separated by hundreds of thousands or even a million nucleotides. For example, mouse chromosomes have about 25 chiasmata per meiosis (12). With a genome of 3×10^9 nucleotide pairs, 0.4 cM of DNA in the mouse would span about 10^6 nucleotide pairs, if meiotic recombination were random. Third, the T15 idiotype appears to be present on at least one type of T cells ("helper T cells") (13), implying that T-cell receptors and B-cell immunoglobulins may share the same V_H repertoire of genes. Thus an analysis of the phosphorylcholine system may provide opportunities to analyze T-cell receptors. Finally, the hybridoma system of Milstein and Köhler (14) has been employed to generate homogeneous antibodies to phosphorylcholine. In collaboration with Dr. Patricia Gearhart, we are analyzing 20 hybridomas to phosphorylcholine in order to broaden our knowledge about the phenotypic diversity patterns of the phosphorylcholine system. The importance of detailed protein sequence studies on the products of complex multigenic systems such as the antibody gene families cannot be overemphasized, for these phenotypic diversity patterns are one of the end results of heavy chain gene organization and rearrangements and any meaningful understanding of this system at the DNA level must account for the resultant diversity of its gene products. Thus we hope the phosphorylcholine system will provide insights into antibody gene diversity and organization and permit us, in time, to begin analyzing the more complex regulatory events of this sophisticated system.

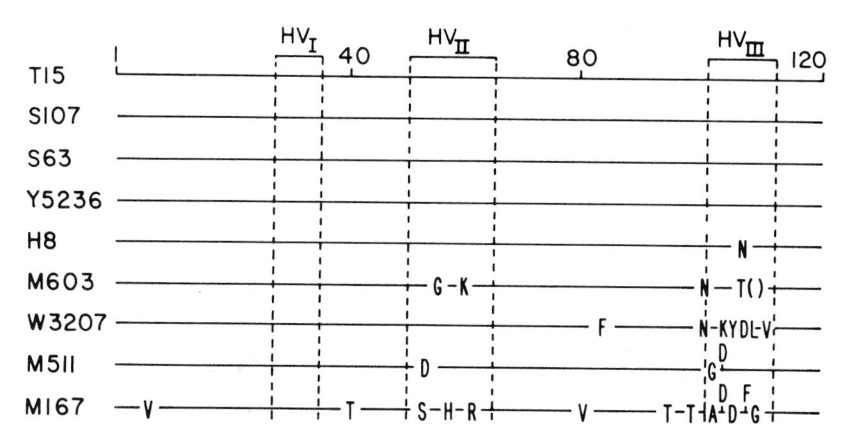

FIGURE 3. The amino acid sequences of V_H regions from immunoglobulins binding phosphorylcholine. Identities of these sequences to the V_H region of T15 are indicated by a straight line. The one letter code of Dayhoff is used to indicate amino acid substitutions (28). Deletions are indicated by brackets. Insertions are denoted by a vertical bar. The three hypervariable regions which fold in three dimensions to constitute the walls of the antigen-binding site of the V domain are designated by HV_I, HV_{II}, HV_{III} and dotted lines.

OUR APPROACH

We have constructed libraries in Charon 4A bacteriophage from partial restriction digests of sperm, embryo, and myeloma DNA (15, 16). The sperm and embryo libraries are a source of undifferentiated DNA. The myeloma library, derived from the tumor MOPC 603 which synthesizes IgA molecules binding phosphorylcholine, represents a terminal stage in the differentiation of an antibody-producing cell. We also have purified mRNA from a variety of myeloma tumors, and used these as templates for the synthesis of double-stranded DNA copies which were then inserted into plasmids (16). Our initial approach has been to compare the genomic organizations of undifferentiated (sperm or embryo) and differentiated (IgA myeloma tumor) DNAs. To this end we have isolated a number of genomic clones from both the M603 library and from a sperm library, using cDNA probes for the complete V_HC_α coding region of myeloma protein S107. The V_H regions of the

S107 and the M603 immunoglobulins are very closely related (Figure 3) and the corresponding mRNAs completely protect one another in S1 nuclease digestion experiments (16). Certain of these initial experiments have recently been published in a paper which describes for the first time a heavy chain genomic clone containing the V_H and C_α gene segments and the presence of intervening sequences within the C_α coding region, probably separating the coding regions for immunoglobulin α domains (16). These results as well as more recent observations are summarized below.

EXPERIMENTAL OBSERVATIONS

The Variable and Constant Regions of α Heavy Chains Appear to be Encoded by Distinct V_H and C_α Gene Segments which are Rearranged During Differentiation. We have analyzed a series of overlapping genomic clones from the M603 library which have the general structures illustrated in Figure 4. The V and the C gene segments are separated by 6.8 kilobases. Furthermore, idiotypic mapping, discussed above, suggests that these regions were separated by hundreds of thousands of nucleotides prior to differentiation of this antibody-producing cell with the concomitant DNA rearrangements. A heteroduplex comparison of a sperm V_H clone with the myeloma M603 clone, which will be discussed subsequently, also provides evidence for the rearrangement of the V_H gene segment in the myeloma DNA. Accordingly, the V_H and C_α gene segments are originally widely separated from one another. As the antibody-producing cell differentiates, DNA rearrangements of antibody V and C gene segments occur over extensive stretches of DNA.

The C_α Gene Segments from the M603 Myeloma Library are Present in Multiple Rearranged Forms. A comparison of Southern blots on sperm M603 DNA using the C_α probe demonstrates that the myeloma DNA has three forms of the C_α gene, none of which are identical to their germ line counterpart (Figure 5). These three forms have been isolated from the M603 library as Charon 4A clones (Figure 6). Restriction enzyme analyses and heteroduplex comparisons demonstrate that, although they share 2.7 or more kilobases of homology just 5' to the C_α gene, each of these three clones is distinct from the others in their more 5' regions.

These observations raise several interesting possibilities. The absence of a germ line-like C_α gene segment in the M603 DNA suggests that the C_α gene segments in both the maternal and paternal chromosomes coding for heavy chain genes have been rearranged. Immunoglobulin-producing cells exhibit allelic exclusion; that is, a particular antibody-

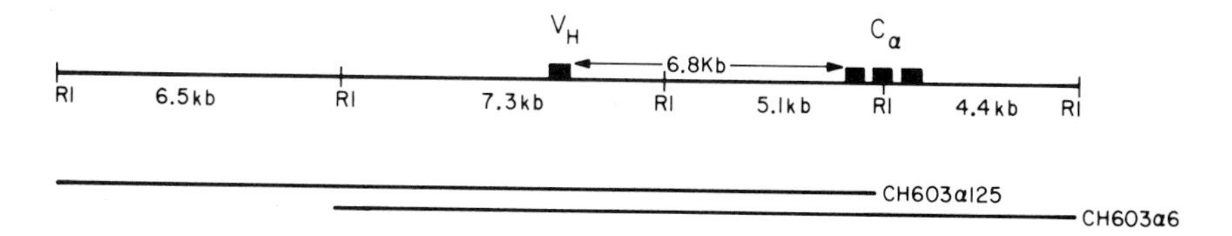

FIGURE 4. The organization of V_H and C_α gene segments from DNA derived from myeloma tumor M603. Kb denotes kilobases. R1 denotes Eco R1 cleavage sites. The distances between Eco R1 sites are indicated. CH603α125 and CH603α6 are two clones derived from the phage library of M603 DNA. The V_H gene segment is separated from the Cα gene segment by 6.8 kilobases of intervening DNA. R-loop mapping and restriction enzyme analyses demonstrate that the C_α segment is divided into three approximately equal segments, presumably coding regions for the three C_α domains, by two small intervening DNA sequences (16).

producing cell may express the maternal or paternal allele for a particular immunoglobulin family, but not both alleles. In the past the phenomenon of allelic exclusion has been explained by suggesting that either the maternal or paternal chromosome does not rearrange at the DNA level and, accordingly, cannot express an immunoglobulin polypeptide. This suggestion has come from Southern blot analyses of myeloma DNAs in which the germ line pattern of constant gene segments for light chains appears to be preserved (17). Our data on the alpha constant region genes of the M603 myeloma DNA suggests that both the maternal and paternal chromosomes undergo rearrangements, but that one of these rearrangements is abortive in the sense no gene product is expressed. It will be interesting to determine whether these abortive DNA rearrangements include V gene segments; or whether only the C gene segment is involved in the rearrangement. Moreover, it will be interesting to analyze carefully the myeloma examples that appear to have germ line C fragments to determine whether the DNA rearrangements have been missed due to technical limitations of the Southern blotting technique, or contamination with somatic DNA. It may be that all myeloma DNAs in fact rearrange both the paternal and maternal chromosomes--one in a productive and the second in an abortive fashion.

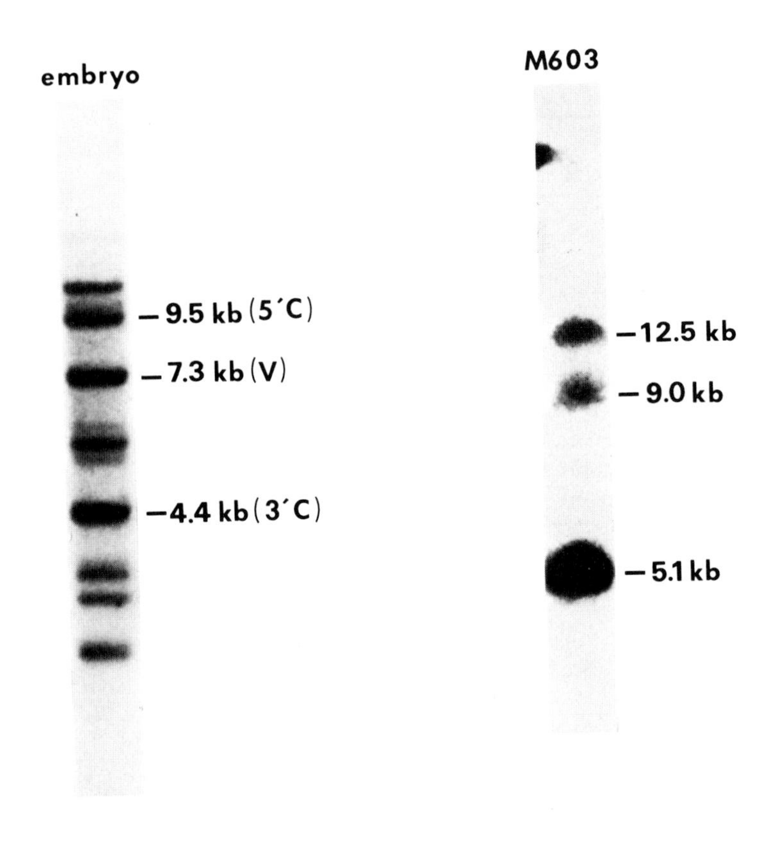

FIGURE 5. Southern blots of embryo (undifferentiated) and myeloma M603 (differentiated) DNAs. The picture on the left is a Southern blot of 13-day embryo DNA after digestion with the Eco R1 enzyme, separation of the DNA fragments on agarose, and hybridization with a cDNA probe derived from mRNA of myeloma tumor S107. This probe contains both the V_H and C_α coding regions. Assignments of the C_α fragments are based on Southern blots with separated V_H and C_α probes (data not shown). The remaining fragments must be V_H gene fragments. Thus there are at least 8-9 germ line V_H genes which cross-hybridize with the V_H probe from myeloma tumor S107. The exposure on the right is a Southern blot of tumor M603 DNA after Eco R1 digestion and hybridization to a plasmid containing the 5' half of the C_α coding region (an R1 site separates the 5' from the 3' half of the C_α gene segment; see Figure 4). The 5' C_α probe gives just one 9.5 kilobase band in the embryo DNA (data not shown) and 5.1, 9.0 and 12.5 kilobase bands in the M603 DNA. Hybridization to the 3' half of the C_α coding region gives a 4.4 kilobase band in both embryo and myeloma DNA (not shown).

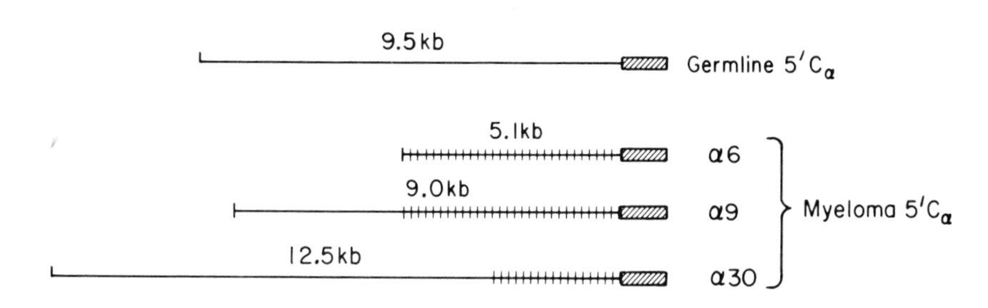

FIGURE 6. Eco R1 genomic fragments including the 5' portion of the C_α gene from myeloma M603 DNA and sperm DNA. The genomic clones α6, α9, and α30 have been derived from the M603 phage library. The structure of the germ line C_α clone comes from a Southern blot analysis of sperm or embryo DNA (Figure 5). The boxes represent the 5' potion of the C_α coding sequence (see Figure 4), whereas the hashmarks represent DNA homologies revealed by heteroduplex analyses.

One surprising observation that is difficult to explain is the presence of three distinct C_α clones in the M603 DNA. Several explanations may be offered, none really satisfactory. First, the germ line may contain two C_α genes, both the same size by Eco R1 restriction analysis. Both of these C_α genes may undergo rearrangements of several different types. Second, perhaps the abortive rearrangement is unstable and may be subject to additional DNA rearrangements. Third, perhaps there are several different M603 cell types in the uncloned tumor from which the DNA was derived. The possibility that the M603 C_α pattern is some aberration of this particular tumor line seems unlikely because at least one other phosphorylcholine binding tumor (H8) has an identical pattern on Southern blots (M. Davis and P. Early, unpublished). Thus in the case of the C_α gene segments, it appears that both the maternal and paternal chromosomes undergo DNA rearrangements, some of which are abortive (nonproductive) while others lead to the expression of one V_H-C_H pair of gene segments.

The V and C Rearrangements in Heavy Chains Resemble Those of Light Chains in Some Respects but Not Others. The V_L and C_L gene segments are rearranged by a fusion at the DNA level of V_L and J_L gene segments with the removal (or rearrangement) of the intervening DNA (Figure 7) (4, 17). Accordingly, the DNA 5' to the V_L gene segment is identical to that of the unrearranged V_L gene and the intervening DNA between the V and C gene segments is derived from the region

Light Chain

V J C
Sperm
VJ C
Myeloma

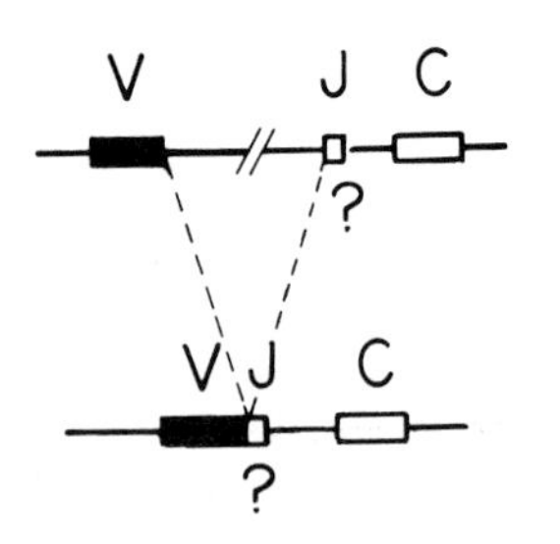

FIGURE 7. A model of the joining of light and heavy chain gene segments. An analysis of λ (4) and κ (18) light chain gene segments indicate that the 3' side of a V segment is fused to the 5' side of a J segment. The intervening DNA sequence between the J segment and the C segment remains unchanged in the DNA rearrangement process. The heavy chain gene segments appear to rearrange in a similar fashion, although the organization of the intervening DNA sequence between the J and C gene segments is altered, presumably because of multiple DNA rearrangements between one V_H gene segment and two (or more) C_H gene segments (see text).

5' to the unrearranged C_L gene. The existence of J_H segments for heavy chains is strongly implied from protein sequence data (18) and has recently been demonstrated by the DNA sequence analysis of a sperm clone containing a V_H segment (P. Early and M. Davis, unpublished observation). Comparison of a sperm V_H clone and the joined V_H and C_α myeloma clone (α6) by DNA heteroduplex analysis demonstrates that those regions 5' to the V segment are homologous and those regions 3' to the V segment are nonhomologous (Figure 8). In this respect the heavy chain variable region gene segment appears to rearrange in a manner similar to its light chain counterparts (Figure 7).

The rearrangement of V_H and C_H gene segments differs from those of the light chains in one important regard. Certain of the intervening sequences between the V_H and C_α gene segments of the α6 clone (Figure 4) are not derived from germ line DNA 5' to the C_α gene. For example, a Southern blot analysis of germ line DNA with a C_α probe shows that the closest Eco R1 site is 9.5 kilobases from the 5' side of the C_α gene segment (Figure 5). However, the α6 clone

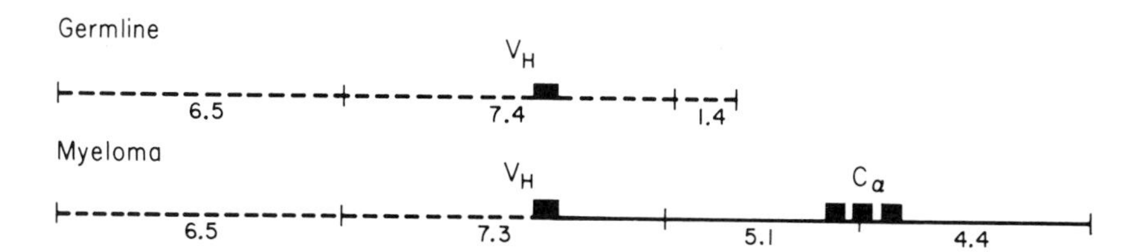

FIGURE 8. Homologies determined by heteroduplex analysis between the flanking sequences of germ line and myeloma V_H clones. The dotted lines indicate germ line sequences. Accordingly, the intervening DNA sequence between the V_H and C_α gene segments is not derived from the sperm V_H clone. The sperm library was constructed by M. Davis and R. Joho.

from the M603 DNA has an Eco R1 site 5.1 kilobases from the 5' end of the C_α gene segment. In addition, as discussed above, this α6 DNA is not homologous to DNA of the sperm V_H clone (Figure 8). Moreover, the Eco R1 site of the α6 clone in the DNA between the V and C gene segments does not seem to have been created by a spurious mutation, since Southern blots of DNA from an independently derived tumor line (H8) show the same C_α Eco R1 fragment. One explanation for the origin of the DNA sequence between V_H and C_α gene segments in the α6 clone containing this Eco R1 site is that it arises from the DNA rearrangement events of an earlier stage in differentiation, in which this V_H gene segment was formerly joined to a different C_H (or J) gene segment. Indeed, during the differentiation of antibody-producing cells, the V_H gene segment appears initially to be joined to a C_μ gene (Figure 2), so we would predict that some of the intervening DNA in the α6 clone between the V_H and C_α gene segments may be derived from the 5' side of a germ line C_μ gene segment. The subsequent joining of this V_H segment to a C_α gene segment later in development might displace or delete (19) the C_μ gene, but not all of its flanking sequences.

Intervening Sequences Appear to Separate the Domains of the C_H Genes. The C_α polypeptide is divided into three discrete molecular domains, each of which encompasses about 110 amino acid residues (20). We initially used R-loop mapping to demonstrate the existence of two small intervening sequences (IVS2, IVS3) which separate the C_α coding region into three roughly equal segments (Figure 4) (16). Subsequent restriction enzyme analyses of the M603 genomic clone (α6) places IVS2 within 30 amino acids of the domain boundary between the $C_{\alpha}1$ and $C_{\alpha}2$ homology units (16; M. Davis, unpublished). Thus it appears likely that the two intervening sequences will separate the C_{α} gene into three distinct coding segments, one for each C_{α} domain (Figure 4). In addition, we have analyzed a μ genomic clone from the M603 library by R-loop mapping. The C_{μ} region has four domains (21) and, as expected, R-loop analysis demonstrates that the C_{μ} coding region is divided by three small intervening sequences into four roughly equivalent segments (K. Calame, P. Early, M. Davis, D. Livant, unpublished observations). The analysis of a genomic γ1 clone has established that intervening sequences separate the three $C_{\gamma 1}$ domains and the hinge region from one another precisely at the interdomain boundaries (22). Therefore it appears reasonable to conclude that intervening sequences will divide all of the immunoglobulin C genes coding into segments for structural domains (see Figure 1).

The function of intervening sequences has generated spirited controversy and discussion. Individual domains of the immunoglobulin molecule carry out discrete and independent functions (20). Accordingly, the immunoglobulin intervening sequences appear to perform the important task of breaking the coding regions into discrete units which may then rearrange independently of one another through recombination at either the DNA level or the nuclear RNA level as proposed by Gilbert (23). Several lines of evidence suggest that the domains of immunoglobulins may be discrete evolutionary units. First, C_H regions with two, three, and four domains are present in vertebrate antibodies. Second, heavy chain disease deletions (24) and spontaneous deletions in tissue culture lines (25) suggest that frequent nonhomologous crossing-over occurs at or between domain boundaries. Perhaps intervening sequences not only separate domains but facilitate recombination as well. It will certainly be interesting to determine the homology relationships, if any, of the various immunoglobulin intervening sequences to one another.

The Germ Line V Gene Segments of Mouse Heavy Chains Appear to be as Diverse as Their V_{κ} Counterparts. The V_H regions derived from myeloma proteins binding phosphorylcholine show a limited range of heterogeneity (Figure 3). We are interested in determining whether these different V_H sequences are germ line or in part derived by somatic mutation. Southern blot analysis of embryo DNA employing the S107 cDNA probe reveals at least 8-9 restriction fragments which hybridize to the S107 V region probe (Figure 5). The PC V_H regions represent a single group of heavy chain variable regions (26). Approximately 20 other groups of

V_H regions have been defined (26). Therefore, if each group is on the average encoded by ~10 germ line genes, the heavy chain gene family may be comprised of approximately 200 V_H genes. Since the amino acid sequence analyses of mouse V_H regions are relatively limited, it appears likely that in time many additional V_H groups will be defined. By similar analyses, the V_κ family of mouse appears to be encoded by 200 or more germ line V genes (3, 27). We have isolated several different PC V_H genes and are now in the process of sequencing them to determine the relative contributions of germ line diversity, somatic mutation, and combinatorial joining of V_H and J_H segments to antibody variability.

The Generality of Nucleic Acid Rearrangements. The intriguing general question posed by the studies on immunoglobulin genes is whether DNA rearrangements are a fundamental aspect of differentiation in other eukaryotic systems. An answer to this question will await more detailed analyses of other gene families, both simple and complex.

ACKNOWLEDGMENTS

The work here is supported by National Science Foundation Grant PCM 76-81546. MD, PE, and DL are supported by National Institutes of Health Training Grant GM 07616. KC is supported by National Institutes of Health Fellowship GM 05442.

REFERENCES

1. Hood, L., Campbell, J. H., and Elgin, S. C. R. (1975). Ann. Rev. Genet. 9, 305.
2. Cohn, M., Blomberg, B., Geckeler, W., Raschke, W., Riblet, R., and Weigert, M. (1974). "The Immune System," ICN-UCLA Symp., p. 89. Academic Press.
3. Weigert, M., Gatmaitan, L., Loh, E., Schilling, J., and Hood, L. (1978). Nature 276, 785.
4. Brack, C., Hirawa, M., Lenhard-Schueller, R., and Tonegawa, S. (1978). Cell 15, 1.
5. Mage, R., Lieberman, R., Potter, M., and Terry, W. (1973). In "The Antigens" (M. Sela, ed.), Vol. I, p. 300. Academic Press.
6. Goding, J. W., Scott, D. W., and Layton, J. E. (1977). Immunol. Rev. 37, 152.
7. Potter, M. (1970). Physiol. Rev. 52, 631.
8. Hood, L., Loh, E., Hubert, J., Barstad, P., Eaton, B., Early, P., Fuhrman, J., Johnson, N., Kronenberg, M., and Schilling, J. (1976). Cold Spring Harbor Symp. Quant. Biol. 41, 817.

9. Hubert, J., Johnson, N., Barstad, P., Rudikoff, S., and Hood, L. In preparation.
10. Rao, D. N., Rudikoff, S., and Potter, M. (1978). Biochemistry 17, 5555.
11. Riblet, R. J. (1977). "Molecular and Cellular Biology," ICN-UCLA Symp., Vol. 6, p. 83. Academic Press.
12. Klein, J. (1975). "The Biology of the Mouse Histocompatibility Complex." Springer-Verlag.
13. Cosenza, H., Augustin, A., and Julius, M. (1977). Cold Spring Harbor Symp. Quant. Biol. 41, 709.
14. Köhler, G., and Milstein, C. (1976). Eur. J. Immunol. 6, 511.
15. Maniatis, T., Hardison, R., Lacy, E., Lauer, J., O'Connell, C., Quon, D., Sim, G., and Efstratiadis, A. (1978). Cell 15, 687.
16. Early, P., Davis, M., Kaback, D., Davidson, N., and Hood, L. (1979). Proc. Nat. Acad. Sci. USA 76, 857.
17. Seidman, J. G., and Leder, P. (1978). Nature 276, 790.
18. Schilling, J., Clevinger, B., Davie, J., and Hood, L. In preparation.
19. Honjo, T., and Kataoka, T. (1978). Proc. Nat. Acad. Sci. USA 75, 2140.
20. Edelman, G. M., Cunningham, B. A., Gall, W., Gottlieb, P., Rutishauser, U., and Waxdal, M. (1969). Proc. Nat. Acad. Sci. USA 63, 78.
21. Beale, D., and Feinstein, A. (1976). Quart. Rev. Biophys. 9, 135.
22. Sakano, H., Rogers, J. H., Huppi, K., Brack, C., Traunecker, A., Maki, R., Wall, R., and Tonegawa, S. (1979). Nature 277, 627.
23. Gilbert, W. (1978). Nature 271, 501.
24. Frangione, B., Lee, L., Haber, E., and Bloch, K. (1977). Proc. Nat. Acad. Sci. USA 70, 1073.
25. Adetugbo, K., Milstein, C., and Secher, D. (1977). Nature 265, 299.
26. Barstad, P., Rudikoff, S., Potter, M., Cohn, M., Konigsberg, W., and Hood, L. (1974). Science 183, 962.
27. Seidman, J., Leder, A., Nau, M., Norman, B., and Leder, P. (1978). Science 202, 11.
28. Dayhoff, M. O. (1972). In "Atlas of Protein Sequence and Structure," Vol. 5, Biomedical Research Foundation, Washington, D.C.

PROBES FOR SPECIFIC mRNAS BY SUBTRACTIVE HYBRIDIZATION: ANOMALOUS EXPRESSION OF IMMUNOGLOBULIN GENES[1]

Frederick W. Alt[2], Vincenzo Enea[3], Alfred L.M. Bothwell[4] and David Baltimore[5]

Department of Biology and Center for Cancer Research
Massachusetts Institute of Technology
Cambridge, Massachusetts 02139

INTRODUCTION

We present here a general method for the purification of DNA complementary to regions of specific immunoglobulin heavy and light chain mRNAs. This procedure requires no prior enrichment for the specific mRNA, but instead relies on a subtractive hybridization step (1) which exploits the observation that myeloma lines usually produce high levels of only a single heavy and a single light immunoglobulin chain (2). This method, which is both easy and efficient, should be generally applicable to the purification of sequences that are characteristic of a particular cellular state of differentiation.

In this report we will describe the purification of several different heavy and light chain-specific cDNA sequences. In addition, we will present some unexpected observations that were made when these specific probes were characterized by hybridization to RNA from a variety of myeloma lines that differ in respect to the immunoglobulins they produce. The results of these experiments indicate that a given line can contain RNA for multiple types of heavy or light immunoglobulin chains.

[1]This work was supported by grant #CA-14051 to S.E. Luria from the National Institutes of Health, N.C.I. and grant #VC-4J from the American Cancer Society.

[2]Chaim Weizmann postdoctoral fellow.

[3]Postdoctoral fellow of the Helen Hay Whitney Foundation.

[4]Postdoctoral fellow of the National Institutes of Health, N.C.I.

[5]American Cancer Society Research Professor of Microbiology.

ISBN 0-12-068350-4

METHODS

The majority of the methods used in these studies including hybridization conditions, S1 nuclease assay, hydroxylapatite (HAP) chromatography and the details of the cDNA purification procedures have been described previously (1). Any modifications of these procedures are described in the text or in appropriate figure legends. The RNA used for both cDNA preparations and hybridizations was total cellular polyA-containing RNA prepared by either the SDS-proteinase K (3) or guanidinium HCl (4) extraction procedures.

The purification of single stranded hybridization probes from cloned immunoglobulin-specific sequences will be described in detail elsewhere (5). Briefly, γ_{2b} (1050 base pairs) or μ (900 base pairs) inserts were enzymatically (Pst) cleaved from the appropriate nick translated (6; specific activity 1-2 x 10^8 cpm/μg) plasmids and then purified by agarose gel electrophoresis. Single stranded hybridization probes were prepared by hybridization of the purified, denatured insert to excess RNA from a myeloma line producing the appropriate complementary RNA sequence and hybrids isolated by equilibrium density gradient centrifugation in guanidinium-CsCl (7) or by preparative S1 nuclease treatment. In both cases RNA was removed from the isolated DNA strand by alkaline hydrolysis.

RESULTS

Figure 1 outlines the basic strategy that we have employed to purify cDNA sequences complementary to RNA sequences encoding immunoglobulins. This two step hybridization procedure exploits the observation that myeloma lines generally produce high levels (up to 5-10 percent) of only a single species of heavy and light immunoglobulin chain (2). In terms of logic and experimental detail this approach is similar to that used previously in the preparation of dihydrofolate reductase-specific cDNA from methotrexate-resistant cell lines (1). To best describe the practical considerations of the method, we will present a detailed description of the purification of cDNA sequences complementary to the α heavy chain mRNA found in the HOPC 2020 (H-2020) myeloma line. Purification of other types of immunoglobulin specific sequences will then be discussed in less detail in following sections.

Purification of α-specific cDNA sequences. The H-2020 myeloma line has been previously described as producing high levels of α heavy chains and λ_1 light chains (2). To purify

FIGURE 1. Purification of heavy chain-specific cDNA by subtractive hybridization.

Step A. Negative (subtractive) selection.

1. *PolyA$^+$ RNA from myeloma 1 (heavy chain A and light chain X)*
   ↓
2. *^{32}P-cDNA*
   ↓
3. *Hybridize to high Rot with large excess of polyA$^+$ RNA from myeloma 2 (heavy chain B and light chain X) plus L cell RNA. Light chain X and vast majority of common cellular sequences are rendered double-stranded.*
   ↓
4. *Recover non-hybridized cDNA by chromatography on hydroxylapatite (enriched for heavy chain A-specific cDNA plus non-hybridizable material).*

Step B. Positive selection (enrichment for hybridizable sequences.

1. *^{32}P-cDNA selected in step A hybridized to excess polyA$^+$ RNA from myeloma 1 (Rot high enough to render majority of cDNA enriched in step A double-stranded).*

2. *Hybridized cDNA recovered by chromatography on hydroxylapatite.*

cDNA complementary to α mRNA, cDNA prepared from H-2020 poly(A)-containing RNA (polyA$^+$ RNA) was hybridized to a R_ot (RNA concentration X time) of 3000 with a 100-fold excess of polyA$^+$ RNA from MOPC 104E (M-104E), a myeloma that produces the same type of light chain (λ_1) as H-2020 but a different class of heavy chain (μ) (Table 1). This procedure should render double-stranded those cDNA sequences complementary to RNA sequences present in both cell lines (presumably the vast majority of all cellular sequences except the specific α heavy chain mRNA). To ensure that general cellular sequences were driven into hybrids, we also included a large excess of polyA$^+$ RNA from a non-immunoglobulin-producing mouse cell (L cells) in the reaction (Table 1). At the end of the reaction, hybridized and non-hybridized material were separated by chromatography on HAP at 60°C. To protect the integrity of the hybridized fraction, 400 mM NaCl was included in the $NaPO_4$ buffers used in this procedure. With the particular batch of HAP used in these experiments, the majority of the single-stranded (S1 nuclease-sensitive) cDNA was eluted at 0.1M $NaPO_4$. Approximately 3% of the total cDNA was recovered in this fraction (Table 1). When tested by RNA excess hybridization, this material was found to be greatly enriched for sequences complementary to RNA species that are very abundant in H-2020 RNA ($R_ot_{1/2}$ = 0.1-0.2, Table 1), but not detectable in M-104E RNA (Table 1) or in L cell RNA (not shown). Thus, by this procedure we have, in essence, subtracted from the total population of H-2020 cDNA sequences all of those sequences complementary to RNA sequences also found in M-104E and L cells. The maximum hybridization of the selected cDNA was approximately 70% (Table 1); therefore, as was observed in the dihydrofolate reductase cDNA purification (1), this step also enriches for non-hybridizable material in the cDNA preparation.

TABLE 1
PREPARATION OF α-SPECIFIC cDNA PROBE

I. Negative selection: 6 μg $^{32}P\text{-}cDNA_{(H\text{-}2020)\ \alpha,\lambda}$ + (600 μg $RNA_{(M\text{-}104E)\ \mu,\lambda}$ and 600 μg L cell RNA) → R_ot 3,000

II. HAP Fractionation
60°, NaPB + 0.4M NaCl

M NaPB	%cpm	$\%S1^R$
0.10	3.1	9
0.12	1.5	76
0.50	94.4	96

III. Analytical Hybridization of 0.1M Fraction

R_ot	RNA H-2020	RNA M-104E
0.1	39	1.5
0.3	59	1.8
1.2	68	4.0
4.7	71	7.5

IV. Positive Selection: 0.1M Fraction + $RNA_{(H\text{-}2020)}$ → R_ot 0.3

V. HAP Fractionation, 60°C, NaPB, 0.4M NaCl

M NaPB	%cpm	$\%S1^R$
0.10	43	7
0.12	3	45
0.50	54	98

For details of cDNA preparation, hybridization conditions, S1 nuclease assay and HAP chromatography see Alt et al. (1).

As a final purification step, the cDNA isolated as described above was hybridized back to the homologous H-2020 RNA and the cDNA which hybridized by a R_0t of 0.3 was isolated by HAP chromatography (Fig. 1, step B; Table 1). This step excludes the non-hybridizable material selected in the previous step. As indicated in Table 1 it was necessary to include a 0.12M $NaPO_4$ wash of the HAP column in order to remove a partially S1 nuclease-resistant cDNA fraction that was not eluted in the 0.1M $NaPO_4$ wash. The nature of this material was not further investigated. Approximately 50% of the cDNA was recovered in the 0.5M $NaPO_4$ (S1 nuclease-resistant) fraction yielding an overall recovery of approximately 1.5% of the initial cDNA. Based on the $R_0t_{1/2}$ data described below, this represents almost a 50% recovery of the expected α cDNA sequences.

The specificity of the purified cDNA was indicated by the kinetics with which it hybridized to excess RNA from a number of α and non-α-producing cell lines (Fig. 2). As expected, the purified cDNA hybridized to abundant RNA sequences in the H-2020 line (Fig. 2, $R_0t_{1/2}$ = 0.15). Importantly, the

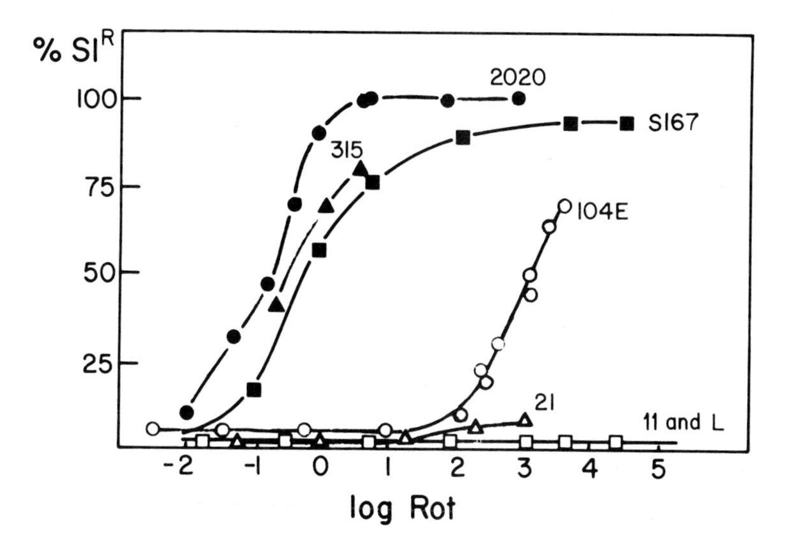

FIGURE 2. Hybridization properties of labeled α-specific cDNA. The cDNA isolated in the 0.5M $NaPO_4$ wash of the positive selection step (Table 1, section V) was treated with 0.3M NaOH for 20 hours to hydrolyze complementary RNA. This material was then hybridized to excess RNA from the indicated lines. Details of the hybridization assay have been described (1).

extent of this reaction approached 100% with pseudo-first order kinetics, suggesting that the purified cDNA preparation consisted mainly of sequences complementary to a single species of abundant RNA. The purified probe also hybridized with similar kinetics to abundant RNA sequences in two other α-producing lines, S167, $R_ot_{1/2}$ = 0.5, and MPC 315, $R_ot_{1/2}$ = 0.5 (Fig. 2). The extent of these reactions (i.e. near completion) is consistent with evidence that the cDNA represented mostly the common 3' constant region sequence of the α mRNA (data not shown). No hybridization, even at high R_ot values was detected with RNA from L cells. Little or no hybridization was detected with RNA from several other non-α-producing myelomas [MPC 11 (γ_{2b}, κ); MOPC 21 (γ_1, κ)]. Unexpectedly, RNA sequences complementary to the purified cDNA were found in the M-104E (μ, λ_1) line that was used for subtractive hybridization (Fig. 1). However, the abundance of these sequences ($R_ot_{1/2}$ = 1000) was about 5000-fold less than that of the hybridizing RNA sequences in H-2020 ($R_ot_{1/2}$ = 0.15). Fortunately, since the excess of M-104E RNA used in the subtractive hybridization procedure was only 100-fold, this low level of presumptive α sequences did not interfere with the purification procedure.

We have used this purified cDNA preparation to identify cloned cDNA sequences derived from H-2020 polyA^+ RNA (5). Cloned DNA that hybridized extensively with the purified cDNA by liquid hybridization analysis was labeled by nick translation and used in the method of Alwine et al. (8) to determine the size of RNA that contains complementary sequence. The DNA hybridized to an RNA species of the size expected for α mRNA (1500 bases) in both of the α-producing H-2020 and MPC 315 lines, but not detectably to RNA from a variety of other non-α-producing and non-immunoglobulin producing cell and tumor lines (data not shown).

Taken together, the data presented above are very strong evidence that the purified cDNA is specifically complementary to α mRNA.

Purification of λ_1-specific cDNA. Purification of cDNA sequences complementary to the λ_1 light chain mRNA of H-2020 (α, λ_1) was performed essentially identically to that of the α-specific cDNA described above except that S167 RNA (α, κ) was used instead of M-104E RNA in the subtractive hybridization step. To provide additional specificity, the positive selection step was performed with RNA from M-104E (μ, λ_1). For further details see the legend to Figure 3. As indicated in Figure 3, cDNA purified in this way appears to be kinetically homogenous and hybridizes to RNA sequences that are quite abundant in λ_1-producing myeloma lines (H-2020, M-104E,

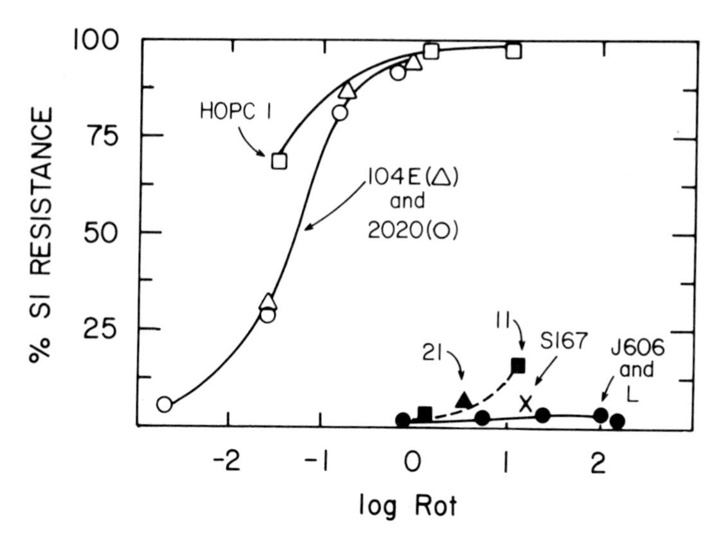

FIGURE 3. Hybridization properties of λ-specific cDNA. 500 ng of cDNA prepared from H-2020 polyA$^+$ RNA was negatively selected (see Fig. 1) with 23 μg of polyA$^+$ RNA from S167 and 100 μg from L cells (R_ot = 2500). The material isolated in this step was positively selected (see Fig. 1) with 34 μg of M-104E polyA$^+$ RNA (R_ot = 0.6) yielding an overall recovery of 3-4 percent. The purified cDNA was then tested for hybridization to excess polyA$^+$ RNA from the indicated lines.

and HOPC-1) but not significantly with RNA from a variety of κ-producing myeloma lines (MPC 11, MOPC 21, J606, S167) or L cells. Recombinant plasmid clones (derived from both H-2020 and M-104E RNA) identified with this probe were shown to be λ_1-specific both by RNA blotting and by restriction mapping analyses (5).

Thus, in summary, we have been able to purify cDNA sequences complementary to either the light or heavy chain mRNA of the H-2020 myeloma by using RNA derived from an appropriate second myeloma line in the basic purification scheme outlined in Figure 1.

Purification of μ-specific cDNA. To purify μ-specific sequences, we prepared cDNA from M-104E RNA (μ, λ_1) and then subtracted those sequences present in the polyA$^+$ RNA of H-2020 (α, λ_1), S167 (α, κ) and L cells. S167 RNA was included in this selection because the M-104E line contains both κ and λ_1 RNA sequences (see below). This procedure led to the recovery of very little hybridizable cDNA (approximately 0.1% recovery). This material was positively selected by hybridization to M-104E RNA ($R_ot = 5$) and then characterized by hybridization to various RNA preparations (for details of purification see legend to Figure 4). As shown in Figure 4, the purified cDNA was essentially kinetically homogenous, hybridizing to nearly 100% with M-104E polyA$^+$ RNA and hybridizing little with RNA from the two myeloma lines used in the negative selection or with RNA from L cells (not shown). However, based on the kinetics of the hybridization reaction ($R_ot_{1/2} = 5$) and assuming an average complexity of 2400 for the reacting sequences (5), the purified cDNA is complementary to RNA sequences which represent as little as 0.1% of the M-104E polyA$^+$ RNA. We expected μ to be a more abundant sequence in M-104E and therefore eventually identified clones containing μ inserts by hybridization to cDNA prepared from purified μ mRNA (a gift of Dr. I. Schechter).

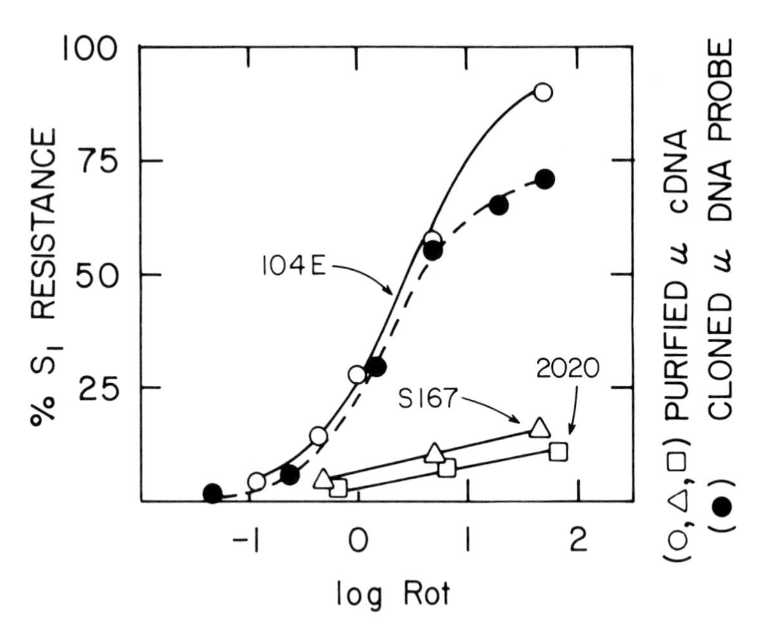

FIGURE 4. Hybridization of μ-specific DNA probes. Approximately 1 μg of M-104E cDNA was negatively selected (Fig. 1) with 60 μg of total polyA$^+$ RNA from both H-2020 and S167 as well as 300 μg from L cells. Isolated material was positively selected (R_ot 5) with 40 μg of M-104E polyA$^+$ RNA. This purified cDNA (O, △,□) and a cloned μ-specific hybridization probe (●) (see Methods) were then tested for hybridization to excess polyA$^+$ RNA from the indicated sources.

However, a single-stranded hybridization probe which was prepared from the cloned μ sequence hybridized to excess M-104E RNA with kinetics identical to those observed with the purified cDNA (Fig. 4). Thus in our M-104E line, μ represents a relatively non-abundant sequence (at least 40-fold lower abundance than the λ_1 sequences in the same line).

The difficulties encountered in the purification of μ-specific cDNA (unexpressed sequences--i.e. κ, see below--in the starting material or low levels of the sequence to be purified) serve well to illustrate some of the problems that might occur in a blind attempt to purify a particular sequence by this method. However, this purification also demonstrates that under appropriate conditions, the procedure is powerful enough to greatly enrich for cDNA sequences specifically complementary to RNA species representing considerably less than 1% of the cellular polyA$^+$ RNA.

Anomalous occurrence of κ sequences in λ_1-producing myelomas. In a number of experiments in which λ_1-producing myelomas were used as sources of cDNA or of RNA for the subtractive hybridization procedures, we experienced difficulties that were eventually related to the occurrence of κ RNA sequences in these lines. Several laboratories have previously reported finding relatively high levels of κ RNA sequences in λ_1-producing myelomas, including M-104E (9, 10, 11). By hybridization with κ cDNA prepared from purified κ mRNA (a gift of Dr. I. Schechter), we have demonstrated significant levels of κ RNA sequences in the three λ_1-producing myelomas (M-104E, H-2020, and HOPC-1) currently carried in our laboratory (Table 2). In two of these lines (M-104E and HOPC-1) the level of κ RNA sequences (we estimate 0.5-1% to polyA$^+$ RNA based on data in Table 2) was nearly as high or higher than the level of κ RNA sequences in several κ-producing myelomas (S167, MPC 11, MOPC 21, Table 2). To our knowledge,

TABLE 2
HYBRIDIZATION OF κ cDNA

	κ-producers				λ_1-producers		
RNA Source	J606	MPC11	MOPC 21	S167	M-104E	HOPC-1	H-2020
$R_0t_{1/2}$	0.05	0.3	0.3	1	0.7	1.2	40

cDNA prepared from purified κ mRNA was hybridized to excess polyA$^+$ RNA from the noted lines. The $R_0t_{1/2}$ values indicate the R_0t at which the hybridization reaction reached 50% of its maximum value (70%). This value is inversely proportional to the abundance of κ sequences in a particular RNA source.

there have been no reports of κ protein production by either the M-104E or HOPC-1 lines. In addition, we have not been able to demonstrate κ synthesis by specific immunoprecipitation of extracts from pulse-labeled M-104E cells (E. Siden, unpublished results). These immunoprecipitation assays have, however, been used to detect κ synthesis in certain Abelson murine leukemia virus (A-MuLV) transformed cell lines (11) in which κ RNA sequences are 10-50-fold less abundant than in M-104E. Furthermore, in both our laboratory and in others (N. Rosenberg, personal communication) M-104E extracts have been used as a negative control in sensitive κ immunocompetition assay. Therefore, it would appear that even though the M-104E myeloma contains high levels of κ sequences (0.5-1% of polyA$^+$ RNA) this line does not produce readily detectable levels of κ chains.

More recently, we have demonstrated by the RNA blotting procedure of Alwine et al. (8) that the major species of κ RNA in the three λ_1-producing myelomas (H-2020, M-104E and HOPC-1) has approximately the same size (950-1100 bases) as the major species of κ RNA found in a number of κ-producing myelomas (MPC 11, MOPC 21 and J606), suggesting that these molecules probably contain both constant and variable region sequences. We have cloned κ-specific sequences from both HOPC-1 and M-104E myelomas and are now in the process of examining the structure of these molecules in detail.

Purification of cDNA sequences complementary to various subclasses of γ heavy chain mRNA. We prepared γ_1, γ_{2b}, and γ_3-specific cDNA sequences essentially as indicated in Figure 1 by subtraction of S167 (α, κ) sequences from cDNA prepared from polyA$^+$ RNA of MOPC 21 (γ_1, κ), MPC 11 (γ_{2b}, κ) and J606 (γ_3, κ), respectively. We did not feel that it was necessary to purify γ_{2a}-specific cDNA because this sequence has considerable homology (60-65%) with the γ_{2b} sequence (T. Honjo, personal communication; F. Alt, A. Bothwell, and V. Enea, unpublished observations). The cDNAs purified by these protocols were all kinetically homogeneous and hybridized to nearly 100% with abundant sequences in the homologous RNA preparations ($R_ot_{1/2}$ of 0.5, 0.5 and 0.2 for the γ_1, γ_{2b} and γ_3-specific probes, respectively). In general, hybridizations at greatly reduced rates (see below) or no hybridization at all, was detected with RNA from myeloma lines producing heterologous types of heavy chains or from L cells (data not shown). All three of these purified γ-specific cDNA sequences have been used to identify cloned sequences which again hybridize to RNA of the expected size and tissue distribution for their respective subclasses of γ chain (F. Alt and V. Enea, unpublished results). Furthermore, the identities of the cloned γ_1 and γ_{2b} sequences have been confirmed by restriction mapping and DNA sequencing (A. Bothwell, unpublished observations).

An unexpected finding of our hybridization analyses was the frequent occurrence of RNA sequences for more than a single heavy chain in a given myeloma line (e.g. α RNA sequences in the RNA from M-104E, Fig. 1). Whereas κ sequences are often present at high levels in λ_1-producing myelomas (Table 2), additional heavy chain sequences are generally present at considerably lower, but variable levels (5% to 0.01% of the level of the major heavy chain sequence). We have observed this phenomenon with a number of heavy chain probes and a number of different myeloma lines. The significance of these findings is generally difficult to interpret because most of our RNA preparations were derived from myeloma tumors which could be invaded by host lymphocytes. However, as shown in Table 3, γ_1 and γ_{2b} sequences are apparently present at relatively high levels (2.5-5% of the level of the major heavy chain sequence) in the cloned MPC 11 (γ_{2b}, κ) and MOPC 21 (γ, κ) cell lines, respectively. In the same set of experiments, the γ_{2b} but not the γ_1 probe was rendered highly S1 nuclease-resistant by hybridization with RNA from the A-MuLV transformed 18-48 cell line (12), suggesting that the result described above was not due to cross hybridization between γ_1 and γ_{2b} sequences (Table 3). Additional support for this conclusion comes from the following experiments. 1) A single-stranded hybridization probe prepared from cloned γ_{2b} C region sequences hybridized to MPC 11 and MOPC 21 RNA with kinetics identical to those observed with purified γ_{2b} cDNA and, furthermore it was rendered almost 100% S1 nuclease resistant by hybridization to the MOPC 21 RNA (Fig. 5). 2) S1 nuclease mapping experiments suggest that MOPC 21 RNA will protect at least 850 contiguous bases from the 3' (C region) end of the γ_{2b} sequence (V. Enea and F. Alt, unpublished observations). Preliminary RNA

TABLE 3

HYBRIDIZATION OF HEAVY CHAIN-SPECIFIC cDNA PROBES

Source of PolyA$^+$ RNA	Hybridization of labeled specific probe					
	α		γ_1		γ_{2b}	
	Ext.	$R_ot_{1/2}$	Ext.	$R_ot_{1/2}$	Ext.	$R_ot_{1/2}$
Cloned MPC11 (γ_{2b}, κ)	<10%	>10,000	>50%	100	>90%	0.5
Cloned MOPC 21 (γ_1, κ)	<10%	>14,000	85%	0.5	70%	20
Cloned 18-48 (A-MuLV transformed)	>85%	20	<10%	>10,000	>60%	100

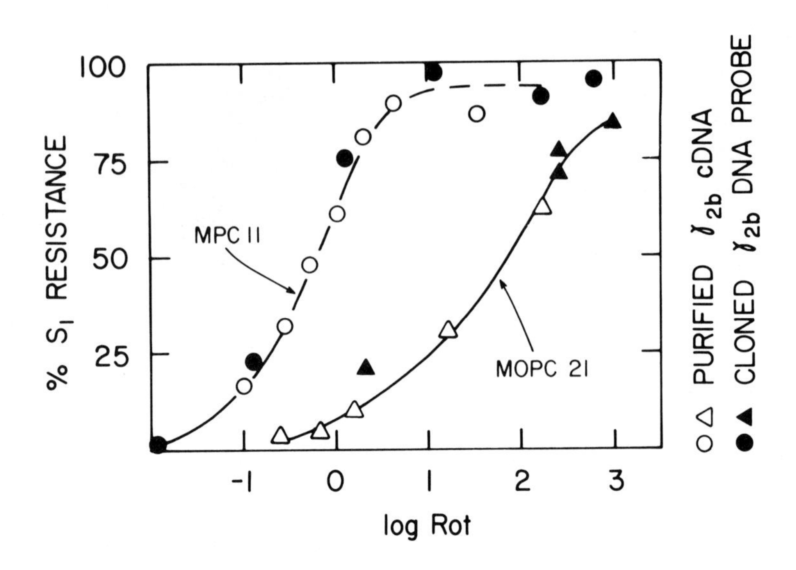

FIGURE 5. Approximately 1 μg of cDNA from MPC 11 was negatively selected with 39 μg of S167 RNA (R_ot = 2500). No L cell RNA was included in this reaction. The enriched material was positively selected with 100 μg of MPC 11 $polyA^+$ RNA (R_ot = 4). The final recovery was approximately 0.5-1% of the initial cDNA mass. The purified cDNA (O, Δ) and a cloned γ_{2b}-specific cDNA hybridization probe (●, ▲) (see Methods) were then tested for hybridization to excess $polyA^+$ RNA from the indicated sources.

blotting experiments indicate that the major forms of MOPC 21 γ_{2b} RNA and MPC 11 γ_{2b} RNA are of approximately the same size (1800 bases), suggesting that the MOPC 21 γ_{2b} sequence may represent a mature heavy chain mRNA (F. Alt and V. Enea, unpublished observation).

It is also interesting to note that the cloned A-MuLV transformed 18-48 cell line, which has previously been characterized as producing κ light chains and no heavy chain (12), contains both α and γ_{2b} RNA sequences (Table 3). We have recently confirmed the presence of γ_{2b} sequences in this line by RNA blotting procedures and in addition have demonstrated μ sequences (of distinctly different size than the γ_{2b} sequences) in the same RNA preparation (F. Alt and V. Enea, unpublished observations). Thus, this cloned cell line contains low, but significant, levels of three different classes of heavy chain RNA sequences.

To understand the significance of these findings we

are currently addressing the following questions. Do the multiple types of heavy chain RNA in a given cell line share identical variable region sequences? If so, are these sequences expressed from the same chromosome? Should these cells produce multiple types of heavy chain with identical variable regions and from the same chromosome, then they would represent an attractive system for studying the mechanism by which a variable region gene can be expressed (concomitantly or sequentially) in association with different constant genes.

REFERENCES

1. Alt, F.W., Kellems, R.E., Bertino, J.R. and Schimke, R.T. (1978). J. Biol. Chem. 253, 1357.
2. Potter, M. (1972). Physiol. Rev. 52, 631.
3. Iarek, M. and Penman, S. (1974). J. Mol. Biol. 89, 327.
4. Strohman, R.C., Moss, P.S., Micon-Eastwood, J., Spectar, D., Przybyla, A. and Paterson, B. (1977). Cell 10, 265.
5. Bothwell, A.L.M., Enea, V., Alt, F.W. and Baltimore, D. (1979). Manuscript in preparation.
6. Rigby, P.W., Diekmann, M., Rhodes, C. and Berg, P. (1977). J. Mol. Biol. 113, 237.
7. Enea, V. and Zinder, N.D. (1975). Science 190, 584.
8. Alwine, J.C., Kemp, D.J. and Stark, G.R. (1977). Proc. Nat. Acad. Sci. U.S.A. 74, 5350.
9. Ono, M., Kawakarni, M., Kataoka, T. and Honjo, T. (1977). Biochem. Biophys. Res. Commun. 74, 796.
10. Storb, U., Hagar, L., Wilson, R. and Putnam, D. (1977). Biochemistry 16, 5432.
11. Rabbits, T.H., Forster, A., Smith, M. and Gillam, S. (1977). Eur. J. Immunol. 7, 43.
12. Siden, E., Rosenberg, N., Clark, D. and Baltimore, D. (1979). Cell 16, 389.

LIGHT CHAIN GENES, TRANSCRIPTION UNITS AND RNA PRECURSORS

I. ISOLATION OF κ GENES FROM MOPC 21 GENOMIC LIBRARIES

Michael Komaromy, Pat Clarke and Randolph Wall

II. NUCLEAR RNA PRECURSORS TO κ mRNAS IN MPC 11 CELLS

Edmund Choi, Michael Kuehl[1] and Randolph Wall

Molecular Biology Institute and the
Dept. of Microbiology and Immunology
UCLA School of Medicine
Los Angeles, California 90024

INTRODUCTION

Immunoglobulin genes, like most eukaryotic genes studied to date (cf. This Symposium Volume) contain noncoding intervening sequences. However, unlike those in other eukaryotic genes, the intervening sequences in immunoglobulin genes occur between DNA sequences coding for regions of defined structural and functional homology in immunoglobulin chains. Thus the immunoglobulin precursor (or signal), variable (V) and constant (C) regions are separated by intervening sequences in the complete λ gene (1,2) and in different κ genes (3,4) being expressed in myeloma cells. Variable and constant regions are also separated by an intervening sequence in an α-heavy chain gene (5). Particularly striking in this regard are the very recent findings that the individual constant region domains in an α-heavy chain gene (5) and the hinge region in a γ_1-heavy chain gene (6) are all separated by intervening sequences.

Immunoglobulin genes appear to be unique in another regard among the eukaryotic genes studied to date. It is now clear that the active configuration of both immunoglobulin λ (1,2) and κ (3,4) genes is generated by somatic rearrangements in DNA. This is in contrast to other eukaryotic genes now analyzed which have the same structure whether or not from cells expressing those genes. The somatic rearrangements generating active immunoglobulin light chain genes join the germline V- region in correct coding phase with the germline J-region (2,3). The molecular mechanism for the joining of germline V- and J-regions is not yet known.

---

[1]Address: Dept. of Microbiology, University of Virginia
Charlottesville, Virginia 22901

ISBN 0-12-068350-4

It is now certain that most mRNAs in eukaryotic cells are derived from larger nuclear RNA precursors which contain the intervening sequences from the active genes (reviewed in 7). RNA splicing has emerged as the most likely mechanism for joining structural gene sequences in the generation of mRNAs from such nuclear RNA precursors. The κ mRNA in MOPC 21 cells is derived from a 10 kb primary transcript through several nuclear RNA processing intermediates larger than κ mRNA (8-10). Furthermore, V- and C- regions appear to be spliced together during the processing of the major intermediate leading to MOPC 21 κ mRNA (10). It is already evident that other κ mRNAsare derived from nuclear RNA precursors which differ from those in MOPC 21 cells. The κ mRNA in MPC-11 myeloma cells appears to be derived from a 5.3 kb nuclear RNA (11). These findings suggest that different κ mRNAs are transcribed from quite different transcription units in DNA. How these are generated in the DNA rearrangements which produce complete κ genes remains an intriguing question.

It appears likely that both DNA rearrangements and RNA splicing represent important steps at which immunoglobulin gene expression could be regulated. We have now constructed cloned genomic libraries from MOPC 21 myeloma cells and isolated κ light chain genes to complement our studies on the transcription and processing of the nuclear RNA precursors to MOPC 21 κ mRNA. In addition we have examined an interesting case involving the production of two κ mRNAs in MPC 11 myeloma cells which seemed to be a possible candidate for the regulation of κ mRNA production through nuclear RNA processing. These subjects are considered below.

I. ISOLATION OF κ GENES FROM MOPC 21 GENOMIC LIBRARIES

We have constructed two genomic libraries from MOPC 21 mouse myeloma tumors. These were generated from a total EcoRI digestion and from size-selected fragments (10-18 kbp) from a total EcoRI digestion. These were screened with a cloned V+C probe from MOPC 21 κ mRNA. Some characteristics of the clones obtained from these two libraries are presented here.

Construction and Screening of Genomic Libraries

High molecular weight DNA was prepared by nitrogen cavitation (12), pronase digestion (13), and CsCl centrifugation. EcoRI was prepared by an unpublished method of T. Landers. λgt WES·λB DNA was prepared by standard techniques. The λB fragment was separated from the arms on

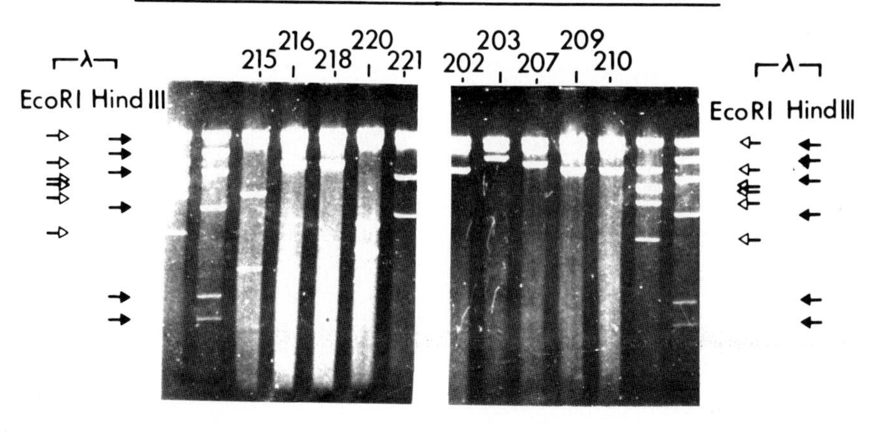

Fig. 1. Sizes of Cloned MOPC 21 κ Gene Fragments

Clones were digested with EcoRI and run on .7% agarose gels. Analysis of band sizes was performed by using a minicomputer to fit higher-order regression curves to the standards.

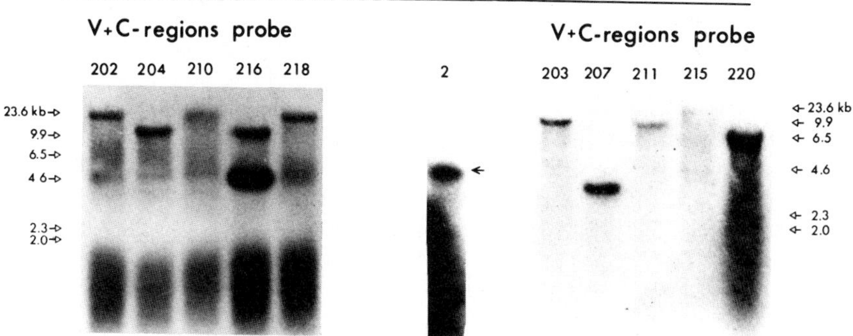

Fig. 2. Sequences in MOPC 21 Clones

Clones were digested with a 10-fold excess of Hind III, run on .7% agarose gels and transferred to nitrocellulose paper using the Southern technique. Duplicate filters were hybridized with either V+C probe or C-only probe and autoradiographed. Only the V+C probe results are shown.

a .4% SeaPlaque agarose (Marine Colloids) gel in E buffer. The gel was melted at 65°C and the agarose removed by phenol extraction. T_4 DNA ligase was prepared by the method of Pamet et al (15). The ligation conditions were those of Sugino, et al (16).

The *in vitro* packaging method used was that of Becker and Gold as described by F. Blattner (personal communication). The average cloning efficiency obtained in 6 trials was approximately 5×10^6 per microgram of insert DNA, as compared to 3×10^7 per microgram of intact λ DNA. Three million plaques were plated on 13 Pyrex pizza dishes (each 30 cm in diameter). Filters were made by the method of Benton and Davis (17) and screened with nick-translated probes. Plasmid clone pL21-1 was used for V+C probe, and pL21-5 was used for C probe (18). It was found that a single round of CsCl/ethidium bromide centrifugation did not suffice to remove enough E. coli DNA to give clean backgrounds in the hybridizations, so all DNA used for nick-translation was additionally purified on SeaPlaque agarose gels.

Characteristics of MOPC 21 light chain Gene Clones

Twenty-one κ-positive plaques were picked from the total primary plating of 3×10^6 plaques. Twelve of these showed 100% κ-positive plaques after three further rounds of plating and plaque hybridization. DNA was made from each of these and the sizes of the inserts were estimated by agarose gel electrophoresis of EcoRI digestions (Fig. 1). The DNAs were also digested with Hind III, run on agarose gels, transferred to nitrocellulose paper by the Southern technique and hybridized to either κ V+C or C-region probes. These data are shown in Fig. 2 and the results are summarized in Table I. From these two restriction digests we have identified seven apparently different V-regions related to the MOPC 21 V-region. None of these contained a C-region. Three of these clones were isolated in duplicate. These findings suggest that a minimum of seven related V-regions make up the $V_{\kappa}15$ subgroup which includes MOPC 21. We have not yet identified a V+C clone in these isolates. However, we have confirmed a C-region clone which does not contain a MOPC 21 V-region (Table 1). This clone has been restriction mapped and appears to be identical to the embryo κ C-region clone reported by others (3,4).

Speculations on the role of Multiple V-region Genes in Antibody Diversity

Two major theories invoking somatic mutations and multiple germline variable region genes have been proposed

20. McKean, D.J., Bell, M. and Potter, M. (1978) Proc. Nat. Acad. Sci. USA 75: 3913.
21. Valbuena, D., Marcu, K.B., Weigert, M. and Perry, R.B. (1978) Nature 276:780.
22. Seidman, J.G., Leder, A., Edgell, M.H., Polsky, F., Tilghman, S.M., Tiemeier, D.C. and Leder, P. (1978) Proc. Nat. Acad. Sci. USA 75:3881.
23. Kuehl, W.M., Kaplan, B.A., Scharff, M.D., Nau, M., Honjo, T., and Leder, P. (1975). Cell 5: 139.
24. Rose, S.M., Kuehl, W.M., and Smith, G.P. (1977) Cell 12: 453.
25. Bailey, J.M., and Davidson, N. (1976). Anal. Biochem. 70: 75.
26. Alwine, J.C., Kemp, D.J., and Stark, G.R. (1977) Proc. Nat. Acad. Sci. USA 74: 5350.
27. Burckhardt, J., and Birnstiel, M.L. (1978). J. Mol. Biol 118: 61.

ON THE ORIGIN OF MUTANTS OF SOMATIC CELLS

Louis Siminovitch

Department of Medical Genetics, University of Toronto
Toronto, Ontario, Canada M5S 1A8

ABSTRACT A large number and variety of mutants which behave recessively in hybrids have been isolated in Chinese hamster ovary and other somatic cell lines. Since somatic cells are presumably diploid, these findings represent a paradox. One hypothesis which could explain these results is that permanent cell lines are functionally hemizygous at several genetic loci. Experiments in our laboratory on mutation frequencies, segregation, and on the RNA polymerase II locus, have provided evidence for this hypothesis. Recent experiments with the recessive thymidine kinase (Tk^-) and emetine resistant (Emt^r) loci have indicated that recessive mutations can be obtained, by means of two independent events, even when two-functional loci are present. One event leads to functional hemizygosity, and occurs at high frequency, and the other consists of a standard type of mutation and occurs at a much lower frequency.

INTRODUCTION

Over the last 10-15 years considerable success has been achieved in the selection of a broad spectrum of mutants of mammalian somatic cells (1). Although some of these isolates behave dominantly in somatic cell hybrids, the large majority show recessive behaviour under such conditions. The types of recessive mutants which have been obtained in one particular cell line, Chinese hamster ovary (CHO) cells, are shown in Table I. Although the list is comprehensive, it is not complete, and is presented mainly to indicate the spectrum and number of isolates which have been selected in this particular cell line. The Table includes auxotrophic, conditional-lethal temperature-sensitive and drug-resistant mutants and, in several cases, there is good evidence that the lesions involve structural gene changes. At a minimum estimate, at least 15 complementation groups have been identified in each of the three major classes - auxotrophs, conditional-lethal temperature-sensitives and resistants.

ISBN 0-12-068350-4

TABLE I
PHENOTYPICALLY RECESSIVE MUTATIONS IN CHO CELLS

Class	Phenotype	Number of Complementation Groups
Auxotroph	Requirement for	
	glycine	4
	proline	1
	adenine	7–8
	glycine, adenine, thymidine	1
	uridine	1
	asparagine	1
	serine	1
	glutamate	1
	alanine	1
	inositol	1
Conditional-lethal temperature sensitive	Amino acyl tRNA synthetase	8–9
	Mitotic apparatus	1
	Undefined	10
Resistant	Resistant to	
	lectins	8–9
	adenosine, toyocamycin, or tubericidin	1
	8 Azaguanine, or 6 Thioguanine	1
	8 Azaadenine, or 6 Mercaptopurine	1
	Emetine, Cryptopleurine, or Tylocebrine	1
	Trichodermin	1
	Chromate	1
	Methotrexate ($MtxR_I^{II}$)	1
	5 Bromodeoxyuridine	1
	2 Deoxygalactose	1

These findings present a paradox. Because of the diploid nature of mammalian somatic cells, it would be anticipated that autosomal recessive mutations would be obtained rarely or not at all at all loci. In an earlier publication (1), I developed an hypothesis which might provide an explanation for the success which has been achieved in the selection of phenotypically recessive mutations in CHO cells. Although CHO cells contain roughly the same amount of DNA as the normal Chinese hamster cell, their karyotype is not normal, and extensive rearrangement has occurred during the evolution of the line. It is conceivable that some of these chromosomal modifications have led to loss of functional transcription of one of the alleles at several genetic loci. There would be little counter selection against such events since one functional allele probably suffices for efficient growth in vitro. According to this hypothesis, therefore, the CHO line is functionally hemizygous over part of its genome, and the isolation of recessive mutants at such loci is not surprising.

Because of the importance of this concept for the understanding of the origin of mutants of somatic cells, our laboratory has sought to obtain evidence for or against it in several different ways. The nature of these experiments is described in this paper.

THE α -AMANITIN RESISTANT (ama^R) LOCUS

Previous studies in our laboratory have shown that α-amanitin-resistant lines of somatic cells contain drug-resistant RNA polymerase II, the target enzyme for α-amanitin, and that the mutation acts codominantly in somatic cell hybrids (2-6). In a resistant diploid cell, it would be expected that equal quantities of resistant and sensitive enzyme should be found. Such a result was obtained in rat myoblasts and human diploid fibroblasts (7-8) but, in contrast, in α-amanitin-resistant CHO cells (5), only the resistant form of the enzyme was found. Hybrids of such CHO resistant cells with wild-type cells contained about 50% each of the resistant and sensitive enzyme as expected (3, 5). These results in themselves provide strong evidence that CHO cells contain only one functional copy of the RNA polymerase II gene.

It was of interest to determine whether this situation was unique to CHO cells or whether α-amanitin-resistant isolates obtained from other Chinese hamster lines would show similar characteristics. Such mutants were selected and the nature of their polymerases examined. In contrast to CHO cells, α-amanitin-resistant M3-1 and CHW cells were found to contain about 50% each of resistant and sensitive RNA polymerase II (9). α-Amanitin-resistant mutants of CHO-K1 (a CHO line related to ours (10)) contained only resistant enzyme. All of these results are summarized in Table 2. These studies on RNA polymerase II thus provide strong evidence for the existence of functional hemizygosity at the ama locus in CHO and CHO-K1 cells and not in other Chinese hamster lines.

TABLE 2

RNA POLYMERASE II ACTIVITIES IN VARIOUS CELL LINES RESISTANT TO α-AMANITIN

Cell Line	Presence of Sensitive Enzyme
Human diploid fibroblasts	+
Rat myoblasts	+
CHO (Toronto strain)	-
CHO-Kl (Denver strain)	-
CHO hybrid (Resistant x Sensitive)	+
CHW (Chinese hamster fibroblast)	+
M3-l (Chinese hamster fibroblast)	+

MUTATION FREQUENCIES FOR RECESSIVE MARKERS IN CHINESE HAMSTER LINES

Campbell and Worton have taken another approach to the study of functional hemizygosity in CHO cells. They have argued that since two events would be required for the expression of a recessive mutation in a diploid cell, it would be expected that functional hemizygosity for any locus in CHO cells would result in higher recessive mutation frequencies at that locus when compared to cells in which such hemizygosity was not present. To test this hypothesis, they compared the mutation frequency at the emt (emetine resistance) locus in CHO cells with that in several other Chinese hamster lines. Previous work in our laboratory had shown that Emt^r cells could be isolated in CHO cells at high frequency and that the mutation behaved recessively in somatic cell hybrids (12-15). The data in Table 3 is taken from Campbell and Worton (16). To control for any differences between the cell lines in mutation frequency per se, the frequencies of a codominant marker, ouabain resistance (Our^R), and an X-linked marker, thioguanine resistance (Thg^r) were also compared. Clearly, because of the nature of the latter two mutations, no systematic differences in these should be observed between the various cell lines, and this result was found. In contrast, however, the induced mutation frequency for Emt^r cells was at least 50-fold higher in CHO cells than in the other Chinese hamster lines. Except for the V79/V6 line, the numbers in the Table represent overestimates, since complementation tests indicated that these latter isolates might not represent mutations at the same locus as the CHO emt^r lines. Although the data is not shown in the Table, it is of interest that one of the lines derived from CHO-Kl cells, glyA, (17), gave frequencies similar to that of our CHO in this test, whereas the frequen-

TABLE 3
EMS-INDUCED FREQUENCIES OF Emt^r CELLS IN CHINESE HAMSTER CELL LINES

Cell Line	Oua^R x10^6	Thg^r x10^6	Emt^r x10^6
CHW	44	365	1.2
F3B	47	320	1.7
G3	149	790	0.38
M3-1	73	240	1.2
V79/V6	72	780	1.1
CHO	186	1,040	50

cy in another CHO-K1 line, glyB, (17), was similar to that of V79/V6.

Similar differences have been observed when mutation frequencies have been compared between CHO cells and other Chinese hamster lines for toyocamycin resistance, a mutation which involves loss of adenosine kinase and which behaves recessively (18). Here, the result is not as clear-cut as for emt^r since the frequencies vary from about 10^{-3} for CHO to 10^{-7} for CHO-K1, with other lines showing frequencies in between. Other studies in progress in our laboratory indicate that several other markers will behave similarly in showing high frequencies of mutation for CHO cells as compared to other cell lines. Although these findings do not, of course, provide formal proof for the existence of functional hemizygosity at these loci in CHO cells, they are consistent with the hypothesis. They indicate as well that mutation to emetine resistance in V79/V6 cells, for example, might involve two events, one at high frequency, and the other at normal mutation frequencies.

SEGREGATION ANALYSIS OF THE emt^r LOCUS IN SOMATIC CELL HYBRIDS

A further approach to the study of functional hemizygosity has been to examine the process of segregation in somatic cell hybrids. By definition, hybrids between cells carrying a recessive mutation involving resistance to a particular drug and wild-type cells will be sensitive to the cytotoxic action of that drug. Resistant cells reselected from these hybrids using the same drug could arise either by mutation at the wild-type locus or by loss of the wild-type allele from the hybrid (segregation). Studies in several laboratories have shown that the frequencies with which resistant cells are obtained from such hybrids are at least 100-fold higher than for the original mutation. Thus they do not arise by mutation.

For at least one locus, the hprt mutation situated on the X chromosome, Farrell and Worton (19) have demonstrated that selection of resistant hybrids is always associated with loss of the whole, or a part, of the short arm of the X chromosome. Whatever the mechanism, measurement of the frequency of the event can provide information on functional hemizygosity for any recessive locus, as described below. Since Emt^r behaves recessively, hybrids between CHO Emt^r and CHO wild-type cells will be sensitive to the drug. Resistant cells selected from these hybrids will have lost the wild-type allele and so if CHO contains only one functional copy for this locus, the frequency of the event should be similar to that for a locus which is present in only one copy (e.g., hprt). However, if emetine-resistant cells are selected from hybrids between CHO Emt^r cells and other Chinese hamster cell lines, and if these lines contain two functional copies of the emt locus, then the frequencies observed should be much lower because two events would be involved. Such studies have been performed in our laboratory using cells which were hybrid between emt^r and $hprt^-$ CHO cells as one parent, and wild-type CHO, or other Chinese hamster cell lines as the other. The results are summarized in Table 4. As expected, assuming functional hemizygosity in CHO cells for the emt locus, the frequencies of Emt^r colonies and Thg^r colonies ($hprt^-$) were similar when examined in CHO X CHO hybrids. Also, as expected, the frequencies of segregants for the hprt locus were similar for all the hybrids. However, for emt the frequencies were much lower for all other Chinese hamster lines, as compared to CHO. A similar result was obtained when segregation rates were compared (20).

These data are, again, consistent with the concept that the locus for Emt^r is functionally hemizygous in CHO cells.

STUDIES ON FUNCTIONAL HEMIZYGOSITY USING LINKED MARKERS

Although the studies described above have provided evidence for functional hemizygosity at some loci in CHO cells, they do not, of course, imply that all loci are in this state in CHO cells. In fact, Siciliano et al. (21) have shown that several loci in CHO-K1 cells are present in two copies and there is good evidence that the aprt locus is also present in two functional copies in CHO cells. We have had considerable difficulty in obtaining thymidine kinase (Tk^-) mutants in CHO cells and we have assumed that this locus is present in two functional copies as well. During the course of studies on this locus, we accidentally found that an α-amanitin-resistant CHO line (Amal) showed a high fre-

TABLE 4
SEGREGATION FREQUENCIES OF Emt^r AND Thg^r FROM HYBRIDS INVOLVING DIFFERENT CHINESE HAMSTER LINES

Cell Lines Used to Form Hybrids with CHO Emt^r Thg^r*	Segregation Frequencies**	
	Emt^r $\times 10^5$	Thg^r $\times 10^5$
CHO Emt^sThg^s	600	480
M3-1 Emt^sThg^s	1.1	2700
GM7 Emt^sThg^s	1.0	550
V79 Emt^sThg^s	19.0	750
CHW Emt^sThg^s	1.2	230
CHL Emt^sThg^s	10.0	930

* The identity of these Chinese hamster lines is described elsewhere (20).

** Except for the CHL cross, all the other frequencies represent the average of two experiments.

quency of Tk^- mutants. The data comparing the frequency of Tk^- isolates from this line and from our wild-type line is shown in Table 5.

TABLE 5
FREQUENCIES OF Tk^- AND Gal^- MUTANTS IN CHO CELLS

Cell Line	Total Cells Plated	No. of Colonies	Frequency*
Frequencies of Tk^- Mutants			
WT5	2.8×10^8	3	1×10^{-8}
Amal	2×10^7	80	4×10^{-6}
Amal	8×10^7	186	2.5×10^{-6}
Frequencies of Galactokinase Mutants			
WT5	2×10^7	None	5×10^{-8}
Tk^- derived from WT	6×10^6	7	3.2×10^{-6}
Amal	8×10^6	104	1.3×10^{-5}

* Corrected for P. E.

Although mutants can be obtained in the latter case, the frequency is much less than in Amal cells. This latter high frequency is probably not related to the α-amanitin mutation since another α-amanitin-resistant isolate gave low Tk^- frequencies, and it is probably not due to an inherently high mutation frequency in the Amal line, since the frequency for emetine resistance is normal.

One immediate explanation for the origin of this line is that it has lost one of the functional Tk alleles and has, therefore, become functionally hemizygous at this locus. Two experiments have been done to examine this question. It is known that the galactokinase gene is linked to Tk in human cells, and we assumed that this linkage was maintained in CHO cells. It has also been shown that galactokinase minus (Gal^-) mutants can be isolated by selecting for resistance to 2-deoxygalactose (22). Thus, if Amal cells yield high Tk^- frequencies because of the previous loss of one functional allele, it seemed possible that they might also be functionally hemizygous for the galactokinase locus. The frequencies of the latter type mutants were, therefore, compared in wild-type and Amal cells. As can be seen in Table 5, the frequencies of galactokinase (Gal^-) mutants were indeed much higher in the Amal cells.

Examination of Table 5 shows that we were successful in obtaining a small number of Tk^- mutants in our mutagenized wild-type line. This frequency was similar to that observed for Emt^r in mutagenized V79/V6 cells and their origin could be explained by assuming that two events are involved - one occurring at a high frequency and leading to functional hemizygosity and the second consisting of a normal mutation for the $Tk^+ \rightarrow Tk^-$ change. If this hypothesis is correct, then the final Tk^- mutant should come from a minority population hemizygous for the section of the chromosome carrying the Tk locus. To test this assumption, we measured the mutation frequency for the galactokinase locus in these Tk^- cells. As can be seen in Table 5, as predicted, the frequencies of galactokinase mutants were much higher than in wild-type cells.

In summary, these experiments provide further indications that hemizygosity for any locus can arise in wild-type cells at relatively high frequency, and that the event is not localized to a single locus but affects a length of the chromosomal segment.

EXPERIMENTS WITH THE <u>emt</u> LOCUS TO EXAMINE THE EVOLUTION OF HEMIZYGOSITY IN V79/V6 CELLS

Our studies with the Tk locus in CHO cells provided evidence that the Emt^r mutants observed in V79/V6 cells involved two separate events. Assuming that the events are independent, one consequence of this assumption is that cultures of V79/V6 cells at any time should contain 1-2% of cells which are hemizygous for the emt locus. We have carried out some preliminary experiments to test this concept.

We have isolated a large number of clones (> 300) of V79/V6 cells, grown them to a population of 10^8 cells, mutagenized each with EMS and then measured the frequency of Emt^r cells in each clone. Most (> 200 clones) gave very low mutation frequencies, characteristic of the parental line. However, in two separate experiments, some clones were observed with Emt^r frequencies very similar to that found in CHO cells.

Cells derived from these clones produced lines of V79/V6 cells which, again, yielded high frequencies of Emt^r colonies.

DISCUSSION

The experiments described in this paper provide evidence for the existence of functional hemizygosity in somatic cells. We have only examined a few markers, but much of the technology could be applied to other recessive loci and to other cell lines. Such studies would allow evaluation of the generality of our findings.

For V79/V6 cells it appears that even if the cells are functionally diploid, recessive mutants can be obtained because of the high frequency of an event which renders the cells susceptible to mutagenesis and selection. Although the nature of the latter process is obscure at present, we believe that it must involve a chromosomal alteration which leads to inactivation of one allele with resultant functional hemizygosity.

It is not clear as yet whether the relatively high frequencies observed in V79/V6 cells for the Emt^r marker presumably present in two copies can be generalized to other parts of the genome. If they can, it may mean that recessive mutants already observed such as some of those shown in Table I owe their origin to two rather than to one genetic event. In addition, if one assumes that the frequency of a bona fide mutation such as a missense or nonsense alteration will be similar from one cell line to another, then success in obtaining any given isolate will depend on whether the cell is already functionally hemizygous for that marker at the outset and, if not, on the frequency of the events leading to functional hemizygosity. These latter frequencies may vary considerably from line to line and may account for difficulties in obtaining a broad variety of mutants in some cell lines.

It is, of course, known that the frequency of mutation for a single locus (presumably in a hemizygous state) can be increased by mutagenesis. However, it may also be possible to increase the probability of rendering the cells hemizygous by specific agents such as those which promote nondisjunction. The methods described in this paper should allow such questions to be examined. The generality of our findings to the field of somatic genetics will depend on the extension of such experiments to other markers and to other cell lines.

ACKNOWLEDGEMENTS

I wish to acknowledge the extensive contribution of the colleagues who contributed to the work described in this paper, particularly Drs. R. S. Gupta, C. Campbell, R. G. Worton and P. Ip, and the assistance of Dr. R. G. Worton in the preparation of the manuscript. Financial assistance was provided by the Medical Research Council of Canada and by the National Cancer Institutes of Canada and of the United States.

REFERENCES

1. Siminovitch, L. (1976). Cell 7, 1.
2. Chan, V. L., Whitmore, G. F., and Siminovitch, L. (1972). Proc. Nat. Acad. Sci. (U. S. A.) 69, 3119.
3. Lobban, P. E., and Siminovitch, L. (1975). Cell 4, 167.
4. Lobban, P. E., Siminovitch, L., and Ingles, C. J. (1976). Cell 8, 65.
5. Ingles, C. J., Guialis, A., Lam, J., and Siminovitch, L. (1976). J. Biol. Chem. 251, 2729.
6. Ingles, C. J., Beatty, B. G., Guialis, A., Pearson, M. L., Crerar, M. M., Lobban, P. E., Siminovitch, L., Somers, D. G., and Buchwald, M. (1976). In: "RNA Polymerase" pp. 835-853. Cold Spring Harbor Laboratory, New York.
7. Somers, D. G., Pearson, M. L., and Ingles, C. J. (1975). J. Biol. Chem. 250, 4825.
8. Buchwald, M., and Ingles, C. J. (1976). Somatic Cell Genet. 2, 225.
9. Gupta, R. S., Chan, D. Y. H., and Siminovitch, L. (1978). J. Cell. Physiol. 97, 461.
10. Worton, R. G. (1978). Cytogenet. Cell Genet. 21, 105.
11. Williams, K. L. (1976). Nature 260, 785.
12. Gupta, R. S., and Siminovitch, L. (1976). Cell 9, 213.
13. Gupta, R. S., and Siminovitch, L. (1978). Somatic Cell Genet. 4, 77.
14. Gupta, R. S., and Siminovitch, L. (1977). Cell 10, 61.
15. Gupta, R. S., and Siminovitch, L. (1978). J. Biol. Chem. 253, 3978.
16. Campbell, C. E., and Worton, R. G. (1979). Somatic Cell Genet.

5, 51
17. Kao, F.-T., Chasin, L., and Puck, T.T. (1969). Genetics 64, 1284.
18. Gupta, R.S., and Siminovitch, L. (1978). Somatic Cell Genet. 4, 715.
19. Farrell, S.A., and Worton, R.G. (1977). Somatic Cell Genet. 3, 539.
20. Gupta, R.S., Chan, D.Y.H., and Siminovitch, L. (1978). Cell 14, 1007.
21. Siminovitch, L., and Thompson, L.H. (1978). J. Cell. Physiol. 95, 361.
22. Thirion, J.-P., Banville, D., and Noel, H. (1976). Genetics 83, 137.

GENE AMPLIFICATION AS A CONCOMITANT TO CHROMOSOME MEDIATED GENE TRANSFER[1]

George Scangos,[2] Lawrence Klobutcher[3]
and Frank H. Ruddle[2,3]

Departments of Biology,[2] Yale University and
Human Genetics,[3] Yale University School of Medicine
New Haven, Connecticut 06520

ABSTRACT We have transferred the gene for thymidine kinase (TK) into the TK deficient mouse cell line LMTK$^-$ using isolated metaphase chromosomes as the vector. Recipient cell lines initially express the transferred gene unstably (lose the gene at a characteristic rate in nonselective medium) and convert to stable expression upon prolonged cultivation. These lines fall into two classes. One class is characterized by large, cytologically detectable chromosomal fragments of donor origin (transgenomes) which are lost at a rate of 3% per day from the unstable state. The transgenomes exist in one copy per cell in both the stable and unstable states. The second class consists of lines which have received a smaller, cytologically undectable fragment. In these lines, the loss rate is approximately 10% per cell per day and there appear to be multiple copies of the transferred gene. Upon stabilization the copy number declines. We have found that transferred markers that were unlinked in the donor are sometimes found to be linked after transfer. In addition, we show that rearrangement occurs during transfer and that this arrangement can occur very close to the selected gene.

[1]This work was supported in part by USPHS grant GM09966 and National Cancer Institute contract No. N01 CP 71060.

ISBN 0-12-068350-4

Purified metaphase chromosomes were first used as vectors for the transfer of genes into mammalian cells by McBride and Ozer (1). Donor cells subjected to mitotic arrest are physically broken, releasing metaphase chromosomes. The chromosomes are mixed with recipient cells, usually in multiplicities of 0.5 to 2 genome equivalents per cell. When conditionally auxotrophic recipient cells (e.g. $HPRT^-$ or TK^-) are treated with chromosomes from a prototrophic donor, cells of the recipient type that express the donor marker can be recovered at a low frequency, typically one in 10^{-6} to 10^{-7}. Two recent technical improvements, co-precipitation of the chromosomes with calcium phosphate, and post-treatment of the recipient cells with dimethyl sulfoxide, have allowed this frequency to be raised as high as 2×10^{-5} (2). In this system, subchromosomal fragments are transferred to the recipient cell. The fragments range in size from large pieces readily detected by light microscopy (2) to pieces carrying no detectable genetic information beyond the selected prototrophic marker itself. We have coined the term "transgenome" to describe these fragments (3).

The transgenome is typically expressed in one of two ways. In nonselective medium, expression of the transferred marker may be lost at a characteristic rate of 3-10% per cell per generation (unstable expression) or may be retained stably. Loss appears to be an all or none phenomenon, in that donor markers that were linked to the prototrophic marker are lost in a concordant fashion (2, 4). Stable sublines that no longer lose the prototrophic marker at a detectable rate can arise in unstable populations (5, 6). In the stable state, the transgenome becomes closely associated with a recipient chromosome (6-9). Indeed, in those cases where a large fragment undergoes stabilization, it can subsequently be detected as a morphologically distinct region of a recipient cell chromosome (2).

A number of chromosome mediated gene transfer (CMGT) experiments have involved the transfer of thymidine kinase (TK, EC 2.7.1.75) into the TK deficient mouse L cell Line $LMTK^-$. In this system, recipient cells expressing the TK gene can be selected by growth in HAT medium (10). We discuss here two sets of experiments involving the transfer of TK into $LMTK^-$ cells. The first involves the transfer of the human form of TK and the cotransfer of the linked markers galactokinase (GalK, EC 2.7.1.6) and type I procollagen (ProCol I). We have used somatic cell genetic analysis to characterize the size of the transgenome and to look for changes in the size of the transgenome and in the expression of cotransferred markers after conversion to stability of the initially unstable lines. A detailed account of this experiment has been published elsewhere (11).

In the second study, we have utilized CMGT to transfer the TK encoded by Herpes Simplex Virus type 1. The TK of HSV can be isolated on a 3.4 kilobase (kb) restriction fragment. This fragment was used to transfect $LMTK^-$ cells to the TK^+ phenotype by Wigler *et al*. (12), who showed that the TK fragment had become stably integrated into high molecular weight recipient DNA. We have utilized a line which possesses the HSV TK fragment as a donor in CMGT experiments and have used the 3.4 kb fragment as a probe to characterize the state of the transgenome in recipient cells. A detailed account of these studies has been published recently (13).

RESULTS

We have transferred the human form of TK from the cell line HeLa S3 to the mouse cell line $LMTK^-$ and have characterized several independent transformant lines. Analysis of one recipient line which possesses a large, detectable fragment has demonstrated several interesting properties of chromosome mediated gene transfer. This line, termed 2TGT4 was a derivative of the TK deficient mouse line $LMTK^-$ to which we have transferred the human form of TK. Cytogenetic analysis utilizing alkaline Giemsa staining (14) demonstrated the presence of a large fragment of human origin, similar in size to the long arm of human chromosome 17 (to which TK has been previously localized (15). This line also expressed the human forms of the chromosome 17 markers GalK and procollagen I. One copy of the fragment was visible in each cell examined, and upon selection against TK, the fragment and the human forms of GalK and ProCol I were lost.

Line 2TGT4 expressed TK unstably. In nonselective medium, TK was lost from the cells at a rate of 3% per cell generation. Nine independent stabilized derivatives were isolated. Each was shown to have the human fragment attached to a mouse chromosome, although the mouse chromosome involved differed from cell to cell. Furthermore, the fragment was found to have decreased in size following stabilization. We were able to correlate the size of the stabilized fragment with the number of human markers retained to generate a deletion map which indicates the order centromere-GalK-(TKprocolI).

Line 2TGT4 appears to be representative of a class of transformants in which the transgenome is large. Such lines appear to have one copy of the transgenome per cell. In the unstable state, the transgenome exists autonomously and stabilization involves the attachment of the transgenome to a recipient cell chromosome. We have generated a number of unstable transformants which differ from the above in several ways. The transgenomes are smaller, always below the limits of cytological detection; the loss rate is much higher, approximately 10% per cell per generation; and the unstable lines appear to have higher enzyme levels than do the stabilized derivatives (Miller and Ruddle unpublished, Scangos and Ruddle unpublished). These findings raise the question of the molecular structure of this type of transgenome: how many copies exist per cell? what is its size range? how is it organized in the unstable state and in the stable state? and what is the molecular basis of stabilization?

In order to obtain information about the fine structure of the transgenome, we utilized the thymidine kinase gene of Herpes Simplex Virus type 1 (HSV). LMTK$^-$ cells treated with live (16) or UV-irradiated (17) HSV-1 give rise to small numbers of cells expressing high levels of TK activity. The TK expressed by these cells is clearly that encoded by HSV-1 and not the murine enzyme, as judged by its isoelectric point, electrophoretic mobility, immunochemical properties, and ability to phosphorylate iododeoxycytidine (18). Recent studies in our own laboratory have shown that in one such stable transformant, the HSV TK phenotype can be correlated with a particular mouse chromosome (18). This study thereby provides somatic cell genetic evidence for the integration of an HSV TK gene at a discrete chromosomal site in the recipient cell.

Recently, Wigler *et al.* (12) have been able to transfer the HSV TK gene to mouse cells after cleaving the HSV genome with any of several restriction endonucleases. When Bam HI was used to cleave the DNA, TK transforming activity was found to reside on a 3.4 kilobase (kb) fragment. Enzymes which cut within the 3.4 kb fragment were found to abolish or drastically reduce TK transforming activity. The purified 3.4 kb fragment was found to transfect cells at a frequency of 1 colony per 10^6 cells per 40 pg DNA. The same group reported biochemical evidence for physical integration of the fragment into host DNA sequences in transformed mouse TK deficient host cells (19) and has further shown that single copies of the 3.4 kb HSV TK fragment integrated in recipient cell genomes can secondarily transform TK deficient mouse cells. The transformation frequency for secondary gene transfer ranges from 1 colony per 10^6 to 1 per 5 x 10^5 cells per 20 μg cellular DNA. Thus, a specific fragment of the HSV-1 genome can be used for the highly efficient transformation of mammalian cells, and the fate of this fragment can be analyzed biochemically.

We have used one line carrying the 3.4 kb fragment of HSV as a donor in chromosome mediated gene transfer. Chromosomes from this line, termed LH7, were isolated and precipitated onto $LMTK^-$ cells. LH7 was kindly provided by Dr. S. Silverstein, Columbia University, College of Physicians and Surgeons. Four recipient lines, termed LHM1, LHM21, LHM22, and LHM23 were obtained and analyzed further. The stability of TK expression was analyzed in each line. It could be shown that all four lines were initially unstable, and became TK deficient at a rate of approximately 10% per cell per day. As expected, the donor, LH7, expressed TK stably. Line LHM1 converted to stable expression very quickly and a stable derivative was carried through subsequent analyses.

DNA from each of the recipient lines was isolated and digested with a number of restriction enzymes. Equal amounts of DNA from each line were loaded onto agarose gels, blotted onto nitrocellulose filters, and hybridized with a radioactively labeled 3.4 kb TK containing fragment of HSV (20). In each case, the state of the TK fragment in lines LHM1, LHM21, and LHM22 was indistinguishable from that of the donor, LH7, while that of LHM23 differed, indicating that a rearrangement has occurred during the transfer into LHM23. The size of the Xba I derived fragment containing the TK gene is over 20 kb in LH7, and larger in LHM 23, indicating that the unstable transgenomes are at least that large.

Eco RI recognizes two sites within the 3.4 kb TK fragment and so cleaves a 2.2 kb fragment from the middle and leaves two roughly equal tails (12). The prominent internal band of 2.2 kb could be visualized in line LH7, and one of the tails (now linked to cellular DNA) was seen as a high molecular weight band. The other tail, previously measured as 0.9 kb, was not detected. Identical patterns were observed in lines LHM1, 21, and 22. In line LHM23, the 2.2

kb internal fragment had disappeared and was replaced by a prominent high molecular weight band. These data indicate that one of the internal Eco RI sites has been deleted during CMGT into LHM23.

The enzyme Hinc II cuts once within the TK gene to yield fragments of 2.4 and 0.9 KB (12). Hinc II digestion of the DNA of LH7 gave two fragments of 4.5 and 1.6 kb. Again, lines LHM1, 21, and 22 displayed patterns indistinguishable from LH7, whereas in line LHM23, the 4.5 kb fragment has been altered. Since the 2.4 kb portion of the TK fragment is included in the 4.5 kb LH7 fragment (S. Silverstein, personal communication), the Eco RI site that has been deleted must be that included in the large Hinc II fragment. These data localize more precisely the TK gene within the 3.4 kb fragment.

These data indicate that a rearrangement has occurred during the CMGT from LH7 into LHM23. Digestion with both Eco RI and Hinc II demonstrates that sequences to one side of the TK fragment in LHM 23 are indistinguishable from those of LH7 while sequences on the opposite side are altered very close to the TK gene. Altered restriction patterns were also seen with Bam HI, XbaI, and HindIII. These data are consistent with a deletion to one side of the TK gene or with breakage and joining to another fragment. Since the size of the restriction fragment was altered with all the above enzymes, a deletion would have to encompass recognition sites for all of them and so would have to be very large. We think it more likely that the chromosomal fragment bearing the TK fragment underwent breakage and fusion to a second, originally unlinked fragment during the transfer process. That breakage can occur so close to the selected gene indicates further that CMGT is likely to be an important tool for fine structure genetic analysis.

Each of the three unstable lines was grown in non-selective medium for 30 days and then replated in selective medium. Subclones of each were obtained and their stability was characterized. In each case stable derivatives were obtained.

The stable derivatives, termed LHM21A, 22A, and 23A, were subjected to filter hybridization analysis. No changes were seen in the restriction pattern following stabilization, indicating that if any loss of material was involved during stabilization, it did not occur close to the selected gene.

We did find, however, that the intensity of the bands decreased dramatically following stabilization, as if the stable lines possessed fewer copies of the TK gene than did their unstable parents. This result was surprising since in the stable donor LH7, the intensity of the bands appeared to be roughly equivalent to that of the unstable lines, and significantly greater than the stabilized derivatives.

LH7 has been rigorously characterized by Wigler _et al._ (12) and shown to have one copy per cell of the TK gene. Since each stable line in this study must possess at least one copy of the gene per cell to survive in HAT medium, it appears that LH7 has undergone amplification of the TK gene during a year of continuous growth in HAT medium. Our data also indicate that each of the unstable lines possesses multiple copies of the TK gene and that upon stabilization, the copy number decreases. The same intensity pattern was found when the DNA was digested with Eco RI, Bam HI, XbaI or HindIII and thus does not appear to be a technical artifact.

In addtion, we have performed analyses to determine the level of TK synthesized by LH7 and each of the transformants. The levels of TK activity correlated with the intensity of the TK bands, supporting our conclusion that multiple copies of the TK gene exist in these unstably transformed lines.

DISCUSSION

Overproduction of the selected gene product in unstably transformed lines has been previously reported (5). It was not possible at that time to determine whether the overproduction was due to multiple copies of the gene or to some characteristic of the unstably maintained transgenome which leads to a higher level of enzyme synthesis. We have presented data which suggest that, at least in our system, a high enzyme level is due to multiple copies. We can not distinguish between tandem duplication of the TK gene or amplification of the number of chromosomes or chromosomal fragments carrying the gene in either LH7 or the unstable lines. One interpretation is that in LH7, a tandem duplication has occurred and that a fragment containing the duplicated region has been transferred to each of the recipients. Upon stabilization most of the gene copies are lost. Since stabilization almost surely involves attachment to a recipient chromosome, however, it is difficult to envision a plausible stabilization process which would delete most copies of the TK gene but not change the restriction pattern from that of the unstable parent.

We think it more likely that in LH7, the number of copies of the TK fragment was amplified through an increase in the copy number of the chromosome into which it is integrated. During the transfer process, a fragment of one such chromosome was transferred and in the unstable state was amenable to duplication. Stabilization involves the integration of one copy of such a transgenome into a recipient chromosome followed by loss of the unstable copies. A schematic representation of our view of chromosome mediated gene transfer is given in Fig. 1.

This model predicts that continued cultivation of our stable gene transformants in HAT medium should lead to an increase in both the level of enzyme produced and in the number of copies of the TK gene. In addition, if the amplification is the result of an increase in the number of chromosomes, it may be possible to identify the chromosome involved and analyze its number through cytogenetic analysis. We believe that the amplification is the result of selective conditions applied by continued cultivation in HAT medium, in which HSV TK, expressed at levels below that of the wild type cellular enzyme (17), may be involved in a limiting step and therefore subject to amplification.

We have performed analyses of the enzyme levels of the mouse cell line A9 to which we have transferred the human form of HPRT by CMGT. Several unstable lines were found to have elevated HPRT levels with respect to their stable derivatives (Miller and Ruddle, unpublished). It now seems probable that these lines too possess multiple copies of the unstably expressed transgenome, and that amplification following gene transfer may be a common event.

Most recipient lines express the transferred marker unstably for a period immediately after transfer (2; Klobutcher & Ruddle, unpublished). It appears that the unstable transgenomes fall into two classes. One class, of which LHM21, 22, and 23 are examples, possesses small fragments, has a high loss rate, elevated enzyme levels, and apparently multiple copies of the transgenome. A second class normally possesses larger, cytologically detectable fragments, and is

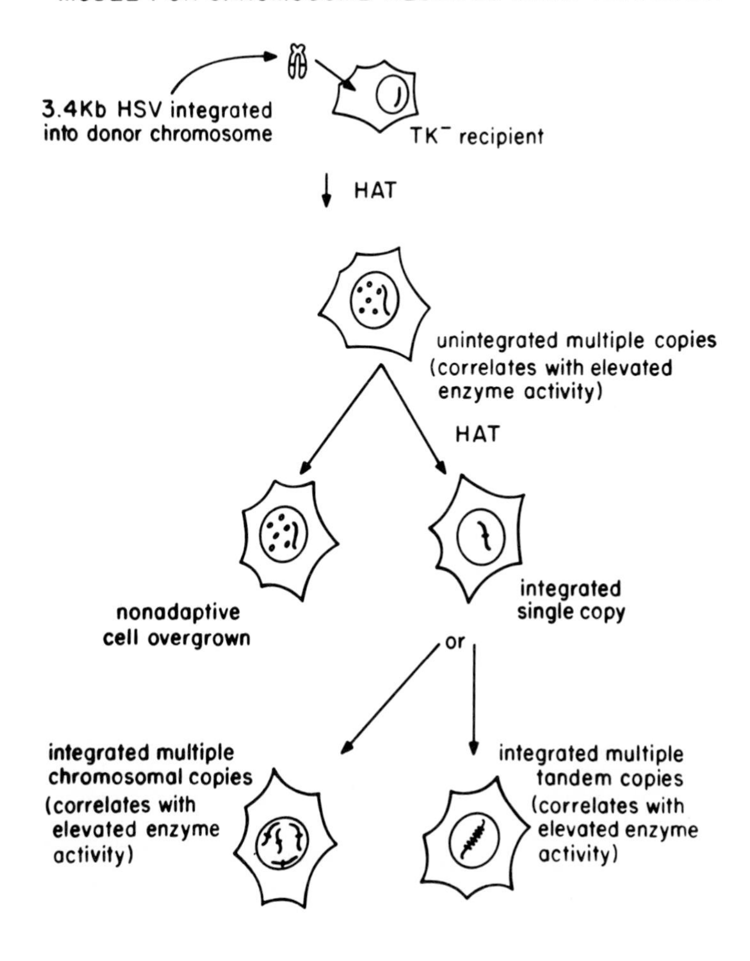

Figure 1. This diagram represents our current model of the chromosome mediated gene transfer process.

characterized by a slower loss rate. We believe that this second class of transgenomes possesses donor centromeric function. Donor centromere-like regions are often visible, and the rate of loss is the same at which chromosomes are lost from interspecific cell hybrids (11). The first class presumably lacks centromeric function and may not be subject to the same type of replication control. Since the transgenomes are small, they may be able to replicate more than once during a cell cycle. The presence of multiple copies coupled with a loss rate of 10% per cell per generation indicate that transgenomes of this type segregate poorly at cell division. It may be that those lines which possess only one copy of a transgenome of this type lose it too rapidly to be detected as viable colonies in selective medium.

We have found other instances of gene amplification following gene transfer. Three primary gene transformants (transfected directly with the 3.4 kb HSV fragment) were subjected to selection against TK by growth in the toxic analog 5-bromodeoxyuridine (BUdR). Cells were diluted into BUdR medium, exposed twice to UV light, and grown in BUdR until a total of 5 x 10^8 cells were obtained. An aliquot of cells was frozen and the DNA was isolated from the remainder of the cells. A back selectant of one line, termed H3, was found to have multiple intense bands in filter hybridization analysis, indicative of many copies of the TK gene at several loci. When cells were thawed and reanalyzed, the TK containing bands were found to have disappeared (Scangos & Ruddle, unpublished). It thus appears that following selection against the gene, a transient amplification of the sequences can occur, followed by the eventual loss of the gene sequences or of cells bearing them.

It is interesting to note that line LHM23 possesses a rearrangement of the transgenome. Since this line is unstable, it is unlikely that the rearrangement reflects integration of the transgenome into a host chromosome. We feel that this phenomenon represents breakage and joining of a chromosomal fragment to a second donor fragment present in the same recipient cell.

This interpretation is consistent with our observation that in approximately 5-10% of the CMGT lines screened, we have detected the cotransfer of unlinked markers along with the selected marker (Klobutcher and Ruddle; and Miller and Ruddle, unpublished). In such lines, selection against the transferred marker results, in many instances in the loss of the cotransferred marker as well, indicating that they were linked in the recipient cell. Four different unlinked enzymes have been cotransferred, indicating that this phenomenon is not the result of novel linkage groups in the donor. Rather, we think that the new linkage groups were generated during the transfer process, perhaps by the same mechanism which led to the alteration of the restriction patterns in line LHM23.

A picture is beginning to emerge of CMGT as a fluid process in which gene rearrangements and amplification can occur with some frequency. We feel that both of these phenomena will be useful to fine structure gene mapping and for studies of gene regulation. In particular, we feel that chromosome mediated gene transfer is a useful tool for the analysis of genome organization on a larger scale than that permitted by cloning and restriction analysis, but in considerably more detail than can be provided by conventional somatic cell genetic analysis.

ACKNOWLEDGMENTS

GAS is a postdoctoral fellow of the Jane Coffin Childs Memorial Fund for Medical Research.

We would like to thank Dr. Saul Silverstein for making available his cell lines and expertise. We acknowledge the excellent technical assistance of Ms. Clemencia Colmenares and Ms. Elizabeth Nichols and thank Mrs. Marie Siniscalchi for help with the preparation of the manuscript.

REFERENCES

1. McBride, O., and Ozer, H. (1973). Proc. Natl. Acad. Sci. USA 70, 1258.
2. Miller, C., and Ruddle, F. (1978). Proc. Natl. Acad. Sci. USA 75, 3346.
3. Ruddle, F., and Fournier, R. (1977). In "Genetic Interaction and Gene Transfer" (C. Anderson, ed.), Brookhaven Symposia in Biology 29, p. 96.
4. McBride, O., Burch, J., and Ruddle, F. (1978). Proc. Natl. Acad. Sci. USA 75, 914.
5. Degnen, G., Miller, I., Eisenstadt, J., and Adelberg, E. (1976). Proc. Natl. Acad. Sci. USA 73, 2838.
6. Willecke, K., Mierau, R., Krüger, A., and Lange, R. (1978). Molec. Gen. Genet. 161, 49.
7. Athwal, R., and McBride, O. (1977). Proc. Natl. Acad. Sci. USA 74, 2943.
8. Willecke, K., Mierau, R., Krüger, A., and Lange, R. (1976). Cytogen. Cell Genet. 16, 405.
9. Fournier, R., and Ruddle, F. (1977). Proc. Natl. Acad. Sci. USA 74, 3937.
10. Littlefield, J. (1964). Science 145, 709.
11. Klobutcher, L., Church, R. , and Ruddle F. (1979). Nature, in press.
12. Wigler, M., Silverstein, S., Lee, L., Pellicer, A., Cheng, Y., and Axel, R. (1977). Cell 11, 223.
13. Scangos, G., Silverstein, S., Huttner, K., and Ruddle, F. (1979). Proc. Natl. Acad. Sci. USA, in press.
14. Friend, K., Dorman, B., Kucherlapati, R., and Ruddle, F. (1976). Exp. Cell Res. 99, 31.
15. Elsevier, S., Kucherlapati, R., Nicholas, E. A., Creagan, R., Giles, R., Ruddle, F., Willecke, K., and McDougall, J. (1974). Nature 251, 633.
16. Kit, S., and Dubbs, D. (1963). Biochem. Biophys. Res. Commun. 11, 55.
17. Munyon, W., Kraiselburd, E., Davis, S., and Mann, J. (1971). J. Virol. 7, 813.
18. Smiley, J., Steege, D., Juricek, D., Summers, W., and Ruddle, F. (1978). Cell 15, 455.
19. Pellicer, A., Wigler, M., Axel, R., and Silverstein, S. (1978). Cell 14, 133.
20. Southern, E. (1975). J. Mol. Biol. 98, 503.

TRANSFORMATION OF MAMMALIAN CELLS WITH PROKARYOTIC AND EUKARYOTIC GENES[1]

Michael Wigler[2], Raymond Sweet, Gek Kee Sim, Barbara Wold, Angel Pellicer, Elizabeth Lacy*, Tom Maniatis*, Saul Silverstein, and Richard Axel

College of Physicians and Surgeons,
Columbia University, New York, N.Y. 10032

*Division of Biology, California Institute of Technology, Pasadena, Calif. 91125

ABSTRACT Cellular genes coding for selectable biochemical functions can be stably introduced into cultured mammalian cells by DNA-mediated gene transfer (transformation). Biochemical transformants are readily identified by the stable expression of a gene coding for a selectable marker. These transformants represent a subpopulation of the competent cells which integrate other physically unlinked genes for which no selective criteria exist. In this manner, we have used a viral thymidine kinase gene as a selectable marker to isolate mouse cell lines which we have stably transformed with the tk gene along with bacteriophage ΦX 174, plasmid pBR 322, or the cloned chromosomal rabbit β-globin gene sequences. ΦX co-transformants were studied in greatest detail. The frequency of co-transformation is high, 15 of 16 tk^+ transformants contain the ΦX sequences. Further, from one to more than fifty ΦX sequences are stably integrated into high molecular weight nuclear DNA isolated from independent clones. The introduction of cloned eukaryotic genes now provides an *in vivo* system to study the functional significance of various features of DNA sequence organization. We have analyzed the ability of the mouse fibroblast transformant to transcribe and process the heterologous rabbit β-globin gene. Hybridization experiments indicate that in at least one transformant, rabbit β-globin sequences are expressed in the cytoplasm as a discrete 9S species, suggesting that mouse fibroblast may contain the enzymes necessary to transcribe and correctly process a rabbit gene whose expression is usually restricted to erythroid cells. These studies demonstrate the potential value of co-transformation systems in the analysis of eukaryotic gene expression.

ISBN 0-12-068350-4

INTRODUCTION

Specific genes can be stably introduced into cultured mammalian cells by DNA-mediated gene transfer. The process of transformation results in a change in the genotype of the recipient cell and provides a unique opportunity to study the function and physical state of exogenous genes in the transformed host. In our laboratories, we have developed transformation systems which may allow the introduction of virtually any defined gene into cultured cells. We have therefore performed a series of transformation experiments with a variety of different eukaryotic and prokaryotic genes: 1) to develop *in vivo* systems to study the functional significance of various features of DNA sequence organization; 2) as a means for gene purification where now classical routes involving recombinant DNA technology and molecular hybridization are inapplicable; 3) to examine the fluidity and promiscuity of the eukaryotic chromosome.

In initial studies, we developed a transformation system for the thymidine kinase (tk) gene of herpes simplex virus (HSV-1). Through a series of electrophoretic fractionations in concert with transformation assays, we isolated a unique 3.4 kb fragment of viral DNA which is capable of efficiently transferring tk activity to mutant Ltk^- cells (Wigler *et al.*, 1977). Extension of these studies to unique cellular genes has resulted in the stable transfer of genes coding for thymidine kinase, adenine-phosphoribosyl transferase and a methotrexate resistant mutant of dihydrofolate reductase to mouse fibroblasts (Wigler *et al.*, 1978, 1979a).

The methods we have used to transfer these genes can, in principle, be applied to any gene for which conditional selection criteria are available. The isolation of cells transformed with genes which do not code for selectable markers, however, is problematic, since current transformation procedures are highly inefficient. We have recently demonstrated the feasibility of co-transforming cells with two physically unlinked genes (Wigler *et al.*, 1979b). Co-transformed cells can be identified and isolated when one of these genes codes for a selectable marker. We have used the viral tk gene as a selectable marker to isolate mouse cell lines which contain the tk gene along with either bacteriophage ΦX 174, plasmid pBR 322, or the cloned rabbit β-globin gene sequences stably integrated into cellular DNA. We have further demonstrated that the gene coding for the rabbit β-globin in transformed mouse fibroblasts is properly recognized by the transcriptional and processing enzymes of the mouse cell to generate RNA indistinguishable from the mature globin mRNA of the rabbit erythroblast (Wold *et al.*, 1979).

These studies demonstrate the value of co-transformation systems in the analysis of eukaryotic gene expression.

RESULTS

Co-Transformation of Mouse Cells with ΦX-174 DNA. The addition of the purified thymidine kinase gene from herpes simplex virus to mutant mouse cells lacking tk results in the appearance of stable transformants expressing the viral gene which can be selected by their ability to grow in HAT (Maitland and McDougall, 1977; Wigler et al., 1977). To obtain co-transformants, cultures are exposed to the tk gene in the presence of vast excess of a well-defined DNA sequence for which hybridization probes are available. Tk^+ transformants are isolated and scored for the co-transfer of unselectable DNA sequences by molecular hybridization.

We initially used ΦX DNA in co-transformation experiments with the tk gene as the selectable marker. ΦX replicative form DNA was cleaved with Pst I, which recognizes a single site in the circular genome (Fig. 1) (Sanger et al., 1977). Purified tk gene (500 pg) was mixed with 1-10 μg of Pst-cleaved ΦX replicative form DNA. This DNA was then added to mouse Ltk^- cells using the transformation conditions previously described (Wigler et al., 1979a). After two weeks in selective medium (HAT), tk^+ transformants were observed at a frequency of one colony per 10^6 cells per 20 pg of purified gene. Clones were picked and grown into mass culture.

We then asked whether tk^+ transformants contained ΦX DNA sequences. High molecular weight DNA from the transformants was cleaved with the restriction endonuclease Eco RI, which recognizes no sites in the ΦX genome. The DNA was fractionated by agarose gel electrophoresis and transferred to nitrocellulose filters, and these filters were then annealed with nick-translated ^{32}P-ΦX DNA (blot hybridization) (Southern, 1975; Botchan et al., 1976; Pellicer et al., 1978).

These annealing experiments indicate that 15 of 16 transformants acquired bacteriophage sequences. Results with two representative clones, ΦX 4 and ΦX 5 are shown in Figure 2. Since the ΦX genome is not cut with the enzyme Eco RI, the number of bands observed reflects the minimum number of eukaryotic DNA fragments containing information homologous to ΦX. The clones contain variable amounts of ΦX sequences: 4 of the 15 positive clones reveal only a single annealing fragment while others reveal at least fifty ΦX-specific fragments.

It should be noted that none of 15 clones picked at random from neutral medium, following exposure to tk and ΦX DNA,

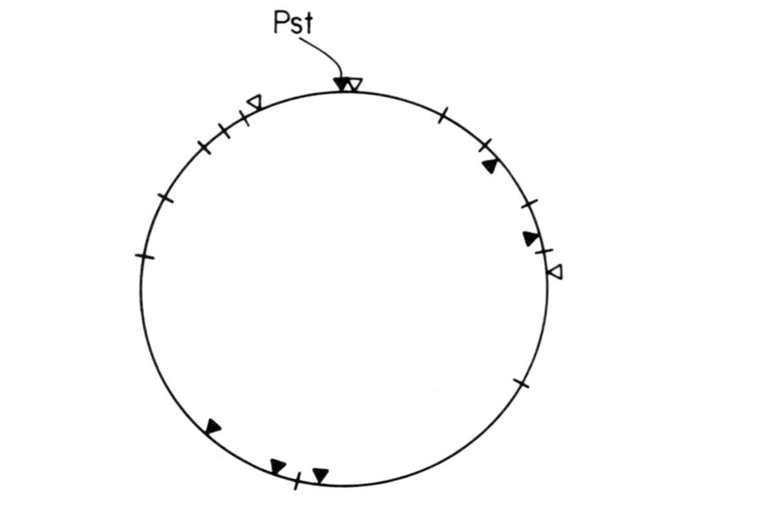

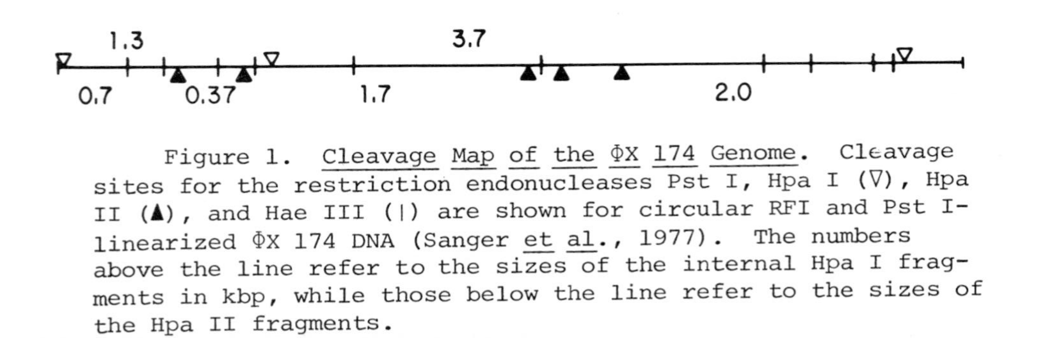

Figure 1. Cleavage Map of the ΦX 174 Genome. Cleavage sites for the restriction endonucleases Pst I, Hpa I (▽), Hpa II (▲), and Hae III (|) are shown for circular RFI and Pst I-linearized ΦX 174 DNA (Sanger et al., 1977). The numbers above the line refer to the sizes of the internal Hpa I fragments in kbp, while those below the line refer to the sizes of the Hpa II fragments.

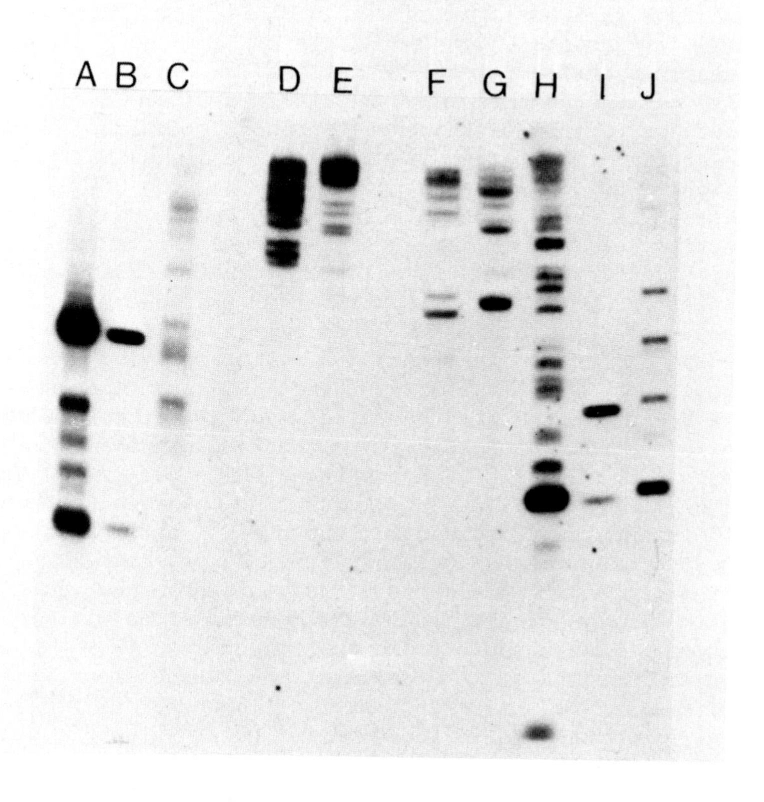

Figure 2. Extent of Sequence Representation in ΦX Co-Transformants. High molecular weight DNA from co-transformant clones ΦX 4 and ΦX 5 was digested with either Eco RI, Bam HI, Hpa I or Hpa II and analyzed for the presence of ΦX sequences as described. (Lanes B and I) 50 pg (4 gene equivalents) of ΦX RFI DNA digested with Hpa I and Hpa II, respectively. (Lanes A, D, E and H) 15 μg of clone ΦX 4 DNA digested with Hpa I, Eco RI, Bam HI and Hpa II, respectively, and analyzed for ΦX sequences by blot hybridization. (Lanes C, F, G and J) 15 μg of clone ΦX 5 DNA digested with Hpa I, Eco RI, Bam HI or Hpa II, respectively.

contain ΦX information. Transformation with ΦX therefore is restricted to a subpopulation of tk^+ transformants. The addition of a selectable marker therefore facilitates the identification of co-transformants.

ΦX Sequences Are Integrated Into Cellular DNA. Cleavage of DNA from ΦX transformants with Eco RI (Fig. 2) generates a series of fragments which contain ΦX DNA sequences. These fragments may reflect multiple integration events. Alternatively, these fragments could result from tandem arrays of complete or partial ΦX sequences which are not integrated into cellular DNA. To distinguish between these possibilities, transformed cell DNA was cut with Bam HI or Eco RI, neither of which cleaves the ΦX genome. If the ΦX DNA sequences were not integrated, neither of these enzymes would cleave the ΦX fragments. Identical patterns would be generated from undigested DNA and from DNA cleaved with either of these enzymes. If the sequences are integrated, then Bam HI and Eco RI should recognize different sites in the flanking cellular DNA and generate unique restriction patterns. DNA from clones ΦX 4 and ΦX 5 was cleaved with Bam HI or Eco RI and analyzed by Southern hybridization (Fig. 2, clone 4, lanes D and E; clone 5, lanes F and G). In each instance, the annealing pattern with Eco RI fragments differed from that observed with the Bam HI fragments. Furthermore, the profile obtained with undigested DNA reveals annealing only in very high molecular weight regions with no discrete fragments observed (data not shown). Similar observations were made on clone ΦX 1 (data not shown). Thus, most of the ΦX sequences in these three clones are integrated into cellular DNA. Experiments with subcellular fractions demonstrate that over 95% of the ΦX sequences are localized in the high molecular weight fraction of nuclear DNA in transformants.

Extent of Sequence Representation of the ΦX Genome. The annealing profiles of DNA from transformed clones digested with enzymes that do not cleave the ΦX genome provide evidence that ΦX sequences are integrated and allow us to estimate the number of ΦX sequences integrated. Annealing profiles of DNA from transformed clones digested with enzymes which cleave within the ΦX genome allow us to determine what proportion of the genome is present and how these sequences are arranged following integration. Cleavage of ΦX with the enzyme Hpa I generates three fragments for each integration event (see Fig. 1): two "internal" fragments of 3.7 and 1.3 kb which together comprise 90% of the ΦX genome, and one "bridge" fragment of 0.5 kb which spans the Pst I cleavage site. The annealing profile observed with clone ΦX 4 digested with Hpa I is shown in Fig. 2, lane A. Two intense bands are observed at 3.7 and 1.3 kb. A less intense series of bands of higher molecular weight is also observed, some of which probably represent ΦX sequences adjacent to cellular DNA. These results indicate that at least 90% of the ΦX genome is present in these cells.

It is worth noting that the internal 1.3 kb Hpa I fragment is bounded by an Hpa I site only 30 bp from the Pst I cleavage site. Comparison of the intensities of the internal bands with known quantities of Hpa I-cleaved ΦX DNA suggests that this clone contains approximately 100 copies of the ΦX genome (Fig. 2, lanes A and B). The annealing pattern of clone 5 DNA cleaved with Hpa I is more complex (Fig. 2, lane C). If internal fragments are present, they are markedly reduced in intensity; instead, multiple bands of varying molecular weight are observed. The 0.5 kb Hpa I fragment which bridges the Pst I cleavage site is not observed in either clone ΦX 4 or clone ΦX 5. Similar analyses of DNA from these clones with the enzymes Hpa II and Hae III confirm our observation that although, in general, the "internal" fragments of ΦX are found in these transformants, "bridge" fragments which span the Pst I site are reduced or absent.

The data allow us to make some preliminary statements on the nature of the integration intermediate. These experiments attempted to distinguish between the integration of a linear or circular intermediate. If either precise circularization or the formation of linear concatamers had occurred at the Pst I cleavage site, and if integration occurred at random points along this DNA, we would expect cleavage maps of transformed cell DNA to mirror the circular ΦX map. The bridge fragment, however, is not observed or present in reduced amounts in digests of transformed cell DNA with three different restriction endonucleases. The fragments observed agree with a model in which ΦX DNA integrates as a linear molecule. Alternatively, it is possible that intramolecular recombination of ΦX DNA occurs, resulting in circularization with deletions at the Pst termini (Lai and Nathans, 1974). Random integration of this circular molecule would generate a restriction map similar to that observed for clones ΦX 4 and ΦX 5. Other more complex models of events occurring before, during or after integration can also be considered. Whatever the mode of integration, it appears that cells can be stably transformed without significant loss of donor DNA sequences.

Stability of the Transformed Genotype. Our previous observations on the transfer of selectable biochemical markers indicate that the transformed phenotype remains stable for hundreds of generations if cells are maintained under selective pressure. If maintained in neutral medium, the transformed phenotype is lost at frequencies which range from <0.1 to as high as 30% per generation (Wigler et al., 1977; 1979). The use of transformation to study the expression of foreign genes depends upon the stability of the transformed genotype. This is an important consideration with genes for which no selective criteria are available. We assume that the presence of ΦX DNA in our transformants confers no selective advantage on the recipient cell. We therefore examined the stability of the ΦX genotype in the descendants of two clones after numerous generations in culture. Clones ΦX 4 and ΦX 5, both containing multiple copies of ΦX DNA, were subcloned and six independent subclones from each clone were picked and grown into mass culture. DNA from one of these subclones was then digested with either Eco RI or Hpa I, and the annealing profiles of ΦX-containing fragments were compared with those of the original parental clone. The annealing pattern observed for four of the six ΦX 4 subclones is virtually identical to that of the parent (Fig. 3). In two subclones, an additional Eco RI fragment appeared which is of identical molecular weight in both. This may have resulted from genotypic heterogeneity in the parental clone prior to subcloning. These data indicate that ΦX DNA is maintained within the subclones examined for numerous generations without significant loss or translocation of information.

Transformation of Mouse Cells with the Rabbit β-Globin Gene. Transformation with purified eukaryotic genes may provide a means for studying the expression of cloned genes in a heterologous host. We have therefore performed co-transformation experiments with the rabbit β major globin gene which was isolated from a cloned library of rabbit chromosomal DNA (Maniatis et al., 1978). One β-globin clone, designated RβG-1 (Lacy et al., 1978), consists of a 15 kb rabbit DNA fragment carried on the bacteriophage λ cloning vector Charon 4A. Intact DNA from this clone (RβG-1) was mixed with the viral tk DNA at a molar ratio of 100:1, and tk^+ transformants were isolated and examined for the presence of rabbit globin sequences. Cleavage of RβG-1 with the enzyme Kpn I generates a 4.7 kb fragment which contains the entire rabbit β-globin gene. This fragment was purified by gel electrophoresis and nick-translated to generate a probe for subsequent annealing experiments. The β-globin genes of mouse and rabbit are partially homologous although we do not observe annealing of the rabbit β-globin probe with Kpn-cleaved mouse DNA, presumably because Kpn generates very large globin-specific fragments (Fig. 4, lanes C, D and G). In contrast, cleavage of rabbit liver DNA with Kpn I generates the expected 4.7 kb globin

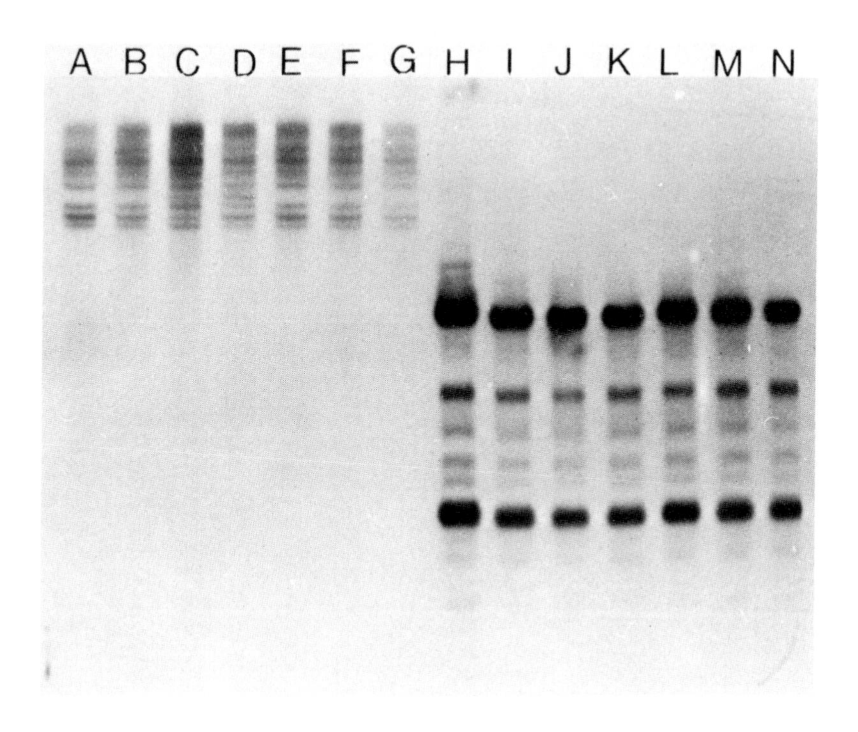

Figure 3. Stability of ΦX Sequences in Subclones of Transformants. Annealing profiles of DNA from parental clone ΦX 4 digested with Eco RI (lane A) and Hpa I (lane H) are compared with DNA from six independent subclones digested with either Eco RI (lanes B-G) or Hpa I (lanes I-N).

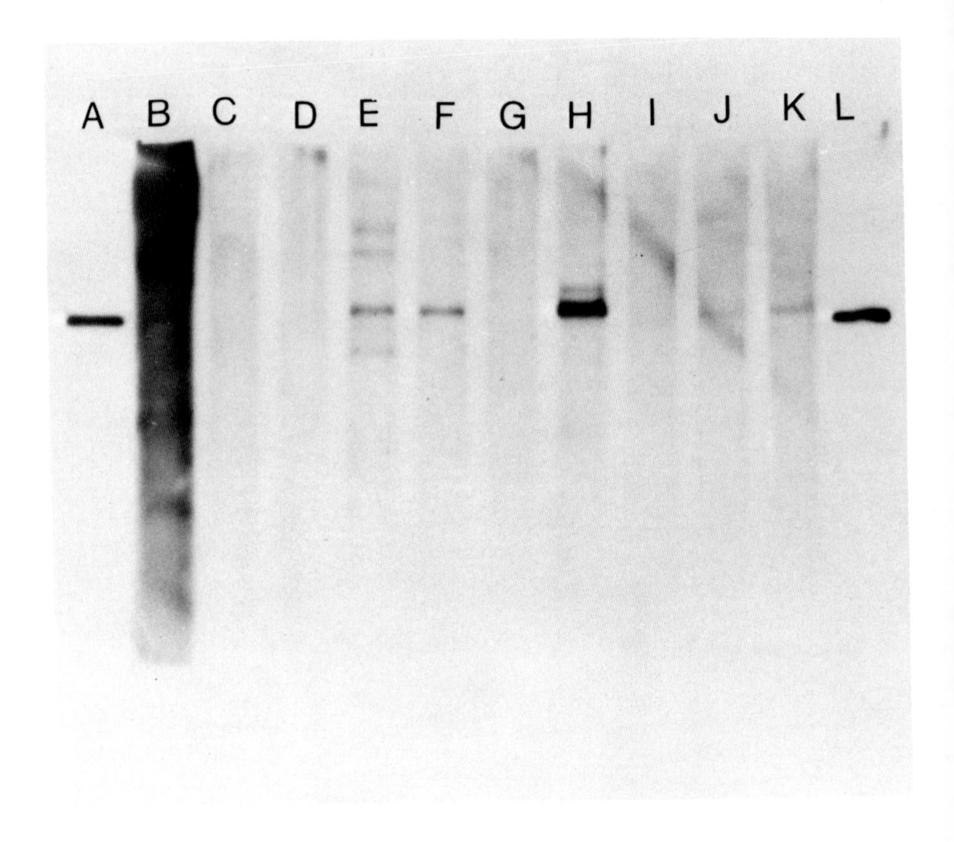

Figure 4. The Rabbit β-Globin Gene Is Present in Mouse DNA. Cells were exposed to RβG-1 DNA and the viral tk gene and selected in HAT. High molecular weight DNA from eight independent clones was digested with Kpn I and electrophoresed on a 1% agarose gel. The DNA was denatured *in situ* and transferred to nitrocellulose filters, which were then annealed with a ^{32}P-labeled 4.7 kbp fragment containing the rabbit β-globin gene. (Lanes A and L) 50 pg of the 4.7 kbp Kpn fragment of RβG-1; (lane B) 15 μg of rabbit liver DNA digested with Kpn; (lane C) 15 μg of Ltk^-aprt^- DNA; (lanes D-K) 15 μg of DNA from each of eight independently isolated tk^+ transformants.

band (Fig. 4, lane B). Cleavage of transformed cell DNA with the enzyme Kpn I generates a 4.7 kb fragment containing globin-specific information in six of the eight tk^+ transformants examined. The number of rabbit globin genes present in these transformants is variable. In comparison with control lanes, some of the clones contain a single copy of the gene, while others may contain as many as 20 copies of this heterologous gene.

<u>Rabbit β-Globin Sequences are Transcribed in Mouse Transformants</u>. The co-transformation system we have developed may provide a functional assay for cloned eukaryotic genes if these genes are expressed in the heterologous recipient cell. Six transformed cell clones were therefore analyzed for the presence of rabbit β-globin RNA sequences (Wold <u>et al</u>., 1979). In initial experiments, we performed solution hybridization reactions to determine the cellular concentration of rabbit globin transcripts in our transformants.

A radioactive cDNA copy of purified rabbit α and β globin mRNA was annealed with a vast excess of total cellular RNA from transformants under experimental conditions such that rabbit globin cDNA does not form a stable hybrid with mouse globin mRNA but will react completely with homologous rabbit sequences. A summary of these hybridization reactions is shown in Table 1. Total RNA from transformed clone 6 protects 44% of the rabbit cDNA at completion, the value expected if only β gene transcripts are present. This reaction displays pseudo-first-order kinetics with an $R_ot_{1/2}$ of 2×10^3. A second transformant (clone 2) reacts with an $R_ot_{1/2}$ of 8×10^3. No significant hybridization was observed with total RNA preparations from four other transformants. Further analysis of clone 6 demonstrates that virtually all of the rabbit β-globin RNA detected in this transformant is polyadenylated and exists at a steady state concentration of about five copies per cell with greater than 90% of the sequences localized in the cytoplasm.

<u>Globin Sequences Exist as a Discrete 9S Species in Transformed Cells</u>. In rabbit erythroblast nuclei, the β-globin gene sequences are detected as a 14S precursor RNA which reflects transcription of two intervening sequences which are subsequently spliced from this molecule to generate a 9S messenger RNA (Flavell, manuscript in preparation; Lacy and Maniatis, unpubished results). Our solution hybridization experiments only indicate that polyadenylated rabbit globin RNA sequences are present in the mouse transformant. It was therefore of interest to determine whether the globin transcripts we detect exist as a discrete 9S species, which is

TABLE 1

Rabbit Globin Gene Transcripts in Total RNA of Ltk^+ RβG-1 Transformants

L-Cell Transformant	Lane in Figure 4	$R_o t_{1/2}$ of Total RNA and Rabbit Globin cDNA	Transcripts of Rabbit β-Globin per Cell
1	D	$> 2.0 \times 10^4$	< 0.2
2	E	$\sim 8.0 \times 10^3$	~ 1.0
3	F	$> 1.2 \times 10^4$	< 0.2
4	K	$> 3.0 \times 10^4$	< 0.1
5	J	$> 1.0 \times 10^4$	< 0.5
6	H	2.0×10^3	~ 5.0

likely to reflect appropriate splicing of the rabbit gene transcript by the mouse fibroblast. Cytoplasmic poly A-containing RNA from clone 6 was denatured by treatment with 6 M urea at 70°C, and electrophoresed on a 1% acid-urea-agarose gel and transferred to diazotized cellulose paper (B. Seed, personal communication). Following transfer, the RNA filters were hybridized with DNA from the plasmid RβG-1 containing rabbit β-globin cDNA sequences (Wold et al., 1979). Using this ^{32}P-labeled probe, a discrete 9S species of cytoplasmic RNA is seen which co-migrates with rabbit globin mRNA isolated from rabbit erythroblasts (Fig. 5). Hybridization to 9S RNA species is not observed in parallel lanes containing either purified mouse 9S globin RNA or polyadenylated cytoplasmic RNA from a tk^+ transformant containing no rabbit globin genes.

We are unable in these experiments to detect the presence of a 14S precursor in nuclear RNA populations from the transformant. This is not surprising, since the levels expected in nuclear RNA, given the observed cytoplasmic concentration, are likely to be below the limits of detection of this technique. Nevertheless, our results with cytoplasmic RNA strongly suggest that the mouse fibroblast is capable of processing a transcript of the rabbit β-globin gene to generate a 9S polyadenylated species which is indistinguishable from the β-globin mRNA in rabbit erythroblasts.

Rescue of pBR 322 DNA from Transformed Mouse Cells. We have extended our observations on co-transformation to the EK-2 approved bacterial vector, plasmid pBR 322. Using the co-transformation scheme outlined earlier, we have constructed cell lines containing multiple copies of the pBR 322 genome. Blot hybridization analyses indicate that the pBR 322 sequences integrate into cellular DNA without significant loss of plasmid DNA. pBR 322 DNA linearized with either Hind III or Bam HI, which destroy the tetracycline resistance gene, integrates into mouse DNA with retention of both the plasmid replication origin and the ampicillin resistance (β-lactamase) gene. We therefore asked whether these plasmid sequences could be rescued from the mouse genome by a second transformation of bacterial cells.

The experimental approach we have chosen is outlined in Figure 6. Linearized pBR 322 DNA is introduced into mouse Ltk^- cells via co-transformation using the tk gene as a selectable marker. DNA is isolated from transformants and screened for the presence of pBR 322 sequences. Since the donor plasmid is linearized, interrupting the tetracycline resistant gene, transformed cell DNA contains a linear stretch of plasmid DNA consisting of the replication origin

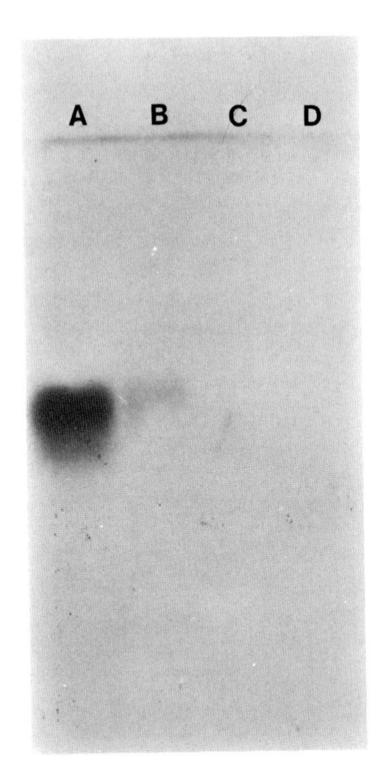

Figure 5. Sizing of Cytoplasmic Polyadenylated Rabbit Globin Transcripts from Clone 6. RNA was electrophoresed in a 1% agarose gel in 6 M urea and the RNA transferred to diazotized cellulose paper. The filter was hybridized with ^{32}P-labeled plasmid DNA ($p\beta G_1$) containing the rabbit β-globin cDNA sequence. Lane A: 2 ng of purified 9S polyadenylated RNA from rabbit reticulocytes, plus 25 μg of carrier chicken oviduct RNA. Lane B: 25 μg of polyadenylated cytoplasmic RNA from clone 6. Lane C: 25 μg of polyadenylated cytoplasmic RNA from a transformant containing no rabbit globin genes. Lane D: 2 ng of purified 9S polyadenylated RNA from mouse reticulocytes plus 25 μg of carrier chicken oviduct RNA.

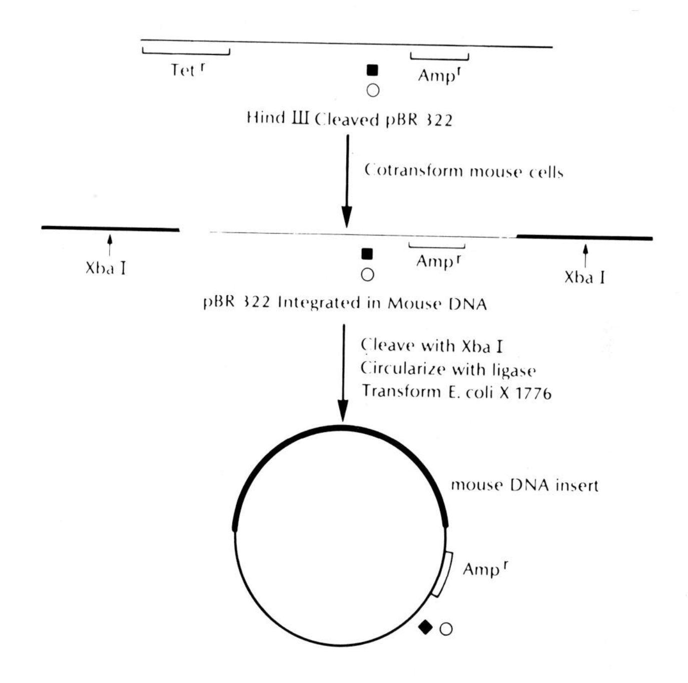

Figure 6. Scheme for the Rescue of Bacterial Plasmids from Transformed Cultured Cells Using Double Selection Techniques. (For explanation, see text).

and the β-lactamase gene covalently linked to mouse cellular DNA. This DNA is cleaved with an enzyme such as Xho I, which does not digest the plasmid genome. The resulting fragments are circularized at low DNA concentrations in the presence of ligase. Circular molecules containing plasmid DNA are selected from the vast excess of eukaryotic circles by transformation of E. coli strain X 1776.

We have carried out this series of experiments and isolated a recombinant plasmid from transformed mouse cell DNA which displays the following properties: 1) The rescued plasmid is ampicillin resistant, but tetracycline sensitive consistent with the fact that the donor pBR 322 was linearized by cleavage within the tetracycline resistance gene. 2) The rescued plasmid is 1.9 kb larger than pBR 322 and therefore contains additional DNA. 3) The rescued plasmid anneals to a single band in blot hybridizations to Eco RI-cleaved mouse liver DNA, suggesting that the plasmid contains an insert of single copy mouse DNA. These observations demonstrate that bacterial plasmids stably integrated into the mouse genome via transformation, can be rescued from this unnatural environment, and retain their ability to function in bacterial hosts.

This result immediately suggests modified schemes utilizing plasmid rescue to isolate virtually any cellular gene for which selective growth criteria are available. The aprt gene of the chicken is not cleaved by Hind III or Xho I and transformation of $aprt^-$ mouse cells with cellular DNA digested with these enzymes results in the generation of $aprt^+$ colonies which express the chicken aprt gene. Ligation of Hind III cleaved chicken DNA with Hind III cleaved pBR 322 results in the formation of hybrid DNA molecules, in which the aprt gene is now adjacent to plasmid sequences. Transformation of $aprt^-$ cells is now performed with this DNA. Transformants should contain the aprt gene covalently linked to pBR 322, integrated into the mouse genome. This transformed cell DNA is now treated with an enzyme which does not cleave either pBR 322 or the aprt gene and the resultant fragments are circularized with ligase. Transformation of E. coli with these circular molecules should select for plasmid sequences from eukaryotic DNA and enormously enrich for chicken aprt sequences. This double selection technique may ultimately permit the isolation of genes expressed at low levels in eukaryotic cells, for which hybridization probes are not readily obtained.

DISCUSSION

In these studies, we have stably transformed mammalian cells with precisely defined prokaryotic and eukaryotic genes for which no selective criteria exist. Our chosen experimental design derives from studies of transformation in bacteria which indicate that a small but selectable subpopulation of cells is competent in transformation (Thomas, 1955; Hotchkiss, 1959; Tomasz and Hotchkiss, 1964; Spizizen et al., 1966). If this is also true for animal cells, then biochemical transformants will represent a subpopulation of competent cells which are likely to integrate other unlinked genes at frequencies higher than the general population. Thus, to identify transformants containing genes which provide no selectable trait, cultures were co-transformed with a physically unlinked gene which provided a selectable marker. This co-transformation system should allow the introduction and stable integration of virtually any defined gene into cultured cells. Ligation to either viral vectors or selectable biochemical markers is not required.

Co-transformation experiments were performed using the HSV tk gene as the selectable biochemical marker. The addition of this purified tk gene to mouse cells lacking thymidine kinase results in the appearance of stable transformants which can be selected by their ability to grow in HAT. Tk^+ transformants were cloned and analyzed by blot hybridization of co-transfer of additional DNA sequences. In this manner, we have constructed mouse cell lines which contain multiple copies of ΦX, pBR 322 and rabbit β-globin gene sequences.

The usefulness of co-transformation will depend to a large extent on its generality. To date, we have limited experience with other cell lines. The use of tk as a selectable marker restricts host cells to tk^- mutants. More recently, we have demonstrated the co-transfer of plasmid pBR 322 DNA into mouse L cells using cellular DNA containing a mutant dihydrofolate reductase gene as donor, and methotrexate resistance as the selectable marker. The use of such dominant acting mutant genes which confer drug resistance may extend the host range for co-transformation to virtually any cultured cell.

The frequency with which DNA is stably introduced into competent cells is high. Furthermore, the co-transformed sequences appear to be integrated into high molecular weight nuclear DNA. The number of integration events varies from one to greater than fifty in independent transformed clones. At present, we cannot make precise statements concerning the nature of the integration intermediate. Although our data with ΦX are in accord with the model in which ΦX DNA integrates as a linear molecule, it is possible that more complex intramolecular recombination events generating circular intermediates may have occurred prior to or during the integration process. Whatever the mode of integration, it appears that cells can be stably transformed with long stretches of donor DNA. We have observed transformants containing contiguous stretches of donor DNA 50 kb long (Wold et al., 1979). Furthermore, the frequency of competent cells in culture is also high. At least one percent of our mouse Ltk⁻ cell recipients can be transformed to the tk⁺ phenotype (Silverstein et al., 1979). Although we do not know the frequency of transformation in nature, this process could have profound physiologic and evolutionary consequences.

The introduction of cloned eukaryotic genes into animal cells provides an in vivo system to study the functional significance of various features of DNA sequence organization. In these studies, we have constructed stable mouse cell lines which contain up to 20 copies of the rabbit β-globin gene and have analyzed the ability of the mouse fibroblast recipient to transcribe and process this heterologous gene. Solution hybridization experiments in concert with RNA blotting techniques indicate that in at least one transformed cell line, rabbit globin sequences are expressed in the cytoplasm as a 9S species indistinguishable from the mature messenger RNA of rabbit erythroblasts. These results suggest that the mouse fibroblast contains the enzymes necessary to transcribe and correctly process a rabbit gene whose expression is normally restricted to erythroid cells. Similar observations have been made using a viral vector to introduce the rabbit globin gene into monkey cells (Leder, personal communication).

These studies indicate the potential value of co-transformation systems in the analysis of eukaryotic gene expression. The introduction of wild type genes along with native and in vitro constructed mutant genes into cultured cells provides an assay for the functional significance of sequence organization. It is obvious from these studies that this analysis will be facilitated by the ability to extend the generality of co-transformation to recipient cell lines, such as murine erythroleukemia cells, which may provide a more appropriate environment for the study of heterologous globin gene expression.

---

[1]This work was supported by grants from the NIH and NSF.

[2]Present address: Cold Spring Harbor Laboratories, Cold Spring Harbor, New York 11724.

REFERENCES

Wigler, M., Silverstein, S., Lee, L.-S., Pellicer, A., Cheng, Y.-c., and Axel, R. (1977) Cell 11, 223.
Wigler, M., Pellicer, A., Silverstein, S., and Axel, R. (1978) Cell 14, 725.
Wigler, M., Pellicer, A., Silverstein, S., Axel, R., Urlaub, G., and Chasin, L. (1979) Proc. Nat. Acad. Sci. USA 76, 1373.
Wigler, M., Sweet, R., Sim, G. K., Wold, B., Pellicer, A., Lacy, E., Maniatis, T., Silverstein, S., and Axel, R. (1979) Cell 16, 777.
Maitland, N. J., and McDougall, J. K. (1977) Cell 11, 233.
Sanger, F., Air, M., Barrell, B. G., Brown, N. L., Coulson, A. R., Fiddes, J. C., Hutchinson, C. A., Slocombe, P. M., and Smith, M. (1977) Nature 265, 687.
Southern, E. M. (1975) J. Mol. Biol. 98, 503.
Botchan, M., Topp, W., and Sambrook, J. (1976) Cell 9, 269.
Pellicer, A., Wigler, M., Axel, R., and Silverstein, S. (1978) Cell 14, 133.
Lai, C., and Nathans, D. (1974) Cold Spring Harbor Symp. Quant. Biol. 39, 53.
Maniatis, T., Hardison, R. C., Lacy, E., Lauer, J., O'Connell, C., Quon, D., Sim, G. K., and Efstradiatis, A. (1978) Cell 15, 687.
Lacy, E., Lawn, R. M., Fritsch, D., Hardison, R. C., Parker, R. C., and Maniatis, T. (1979) In: Cellular and Molecular Regulation of Hemoglobin Switching (New York: Grune & Stratton), in press.
Thomas, R. (1955) Biochim. Biophys. Acta 18, 467.
Hotchkiss, R. (1959) Proc. Nat. Acad. Sci. USA 40, 49.
Thomasz, A., and Hotchkiss, R. (1964) Proc. Nat. Acad. Sci. USA 51, 480.
Spizizen, J., Reilly, B. E., and Evans, A. H. (1966) Ann. Rev. Microbiol. 20, 371.
Wold, B., Wigler, M., Lacy, E., Maniatis, T., Silverstein, S., and Axel, R. (1979) Proc. Nat. Acad. Sci. USA, in press.
Silverstein, S., et. al. (1979) Manuscript in preparation.

SYNTHESIS OF RABBIT β-GLOBIN-SPECIFIC RNA IN MOUSE L CELLS AND YEAST TRANSFORMED WITH CLONED RABBIT CHROMOSOMAL β-GLOBIN DNA.

Ned Mantei, Albert van Ooyen, Johan van den Berg,[1] Jean D. Beggs,[2] Werner Boll, Robert F. Weaver and Charles Weissmann

Institut für Molekularbiologie I, Universität Zürich, 8093 Zürich, Switzerland

ABSTRACT Mouse thymidine kinase negative L cells were transformed with a cloned rabbit chromosomal β-globin gene linked to the cloned thymidine kinase gene of Herpes simplex I. Most thymidine kinase positive cell lines contained one or more copies of rabbit β-globin DNA and produced up to 2200 copies per cell of rabbit β-globin mRNA which was indistinguishable from its authentic counterpart. No mouse β-globin mRNA was detected.

Yeast was transformed with the same rabbit β-globin DNA linked to a hybrid containing the 2 μ-yeast plasmid. Here, too, β-globin-specific transcripts were found; however, they did not have the normal termini and splicing had not occurred.

INTRODUCTION

A promising strategy for elucidating the mechanism of expression and its control in eukaryotic cells involves the cloning and analysis of a gene and its surrounding regions, *in vitro* modification of the cloned DNA and its reintroduction into a eukaryotic cell (1). In this paper we show that a

[1]Present address: Gist-Brocades, Delft, Holland
[2]Present address: Plant Breeding Institute, Cambridge, England

ISBN 0-12-068350-4

cloned chromosomal rabbit β-globin DNA fragment can be introduced and maintained in mouse L cells and that it is transcribed to yield a β-globin mRNA indistinguishable from the rabbit reticulocyte mRNA. In contrast, yeast cells transformed with the rabbit β-globin DNA yield β-globin-specific RNA which is not spliced and has abnormal 5' and 3' termini.

Transformation of Mouse L Cells with Rabbit β-Globin DNA. Fig. 1 shows the nucleotide sequence of the cloned rabbit β-globin gene contained in the hybrid plasmid Z-pCRI/RchrβG-1 (RβG for short) (2), as well as 223 and 109 nucleotides preceding and following it, respectively (3). Comparison with the homologous mouse sequence (4) shows marked similarities in all regions of the gene, except for the internal stretches of the two introns, which are highly divergent (3). Strikingly, segments of 130 nucleotides preceding the mRNA sequences are closely related in mouse and rabbit.

The β-globin DNA-containing plasmid RβG and a hybrid plasmid containing the thymidine kinase (TK) gene of Herpes simplex I virus (5,6) were both cleaved at their single SalI sites and concatenated with DNA ligase. TK negative (TK^-) mouse L cells were transformed (7) with the concatenates and TK

Fig. 1. The nucleotide sequence of a rabbit β-globin gene and its flanking regions.

Position 1 corresponds to the capped nucleotide of the β-globin mRNA. Nucleotides marked by a dot have been deduced from the recognition sequence of a restriction site and the amino acid sequence in that region. The five nucleotides marked by an asterisk have not been reliably determined. "Cap" and "pA" designate the positions of the capped and the polyadenylated nucleotides, respectively. The borders of the introns have not been determined experimentally; they have been placed at the positions predicted by the Chambon rule (23). From A. van Ooyen, J. van den Berg, N. Mantei and C. Weissmann, submitted for publication.

Nucleotide sequence of a chromosomal rabbit DNA segment containing a β-globin gene.

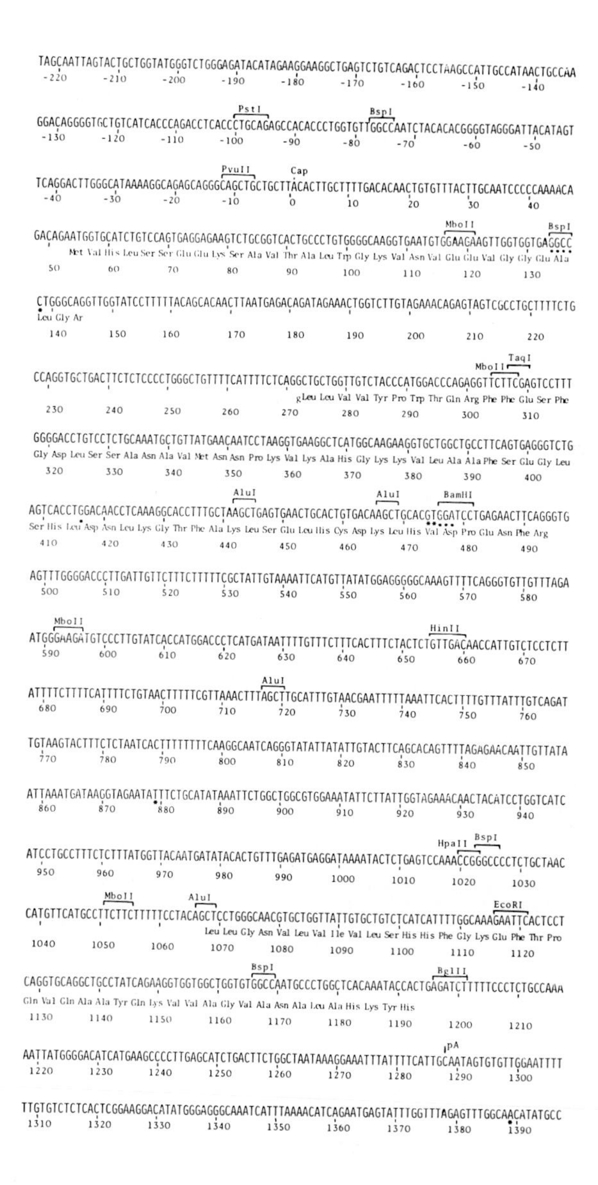

Fig. 1

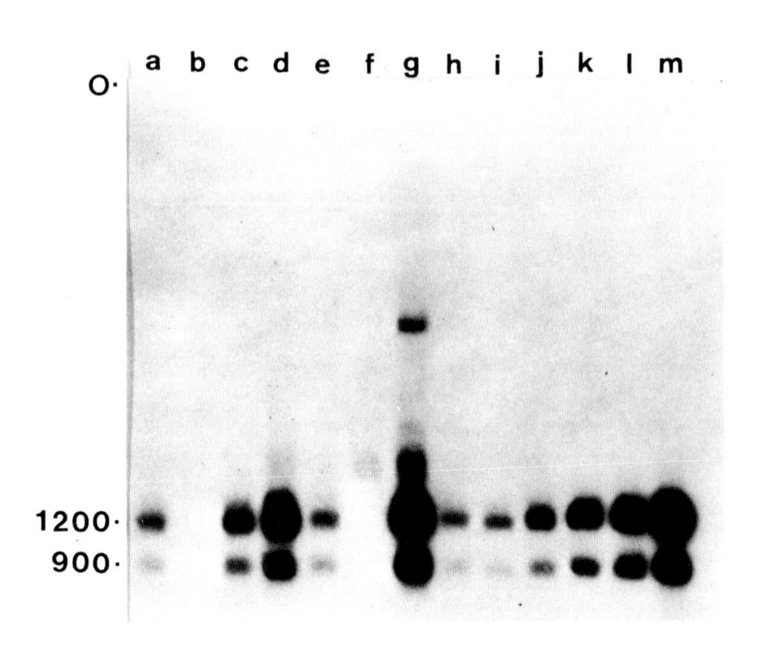

Fig. 2. Detection of chromosomal rabbit β-globin sequences in DNA of transformed mouse L cells.

DNA from mouse L cell lines transformed with the rabbit β-globin hybrid DNA linked to a thymidine kinase (TK) plasmid or with a TK plasmid only, was cleaved with PstI and EcoRI. The fragments were subjected to agarose gel electrophoresis, transferred to a Millipore membrane (8), hybridized to ^{32}P-labeled cloned rabbit β-globin cDNA (13) and located by autoradiography. Lanes b and f, DNA from cell lines P1/S-2 and M2/S-2, transformed only with a SalI-cleaved TK-gene plasmid. Lanes c, d, e and g, DNA from cell lines P1/R-3, P1/R-4, P1/R-5 and M2/R-1, transformed with the rabbit β-globin-containing hybrid RβG linked to a TK-gene plasmid. Lanes a and h to m contain as reference RβG DNA digested with PstI and EcoRI: 1.6, 1.6, 1.6, 3.2, 6.4, 12.8, and 32 pg of DNA hybridizable to the probe (1.8 pg is the equivalent of one β-globin gene). From N. Mantei, W. Boll and C. Weissmann, submitted for publication.

positive (TK^+) clones were selected by growth in HAT medium. To identify cells transformed with β-globin DNA, DNA extracted from 21 clones was cleaved with PstI and EcoRI, the fragments were separated by agarose gel electrophoresis and transferred to a Millipore membrane by the Southern (8) technique. To provide reference markers, the β-globin DNA-containing plasmid RβG was subjected to the same procedures in parallel. Hybridization with a rabbit β-globin-specific ^{32}P-DNA probe demonstrated the presence of the expected 1200 and 900 bp β-globin DNA fragments (Fig. 2) in the 4 samples from the cells transformed with the rabbit β-globin chromosomal-TK DNA concatenates. Altogether, 19 of 21 TK^+ cell lines contained β-globin DNA by this analysis. Control DNA samples, from cells transformed with a TK plasmid only, did not show the characteristic bands (Fig. 2). From the intensities of the β-globin specific bands of Eco-Pst-cleaved cell DNA relative to those of known amounts of similarly restricted RβG DNA run in parallel, we estimate that the cell lines P1/R-3, P1/R-4, P1/R-5 and M2/R-1 contain about 3, 8, 1 and 20 copies of rabbit β-globin DNA per cell.

Identification and Characterization of Rabbit β-Globin Specific Transcripts from Transformed Mouse L Cells. In vivo transcription of the extraneous rabbit β-globin gene was monitored by a modification of the Berk-Sharp procedure (9) developed by Weaver et al. (10), as exemplified in Fig. 3. A specific probe was prepared as follows. The hybrid plasmid RβG contains a BglII site immediately following the 3' end of the β-globin coding sequence. The hybrid was cleaved with BglII and labeled with ^{32}P at the 5' termini. After digestion with PstI, the PstI-BglII (1291/1299) fragment was isolated. The labeled DNA was denatured, hybridized with the RNA sample in 80% formamide, treated with nuclease S_1 (9) and analyzed by polyacrylamide gel electrophoresis in 7 M urea. Only if the labeled 5' terminus of the DNA is protected by complementary RNA is a radioactive fragment recovered. Since the probe is labeled on the minus strand, only plus strand RNA is detected. Mature rabbit β-globin mRNA protects a fragment of 134 nucleotides length, from the BglII site to the position where the sequences of β-globin cDNA and β-globin chromosomal DNA no

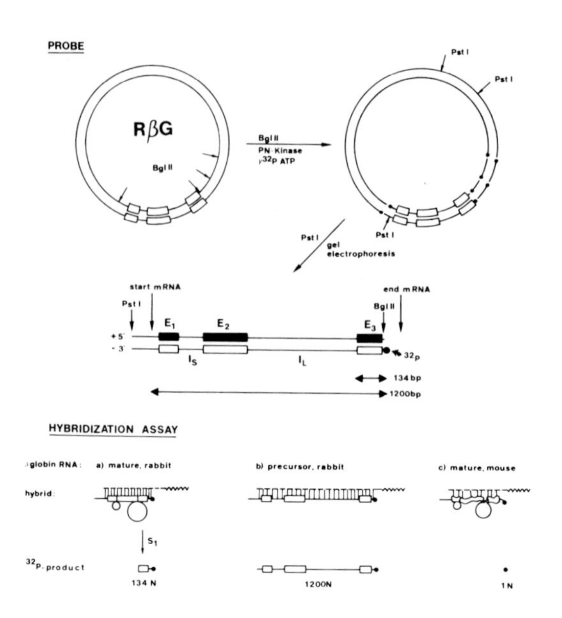

Fig. 3. Application of the Weaver-Berk-Sharp method to distinguish mature and precursor rabbit β-globin mRNA in the possible presence of mouse β-globin mRNA.

The procedure is a modification of the method of Berk and Sharp (9). The probe is a DNA restriction fragment ^{32}P-labeled at one terminus (rather than being uniformly labeled). The denatured DNA is hybridized to the RNA sample in 80% formamide, digested with S_1 nuclease, denatured, and analyzed by polyacrylamide gel electrophoresis in 7 M urea. Only if the labeled terminus of the probe was hybridized to RNA is a labeled DNA fragment recovered. The Pst-BglII (1291/1299) fragment derived from the rabbit chromosomal β-globin plasmid (cf. Fig. 1), labeled at the BglII site, is diagnostic for plus strand β-globin sequences and allows the distinction between rabbit precursor RNA, rabbit mature RNA, and mouse RNA. E, exon; I_S and I_L, small and large intron, respectively.

longer match; this is in principle the 3' edge of the large intron, however, because of a small redundancy, one nucleotide from the 2nd exon may also hybridize to the chromosomal DNA. A precursor containing both introns would protect 1200 nucleotides and a precursor containing only the large intron, 929 nucleotides. The probe discriminates between mouse and rabbit globin mRNA because the rabbit mRNA has a BglII recognition sequence (11) while mouse mRNA does not (N. Mantei, unpublished results and ref. 4), and therefore does not protect the labeled 5' terminus of the probe. Fig. 4 shows that as little as 0.1 ng rabbit β-globin mRNA (lane e) gave the expected 134-nucleotide fragment while 50 ng mouse globin mRNA (lane 1) gave no such band. The RNA from transformed cell lines P1/R-3, P1/R-4, P1/R-5 and M2/R-1 yielded as major product the 134-nucleotide fragment. Thus, the majority of globin-specific RNAs are partly or entirely devoid of the large intron. RNA from cells transformed with only a TK plasmid gave no detectable bands. We estimate that cell line M2/R-1 contains about 2200 rabbit β-globin RNA molecules per cell and P1/R-3, P1/R-4 and P1/R-5 contain about 600, 70 and 400 β-globin transcripts per cell, respectively. Altogether, 12 of 16 globin DNA-containing clones produced β-globin-specific transcripts.

The absence of the small intron was determined by a similar approach, using a probe labeled at the Bam cleavage site (data not shown).

To map the 5' terminus of the β-globin-specific transcripts, a restriction fragment extending from the 5'-^{32}P-labeled MboII cleavage site at position 129/128 to the PstI site at position -95/-99, was used as a probe in the same way as described above. As shown in Fig. 5, hybridization to authentic rabbit β-globin mRNA gave rise to a labeled fragment 128 nucleotides long; RNA from the cell line M2/R-1 gave rise to a strong and P1/R-3 to a weaker band in the same position as the β-globin mRNA, while in the case of cell lines P1/R-4, and P1/R-5 only very weak bands were discernible on the original film (but not on the photograph), probably because of their relatively low content of β-globin-specific RNA. Thus, the only detectable 5' terminus of the β-globin transcripts mapped in the same position as that of authentic β-globin mRNA.

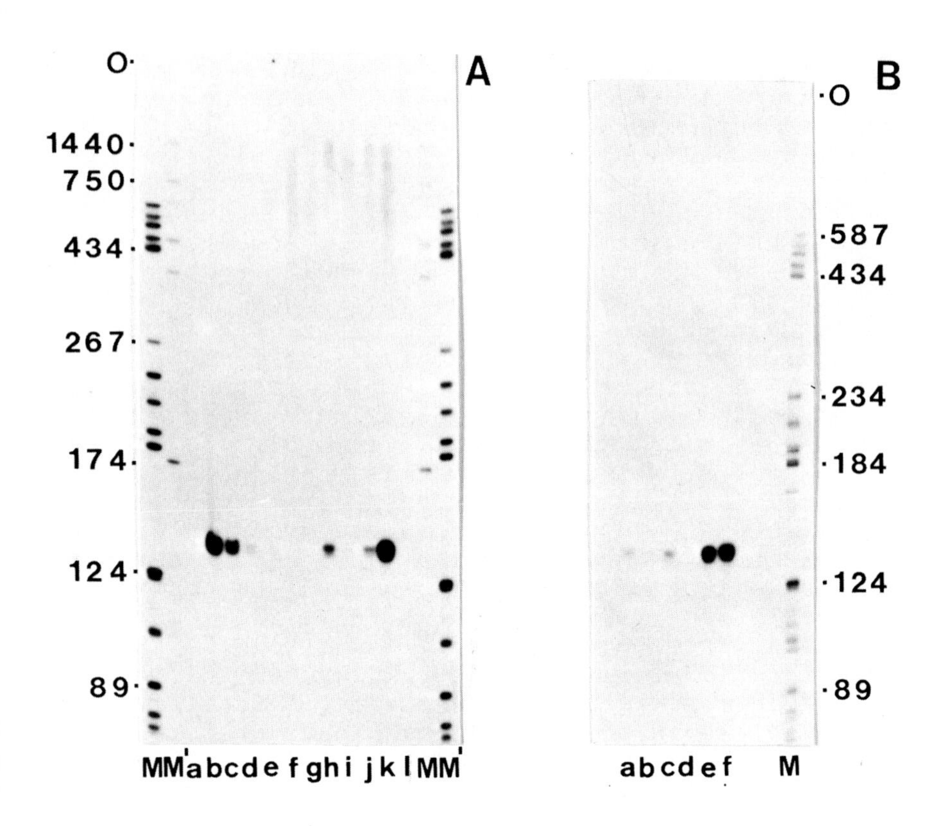

Fig. 4. Detection and characterization of rabbit β-globin RNA in transformed mouse L cells by the Weaver-Berk-Sharp method.

Total cell RNA was prepared from about 10^8 cells as described (24). The probe (described in Fig. 3) was prepared by cleaving Z-pCRI/RchrβG-1 with BglII, 5' labeling with γ-^{32}P-ATP (specific activity 3300 Ci/mmole), cleaving with PstI and isolating the labeled 1291-1299 bp fragment. About 25 ng of probe (13'000 cpm) were denatured and hybridized with up to 50 μg RNA in 80% formamide (9), at 57°C for 16 h. The sample was digested with S_1 nuclease, denatured and electrophoresed through a 5% polyacrylamide gel in 7 M urea. Autoradiography with an intensifier screen was at -70°C for 2-5 days. (A) unfractionated RNA samples. Lanes a to e, 0, 3, 1, 0.3 and 0.1 ng rabbit (α+β) globin mRNA. Lanes f and g, RNA from cell lines P1/S-1 and P1/S-2 (transformed only with TK-DNA plasmid).

The presence of poly(A) on the β-globin-specific transcripts from transformed L cells was demonstrated by fractionating RNA from M2/R-1 on an oligo(dT) column and assaying separately the flow-through fraction (poly(A)$^-$) and the fraction retained at high salt (poly(A)$^+$), using as probe the PstI-BglII (1291/1299) fragment 5'-labeled at the BglII terminus. About 80% of the β-globin transcript was detected in the poly(A)$^+$ fraction (cf. Fig. 4B).

The mobility of the rabbit β-globin-specific RNA present in the cell line M2/R-1 was compared with that of natural rabbit globin mRNA using the "Northern transfer" procedure of Alwine et al. (12). The RNAs were subjected to electrophoresis on an agarose gel under denaturing conditions, transferred to diazobenzyloxymethyl paper and hybridized to ^{32}P-labeled, cloned rabbit β-globin cDNA. Poly(A)$^+$ RNA derived from 150 μg of M2/R-1 RNA gave an autoradiographic band at the same position and with about the same intensity as 8 ng of authentic rabbit (α+β) globin mRNA; no band was found with the poly(A)$^+$ RNA from 150 μg of RNA from cells transformed with thymidine kinase DNA only (Fig. 7A). As far as tested, the β-globin transcript is indistinguishable from the natural mRNA; however, the 3' terminus remains to be mapped.

To determine whether or not transformed L cells produce mouse β-globin mRNA, we again used the Weaver-Berk-Sharp assay, with the 5'-^{32}P-labeled 123-nucleotide Bsp-Bsp minus strand fragment

---

Lanes h to k, RNA from cell lines P1/R-3, P1/R-4, P1/R-5, and M2/R-1 (transformed with rabbit β-globin DNA linked to TK DNA). Lane 1, 50 ng mouse globin mRNA. Lanes M and M', 5'-^{32}P-labeled fragments of plasmid pBR322 (25) as size markers. (B) 50 μg of RNA from cell line M2/R-1 or 3 ng rabbit globin mRNA were passed through a 0.1 ml oligo(dT) cellulose column. Lanes a and b, poly(A)$^+$ and poly(A)$^-$ fractions of M2/R-1 RNA; lanes c and d, poly(A)$^+$ and poly(A)$^-$ fractions of rabbit globin mRNA. Lanes e and f, 20 and 40 μg unfractionated rabbit bone marrow RNA (a gift from R. Flavell).

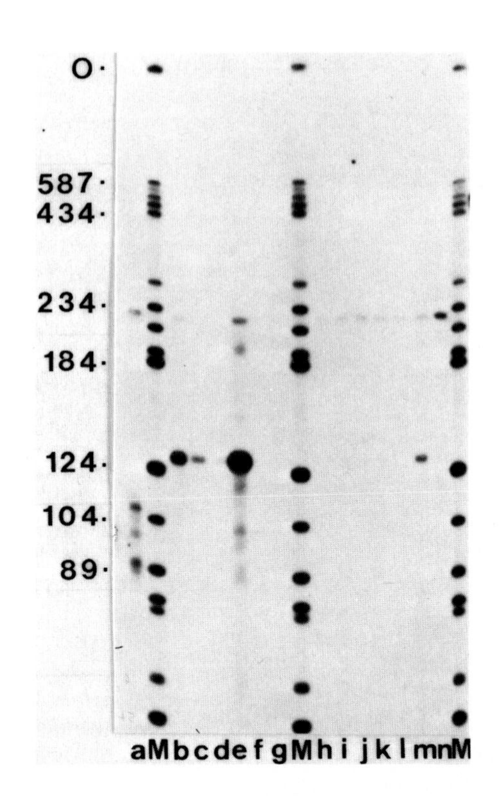
O
587
434
234
184
124
104
89
aMbcdefgMhijklmnM

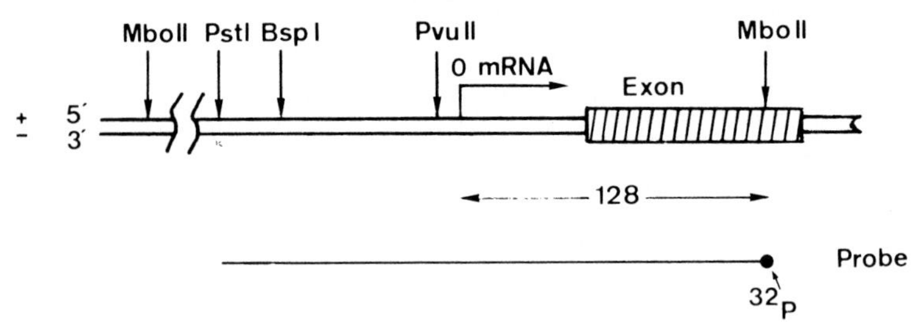
MboII
PstI
BspI
PvuII
MboII
0 mRNA
Exon
+ 5′
− 3′
128
Probe
32P

Fig. 5

of cloned mouse β-globin cDNA (17) as probe. 0.1 ng of mouse β-globin mRNA gave a strong band while 10 ng rabbit (α+β) globin mRNA gave no signal, because there is no sequence complementary to the ^{32}P-labeled terminus of the probe on the rabbit β-globin mRNA. 50 μg total RNA from L cell lines transformed with TK DNA only or with rabbit β-globin-TK DNA concatenates gave no detectable signal (data not shown). From this experiment we estimate that the transformed L cell lines we tested, in particular the one producing 2000 molecules rabbit β-globin mRNA (M2/R-1), contained less than 50 strands per cell of mouse β-globin mRNA.

Transformation of Yeast with Cloned Rabbit Chromosomal β-globin DNA. A hybrid DNA, pJDB219 consisting of pMB9, the 2 μ yeast plasmid and the leu-2 gene of yeast was constructed by Beggs and shown to be maintained predominantly in plasmid form in

---

Fig. 5. Mapping of the 5' terminus of rabbit β-globin-specific RNA from transformed mouse L cells and transformed yeast.

The assay was as described in Fig. 3. The prob probe for mapping the 5' end (cf. scheme) was prepared as follows. Plasmid Z-pCRI/RchrβG-1 was digested with HhaI, the 6.4 kb fragment isolated, cleaved with MboII, and 5'-terminally labeled with ^{32}P (specific activity, 3200 Ci/mmole).After cleavage with PstI, the 223 bp fragment was isolated by agarose gel electrophoresis. Lane a, RNA from yeast cells transformed with Z-pJDB219/ RchrβG-3. Lanes b to d, 3, 1, and 0.3 ng rabbit (α+β) globin mRNA. Lane e, 20 μg rabbit bone marrow RNA (a gift from R. Flavell). Lane f, no RNA. Lane g, 50 ng mouse globin mRNA. Lanes h and i, RNA from cell lines P1/S-1 and P1/S-2, transformed only with a SalI-cleaved TK-gene plasmid. Lanes j to m, RNA from cell lines P1/R-3, P1/R-4, P1/R-5, and M2/R-1, transformed with the rabbit β-globin DNA-containing hybrid linked to a TK-gene plasmid. Lane n, DNA probe not digested with S_1 nuclease. M, size marker: pBR322 digested with BspI and 5' terminally ^{32}P-labeled.

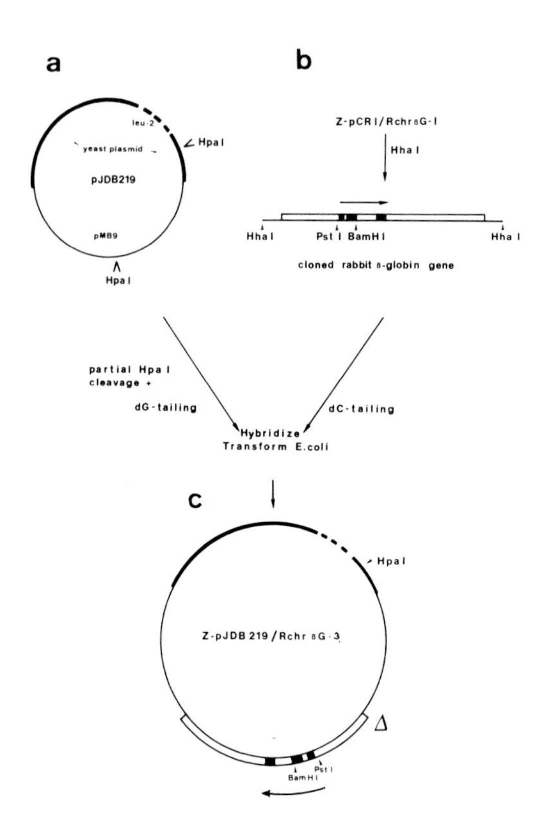

Fig. 6. Construction of hybrids containing rabbit chromosomal β-globin DNA, the 2 μ yeast plasmid, the yeast leu-2 gene and pMB9.

The hybrid plasmid pJDB219 (a) (consisting of pMB9, the 2 μ yeast plasmid and the leu-2 gene (14)) was partially digested with HpaI, the full-length linear molecules were isolated and elongated with dGMP residues. The chromosomal β-globin plasmid Z-pCRI/RchrβG-1 was cleaved with HhaI and the 6.4 kb fragment (b) elongated with dCMP residues. The elongated fragments were joined and transfected into E.coli X1776. Tetracycline-resistant clones were isolated and their plasmids characterized by restriction site analysis. Two plasmids, Z-pJDB219/RchrβG-1 and -3, containing the β-globin inserts in the orientation indicated (c) were used for further work. ZpJDB219/RchrβG-3 had a 1400 bp deletion at the position indicated with **Δ** .

yeast (14). The rabbit β-globin DNA sequence (with some pCRI-specific flanking sequences (cf. Fig. 6) was excised with HhaI and linked into the HpaI site of the pMB9 moiety of pJDB219 by the dG-dC tailing procedure (cf. refs. 15, 16). Two hybrid plasmids, Z-pJDB219/RchrβG-1 and Z-pJDB219/RchrβG-3 (μRβG-1 and μRβG-3 for short), containing the β-globin HhaI fragment in the orientation shown in Fig. 6 were used for the transformation experiments. μRβG-3 has a 1400 bp deletion at the junction with pMB9 (indicated by Δ in the figure).

S. cerevisiae MC16 leu⁻ was transformed with both plasmids and leucine autotrophic colonies were isolated. One clone from each transformation was examined in greater detail. Total DNA from the transformed yeast clones, as well as the original hybrid plasmid DNAs were digested with PstI and BglII and subjected to agarose gel electrophoresis. The DNA fragments were transferred to nitrocellulose filters (8) and hybridized to ^{32}P-labeled, cloned rabbit β-globin cDNA. The pattern of globin-specific bands given by the transformed yeast DNA was the same as that of the original β-globin DNA-containing plasmid Z-pCRI/Rchrβ-1 (data not shown).

Characterization of β-Globin-Specific Transcripts in Yeast. To determine whether globin-specific RNA had been synthesized in the transformed yeast cells, total RNA was subjected to agarose gel electrophoresis under denaturing conditions, transferred to diazobenzyloxymethyl paper (12), and hybridized with ^{32}P-labeled, cloned β-globin cDNA. As shown in Fig. 7B, a strong β-globin-specific band with a mobility corresponding to an 850-nucleotide RNA was found in yeast cells transformed with either μRβG-1 or μRβG-3. This RNA is about 100 nucleotides longer than mature rabbit β-globin mRNA, but 600 nucleotides shorter than the 15 S β-globin precursor RNA (about 1450 nucleotides). By comparing the intensity of the globin-specific RNA band with that of known amounts of β-globin mRNA, we estimate that about 0.01% of the total yeast RNA is β-globin-specific.

To map the 5' terminus of the transcript, the MboII-labeled probe described above (cf. Fig. 5) was used in the Weaver-Berk-Sharp assay. As shown in Fig. 5, natural rabbit β-globin mRNA protected a DNA fragment about 128 nucleotides in length;

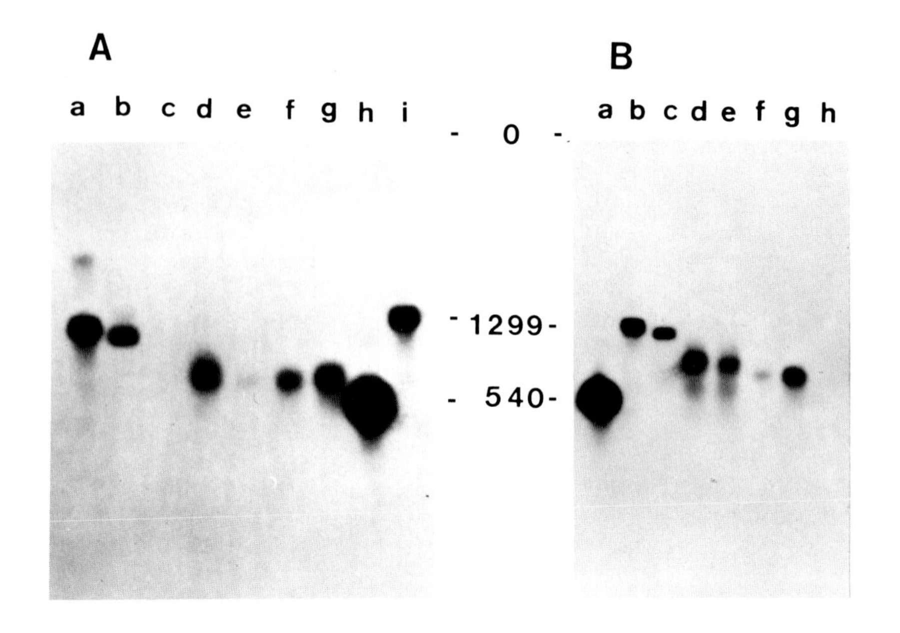

Fig. 7. The electrophoretic mobility of rabbit β-globin RNA from transformed mouse L cells and yeast cells.

Samples were treated with glyoxal in dimethylsulfoxide at 50°C for 60 min (26), subjected to electrophoresis through a 1.5% agarose gel, transferred to diazobenzyloxymethyl paper (12) and hybridized with a ^{32}P-labeled rabbit β-globin cDNA probe (1.6 x 10^8 cpm/μg). Denatured restriction fragments of rabbit β-globin-specific DNA and β-globin mRNA were used as size markers. A. (a) RβG DNA digested with PstI and BglII (1299 bp). (b) RβG DNA digested with PvuII and BglII (1209 bp). (c) Poly(A)$^+$ RNA from 150 μg of P1/S-1 cells RNA (transformed with TK DNA only). (d) Poly(A)$^+$ RNA from 150 μg of M2/R-1 cell RNA (transformed with rabbit β-globin DNA and TK DNA). (e-g) Rabbit (α+β) globin mRNA 1, 4 and 8 ng, respectively (of which about 30% are hybridizable to β-globin DNA). (h) Cloned rabbit β-globin cDNA (13); ca. 540 bp). (i) RβG cleaved with PstI and BglII (1299 bp). All samples contained 20 μg yeast RNA as carrier.

the DNA protected by RNA from β-globin-transformed yeast cells gave five major bands, corresponding to 88, 91, 93, 102 and 110 nucleotides. The β-globin-specific transcripts from yeast apparently have several 5' termini, all of which map downstream from the position corresponding to the 5' terminus of natural β-globin mRNA. The band corresponding to 223 nucleotides is most likely due to self-annealing of the DNA probe (the MboII site has an overhanging 3' end which protects the labeled 5' terminus) since it occurs also in control reactions without RNA.

To determine whether the small intron was present or absent from the β-globin transcripts, a hybridization experiment similar to the one above was carried out. The probe was a Bam-Sal fragment of chromosomal β-globin DNA, labeled at the 5' (minus strand) terminus of the Bam cleavage site. As shown in the scheme of Fig. 8, an RNA molecule lacking the small intron would protect the probe from the 5'-^{32}P-labeled Bam site to the beginning of the small intron, yielding a fragment of 212 nucleotides. If the small intron were present, then the probe would be protected up to the region corresponding to the 5' ends of the transcripts, yielding fragments between 440 and 462 nucleotides. The autoradiogram of Fig. 8 shows the presence of fragments of about 440 nucleotides length and no signal at the position corresponding to 212 nucleotides, proving that the small intron was present in the β-globin-specific RNA.

B. (a) Same as A(h). (b) Same as A(a). (c) Same as A(b). (d and e) RNA from S. cerevisiae Mc16 transformed with Z-pJDB219/RβG-1 (d) or with Z-pJDB219/RβG-3 (e). (f and g) Rabbit (α+β) globin mRNA 2 ng (f) and 10 ng (g). (h) RNA from S. cerevisiae Mc16 transformed with pJDB219.

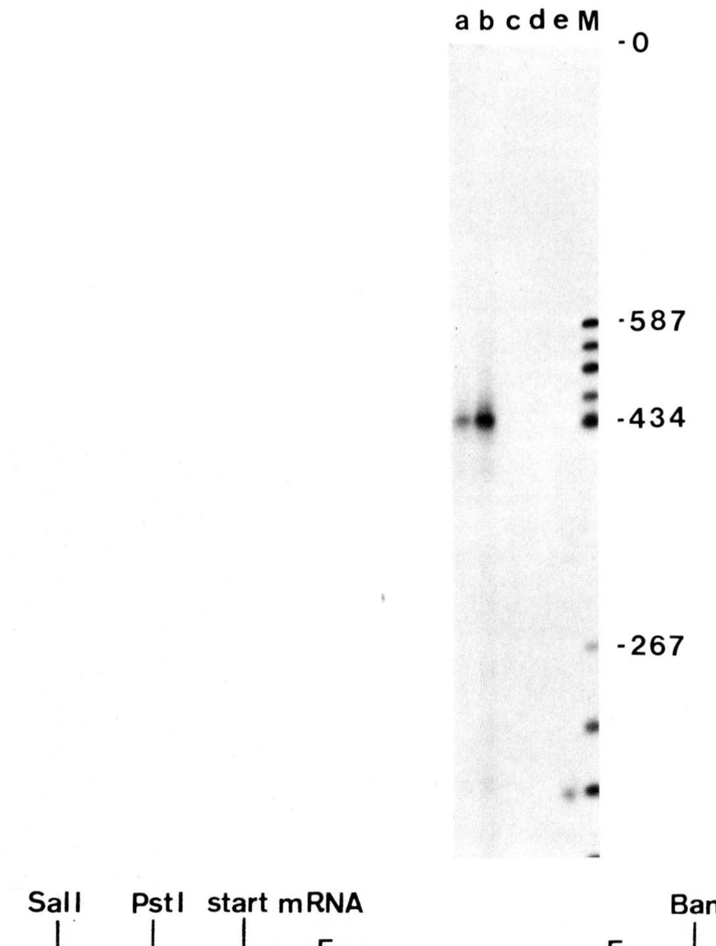
a b c d e M
-0
-587
-434
-267

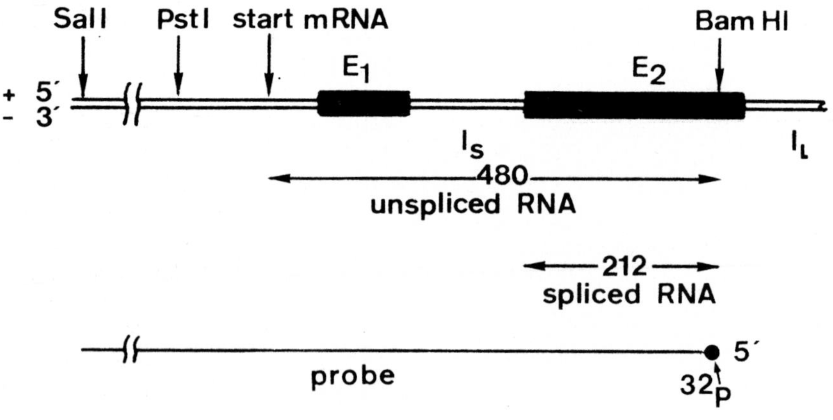
SalI
PstI
start mRNA
Bam HI
E_1
E_2
+ 5′
- 3′
I_S
I_L
480
unspliced RNA
212
spliced RNA
probe
5′
^{32}P

Fig. 8

Finally, the 3' termini of the transcripts were mapped using a Bam-Sal 14'000 bp fragment labeled at the 3' (minus strand) terminus of the Bam site. Using this probe, hybridization with RNA from transformed yeast yielded 4 major bands of 270, 295, 340 and 350 nucleotides (Fig. 9). Since the distance from the Bam site to the 5' border of the large intron is 18 nucleotides, many or all of the β-globin transcripts contain part of the large intron, and most likely terminate within its middle third. This experiment by itself does not exclude that some β-globin RNA lacking the large intron was present, since a 18 nucleotide fragment could not have been detected under the conditions of the experiment. However, since the Northern transfer had not revealed much RNA smaller than 850 nucleotides, the proportion of mature β-globin mRNA could not have been significant.

DISCUSSION

Most mouse L cell lines transformed with rabbit β-globin DNA produced rabbit β-globin-specific RNA. Cell line (M2/R-1), which contained the highest number of gene copies, about 20, also contained the highest number of RNA molecules, namely about 2200 per cell; however, this correlation was not general.

The rabbit β-globin-specific RNA of the line M2/R-1 was indistinguishable from natural mRNA in regard to its electrophoretic mobility, its 5'

Fig. 8. Detection of the small intron in β-globin-specific RNA from transformed yeast cells.

The assay was carried out by the Weaver-Berk-Sharp method explained in Fig. 3; the probe was a BamHI-SalI fragment of Z-pCRI/RchrβG-1 5'-terminally ^{32}P-labeled at the BamHI cleavage site (cf. scheme in this figure). M, marker (BspI-cleaved pBR322, 5'-^{32}P-labeled). (a) to (c), RNA from S. cerevisiae Mc16 transformed with plasmid Z-pJDB219/RchrβG-3 (a), Z-pJDB219/RchrβG-1 (b) and pJDB219 (c). (d) and (e), rabbit (α+β) globin mRNA, 1 and 4 ng, respectively.

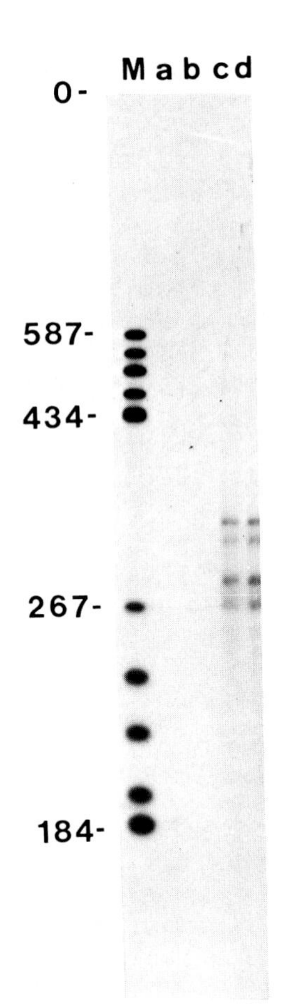
M a b c d
0-
587-
434-
267-
184-

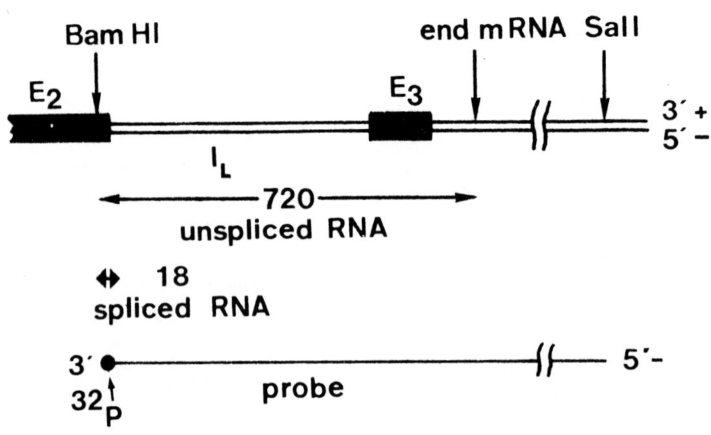
Bam HI
end mRNA
SalI
E2
E3
3′ +
5′ −
IL
720
unspliced RNA
18
spliced RNA
3′
5′-
32P
probe

Fig. 9

terminus, the absence of the large and the small (18) intron and the presence of a 3' terminal poly(A) tail. The β-globin RNAs of the other cell lines, when tested for one or the other criteria, had similar properties.

In the case of the β-globin DNA-transformed yeast cells, substantial quantities of β-globin-specific RNAs were also produced, however they differed considerably from natural β-globin mRNA: they were shorter by 18-40 nucleotides at the 5' end, they contained the small intron part of the large intron, and lacked the normal 3' terminus.

Ziff and Evans (19) have shown that in the case of the adenovirus late mRNA the cap-carrying 5' terminus represents the initiation site of transcription. In the case of mouse β-globin mRNA the 15 S precursor and the mature mRNA have the same 5' terminal sequence and the same cap structures (10, 17). If, as we believe, the 15 S β-globin precursor is the primary transcript (but for a contrary view, see Bastos & Aviv (20)) then we may conclude that the rabbit β-globin RNA produced in transformed mouse L cells originates at the true promotor. We cannot yet, however, exclude the possibility

---

Fig. 9. Mapping of the 3' terminus of β-globin-specific RNA from transformed yeast cells.

The assay was carried out by the Weaver-Berk-Sharp assay as explained in Fig. 3, however the probe was labeled at the 3' terminus of the Bam-cleavage site. The probe was prepared by digesting Z-pCRI/RchrβG-1 with BamHI, filling in the cohesive ends with dTTP, dGTP and α-^{32}P-labeled dATP and dCTP using DNA polymerase I, cleaving with SalI and isolating the larger fragment (cf. map of plasmid in ref. 2). In order to reduce the background resulting from self-annealing of the probe, the fragment was digested with 5' exonuclease prior to denaturation. M, marker (pBR322 cleaved with BspI and 5' terminally ^{32}P-labeled). (a) and (b), rabbit (α+β) globin mRNA 4 ng and 1 ng, respectively. (c) and (d), RNA from S. cerevisiae Mc16 transformed with Z-pJDB219/RchrβG-1 and Z-pJDB219/RchrβG-3, respectively. The rabbit β-globin RNA should give rise to a 18-nucleotide band which could not be detected on this gel.

that the transcript is initiated upstream at a plasmid or a host DNA promotor, and that the 5' terminus arises by processing. In the β-globin DNA-transformed yeast cells transcription either starts downstream from the correct position or a yeast-specified processing enzyme(s) cleaves a longer transcript.

Why do transformed mouse L cells synthesize rabbit but not or very little mouse β-globin mRNA? The simplest explanation is that transcription starts at an unphysiological promotor; if this promotor were located in the host DNA and integration of the globin DNA could occur at various different sites, the varying degrees of expression in different lines of transformed cells could be accounted for. If however the physiological promotor is being utilized, then the mechanism that prevents the transcription of the mouse gene may not be active on the rabbit gene, for example because of sequence differences in the control regions, or because of a position effect if control is cis-directed. The experimental results clearly argue against a simple dosage effect, since several transformed cell lines containing only one or few copies of the rabbit β-globin gene produce detectable amounts of rabbit but not of mouse mRNA.

As far as splicing is concerned, there is little doubt that the introns are accurately excised from the rabbit transcript in mouse L cells. This is not surprising in view of the similarity of the intron-exon junction regions in the rabbit and mouse β-globin gene (2) and of the finding that splicing appears to occur normally in very heterologous systems, such as SV40 transcripts in Xenopus oocytes (21). The fact that β-globin transcripts are not (or incorrectly) spliced in yeast could have one or more causes. (1) Yeast may not contain the appropriate splicing enzymes for dealing with the β-globin transcript, or the enzymes if existent, may not be in the same cellular compartment as the hybrid DNA. (2) The splicing enzymes may be present, but incapable of processing the incomplete transcript formed in the yeast cell, perhaps because of specific secondary or tertiary structure requirements.

Berg and his colleagues (22) have replaced a segment of the SV40 late region by a cloned cDNA copy of rabbit β-globin mRNA. Monkey cells infected by such hybrid DNA underwent a lytic cycle, during the course of which the DNA was greatly amplified. Hybrid mRNA molecules containing the β-globin sequence were transcribed from an SV40 promotor, leading to the accumulation of about 10^4 hybrid mRNA molecules per cell (P. Berg, personal communication); β-globin was efficiently and accurately translated. We have not yet analyzed the proteins produced in our transformed mouse cell lines, but it is likely that β-globin synthesis, if it occurs, will be substantially lower than in the lytic SV40 hybrid system. At this stage it would seem that thymidine kinase-linked transformation of mouse L cells will be useful for the study of transcription initiation and splicing, while the SV40 hybrid system will be advantageous for the investigation of translation.

ACKNOWLEDGMENTS

This work was supported by the Schweizerische Nationalfonds (No. 3.114.77) and the Kanton of Zürich. A.v.O. was supported by grants from EMBO and the Netherlands Organization for the Advancement of Pure Research. J.v.d.B. received grants from EMBO and Koningin Wilhelmina Fonds and J.D.B. had a short-term EMBO Fellowship.

REFERENCES

1. Weissmann, C. (1978). TIBS 3, N109-N111.
2. van den Berg, J., van Ooyen, A., Mantei, N., Schamböck, A., Grosveld, G., Flavell, R.A., and Weissmann, C. (1978). Nature 275, 37-44.
3. van Ooyen, A., van den Berg, J., Mantei, N., and Weissmann, C. (1979) submitted for publication.
4. Konkel, D.A., Tilghman, S.M., and Leder, P. (1978). Cell 15, 1125-1132.
5. Boll, W., Mantei, N., Wilkie, N., Clements, B., Greenaway, P., and Weissmann, C. (1979) Experientia, in press.
6. Weissmann, C., Mantei, N., Boll, W., Weaver, R.F., Wilkie, N., Clements, B., Taniguchi, T., van Ooyen, A., van den Berg, J., Fried, M., and Murray, K. (1979) 11th Miami Winter Symp., in press.

7. Wigler, M., Silverstein, S., Lee, L.-S., Pellicer, A., Cheng, Y.-C., and Axel, R. (1977). Cell 11, 223-232.
8. Southern, E.M. (1975). J. Mol. Biol. 98, 503-517.
9. Berk, A.J., and Sharp, P.A. (1977). Cell 12, 721-732.
10. Weaver, R., Boll, W., and Weissmann, C. (1979). Experientia, in press.
11. Efstratiadis, A., Kafatos, F.C., and Maniatis, T. (1977). Cell 10, 571-585.
12. Alwine, J.C., Kemp, D.H., and Stark, G.R. (1977). Proc. Natl. Acad. Sci. USA 74, 5350-5354.
13. Meyer, F., Heijneker, H., Weber, H., and Weissmann, C. (1979). Experientia, in press.
14. Beggs, J.D. (1978). Nature 275, 104-109.
15. Lobban, P.E., and Kaiser, A.D. (1973). J. Mol. Biol. 78, 453-471.
16. Jackson, D.A., Symons, R.H., and Berg, P. (1972). Proc. Natl. Acad. Sci. USA 69, 2904-2909.
17. Curtis, P.J., Mantei, N., and Weissmann, C. (1977). Cold Spring Harbor Symp. Quant. Biol. 42, 971-984.
18. Mantei, N., Boll, W., and Weissmann, C. (1979). submitted for publication.
19. Ziff, E.B., and Evans, R.M. (1978). Cell 15, 1463-1475.
20. Bastos, R.N., and Aviv, H. (1977). Cell 11, 641-650.
21. De Robertis, E.M., and Mertz, J.E. (1977). Cell 12, 175-182.
22. Mulligan, R.C., Howard, B.H., and Berg, P. (1979). Nature 277, 108-114.
23. Breathnach, R., Benoist, C., O'Hare, K., Gannon, F., and Chambon, P. (1978). Proc. Natl. Acad. Sci. USA 75, 4853-4857.
24. Curtis, P.J., and Weissmann, C. (1976). J. Mol. Biol. 106, 1061-1075.
25. Sutcliffe, J.G. (1978). Nucl. Acids Res. 5, 2721-2728.
26. McMaster, G.K., and Carmichael, G.G. (1977). Proc. Natl. Acad. Sci. USA 74, 4835-4838.

Structure and Localization of Dihydrofolate Reductase Genes in Methotrexate-Resistant Cultured Cells

Robert T. Schimke, Peter C. Brown, Randal J. Kaufman and Jack H. Nunberg

Department of Biological Sciences, Stanford University, Stanford, CA 94305

ABSTRACT

Methotrexate resistance in a number of cultured mammalian cell lines obtained by step-wise selection in increasing concentrations of methotrexate is associated with increased dihydrofolate reductase levels and results from a selective multiplication of DNA sequences coding for dihydrofolate reductase. Based on restriction analyses employing cloned DNA synthesized from mRNA, the dihydrofolate reductase sequences are spread over at least 40 kilobases in the genome with 5 intervening sequences.

High dihydrofolate reductase gene copy number occurs in both a stable and an unstable state in different resistant cell variants. In various stably amplified cell lines the dihydrofolate reductase genes are localized to an expanded region of a specific chromosome as judged by *in situ* hybridization. In all unstably amplified cell lines studied, metaphase chromosomes have small, paired chromosomal elements denoted "double minute chromosomes," which we have tentatively identified as containing the amplified dihydrofolate reductase genes.

INTRODUCTION

We have reported previously that in a number of methotrexate (MTX)-resistant cultured cell lines of hamster and mouse origin, high levels of dihydrofolate reductase (DHFR) are associated with a corresponding amplification of DHFR genes (1-4). Such cell lines are characteristically obtained by step-wise selection in increasing concentrations of MTX in the medium, and we have proposed that amplifications of genes occur randomly in cells and are detected only when selection pressures are imposed (5). In some cell variants MTX resistance and amplified genes are stable in the absence of selection pressure, whereas in others, these parameters are unstable (3). In this report we describe our current studies on genome organization and localization of the amplified genes in stable and unstable cell variants.

ISBN 0-12-068350-4

RESULTS

Genomic Representation of Dihydrofolate Reductase mRNA Sequences

In collaboration with Stanley Cohen and A.C.Y. Chang we have cloned DHFR mRNA-derived DNA sequences that are expressed in E. coli strain X2282, a thy+ derivative of X1776 (6). As depicted in Fig. 1 the mRNA-derived sequence contains approximately 950 non-coding nucleotides at the 3' end, and at least 80 non-coding nucleotides at the 5' end, as well as the 558 coding nucleotides. We have employed this recombinant DNA sequence in restriction map analyses of the genomic representation of this sequence. Fig. 1 shows our tentative conclusions. The entire 3' non-coding sequences, as well as approximately 70 3'-ended coding nucleotides are present on a colinear genomic DNA sequence. There are 4 intervening sequences in the coding region and one intervening sequence in the 5' non-coding sequence. The genomic representation of the mRNA sequence is spread over at least 40 kb. The same restriction patterns are obtained with all mouse cell lines with amplified genes (L1210, L5178Y, 3T6, S-180), as well as with sensitive cell lines and mouse liver DNA (7). Thus our findings of the DHFR gene organization are not related to the amplification process, but occur in normal mouse DNA as well.

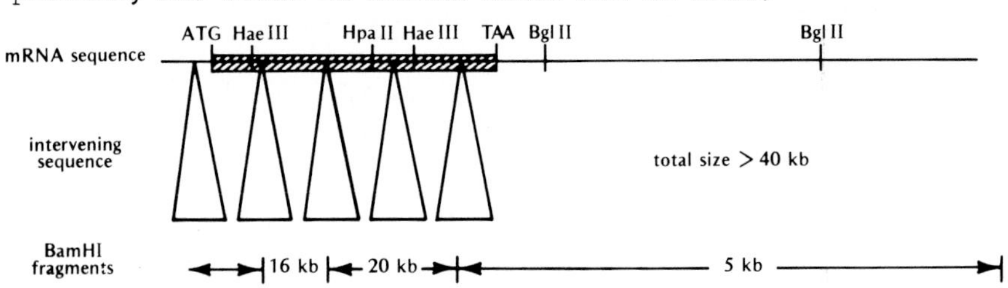

Figure 1 Summary of DHFR Sequences in mRNA and Mouse Genome

Stable and Unstable Amplification of Dihydrofolate Reductase Genes

Fig. 2 shows the results of growth of two S-180 cell lines originally derived from the same clone of resistant AT-3000 cells in the absence of MTX. In one cell line high resistance, rate of DHFR synthesis, and DHFR gene copy number decline rapidly, such that after 20 cell doublings the population of cells has lost approximately 60% of the DHFR genes. In another clone, grown continuously for two years in high MTX concentrations, high resistance and gene copy number declines over 40 cell doublings and then the

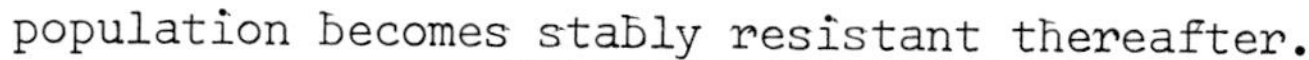
population becomes stably resistant thereafter.

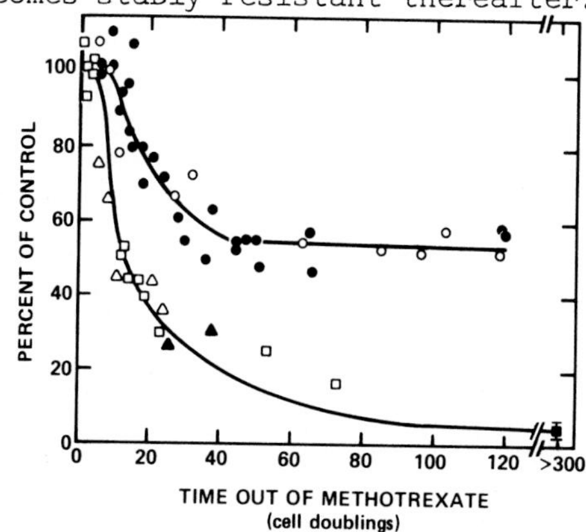

Figure 2 Loss of Elevated DHFR upon Growth in the absence of MTX. The murine S-180 MTX-resistant clone R_1 has been propagated for increasing numbers of cell doublings in the absence of MTX and assayed for either DHFR specific activity (closed symbols) or relative rates of DHFR synthesis (open symbols) (1). Squares are the cells studied 2 years ago. Triangles are the same cells that have been frozen in the interim and studied in the last 6 months. Circles are R_1 cells grown for more than 2 years continually in the presence of MTX prior to removal from MTX.

In order to analyze these two cell populations in more detail, we have made use of the Fluorescence-Activated Cell Sorter (FACS) and the ability of a fluorescein-derivative of MTX to bind quantitatively to DHFR enzyme to determine the behavior of the cell population with respect to declining enzyme content and gene copy number (8). Fig. 3 (right panel) shows the distribution of DHFR in cells from an unstably resistant population (Fig. 2) 3 cell doublings following removal of MTX from the medium. This population has been sorted to obtain three sub-populations with different fluorescence/cell (FU/cell) values (left panel). That this population is heterogeneous with respect to resistance, enzyme synthesis and gene copy number is shown in Fig. 3 (table). More extensive studies of a number of cell populations sorted for FU/cell show that there is a direct relationship between FU/cell and DHFR synthesis and gene copy number (9). Hence we conclude that in our sorting studies FU/cell is a measure of gene copy number.

Fig. 4 shows the change in distribution of unstably resistant cells sorted for different gene copy numbers after 2, 7, and 25 cell doublings in the absence of MTX.

SORTING OF UNSTABLE RESISTANT CELLS WITH THE FACS

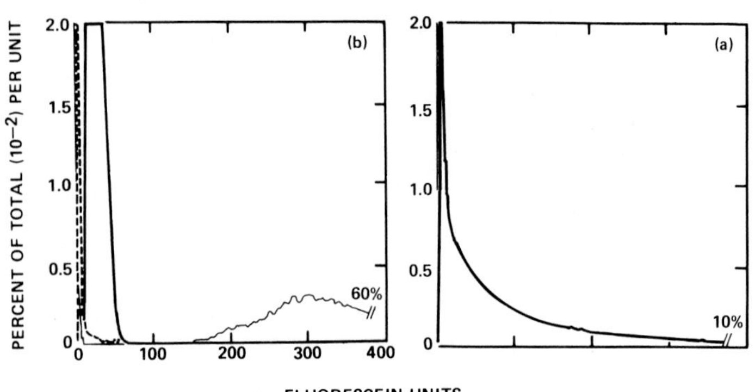

Cells	MF/Cell	Specific Activity	Gene Copy Number
S_3	1	1	1
R_2	105	50	44
R_2^{++}	300	160	-
R_2^{+}	33	23	20
R_2^{-}	38	3	3

Figure 3 Heterogeneity in DHFR Gene Copy Number and MTX Resistance in an Unstably Resistant Cell Line.

Cellular DHFR in S-180 unstably MTX resistant cloned line R_2(9) has been saturated with MTX-F and analyzed for fluorescence emission with the FACS(8). Panel B shows the original R_2 population as a fluorescence histogram with 5% of cells having greater than 400 fluorescence units. One fluorescence unit equals 2×10^6 DHFR molecules (8). Three subpopulations were sorted sterily. Panel A shows these populations immediately after the sort. DHFR specific activities and MTX resistances (the MTX concentration which reduces growth by 50% after 5 days) were determined after subculturing cells 4 cell doublings in the absence of MTX after the sort. DHFR gene copy number was determined by Cot analysis after growing the cells 12 cell doublings in the absence of MTX. The results are shown in the table (9).

Cells with an initial high gene copy number progressively lose genes, and after 25 cell doublings have become a skewed distribution still spreading out to high FU/cell. More instructive is the pattern of cells sorted for intermediate gene copy number. Within a few cell doublings the distribution widens, with cells not only losing genes, but also some gaining genes. After 20 cell doublings the distribution of cells sorted for both high and intermediate gene copy number are similar. We conclude that gene loss in the population occurs gradually, and that there appears to be an extremely rapid randomization of genes in daughter cells, as evidenced by the rapid widening of distribution of FU/cell in the cells sorted for intermediate gene copy number.

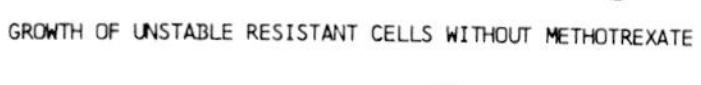

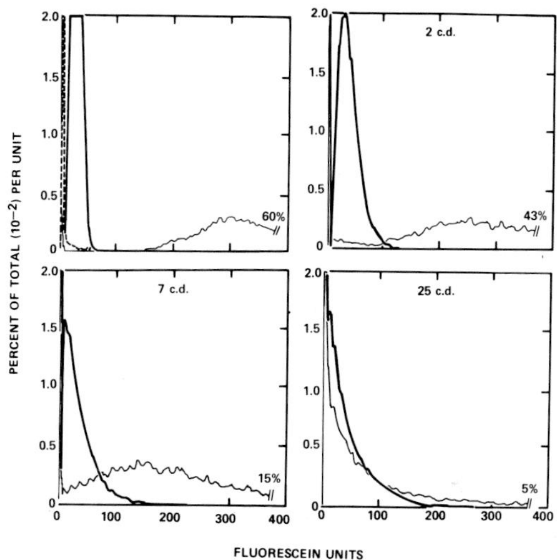

Figure 4 Expansion and Contraction of DHFR Levels in Individual Cells.

Two subpopulations obtained in Fig. 3 have been grown in the absence of MTX and cells analyzed for fluorescence after saturation with MTX-F. Shown here are fluorescence histograms after 2 (panel A), 7 (panel B), and 25 (panel C) cell doublings. The heavy line indicates cells sorted for intermediate DHFR levels and the light line indicates cells sorted for high DHFR levels (9).

Fig. 5 (upper panel) shows the original distribution of the partially stable cell line (Fig. 2). When this line is subsequently grown 80 cell doublings in the absence of MTX, there emerge two non-overlapping populations with approximately 10 and 50 gene copies per cell. These two populations have been stably resistant for over 2 years in the absence of MTX. We have also cloned individual cells having more than 200 fluorescence units per cell from the original partially stable cell line and find that each cell progressively loses genes, and that some clones become only the high gene copy distribution, whereas in other clones, both high and low amplified populations emerge. Hence each cell has some unstably amplified genes, and, in addition, can revert to one or both of the stably amplified populations (9).

GROWTH OF PARTIALY STABLE RESISTANT CELLS WITHOUT METHOTREXATE

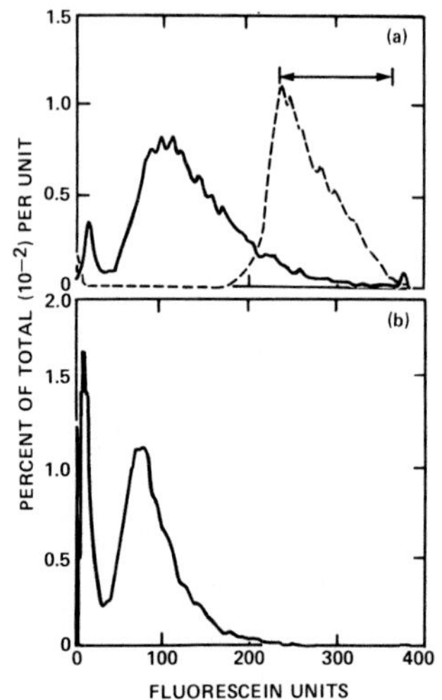

Figure 5 Stabilization of Elevated DHFR Levels.

The S-180 stably MTX-resistant R_1 clone depicted in Fig. 2 has been analyzed with the FACS for DHFR content 3 cell doublings (panel A) and 80 cell doublings (panelB) after growth in the absence of MTX.

Localization of Amplified Dihydrofolate Reductase Genes

In order to understand the behavior of the stably and unstably amplified cell populations we have undertaken a series of studies to localize the DHFR genes in both cell populations. We have examined three different cell lines in which sensitive, unstable and stable resistant variants are avaliable: S-180 (see below), murine L5178Y, and hamster CHO. In collaboration with Laurence Chasin we have shown that in a stably resistant CHO cell line there is a characteristic expansion of the identifiable chromosome 2 that is not present in the sensitive cells (Fig. 6) (11). *In situ* hybridization of cDNA from murine mRNA shows that the 150 amplified genes are localized to this so-called "homogeneously staining region" (HSR) (11). In collaboration with Joseph Bertino we have also examined a L5178Y cell line with

300 stably amplified DHFR genes (12). This cell line also contains a HSR, which by in situ hybridization, contains (at least the majority of) the DHFR genes (Fig. 6). As will be shown below, in the stably amplified S-180 cell line with 50 DHFR genes, the genes are also localized on a large chromosome. Hence we conclude from this limited number of three stably amplified cell lines that the genes are chromosomal, and most likely present in a tandemly repeated organization. We have also made calculations of the probable length of the duplicated DNA sequence in the CHO and L5178Y cell lines and come to the conclusion that it is of the order of 500 to 1000 kb in both cell lines (11, 12). Thus the amplified sequence is clearly much larger than that required only for the coding of DHFR, i.e. some 40 kb.

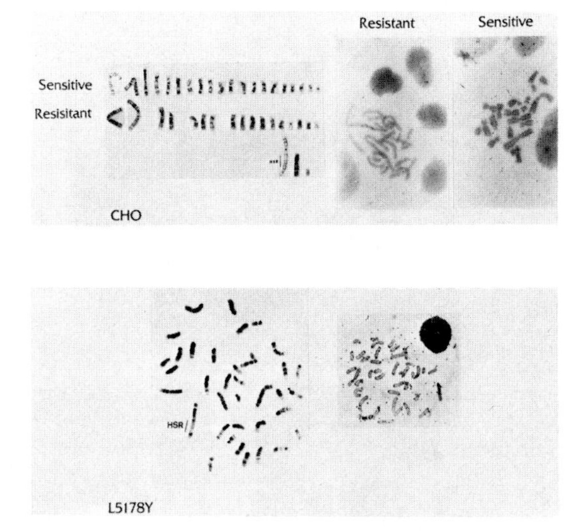

Figure 6 In situ Hybridization to Localize DHFR Genes

^{3}H cDNA to mouse DHFR mRNA has been hybridized to chromosomes from stable CHO and L5178Y MTX resistant cells. Top left shows Giemsa-trypsin banded chromosomes from CHO. The sensitive CHO karyotype is presented on the top and the resistant below. Three chromosomes which are altered are shown below, including the HSR and two chromosomes with a translocation. The karyotype of the resistant L5178Y is shown in the left panel with the identifiable HSR. Results from in situ hybridizations are depicted on the right for the sensitive and resistant CHO (top panel) and for the resistant L5178Y (lower panel).

In the three cell lines with unstably amplified genes we observe small, paired chromosomal elements called "double minute chromosomes" (DMs) (13). Two examples of these are shown in Fig. 7. The top left is a normal metaphase spread of unstably amplified S-180 cells, paired on the right with a sensitive S-180 cell. Note the large number of DMs (these cells were sorted for high gene copy number), as well as the fact that they do not participate in the normal spindle formation process. In the lower part of Fig. 7 (left) is a colcemid-arrested unstably resistant L5178Y chromosome pattern stained with Ethidium Bromide, as well as the parental sensitive cell (right). We have found DMs in all unstably amplified cell lines. In data not shown we have made a correlation between the number of double minute chromosomes and the gene copy number in cells sorted for differing FU/cell.

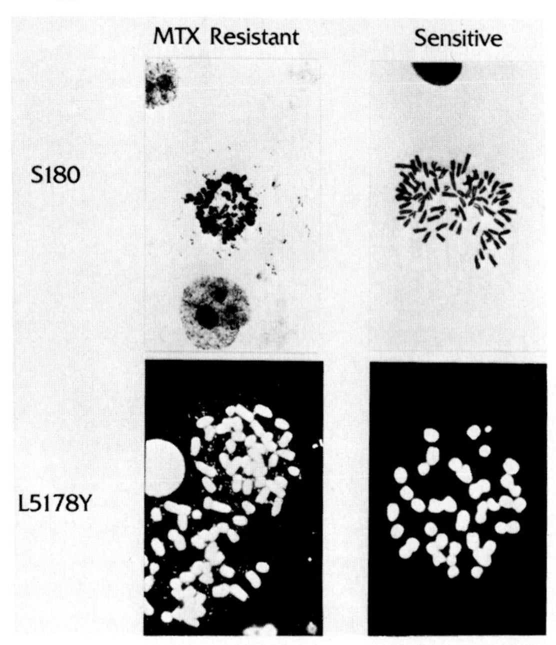

Figure 7 Unstably MTX-Resistant Cells Contain Double Minute Chromosomes

Metaphase spreads have been prepared from logarithmic phase S-180 unstably MTX-resistant (R_2) and sensitive cell lines and from 20 min colcemid arrested MTX-resistant and sensitive L5178Y cells. The S-180 metaphases are stained with Giemsa and photographed with visible light; the L5178Y are stained with ethidium bromide and photographed with fluorescence microscopy.

To further identify DMs as carriers of the DHFR genes we have separated chromosomes on the basis of size by sedimentation through sucrose gradients. Fig. 8 shows the results with stably and unstably resistant S-180 cells. The localization of the DHFR sequences was obtained by EcoRl digestion of a constant percentage of DNA from the fractions and subsequent Southern blot analysis using the cloned DHFR cDNA as probe. The distribution of chromosomes, indicated by DNA content, was similar in stably and unstably resistant cells. In the stably resistant cells, the DHFR genes are present on a chromosome that is larger than the bulk of mouse chromosomes. In contrast, in the unstably resistant cell line, the DHFR genes are present in a fraction highly enriched for DMs as determined by light microscopy. We conclude that in unstably resistant cell lines the DHFR genes are not, for the most part, integrated into chromosomes, and most likely constitute part of the sequences present in the DMs.

DISCUSSION

The finding that step-wise selection for increasing MTX resistance results from gene amplification suggests that this process occurs with a higher frequency than other possible mechanisms for increased mRNA content, including promoter mutations or alterations in mRNA stability. It is also interesting to note that there is a good correspondence between relative enzyme levels and gene copy number in various cell lines (4). However, expression of the amplified genes can vary. We reported that in resistant S-180 cells the rate of enzyme synthesis is high in early logarithmic growth and low in stationary cells (1). More recently we have reported that infection of MTX resistant 3T6 cells with polyoma results in a 5 fold increase in DHFR mRNA content within 24-48 hrs and a comparable increase in DHFR synthesis (14).

Our findings that stably amplified DHFR genes are integrated in a single site of a chromosome, and that unstably amplified genes are extrachromosomal, i.e. tentatively present as double minute chromosomes, assists in our understanding of the phenomenon of instability (see ref 13, 15, 16) for discussions of HSRs and DMs. The rapid segregation of DHFR genes in the unstable state between daughter cells at mitosis (Fig. 3) can be understood on the hypothesis that the DMs do not participate in spindle formation and will, hence, segregate independently and randomly. This hypothesis is supported by the finding (Fig. 7) as pointed out previously. There are several explanations for the loss of DHFR genes from the population. These include: altered replication rates of DMs, extrusion of DMs from cells, and preferential growth

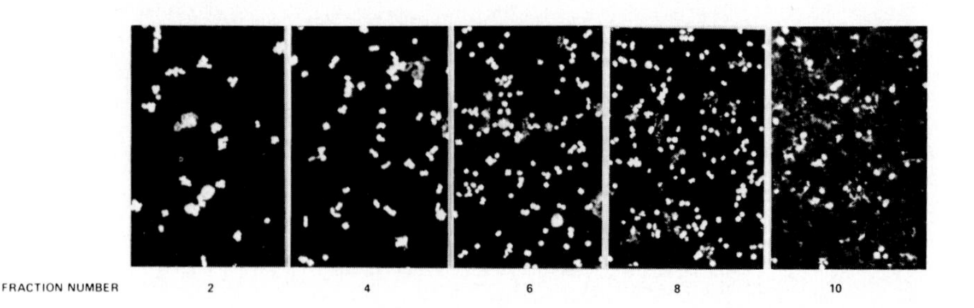

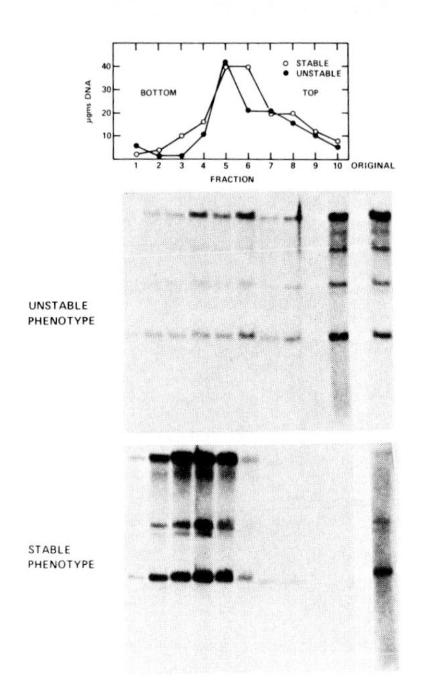

Figure 8 Dihydrofolate Reductase Genes Co-sediment with Double Minute Chromosomes in Unstably Resistant Cells

Chromosomes fro 5×10^7 to 5×10^8 unstably MTX- resistant (R_2) and stably resistant (R_1) mitotic cells were separated on a 20-50% linear sucrose gradient (10). Panel A shows representative photographs taken from fractions 2,4,6, and 10. Panel B indicates the amount of DNA isolated in each fraction (stably resistant————), (unstably resistant --------). The results of Southern blot analysis of EcoR1 restriction patterns is shown in Panel C for the unstably resistant R_2 and in Panel D for the stably resistant R_1 cells. Each lane contains an equal percentage of DNA from each fraction. Also indicated at the far right are Southern blot hybridizations of DNA isolated from the total chromosome preparation.

of cells with low DHFR gene copy number (10). The extent to which these mechanisms explain DHFR gene loss remains for further study.

The mechanism for the amplification process is not understood at present. A number of possibilities exist. Initial duplication could occur by unequal crossing over among sister chromatids, resulting in chromosomal localization of the repeated DNA sequences. Chromosomal localization of repeated sequences could also arise by some form of saltatory replication (17). Extrachromosomal DNA sequences could occur by an excision-replication process analogous to rDNA in *Xenopus* oocytes (18), or as proposed by Botchan (19). Extrachromosomal sequences could also arise by uptake of a DNA segment from lysed cells containing a replication origin and the DHFR gene sequences (20). Potentially, such sequences could integrate into chromosomes to generate the chromosomal HSR regions. Any or all of the different mechanisms could be occuring in individual cells or in different cell lines.

ACKNOWLEDGEMENTS

This work was supported by research grants from the National Institutes of Health (CA 16318) and the American Cancer Society (NP148).

REFERENCES

1. Alt, F., Kellems, R., and Schimke, R.T. (1976). The Journal of Biol. Chem. 251:3063.
2. Kellems, R., Alt, F., and Schimke, R.T. (1976). The Journal of Biol. Chem. 251:6987.
3. Alt, F.W., Kellems, R., Bertino, J.R., and Schimke, R.T. (1978) The Journal of Biol. Chem. 253: 1357.
4. Schimke, R.T., Kaufman, R.J., Nunberg, J.H., and Dana, S.L. (1978). Cold Spring Harbor Symp. Quant. Biol. 43: (in press).
5. Schimke, R.T., Kaufman, R.J., Alt, F.W., and Kellems, R.E. (1978) Science 202: 1051.
6. Chang, A.C.Y., Nunberg, J.H., Kaufman, R.J., Erlich, H. A., Schimke, R.T., and Cohen, S.N. (1978). Nature 275: 617.
7. Nunberg, J.H., Kaufman, R.J., Chang, A.C.Y., Cohen, S.N., and Schimke, R.T. (1979) to be submitted.
8. Kaufman, R.J., Bertino, J.R., and Schimke, R.T. (1978) The Journal of Biol. Chem. 253: 8952.
9. Kaufman, R.J., and Schimke, R.T. (1979) to be submitted.

10. Brown, P., Kaufman, R.J., and Schimke, R.T. (1979) to be submitted.
11. Nunberg, J.N., Kaufman, R.J., Schimke, R.T., Urlaub, G., and Chasin, L.A. (1978) Proc. Natl. Acad. Sci. 75: 5553.
12. Dolnick, B.J., Berenson, R.J., Bertino, R.J., Kaufman, R.J., Nunberg, J.H., and Schimke, R.T. (1979) to be submitted.
13. Balaban-Mazenbaum, G., and Gilbert, F. (1977) Science 198: 739
14. Kellems, R.E., Morhenn, V.B., Pfendt, E.A., Alt, F.W., and Schimke, R.T. (1979) The Journal of Biol. Chem. 254: 309.
15. Barker, P.C., and Hsu, T.C. (1978). Exp. Cell Res. 113:457.
16. Beidler, J.L., and Spengler, B.A. (1976) Science 191:185.
17. Britten, R.J., and Kohne, D.E. (1967) Science 161:529.
18. Hourcade, D., Dressler, D., and Wolfson, J. (1974) Cold Spring Harbor Symp. Quant. Biol. 38:53.
19. Botchan, M., Topp, W., and Sambrook, J. (1978) Cold Spring Harbor Symp. Quant. Biol. 43: in press.
20. Pellicer, A., Wigler, M., Axel, R., and Silverstein, S. (1978) Cell 14: 133.

The *IN VITRO* Transcription of Xenopus 5S DNA

D. D. Brown, L. J. Korn, E. Birkenmeier,
R. Peterson, and S. Sakonju

Department of Embryology, Carnegie Institution
of Washington, Baltimore, Maryland 21210

ABSTRACT Cloned Xenopus 5S DNA supports accurate transcription of 5S RNA in an extract of Xenopus oocyte nuclei. By analyzing 5S DNA variants we have begun to catalogue some of the nucleic acid sequences required for accurate initiation and termination of transcription. Comparison of the forty nucleotides that flank the 5' end of the gene in five different multigene families of 5S DNA from two species of Xenopus revealed three conserved oligonucleotides. The most highly conserved oligonucleotide, AAAAG, is found between positions -13 and -19; the likelihood of this occurring by chance in all five gene families is about one in 10^{10}. Yet, deletions to within 11 nucleotides of the gene that have lost all three conserved oligonucleotides still support 5S RNA synthesis.

Initiation of transcription in this *in vitro* system has been studied using γ-^{32}P-ribonucleoside triphosphate precursors. Transcription by RNA polymerase form III *in vitro* begins with purines. Termination of 5S RNA transcription requires a cluster of T-residues on the noncoding strand. A single T to C change in the fourth T of the termination cluster reduces by more than half the number of molecules that terminate accurately.

INTRODUCTION

The genome of Xenopus contains at least two kinds of multigene families encoding for 5S ribosomal RNA. "Somatic" 5S DNA (Xls and Xbs)* is expressed constitutively in all cell types. "Oocyte" 5S DNA (Xlo, Xlt and Xbo)* functions

*Abbreviations: Xlo, Xlt and Xls are *Xenopus laevis* principal oocyte-type, trace oocyte, and somatic 5S DNAs; Xbo and Xbs are *Xenopus borealis* oocyte and somatic 5S DNAs respectively. In recombinant form with a plasmid they are referred to with the prefix "p" (pXlo, pXbs, etc).

ISBN 0-12-068350-4

exclusively in oocytes, accounting for most of the enormous amount of 5S ribosomal RNA that is synthesized in these cells. We study this "dual" gene system because of its simplicity and developmental control. The primary transcript from these genes is 5S RNA and, RNA polymerase form III transcribes these genes in the living cell. Recently, an assay system has been developed that enables us to test single cloned repeating units of 5S DNA for faithful initiation and termination of transcription (1). This has permitted an analysis of nucleotides responsible for accurate transcription.

RESULTS

Initiation of 5S RNA Synthesis. Representative repeating units of five different 5S DNA multigene families purified from the genomic DNA of two species of Xenopus are diagramed in Figure 1. At least one repeating unit of each of these

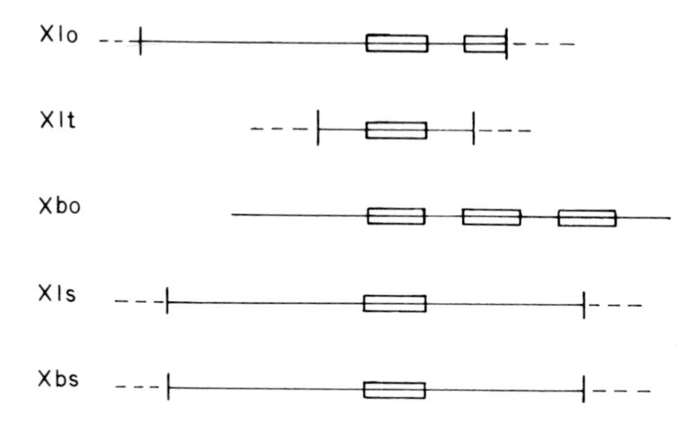

FIGURE 1. Schematic diagram of repeating units of five 5S DNA multigene families from two species of Xenopus. They are drawn to scale and the rectangles refer to the gene. For reference, the repeat length of Xlo 5S DNA is about 700 base pairs.

5S DNA families has been cloned using recombinant DNA technology and sequenced in our laboratory. Computer analysis (2) of the sixty nucleotides flanking the 5' end of the gene has revealed three regions of homology common to the five different 5S DNAs (Figure 2, homologies underlined). The

```
            -40        -30        -20        -10       -1
Xlo  TCCACAGTGCCGCTGACAAGTCAAGAAGCCGAAAAGTGCCGCTGTTCATC
Xlt  TACACAGCACCGCCCACAGGTCCAAAAGGCCCAAAGTGCCAAGCTTCATC
Xbo  ACTTTTTTCGTCAAAGTCTTCATAGAAGCGTCAAAAGTCTTCACTCTGAT
Xbs  GGCCCCCCCCAGAAGGCAGCACAAGGGGAGGAAAAGTCAGCCTTGTGCTC
Xls  CTGGGCCCCAAGAAGGCAGCACAAGAGGAGGAAAAGTCAGCCTTGTGTTC
```

FIGURE 2. The 50 nucleotides upstream from the 5' end of the 5S RNA genes. They are aligned at the genes. The three conserved regions are underlined.

probabilities that these homologies occurred in the same locations relative to the gene by chance alone are 2×10^{-1}, 1×10^{-3} and 1×10^{-10} for the three regions from left to right, respectively.

All five gene families in recombinant form support accurate transcription in an *in vitro* extract of oocyte nuclei (1). This has provided an assay system to test the effect on transcription when the nucleotide sequence of individual repeating units is altered. N. Fedoroff (3) has shown that deletion of essentially all of the AT-rich spacer of *X. laevis* oocyte 5S DNA to 55 nucleotides before the gene does not reduce its ability to support 5S RNA synthesis. We have used the plasmid pXbs1 which contains one repeating unit of *X. borealis* somatic 5S DNA (4), to create progressively increasing deletions of the 5' flanking region. This cloned gene (pXbs1) supports a large amount of 5S RNA synthesis; about 80% of the RNA made *in vitro* is 5S RNA (Figure 3, left). By optimizing transcription with this

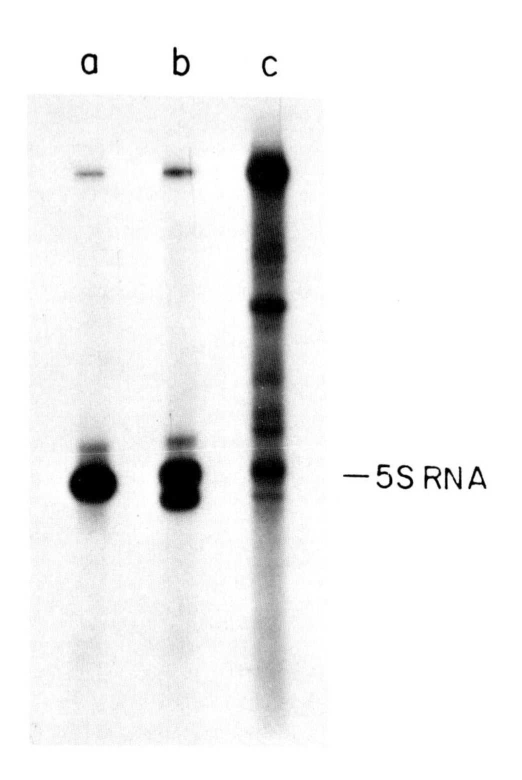

FIGURE 3. Radioautogram of a gel in which radioactive RNA has been electrophoresed. The RNA was transcribed in the *in vitro* nuclear extract (1) from the plasmid a) pXbs1, b) pXbo1, and c) pXlo31.

plasmid we can achieve synthesis of about ten molecules of 5S RNA per gene per hour, a value we estimate to be about 5% of the *in vitro* rate of synthesis.

Deletions have been prepared by exonucleolytic cleavage, trimming the single stranded regions with S1 nuclease, joining the flush ends with synthetic linker DNA and recloning the mixture of deleted molecules. Although the analysis of these deletions is still in progress, one deletion has been confirmed by DNA sequencing to have only the first eleven nucleotides flanking the 5' end of the 5S RNA gene

```
-50        -40        -30        -20        -10
GCCGCCCCCCAGAAGGCAGCACAAGGGGAGGAAAAGTCAGCCTTGTGCTC|GCCTA
GAATTCTCATGTTTGACAGCTTATCATCGATAAGCTTGG---------------
```

FIGURE 4. The sequence flanking the 5' end of the gene in pXbs1 is shown on the top line. Below it is the new sequence adjoining the deleted DNA (pXbs-Δ510). The line means an identical sequence. The new sequence beginning at -12 is a synthetic Hind III linker (8 bases) and then the adjoining sequence of pBR322 is given.

(Fig. 4). All of the remaining spacer sequences upstream from the gene including the three homology regions have been deleted. It is joined by a synthetic DNA linker to the plasmid pBR322. Yet this DNA supports 5S DNA synthesis in the *in vitro* extract (Fig. 5). No change in the amount of

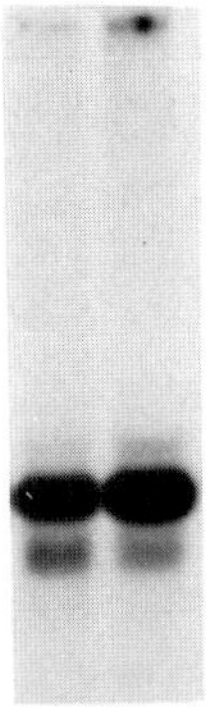

FIGURE 5. Radioautogram of an electrophorogram of (^{32}P)-RNA transcribed from pXbs, (right) and the deletion mutant pXbs1 -Δ510 (left).

5S RNA synthesis was detected, but the assay system is not designed to measure slight changes in rates of synthesis. Several other deleted DNAs that retain more than eleven nucleotides preceding the 5S RNA gene all synthesize 5S RNA. Shorter deletions have not yet been tested.

An Initiation Assay. The determination of faithful transcription in the *in vitro* system is made by eluting an RNA band of the correct size from a gel and fingerprinting it. Since, a positive result requires the production of a full length transcript, transcription must both initiate and terminate properly. We have developed a simple assay for faithful initiation of transcription that does not require proper termination. The assay uses gamma labeled-^{32}P-nucleoside triphosphate to label the first nucleotide in each transcript. We found that incubation of γ-^{32}P-ATP with the nuclear extract randomizes the label between all four nucleoside triphosphates, but it never enters the alpha phosphate since no internal labelling of RNA occurs. Reactions are begun by incubating the DNA to be tested with the complete system minus the nucleoside triphosphates. The reason for the preincubation is that a 30 minute lag is known to occur before 5S RNA synthesis is linear (1). The γ-^{32}P-ATP and four nucleoside triphosphates are added after thirty minutes and incubation continued for 1 hr. Total RNA is purified as the excluded fraction from a Sephadex G-50 column. The excluded fraction is precipitated with ethanol, digested with pancreatic RNase, and the oligonucleotides separated by chromatography in two dimensions on PEI-sheets (Fig. 6) (5). The pattern of

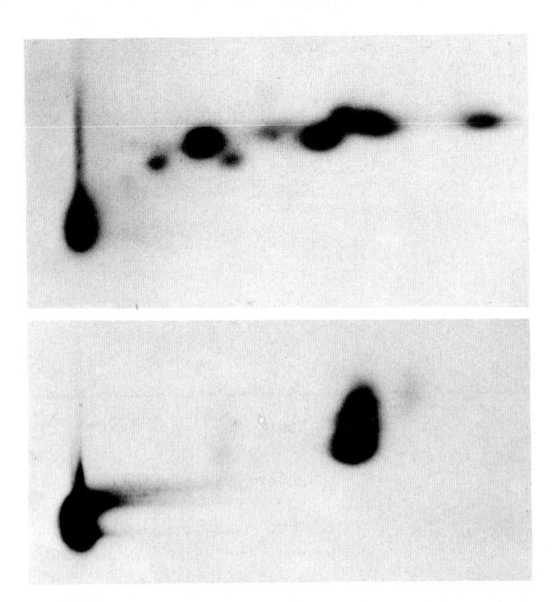

labelled 5' terminal oligonucleotides using plasmid DNA (pMB9 or pBR322) as template is heterogeneous. When a repeating unit of 5S DNA is inserted into the plasmid, the initiation sites change. In the case of pXbs1, plasmid initiation events are suppressed and the majority of radioactivity is in a single oligonucleotide that we presume is pppGpCp, the expected product from the 5' end of Xbs 5S RNA. When only the first nucleotide needs to be identified a second digestion with S_1 nuclease gives the 5' nucleotide in its phosphorylated form. It can be identified readily by including known standards in the chromatogram. When RNA transcribed from the plasmid pBR322 by the nuclear extract is digested with pancreatic RNase and S_1 nuclease, the only labelled nucleotides are the purines ATP and GTP. In these experiments RNA polymerase form III is the principal, perhaps the only enzyme synthesizing RNA. We can conclude that this enzyme initiates at purine residues.

Initiation on a variant 5S RNA gene occurs one nucleotide upstream. One 5S RNA gene that is known to differ in its 5' nucleotide and in residues flanking the 5' end of the gene is the third gene in the cloned DNA referred to as pXbo1 (6). This cloned fragment of *X. borealis* oocyte 5S DNA has three genes separated by about 80 nucleotides. The third gene differs from the other two in 15 of its 120 residues. The sequence of the 5' end of the third gene is shown in Fig. 7. The nucleotides in gene 3 that correspond

↓
AAAAGTCTTCACTCCGATGCCTACGGCCACACC

AAAAGTCAGCAAACCTACCCTGCGGCTACACC
↑

FIGURE 7. The sequence preceding gene two (upper) and gene three (lower) of pXbo1. The arrows denote where transcription begins.

FIGURE 6. Two dimensional thin layer chromatography on PEI-sheets (5) of 5' labelled RNA transcribed from pBR322 (top) and pXbs1. The RNA was digested with pancreatic RNase before chromatography.

to the first nucleotide of most of the 5S RNA genes in Xbo 5S DNA and the one preceding it (-1) have been mutated from G to C and C to A respectively. Previous experiments demonstrated that gene three in pXbo1 is transcribed. Using the initiation assay, we have shown that the third gene initiates transcription at the A residue which is one nucleotide upstream from the usual initiation site. This is further evidence for the predelection of enzyme III to initiate at purines.

A Termination Mutant. The features of nucleotide sequences at or near the termination site of Xenopus 5S DNAs have been summarized previously (6). They include a cluster of four or more T residues at the end of the transcription unit with termination occurring within this cluster. There is imperfect dyad symmetry near the 3' end of the gene, and the sequence preceding the T cluster is GC-rich. The spacer regions that follow the gene have homologies between the five gene families. These include one or more additional T-clusters usually spaced at 10 nucleotide intervals following the gene.

When cloned single repeating units of *X. laevis* oocyte 5S DNA are transcribed, transcription continues beyond the normal transcription termination site more often than when single repeating units of other 5S DNA multigene families are transcribed (see Fig. 3). One such cloned repeat (called pXlo6) was found to accumulate more read through transcripts than did the other cloned repeats (7). This variant has the fourth T-residue in the cluster at the termination site changed to a C-residue. We estimate that this single base change reduces the number of termination events at nucleotide 120 by 80%.

DISCUSSION

We have emphasized here the usefulness of the dual 5S RNA gene system in Xenopus for determining the nucleotide sequences required for accurate initiation and termination of transcription by RNA polymerase form III. Single repeating units of 5S DNA joined to plasmid molecules support accurate transcription of 5S RNA. Some naturally occurring variants that affect initiation or termination have already been analyzed. We have also begun the systematic alteration *in vitro* of sequences in various parts of the repeating unit. An initiation assay permits the detection of accurate initiation in the absence of specific termination.

The most surprising result to date is the synthesis of 5S RNA from DNA molecules whose entire spacer, except for eleven bases preceding the 5S RNA gene, have been deleted. All three conserved regions (see Fig. 2) have been lost in this deletion. The new sequence adjoining the 5S DNA insert consists of a synthetic DNA linker and plasmid. We cannot say at present whether we have recreated fortuitously another promoter region for enzyme III. We must now consider the possibility that "promoter" function for these genes may have a very different location relative to the gene than the analogous control region in prokaryotes. A logical candidate would be the 5' end of the gene itself.

ACKNOWLEDGEMENTS

We are grateful to Mrs. E. Jordan for her excellent technical assistance. The manuscript was improved by the helpful criticism of Dr. D. Bogenhagen.

This work was supported in part by NIH Grant GM22395.

REFERENCES

1. Birkenmeier, E. H., Brown, D. D. and Jordan, E. (1978). Cell *15*, 1077.
2. Korn, L. J., Queen, C. L. and Wegman, M. N. (1977). Proc. Nat. Acad. Sci. USA *74*, 4401.
3. Fedoroff, N. V. (1978). Carnegie Inst. of Washington Yearbook *77*, 131.
4. Doering, J. L. and Brown, D. D. (1978). Carnegie Inst. of Washington Yearbook *77*, 128.
5. Miller, J. S. and Burgess, R. R. (1978). Biochemistry *17*, 2054.
6. Korn, L. J. and Brown, D. D. (1978). Cell *15*, 1145.
7. Brown, D. D. and Gurdon, J. B. (1978). Proc. Nat. Acad. Sci. USA *75*, 2849.

FACTORS INVOLVED IN THE TRANSCRIPTION OF PURIFIED GENES BY RNA POLYMERASE III

R.G. Roeder, D.R. Engelke, S. Ng, J. Segall, B. Shastry, and P.A. Weil

Departments of Biological Chemistry and Genetics, Washington University School of Medicine, St. Louis, Missouri 63110

ABSTRACT When purified DNA templates are incubated with cell free extracts derived from various human or amphibian cell types a variety of homologous and heterologous class III genes (in their respective purified DNA templates) are accurately transcribed. These include the VA RNA genes in adenovirus 2, *Xenopus* and *Drosophila* 5S RNA genes, and most likely both *Xenopus* and *Drosophila* tRNA genes. However, some discrimination is evident since all genes do not appear to be transcribed at the same relative rates in human versus amphibian extracts. Moreover, some genes (*Xenopus* oocyte-type 5S) are expressed in extracts derived from cell types (*Xenopus* kidney) which contain but do not express these genes.

Various chromatographic procedures have been used to isolate, from human cells and from *Xenopus* oocytes, subcellular fractions which are devoid of RNA polymerase III activity but which contain other components essential for accurate transcription of class III genes by a purified RNA polymerase III. The most highly purified fraction from KB cells directs the accurate transcription of the adenovirus VA RNA and cloned tRNA genes, but not cloned 5S RNA genes. However, an accurate transcription of 5S RNA genes is observed in the presence of this fraction and another chromatographically distinct fraction. Similarly, the oocyte extract has been resolved into two chromatographically distinct fractions which are both necessary for transcription of 5S RNA genes by a purified RNA polymerase III. In addition, the oocyte-derived fractions markedly enhance transcription of exogenous 5S RNA genes by the endogenous RNA polymerase III in extracts derived from transcriptionally inactive unfertilized eggs.

ISBN 0-12-068350-4

INTRODUCTION

A detailed understanding of how transcriptional processes are regulated must ultimately involve the ability to reproduce natural transcription events in cell-free systems reconstituted with purified components. Progress toward this goal has been most rapid for those genes transcribed by the class III RNA polymerases. These genes include those which specify the cellular 5S and tRNAs and the adenovirus 2 VA RNA_I and VA RNA_{II} species synthesized in lytically infected cells (1). These investigations have been facilitated by the availablility of well characterized viral and cloned cellular DNAs containing these genes (for review and references see 2-4) and by the availablility of purified (homologous) class III RNA polymerases (1,5,6). A particular advantage of these genes for assessing the fidelity of transcription _in vitro_ is that the primary gene products are in many cases simple and well characterized and closely related to the mature transcripts (reviewed in 2-4).

In earlier studies 5S RNA, tRNA, and VA RNA genes in chromatin or nuclear templates were shown to be accurately transcribed by purified class III RNA polymerases (7-9). This provided the first indication that specific transcription events in living cells could be duplicated in reconstituted cell free systems. In contrast, these genes were randomly transcribed when the same nucleoprotein templates were transcribed by RNA polymerases other than the class III enzymes or when purified DNA templates were transcribed by the class III enzymes alone (5-7). These findings indicated that in addition to RNA polymerase III, chromatin or nuclear template-associated factors are essential for the accurate transcription of these genes.

The further identification of these other transcription components has now been facilitated by the isolation from amphibian oocytes and from cultured animal cells of soluble subcellular fractions in which various class III genes are accurately transcibed by endogenous class III RNA polymerases (2-4, 10-12). In the present report we have extended our earlier preliminary observations (3,4) on the isolation, from these systems, of factors which direct the transcription of these genes by purified (exogenous) class III RNA polymerases.

METHODS

The preparations of soluble cell-free extracts (S-100 fractions), purified DNAs, and purified RNA polymerases were as described previously (3,4). The transcription reaction conditions and the isolation and electrophoretic analyses of $\alpha[^{32}P]$nucleoside triphosphate-labeled transcripts have also been reported (3,4). Standard chromatographic procedures, analogous to those employed for isolation of the RNA polymerases, were employed for fractionation of the cell-free extracts and will be detailed elsewhere.

RESULTS

Transcription of Specific Genes by Endogenous Class III RNA Polymerases in Homologous and Heterologous Cell-Free Systems. One of the most important findings from recent studies with the soluble cell-free transcription systems is that a variety of class III genes (those transcribed by RNA polymerase III *in vivo*) are accurately transcribed by endogenous class III RNA polymerases in crude extracts derived from both homologous and heterologous cell types. In our own studies we have used extracts prepared from cultured human, murine and amphibian cells and from amphibian oocytes to analyze the transcription of the following purified DNA templates: (1) intact adenovirus 2 DNA which contains the tandemly arranged VA RNA_I and VA RNA_{II} genes at about map position 30 on the linear map (13); both of these genes are expressed in lytically infected human cells at a ratio of 5 or 10 to 1; (2) the plasmid pXbo3 which contains a 2700 base-pair fragment of *Xenopus borealis* oocyte-type 5S DNA with six 5S genes; at least two of these genes (type I) encode the predominant oocyte 5S RNA while at least three (type III) appear to contain a slightly divergent sequence and are apparently not expressed *in vivo*; (see references 2-4, J. Doering, personal communication); (3) the plasmid pCIT19 (pDm5S) which contains a 2600 base pair fragment of *Drosophila melanogaster* DNA with three 5S genes (ref. 15); (4) the recombinant bacteriophage λt210, which contains a 3100 base-pair fragment of *Xenopus laevis* DNA with several tRNA genes (ref. 14); S.G. Clarkson, personal communication); and (5) the plasmid pCIT 12 (pDmt) which contains a 9340 base pair fragment of *Drosophila* DNA with several tRNA genes (16, personal communications from N. Davidson and D. Söll).

The autoradiograms in Figure 1 show the electrophoretic resolution of the transcripts generated in response to several of these templates in extracts from cutured human KB

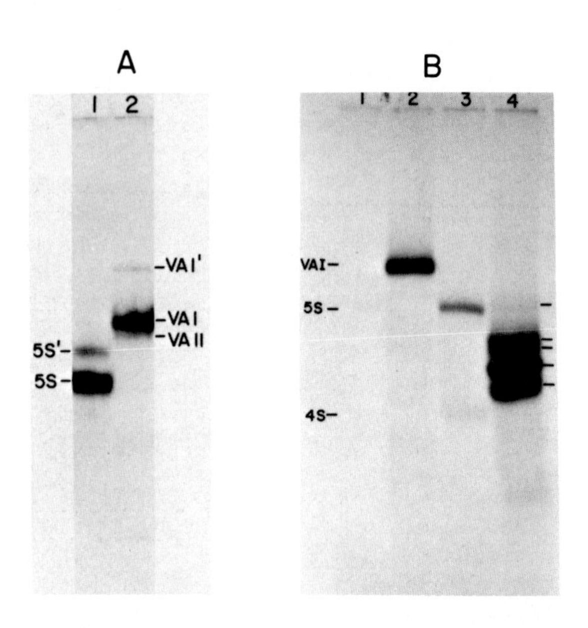

FIGURE 1. Polyacrylamide gel electrophoresis of RNAs synthesized in cell-free extracts with purified DNA templates. Panel A, RNAs synthesized by a KB cell S-100 fraction with pXbo3 (lane 1) or adenovirus 2 DNA (lane 2) as template. Panel B, RNAs synthesized by an *X. laevis* kidney cell S-100 fraction with adenovirus 2 DNA (lane 2), pXbo3 (lane 3), or λ t210 (lane 4) as template, or in the absence of exogenous DNA (lane 1). All procedures were as described elsewhere (3,4). The autoradiogram in panel A is taken from Weil et al (4).

and amphibian kidney cells. Except for the cloned tRNA gene transcripts (lane 4 in Figure 1B) all the discrete RNA bands have been further characterized by fingerprint analyses which have led to the following conclusions (3,4). In the KB cell extract (Panel A) both the type I and type II VA RNA genes (of adenovirus 2) and both type I and type III 5S RNA genes are transcribed. (The single 5S band in lane 1 of Figure 1A is heterogenous and contains both type III and type I 5S RNAs.) Additionally, type I 5S and type I VA gene transcripts with elongated 3' terminii, designated 5S' and VA I', are accumulated in the KB extract and apparently reflect incomplete termination at the normal termination site (for further discussion see ref. 4). Identical results were obtained when these same genes were transcribed in murine cell extracts (4). However, in the amphibian kidney cell extract (Panel B), only the type I 5S RNA gene and only the type I VA RNA gene are transcribed to significant (detectable) extents. (The single 5S band in lane 3 of Figure 1B is homogeneous and contains type I 5S RNA; and no VA II or VA I' species are detected, even with much longer exposures, in lane 2.) Similar results were observed when pXbo3 was transcribed in extracts from mature amphibian oocytes (3). Thus, the homologous (amphibian) extracts transcribe mainly that 5S gene (type I) which encodes the predominant oocyte 5S RNA whereas the mammalian extracts transcribe this same gene and another distinct 5S gene which apparently is not expressed significantly in oocytes. On the other hand, the homologous (human) extract transcribes both the adenovirus VA I and VA II genes, both of which are expressed in infected cells, whereas the heterologous (amphibian) extract appears to transcribe only the VA I gene.

In both the amphibian cell extract (lane 4 in Figure 1B) and the human cell extract (data not shown) the *X. laevis* tRNA gene containing template directs the synthesis of a number of discrete RNA species (indicated by unlabeled horizontal lines) which are of the size expected for pre-tRNAs and which presumably reflect transcription of several of the tRNA genes contained in this DNA (S. Clarkson, personal communication). Similar results have been observed with the *Drosophila* tRNA gene-containing templates (below).

Factors from Human Cells Implicated in the Transcription of tRNA and Adenovirus VA RNA Genes. Previous studies had indicated that factors other than RNA polylmerase III are present in excess in the human cell extract since purified (exogenous) RNA polymerase III greatly stimulated 5S and VA

RNA gene transcription at high dilutions of the extract (4). To further fractionate the system the S-100 extract was loaded onto a phosphocellulose column at 0.05 M KCl and the column was subsequently eluted with buffers containing 0.30, 0.55, and 1.0 M KCl. When assayed in the presence of exogenous RNA polymerase III, only the 0.55 M KCl step fraction showed a capacity for specific transcription of the VA genes in adenovirus 2 DNA. However, this transcription was independent of exogenous RNA polymerase III as shown in Figure 2A, indicating the presence of both RNA polymerase III and other transcription factors in this fraction. When this phosphocellulose fraction was subsequently subjected to chromatography on a Biogel A-1.5m column, the factor(s) necessary for accurate transcription of the adenovirus VA RNA genes was found in the included volume (lane 2 in Figure 2B). The factor was also separated from the RNA polymerase III, since VA gene transcription now required exogenous RNA polymerase III (lanes 1 and 2 in Figure 2B). In agreement with earlier results, no specific transcription of the VA genes (as evidenced by discrete transcripts) was evident when the DNA was transcribed by RNA polymerase III alone (lane 3 in Figure 2B) indicating the requirement for the other factor(s) retained by the Biogel column.

To determine whether the partially purified Biogel factor(s) would direct transcription of other class III genes, several tRNA gene and 5S gene templates were employed. As shown in Figure 2C (lane 3) a *Drosophila* tRNA gene-containing plasmid directs the synthesis of several discrete RNA species (indicated by unlabeled horizontal lines) which are between 4 and 5S in size and which presumably represent pre-tRNAs. Several of these correspond in size to discrete RNA species observed when the same plasmid is transcribed in the original S-100 extract (lane 1) or in the presence of the phosphocellulose step fraction (lane 2). Whether the slight differences in the patterns observed represent variations in processing or transcription in the various fractions is not yet clear. Very similar results have been observed with the *X. laevis* tRNA gene-containing bacteriophage DNA template (data not shown), leading to the conclusion that the Biogel factor(s) suffices, along with an exogenous RNA polymerase III, for tRNA gene transcription. Whether these factors suffice for transcription of all the tRNA genes in each cloned DNA fragment (in excess of seven in each case; personal communications from S. Clarkson, from N. Davidson, and from D. Söll) is not clear. The tentative conclusions reached here are supported by the previous finding (12) that cloned tRNA genes from both *Drosophila* and yeast are

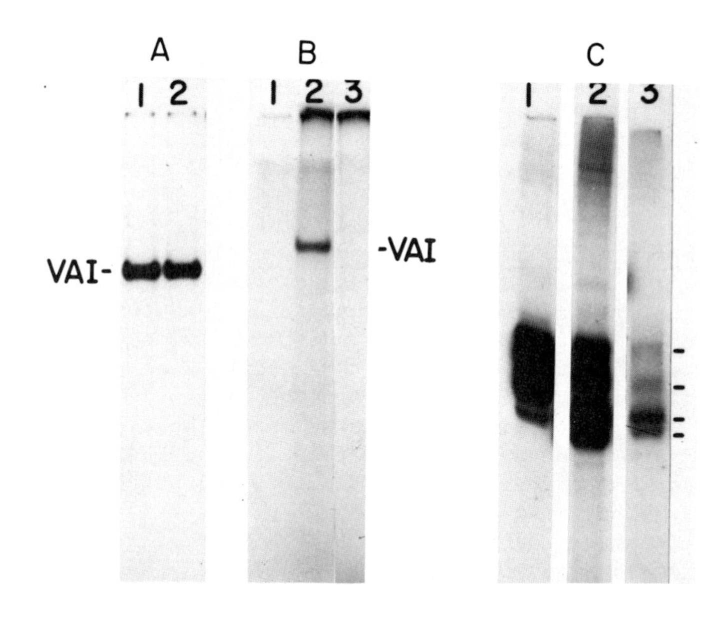

FIGURE 2. Polyacrylamide gel electrophoresis of RNAs synthesized in the presence of partially purified fractions derived from a KB S-100 extract. A KB S-100 extract was loaded onto a phosphocellulose column in a buffer containing 0.05 M KCl. Material bound to the column was subsequently eluted with buffered solutions containing 0.30, 0.55, and 1.0 M KCl. The 0.55 M KCl step fraction was concentrated and subjected to chromatography on a Biogel A-1.5m column. Panel A, RNAs synthesized with adenovirus DNA as template in the presence of the following: phosphocellulose 0.55 M KCl step fraction (lane 1); the phosphocellulose 0.55 M KCl step fraction and purified KB RNA polymerase III (lane 2). Panel B, RNAs synthesized with adenovirus DNA as template in the presence of the following: the Biogel included fraction (lane 1); the Biogel included fraction and purified KB RNA polymerase III (lane 2); purified RNA polymerase III alone (lane 3). Panel C, RNAs synthesized with pDmt as template in the presence of the following: a KB S-100 extract (lane 1); the phosphocellulose 0.55 M KCl step fraction and purified KB RNA polymerase III (lane 2); the Biogel included fraction and purified KB RNA polymerase III (lane 3).

accurately transcribed in cell-free extracts from Xenopus oocyte nuclei.

In contrast to these results, no discrete 5S RNA transcripts are observed when either the Xenopus or the Drosophila 5S DNA templates are transcribed by RNA polymerase III in the presence of those Biogel fractions which suffice for transcription of the VA and tRNA genes (data not shown, see below).

Factors from Human Cells Implicated in the Transcription of 5S Genes. The above studies suggested that at least some of the factors necessary for accurate 5S gene transcription failed to copurify, through the phosphocellulose and Biogel chromatography steps, with the tRNA and VA gene transcription factor(s). To further investigate the fate of the presumptive 5S gene factors the various phosphocellulose fractions were assayed alone or in combination with 5S DNA (pXbo3) and exogenous RNA polymerase III. Figure 3 shows the results of this experiment. As seen in panel A of Figure 3, and as indicated above, the adenovirus VA RNA specificity factors are contained predominantly in the 0.5 M KCl step fraction (lane 4) with only trace levels of activity in the other fractions; and the activity in this fraction is not significantly altered by the presence of the other fractions in mixing experiments (lanes 6-8). However, as seen in panel B of Figure 3, neither the non-bound fraction (lane 2) nor the 0.3 M KCl step fraction (lane 3), nor the 0.5 M KCl step fraction (lane 4) contains factors which alone suffice for specific 5S gene transcription, although it should be noted that several uncharacterized high molecular weight RNA species are present near the top of the gel in the latter case. The mixing experiments show that a combination of the non-bound fraction and the 0.5 M KCl step fraction leads to the synthesis of discrete 5S RNA transcripts in the absence (lane 6) or presence (lane 8) of the 0.3 M KCl step fraction. Results identical to these have been observed with the Drosophila 5S DNA plasmid (data not shown). This result explains the inability of the Biogel purified fraction (above), derived from a phosphocellulose 0.5 M KCl step fraction, to transcribe the 5S genes.

Preliminary experiments have also shown that 5S gene transcription factors can be separated by chromatography of the crude S-100 fraction on DEAE-Sephadex. In this experiment the crude fraction was loaded at 0.15 M KCl and the column eluted with a buffer containing 0.5 M KCl. The

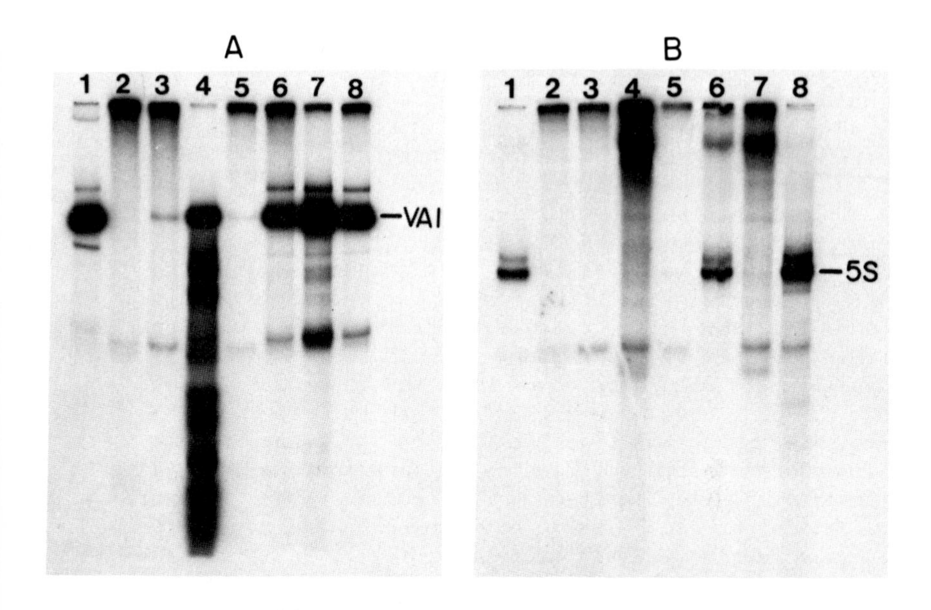

FIGURE 3. Polyacrylamide gel electrophoresis of adenovirus 2 DNA and 5S DNA transcripts synthesized in the presence of recombined KB S-100 components separated by chromatography on phosphocellulose. A KB S-100 extract was subjected to chromatography on phosphocellulose as described in the legend to Figure 2. The autoradiogram shows the RNAs synthesized in the presence of adenovirus 2 DNA (panel A) or pXbo3 DNA (panel B), KB RNA polymerase III (all lanes), and the following components (the same for the corresponding lanes in both panels): KB S-100 (lane 1); the phosphocellulose breakthrough fraction (lane 2); the phosphocellulose 0.3 M KCl step fraction (lane 3); the phosphocellulose 0.55 M KCl step fraction (lane 4); a mixture of the phosphocellulose breakthrough and 0.30 M KCl step fractions (lane 5); a mixture of the phosphocellulose breakthrough and 0.55 M KCl step fractions (lane 6); a mixture of the phosphocellulose 0.30 and 0.55 M KCl step fractions (lane 7); or a mixture of the phosphocellulose breakthrough, 0.30 M KCl step, and 0.55 M KCl step fractions (lane 8). (The low molecular weight RNAs below VA_I in lane 4, panel A are a result of degradation during the preparation of the sample for electrophoresis and are not routinely seen.) The positions of the predominant VA RNA_I and 5S RNA species are indicated.

control experiment with adenovirus 2 DNA as template (Figure 4, lanes 1-4) shows that the 0.5 M KCl fraction results in an accurate transcription of the VA I gene in the presence or absence of the non-bound fraction, which alone has no activity. In contrast, when either the pXbo3 (lanes 5-8) or pDm5S (lanes 9-12) DNA templates are employed, neither the non-bound fraction nor the 0.5 M KCl step fraction alone yields discrete 5S RNAs, although the latter results in synthesis of RNA which does not enter the gel. However, in each case these two fractions in combination result in synthesis of discrete 5S RNA species (lane 8 and 12). In all these experiments reactions were conducted with exogenous RNA polymerase III (the endogenous enzyme is retained by the column). However, very little synthesis was observed in the presence of the unbound fraction under the conditions employed. It should be noted that the major transcript from the pDm5S template is larger than the mature 5S RNA. That this larger RNA (labeled $5S^{+}$) is in fact a 5S gene transcript has been verified by fingerprint analysis (data not shown). This elongated 5S RNA contains about 25 extra nucleotides at the 3' end of the molecule and may represent the true *in vivo* primary transcript (17).

The above experiments clearly suggest the presence of at least two separable 5S gene transcription factors. It might be argued that one of these, namely that retained by the ion-exchange columns and subsequently eluted by high salt, is simply RNA polymerase III. This is unlikely since exogenous RNA polymerase III, in combination with the non-bound fraction, does not accurately transcribes the 5S gene. However, to further investigate this question we have examined transcription of the 5S DNA templates by a purified RNA polymerase III in the presence of the non-bound DEAE-Sephadex fraction and the Biogel fraction described in the preceeding section. Neither of the latter two fractions have detectable RNA polymerase III activity. However, as shown in Figure 5 for the *Xenopus* (lane 8) and the *Drosophila* (lane 12) DNA templates this combination results in specific 5S gene transcription. Since neither the Biogel fraction alone (lanes 7 and 11 in Figure 5) nor the non-bound DEAE-Sephadex fraction alone (preceeding paragraph) result in specific transciption in the presence of exogenous RNA polymerase III we tentatively conclude that there are at least two separable components, other than those present in a purified RNA polymerase III, necessary for 5S gene transcription. One of those could be the same as the VA and tRNA gene transcription factor(s) present in the Biogel fraction, but this is not yet estab-

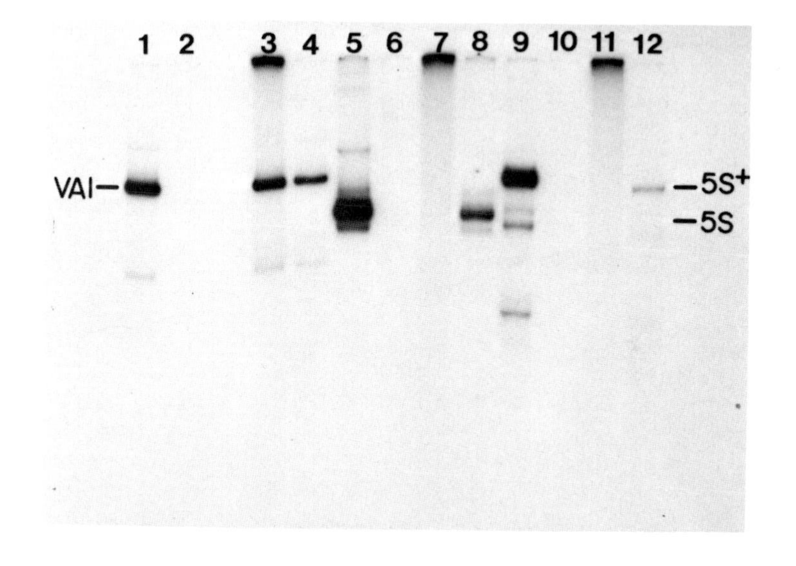

FIGURE 4. Polyacrylamide gel electrophoresis of adenovirus 2 DNA and 5S DNA transcripts synthesized in the presence of recombined KB S-100 components separated by chromatography on DEAE-Sephadex. A KB S-100 extract was loaded onto DEAE-Sephadex (A-25) in a buffer containing 0.15 M KCl and subsequently eluted with a buffer containing 0.50 M KCl. The autoradiogram shows the RNAs synthesized in the presence of adenovirus 2 DNA (lanes 1-4) or pXbo3 DNA (lanes 5-8) or pDm5S DNA (lanes 9-12), KB RNA polymerase III (lanes 1-12) and the following components: KB S-100 (lanes 1,5,9); the DEAE-Sephadex breakthrough fraction (lanes 2,6,10); the DEAE-Sephadex step fraction (lanes 3,7,11); and a mixture of the DEAE-Sephadex breakthrough and step fractions (lanes 4, 8,12). The position of the VA RNA_I and the *Xenopus* 5S RNA species are indicated, as is that of the elongated *Drosophila* 5S RNA species (designated $5S^+$, see text).

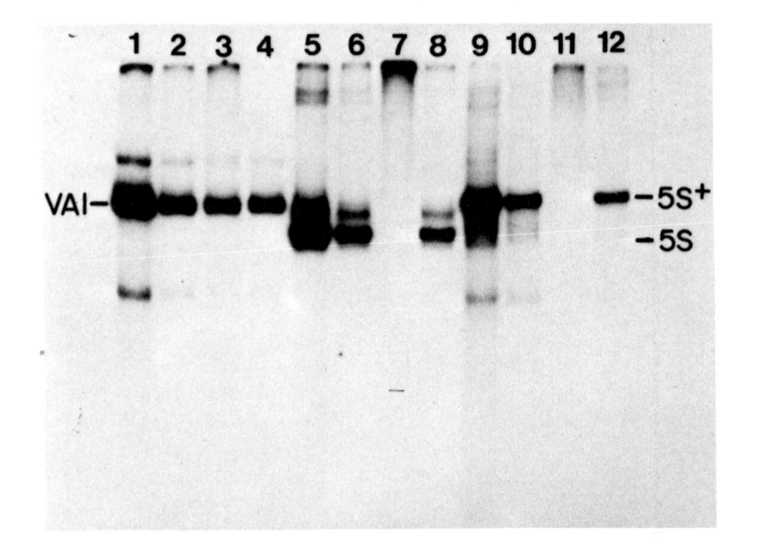

FIGURE 5. Polyacrylamide gel electrophoresis of adenovirus 2 DNA and 5S DNA transcripts synthesized in the presence of recombined KB S-100 components separated by various chromatographic procedures. The autoradiogram shows the RNAs synthesized in the presence of adenovirus 2 DNA (lanes 1-4) or pXbo3 DNA (lanes 5-8) or pDm5S DNA (lanes 9-12), KB RNA polymerase III (lanes 1-12) and the following components: KB S-100 (lanes 1,5,9); a mixture of the DEAE-Sephadex breakthrough and step fractions of Figure 4 (lanes 2,6,10); the included Biogel fractions of Figure 2 (lanes 3,7,11); and a mixture of the DEAE-Sephadex breakthrough and the included Biogel fractions (lanes 4,8,12).

lished. Figure 5 also shows a control experiment with an adenovirus 2 DNA template. As shown earlier, the Biogel factor suffices for VA RNA_I synthesis and this synthesis is unaffected by the presence of the unbound DEAE-Sephadex fraction.

Factors from Amphibian Oocytes Implicated in the Transcription of 5S RNA Genes. Earlier experiments from this laboratory (3) reported the isolation from immature oocytes of a subcellular fraction which was devoid of RNA polymerase III activity but which directed the accurate transcription of purified 5S genes by a purified RNA polymerase III. To pursue the characterization of these factors we turned to the more abundant large oocytes. Crude extracts derived from these oocytes were found to contain all the components, including RNA polymerase III, necessary for transcription of exogenous 5S DNA (3). For further fractionation, this extract was first loaded onto DEAE-cellulose at low salt (0.05 M ammonium sulfate) and eluted with a high salt (0.13 M ammonium sulfate) buffer. As shown in Figure 6A, the components which are necessary for 5S RNA synthesis are bound, including RNA polymerase III (lanes 1-3). When the high salt step fraction is passed through DEAE-Sephadex the endogenous RNA polymerase III is bound to the column whereas the non-bound fraction contains components which result in accurate 5S gene transcription in the presence of an exogenous RNA polymerase III (lanes 4 and 5). When this flow-through fraction was subsequently subjected to chromatography on phosphocellulose two kinds of results were obtained. In some cases when the fraction was loaded at a low salt concentration (Figure 6B), the bound material (eluted with a high salt concentration) was found to yield discrete RNA transcripts when assayed with exogenous RNA polymerase III (lane 3) whereas the non-bound fraction did not (lane 1). In other cases (Figure 6C) where the DEAE-Sephadex flow-through fraction was loaded at a higher salt concentration, neither the bound material (eluted at a high salt concentration) nor the non-bound material resulted in a high level of 5S RNA transcripts in the presence of exogenous RNA polymerase III (lanes 1 and 2). However, in several cases the non-bound and the bound phosphocellulose fractions together resulted in a greatly increased level of synthesis of discrete 5S RNA transcripts in the presence of exogenous RNA polymerase III (lane 3). While there has been considerable variability in these experiments (for unknown reasons), these preliminary data are similar to those reported above for the human cell

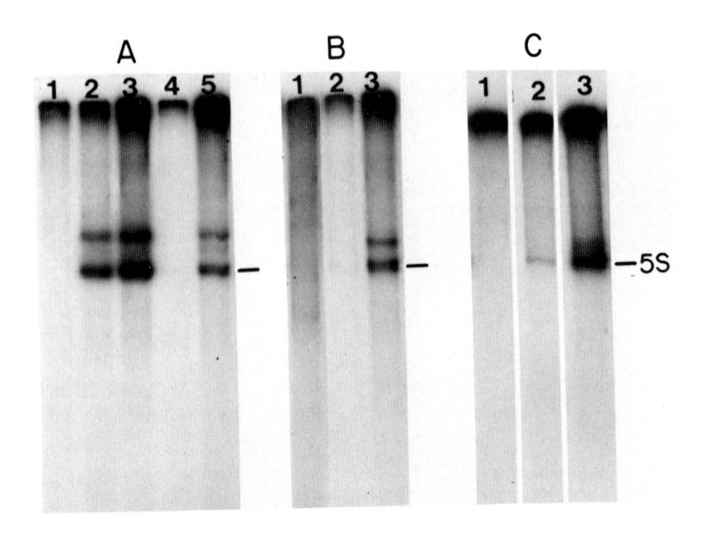

FIGURE 6. Polyacrylamide gel electrophoresis of 5S DNA transcripts synthesized in the presence of chromatographically separated components derived from an *X. laevis* oocyte S-100 extract. An *X. laevis* oocyte S-100 was precipitated with solid ammonium sulfate (0.30 g/ml). The resuspended pellet was loaded onto a DEAE-cellulose (DE52) column at 0.05 *M* ammonium sulfate and eluted with buffer containing 0.13 *M* ammonium sulfate. The 0.13 *M* salt step fraction was passed through a DEAE-Sephadex (A25) column at the same salt concentration. The flow-through fraction was loaded onto a phosphocellulose column at either 0.05 *M* (experiment 1) or 0.08 *M* (experiment 2) ammonium sulfate and eluted with 0.60 *M* ammonium sulfate. *Panel A*; RNAs synthesized with pXbo3 in the presence of the following (see above): the DE52 flow-through and purified *X. laevis* RNA polymerase III (lane 1); the DE52 step fraction (lane 2); the DE52 step fraction and RNA polymerase III (lane 3); the DEAE-Sephadex flow-through fraction (lane 4); the DEAE-Sephadex flow-through fraction and RNA polymerase III (lane 5). *Panel B*; RNAs synthesized with pXbo3 template in the presence of the following: the phosphocellulose flow-through from experiment 1 plus purified RNA polymerase III (lane 1); the phosphocellulose step fraction from experiment 1 (lane 2); and the phosphocellulose step fraction from experiment 1 plus RNA polymerase III (lane 3). *Panel C*; RNAs synthesized with pXbo3 in the presence of purified RNA polymerase III and the following components from experiment 2: the phosphocellulose flow-through (lane 1); the phosphocellulose step fraction (lane 2); and a mixture of the phosphocellulose flow-through and step fractions (lane 3).

5S DNA transcription studies, and suggest the presence of separable 5S RNA transcription factors in amphibian oocytes.

We have also begun to investigate the possibility that the transcription components identified here might be deficient in unfertilized eggs which are transcriptionally inactive. Previous experiments showed that unfertilized eggs contain levels of RNA polymerase III comparable to those found in oocytes and ruled out the absence of RNA polymerase III *per se* as an explanation for the absence of transcription (18). For these studies, crude S-100 extracts were prepared from unfertilized eggs in parallel with extracts from mature ovaries. When assayed with exogenous 5S DNA, the control oocyte extract showed the expected 5S RNA synthesis (Figure 7, lane 7), but the egg extracts repeatedly failed to show signifcant levels of 5S RNA synthesis in the absence (lane 2) or presence (lane 3) of exogenous RNA polymerase III. However, transcription of the 5S DNA templates by the endogenous RNA polymerase III in the egg extract was markedly enhanced when the system was supplemented with an oocyte-derived phosphocellulose fraction which contains 5S DNA transcription factors (Figure 7, lane 5). (When incubated alone with 5S DNA this phosphocellulose fraction showed no 5S gene transcription although there was a significant level of RNA synthesis due to contamination with RNA polymerase I (lane 6).) The discrete 5S RNAs synthesized in the supplemented egg extract clearly result from 5S gene transcription by (endogenous) RNA polymerase III as shown by the sensitivity to high but not low concentrations of α-amanitin (lanes 4 and 5). These studies support the idea that the egg extracts are deficient in at least one functional 5S gene transcription factor. That these observations do not result from the loss or inactivation of such factors during preparation of the extracts is suggested by the following controls. Extracts prepared from equal mixtures of oocytes and eggs (prior to disruption) or by mixing egg and oocyte extracts (prepared independently) show 5S RNA synthesis when incubated with 5S DNA (data not shown).

DISCUSSION

A most significant development for further investigations of transcriptional controls in eukaryotes has been the establishment of soluble cell-free systems in which specific class III genes in purified DNA templates are accurately transcribed by endogenous class III RNA polymerases. Detailed sequence analyses of the transcripts gener-

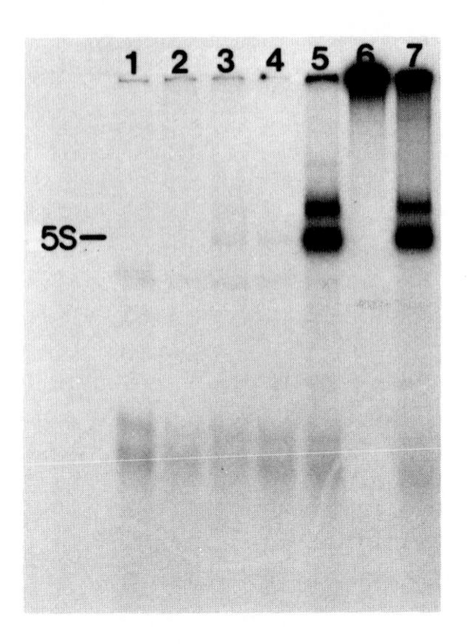

FIGURE 7. Polyacrylamide gel electrophoresis of pXbo3 transcripts synthesized in the presence of an unfertilized egg S-100 extract complemented with oocyte S-100 factors. S-100 extracts were prepared from unfertilized *X. laevis* eggs and from *X. laevis* oocytes in parallel. Phosphocellulose 0.60 *M* ammonium sulfate step fractions were prepared from an oocyte S-100 as described in Figure 6. The autoradiogram shows the RNAs synthesized in reactions containing the following components: egg S-100 (lane 1); egg S-100 plus pXbo3 (lane 2); egg S-100, pXbo3 and purified *X. laevis* RNA polymerase III (lane 3); egg S-100, pXbo3, oocyte phosphocellulose fraction, and 200 μg α-amanitin per ml (lane 4); egg S-100, pXbo3, oocyte phosphocellulose fraction and 0.5 μg α-amanitin per ml (lane 5); pXbo3 and the oocyte phosphocellulose fraction (lane 6); and oocyte S-100 and pXbo3 (lane 7). The experiments shown here utilized the phophocellulose step fraction prepared as in experiment 1 of Figure 6 but a phosphocellulose step fraction prepared as in experiment 2 of Figure 6 also gave the same results (i.e. markedly stimulated 5S RNA synthesis in the egg S-100). The position of the major 5S RNA band is indicated. The bands near the bottom of the gel in the lanes containing egg S-100 reflect end labeling of tRNAs present in the extract.

ated in these systems argue strongly for accurate initiation and termination events, as opposed to random transcription followed by specific processing events (2-4, 10-12). The various studies reported thus far, including those summarized here, also point to a general "permissiveness" in these systems in that a variety of class III genes are accurately transcribed, and often at comparable rates, in extracts from both homologous and heterologous cell types. Thus, *Drosophila*, *Xenopus*, and human (adenovirus) genes are transcribed in extracts from amphibian oocytes and in extracts from cultured amphibian and mammalian cells (2-4, 10,11; present data). Even genes from eukaryotes as simple as yeast are transcribed in such systems (12). These results in themselves suggest that structural features of the transcription components, at least those necessary for the reactions observed, are highly conserved. This is substantiated in part by studies of the DNA sequences surrounding the class III genes (2) and by studies of the class III RNA polymerases from various organisms (1,5,6).

It is clear that the crude transcription systems can be used to advantage to study the functionality of various purified genes (including variants), and the importance of specific base sequences for gene transcription (2). However, as shown by the results presented here and elsewhere (3,4), not all genes are uniformly transcribed in the various heterologous cell types. Thus, under the conditions employed, the amphibian cell-free extracts show a much greater discrimination between two amphibian oocyte-type 5S genes than does the human cell extract; and the human cell extract transcribes both adenovirus VA genes, whereas the amphibian cell extract apparently transcribes only that VA gene (type I) which is most active in virus-infected cells. The exact significance of these results is not yet clear, but since purified DNA templates are employed in the cell-free systems, the variations in relative transcription rates for individual genes within a related set (e.g. VA I versus VA II) most likely reflect sequence variations within recognition sites and consequently promotor strengths. If there is simply a competition for transcription factors, then it may be possible to enhance the transcription of the poorly transcribed genes by altering the reaction conditions or by separating these genes from the others. Such studies might provide some insight into the various parameters which determine overall rates of transcription of various class III genes, especially if done in conjunction with individual transcription factors (below).

Despite the utility of the crude systems for a variety of studies of (purified) gene function, such systems must be further fractionated to more clearly define the nature and mechanisms of action of other components involved in gene activiation and regulation. In this regard it is significant that we have been able to isolate, from these extracts, factors (or fractions) which are necessary and sufficient for the accurate transcription of purified genes by purified RNA polymerases. The ability to perform such operations (i.e. the chromatographic fractionations) without losing the ability to reconstitute specificity makes the further fractionation and purifications seem plausible. Until this is done our conclusions regarding the various factors must be tentative. However the present data indicate that the factor(s) necessary for adenovirus VA and cellular tRNA gene transcription copurify through several steps and raise the possibility that they may be the same. This possibility is not unreasonable since the VA RNA genes can be transcribed *in vitro* (4,10; this report) and *in vivo* (13) solely by cellular components.

The present data also suggest that there are at least two separable components, not present in purified class III RNA polymerases, which are necessary for 5S gene transcription in both mammalian cells and amphibian oocytes. The data do not yet exclude the possibility that one of these may be the same as the factor(s) which is implicated in VA and tRNA gene transcription. Should this be the case it seems somewhat more likely that the common factor would be an initiation component. This speculation is made solely on the basis of the observation that significant levels of synthesis of high molecular weight RNAs are observed when the 5S DNA plasmids are transcribed with the single fraction which suffices for VA and tRNA gene transcription. Conceivably, such high molecular weight RNAs could represent 5S gene transcripts which were correctly initiated but improperly terminated (i.e. readthrough products); this possibility is currently being tested. As already stated, the ultimate resolution of these and other questions about the diversity of class III gene transcription factors must await the further purification of the factors identified here and a further search for other factors which may be detected when these systems are screened with yet other class III genes. An important aspect of these studies will be analyses of individual tRNA genes.

We presently have little information on the site(s) and mechanism(s) of action of the factors whose effects are manifested here. However, it does not seem probable on the basis of the purification schemes employed, that the factors are, or require, the presence of histones, at least in the case of the VA and tRNA genes. Another reasonable hypothesis is that the factors identified here represent *general* initiation and/or termination factors which might be sufficient, along with an RNA polymerase, for transcription of purified (derepressed) DNA templates. When present in chromatin, the transcription of these genes might well require other, possibly gene specific, components which could mediate chromatin structural modifications (see, for example, ref. 19). This could explain the observed "permissiveness" of the cell-free systems and the transcription of purified genes in extracts from cells which contain but do not express these genes. Another interesting possibility raised by the unfertilized egg studies presented here is that a general regulation of gene activity (e.g. for 5S genes) could be mediated via the functional modification or inactivation of one or more of the factors identified here. The further analysis and resolution of these questions should be facilitated by the purification and characterization of the factors thus far identified.

ACKNOWLEDGEMENTS

This work was supported by research grants CA 16640 and CA 23615 (to R.G. R.) from the NCI and in part by Cancer Center Support Grant PBO CA 16217 (to Washington University) from the NCI. R.G.R. is a Camille and Henry Dreyfus Teacher-Scholar awardee; P.A.W. was supported by a fellowship (CA 05860) from the NCI; J. S. by a fellowship from the Damon Runyon-Walter Winchell Cancer Fund; and S. N. by a fellowship from the Sigma Chemical Company. We thank Donald Brown, Stuart Clarkson, and Norman Davidson for the various recombinant DNA plasmids and bacteriophage employed here and Becky Teitze for technical assistance.

REFERENCES

1. Roeder, R.G. (1976). In "RNA Polymerase" (R. Losick and M. Chamberlin, eds.), pp. 285-329. Cold Spring Harbor Laboratory, New York.

2. Korn, L.J., and Brown, D.D. (1978). Cell 15, 1145.

3. Ng. S.Y., Parker, C.S., and Roeder, R.G. (1979). Proc. Natl. Acad. Sci. U.S.A. 76, 136.

4. Weil, P.A., Segall, J., Harris, B. Ng, S.Y., and Roeder, R.G. (1979). J. Biol. Chem., in press.

5. Parker, C.S., Ng, S.Y., and Roeder, R.G. (1976). In "Molecular Mechanisms in the Control of Gene Expression (D. P. Nierlich, W.J. Rutter, and C.F. Fox, eds.), pp. 223-242, Academic Press, New York.

6. Jaehning, J.A., Woods, P.S., and Roeder, R.G. (1977). J. Biol. Chem. 252, 8762.

7. Parker, C.S., and Roeder, R.G. (1977). Proc. Natl. Acad. Sci. U.S.A. 74, 44.

8. Jaehning, J.A., and Roeder, R.G. (1977). J. Biol. Chem. 252, 8753.

9. Sklar, V.E.F., and Roeder, R.G. (1977). Cell 10, 405.

10. Wu, G.J. (1978). Proc. Natl. Acad. Sci. U.S.A. 75, 2175.

11. Birkenmeier, E.H., Brown, D.D., and Jordan, E. (1978). Cell 15, 1077.

12. Schmidt, O., Mao, J., Silverman, S., Hovemann, B., and Söll, D. (1978) Proc. Natl. Acad. Sci. U.S.A. 75, 4819.

13. Mathews, M.B., and Pettersson, U. (1978). J. Mol. Biol. 119, 293.

14. Clarkson, S.G., Kurer, V., and Smith, H.O. (1978). Cell 14, 713.

15. Hershey, N.D., Conrad, S.E., Sodja, A., Yen, P.H., Cohen, M., Davidson, N., Ilgen, C., and Carbon, J. (1977). Cell 11, 585.

16. Yen, P.H., Sodja, A., Cohen, M., Conrad, S.E., Wu, M., Davidson, N., and Ilgen, C. (1977). Cell 11, 763.

17. Jacq, B., Jourdan, R., and Jordan, B.R. (1977). J. Mol. Biol. 117, 785.

18. Roeder, R.G. (1974). J. Biol. Chem. 249, 249.

19. Weisbrod, S. and Weintraub, H. (1979). Proc. Natl. Acad. Sci. U.S.A. 76, 630.

THE INTERNAL ORGANIZATION OF THE NUCLEOSOME

G. Felsenfeld, J. McGhee, B. Sollner-Webb
and P. Williamson

National Institute of Arthritis, Metabolism
and Digestive Diseases, Bethesda, Md. 20205

Abstract Nuclease digestion studies have suggested that a considerable part of the DNA of the nucleosome core may be accessible to some nuclease. We have used a variety of enzymic and chemical probes to investigate the accessibility of DNA in the nucleosome core. We show here that the N7 of guanine in the large groove of nucleosome DNA is largely accessible to the chemical probe, dimethyl sulfate, and that the N3 of adenine, in the small groove, also reacts similarly in the DNA of the nucleosome and in naked DNA. We further report that a nucleosome-like particle can be formed by complexing histones with phage T4 DNA, in which the large groove is occupied by glucose residues. This result suggests that nucleosome structure does not depend upon extensive contacts between histones and sites in the large groove of DNA. Finally, we have examined the template properties of 'chromatin' made by reconstituting the core histones onto phage T7 DNA. We show that under appropriate conditions, this nucleoprotein complex is indistinguishable from naked DNA as a template for *E. coli* RNA polymerase. These results all suggest that the histones of the nucleosome may be so designed that they can cause compaction without severely blocking the chemical and biological reactivity of the DNA.

There is now ample evidence that the DNA of the nucleosome core particle is on the "outside" of the structure (1), but the real meaning of this description, in terms of the chemical accessibility of the DNA, remains to be established. We discuss in this paper the results of a variety of chemical probe experiments which suggest that nucleosome DNA is remarkably reactive.

DIGESTION BY NUCLEASES

The first class of experiments involves the use of nucleases to digest the DNA of the nucleosome core. (The core contains about 145 nucleotide pairs of DNA, and two molecules each of the core histones H2A, H2B, H3 and H4. The spacer DNA separating the neighboring core particles has been digested

ISBN 0-12-068350-4

away.) It is well known (1-4) that digestion of core particles by staphlyococcal nuclease, pancreatic DNase, spleen acid DNase, and other nucleases, gives rise to a series of DNA fragments of discrete size, approximate multiples of ten nucleotides in length. We have asked whether all these nucleases cut at the same sites on the DNA (5). The experimental results show that the cutting sites for the enzymes, though clustered in repeats separated by 10 nucleotides, vary according to the identity of the enzyme (5,6). The positions of the average potential cutting sites are shown in Figure 1.

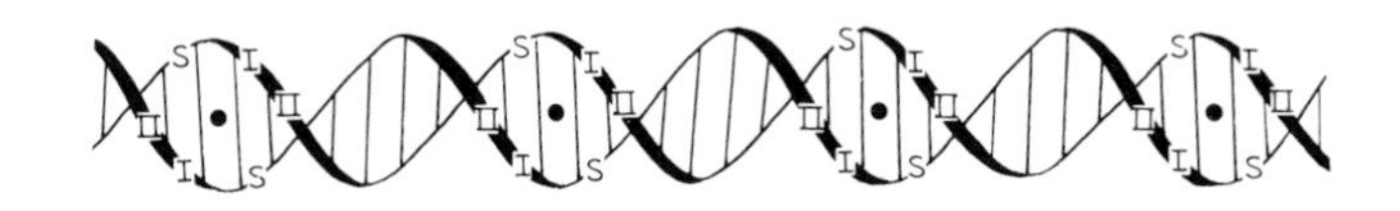

FIGURE 1. Representation of potential average cutting sites positions for DNase I, DNase II, and staphylococcal nuclease (s) on a region of nucleosome core DNA. From reference (5).

REACTION WITH DIMETHYL SULFATE

The above results suggested to us that a considerable proportion of the nucleosome DNA is in some way accessible even to these very large enzymic probes. To test this idea further, we chose a very small chemical probe, dimethyl sulfate. This molecule has been employed by Gilbert and his collaborators to measure the sites of interaction between *lac* operator and *lac* repressor (7). The method depends upon the lability induced in the polynucleotide backbone by methylation of the bases guanine or adenine. In our experiments, nucleosome cores isolated from chicken erythrocyte nuclei (and thus containing DNA sequences representing the entire genome) were labelled at their 5' termini using T4 polynucleotide kinase. The particles were then treated with dimethyl sulfate in such a way as to produce about one or two methylated bases per DNA strand. By appropriate choice of conditions, strand breakage was induced exclusively at the sites occupied by 7 methyl guanine (7), and the distribution of such sites displayed on an appropriate electrophoretic gel.

When the cleavage pattern of DNA in methylated nucleosomes is compared with the pattern obtained with DNA that has been freed of histones before reaction, the similarity is striking (Fig. 2). With the exception of a band in the nucleosome DNA

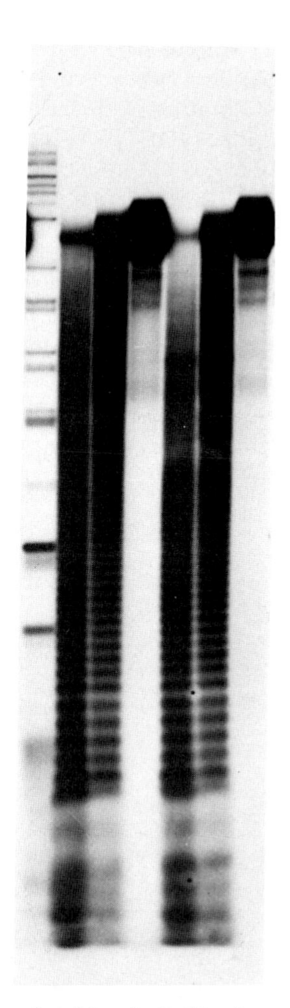

Figure 2. 5' Terminally labelled nucleosome core particles or control of protein-free DNA were reacted at 20° C with 50 mM dimethyl sulfate for varying times. After appropriate treatment to purify the DNA and cleave it at the 7-methyl guanine residues, the cleavage products were electrophoresed on 10% polyacrylamide - 10 M urea denaturing gels. From the right, the slots contain nucleosome cores treated for 0, 10 and 30 minutes, followed by three DNA control samples treated for 0, 10 and 30 minutes. The left hand slot contains size standards generated by digestion of plasmid pBR322 with restriction enzymes HaeIII and HpaII. From reference (8).

at about 62 nucleotides from the 5' terminus, there is no evidence for any variation in reactivity (8).

We found this result somewhat surprising, since we had expected that there might be some periodic variation in the accessibility of the DNA to dimethyl sulfate. The N7 of guanine is in the large groove of DNA, and as the DNA twists the large groove must point alternately inward toward the histones of the core and outward toward the solvent. We therefore expected to find some periodic modulation of reaction, with a period of about 10 nucleotides as a function of distance from the 5' end.

A careful search was made for such periodicity by digitizing the gel data and making a point-by-point comparison of the band intensities for nucleosome DNA and naked DNA (Fig. 3).

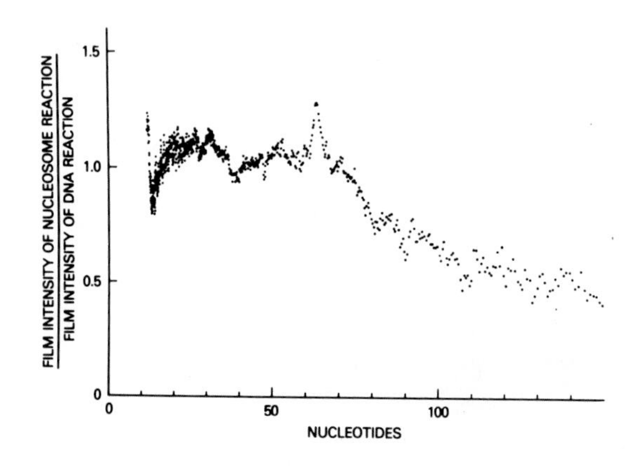

Figure 3. Ratio of corected film intensity in the nucleosome methylation reaction to that in the DNA reaction, plotted against fragment size in nucleotides. From reference (8).

No significant variation was found, except for the reactive region at 62 nucleotides mentioned above. Detailed comparison with various model calculations (8) shows that we would be unlikely to overlook a periodic protection in which even one base in ten was completely protected with respect to its neighbors.

Before these results can be interpreted, it is important to show that the nucleosome remains intact and that the DNA remains fixed relative to the nucleosome surface during the reaction with dimethyl sulfate. To assure the integrity of the nucleosome structure, we reisolated methylated nucleosomes as 11 S particles on sucrose gradients after the initial reaction. To show that no sliding of the DNA had occurred, we digested the terminally labelled core particles with pancreatic DNase under the conditions used for methylation, and observed the pattern of sharp bands at discrete fragment lengths characteristic of the unperturbed nucleosome core. Other controls ruled out the possibility of DNA exchange between histone cores, and showed that methylation of histones occurred at too low a level to have a major effect on the results (8).

The resolution of the gels falls off at sizes larger than 70 nucleotides, and for this reason, we repeated the experiment with nucleosomes labelled at the 3' terminus with terminal deoxynucleotidyl transferase. The results were essentially identical to those shown in Fig. 3: in the region beyond 70 nucleotides from the 5' end, there is also no detectible periodic protection. Thus these results reveal that the DNA in the large groove is equally accessible almost everywhere within the nucleosome. Only the single reactive region 62 nucleotides from the 5' end could be detected.

We have carried out similar experiments under conditions which favor reaction of the N3 of adenine in the small groove. Once again, no periodic protection was observed, but there was enhanced reactivity at a point 65 nucleotides from the 5' end, quite close to that described above at 62 nucleotides in the large groove. Such points of greater reactivity have been observed in the lac repressor-operator complex, and tentatively ascribed to the presence of hydrophobic regions in the protein bound to DNA (7). Since the phosphodiester linkage at 64 nucleotides is one of those cut only weakly by pancreatic DNase, is reasonable to suppose that this region is complexed to histone. On the other hand, it should be noted that other regions of the nucleosome that are protected against DNase attack, such as those about 30 and 80 nucleotides from the 5' end, show neither increased nor decreased reactivity with dimethyl sulfate.

We have also investigated the accessibility of the DNA large groove in nucleosome structure by using T4 phage DNA, in which this groove is occupied by glucose residues bonded to hydroxymethyl cytosine. The core histones can be reconstituted onto such DNA and digested with staphylococcal nuclease to give particles sedimenting at about 11.6 S, and containing between 140 and 150 base pairs of DNA. Digestion of the reconstituted complex with pancreatic DNase gives a pattern quite similar to that found for normal chromatin (8).

These results suggest that much of the large groove of DNA cannot be the site of interactions with histones that are important in preserving nucleosome structure. The interactions that are important in preserving nucleosome structure may very well be electrostatic in nature, involving contact between the phosphodiester groups and basic amino acids of the histones. If that is the case, the important amino acid residues do not lie in the amino terminal portions of the histones, since it has been shown (9) that many aspects of nucleosome structure are very little perturbed when these amino terminal tails are removed by trypsin. Furthermore, it is not clear that a large number of histone-DNA interactions would be necessary to fold the DNA. It may be that the DNA and protein are tightly bound to each other only at a few points, and that these contacts are sufficient to keep the DNA in place.

REACTION WITH *E. coli* RNA POLYMERASE

RNA polymerase can also be used as a probe of chromatin structure, though it clearly responds to different aspects of structure than either nucleases or dimethyl sulfate. The choice of RNA polymerase as a probe was prompted by the

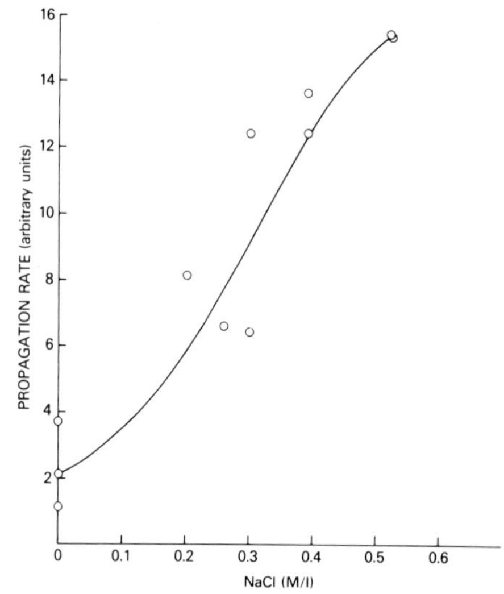

FIGURE 4. Propagation rates on T7 chromatin as a function of ionic strength. In each experiment, the initiation step was performed under identical conditions using low ionic strength and three nucleoside triphosphates. At the end of the initiation period, rifampicin was added, followed by the fourth nucleoside triphosphate and sufficient NaCl to give the final concentration shown. The propagation rate is calculated from the slope of a graph of [^{3}H]UTP incorporation vs. time for a 0-30 min period of propagation. From reference (10).

considerable evidence that histones are present on DNA regions that are active *in vivo* as templates for transcription. [For a review, see Ref. (1)]. We asked the question: Does the presence of the core histones on DNA interfere with the passage of a polymerase molecule?

To answer this question, we made a "chromatin" consisting of core histones reconstituted onto T7 bacteriophage DNA, and tested its activity as a template for *E. coli* RNA polymerase (10). The experiments were carried out in such a way that no new chains could be initiated, so that only chain elongation was measured. If the chain elongation rate on such a template is measured as a function of NaCl concentration, the results shown in Fig. 4 are obtained. At low salt concentrations, elongation proceeds slowly, but as the salt concentration is raised the rate of elongation increases markedly. At a salt concentration of 0.45 M, the rate on the nucleoprotein template is about the same as on naked T7 DNA, although it can be shown that the histones are still bound to the DNA of the nucleoprotein. The similarity of the elongation rates on the two templates at 0.45 M NaCl is confirmed by direct measurement of the size of the transcripts obtained when nucleoprotein and DNA are transcribed in parallel experiments (Fig. 5).

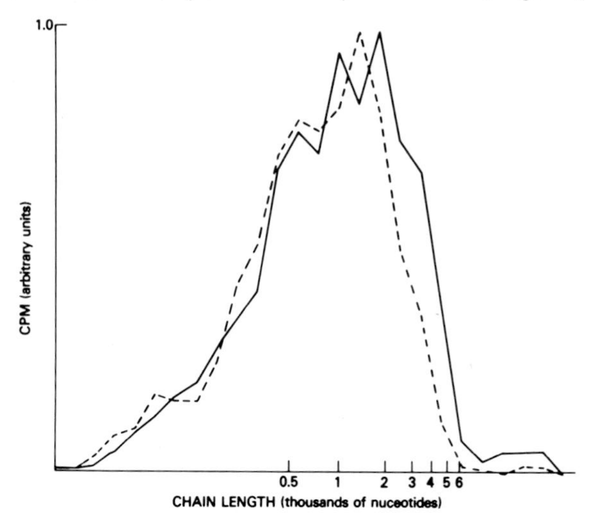

Figure 5. Electrophoretic analysis, carried out in 1.5% agarose slab gels containing methyl mercury hydroxide, of RNA chains from propagation reactions in 0.45 M NaCl on T7 chromatin (——), and T7 DNA (---). From reference (10).

We conclude that the core histones do not necessarily provide a serious barrier to the passage of *E. coli* RNA polymerase. Similar observations have been made in other laboratories (11,12). This does not mean that modification of histones or of nucleosome structure may not be necessary for transcriptional function *in vivo*, but it suggests that even in an unmodified form the histones of the nucleosome core allow the polymerase to traverse the DNA. We do not yet know the mechanism which allows the nucleosome to accomodate the passage of the polymerase, but we have discussed possible mechanisms elsewhere (10).

DISCUSSION

We have presented data from several kinds of experiments that probe nucleosome core structure. Nuclease digestion studies reveal that different nucleases may cut at different places within the core, and suggest that a considerable proportion of the DNA is available even to these very large molecules. Studies with the small chemical probe, dimethyl sulfate, show that most of the nucleotides of the nucleosome core DNA react at identical rates; there are not two major classes consisting of "reactive" and "protected" sites. (Similar results have been obtained in our laboratory with the larger probe, diethyl sulfate.) Our experiments with T4 DNA-histone complexes appear to confirm that the large groove of DNA is not the site of extensive intimate contact between histones and nucleotides. Finally, the *in vitro* transcription studies provide an indication of the accessibility of nucleosome DNA for a biological function though the experimental system is probably far removed from the conditions *in vivo*.

All of these experiments indicate that the DNA of the nucleosome core is remarkably reactive. Of course it would be an exaggeration to suggest that the histones do not alter the physical properties of the DNA, but it is surprising that the binding of a gram of basic proteins per gram of DNA has such a modest chemical effect. This leads us to speculate that the accessibility of DNA is an important design feature of nucleosomes, and that the requirement that the DNA be available to react with enzymes and other molecules places severe restrictions on nucleosome structure and the amino acid sequences of its component histones.

Nucleosomes form high order structures, and in these structures the DNA is possibly much less accessible. Furthermore, the histones undergo specific modifications and reactions with other molecules that are undoubtedly of biological importance in modulating the reactivity of the DNA. It may be, however, that the histones are constructed to have the minimum possible effect on DNA at the level of the individual nucleosome.

REFERENCES

1. Felsenfeld, G. (1978). Nature 271, 115.
2. Noll, M. (1974). Nucl. Acids Res. 1, 1573.
3. Axel, R., Melchior, W., Sollner-Webb, B. and Felsenfeld, G. (1974). Proc. Natl. Acad. Sci. USA 71, 4101.
4. Camerini-Otero, R. D., Sollner-Webb, B. and Felsenfeld, G. (1976). Cell 8, 333.
5. Sollner-Webb, B., Melchior, W. and Felsenfeld, G. (1978). Cell 14, 611.
6. Sollner-Webb., B. and Felsenfeld, G. (1977). Cell 10, 537.
7. Gilbert, W., Maxam, A. and Mirzabekov, A. D. (1976). In "Control of Ribosome Synthesis", pp 139-148. The Alfred Benzon Symposium IX, Munksgaard, Copenhagen.
8. McGhee, J. and Felsenfeld, G. (1979). Proc. Natl. Acad. Sci. USA, in press.
9. Whitlock, J. P. Jr., and Simpson, R. T. (1977). J. Biol. Chem. 252, 6516.
10. Williamson, P. and Felsenfeld, G. (1978). Biochemistry 17, 5695.
11. Meneguzzi, G., Pignatti, P., Barbanti-Brodano, G., and Milanesi, G. (1978). Proc. Natl. Acad. Sci. USA 75, 1126.
12. Brooks, T. L., and Green, M. H. (1977). Nucleic Acids Res. 12, 4261.

THE MECHANISM OF NUCLEOSOME ASSEMBLY

R.A. Laskey, B.M. Honda, A.D. Mills, R.M. Harland, and W.C. Earnshaw.

MRC Laboratory of Molecular Biology, Hills Road, Cambridge CB2 2QH, England.

ABSTRACT Fractionation of a cell-free nucleosome assembly system from eggs of *Xenopus laevis* (1) yields an acidic, thermostable protein which promotes nucleosome assembly (2). Further properties of the protein are described including some similarities to the HMG non-histone nuclear proteins. An alternative pathway for nucleosome assembly *in vitro* by purified nicking-closing enzyme has recently been described (3). This paper considers the relationship between the two *in vitro* assembly systems and discusses their possible relevance to the cellular mechanism of nucleosome assembly.

INTRODUCTION

We have previously described a cell-free system from eggs of *Xenopus laevis* which converts purified DNA into nucleosomes at physiological ionic strength using either an endogenous histone pool or purified exogenous histones (1). Fractionation of egg homogenates yielded a thermostable acidic protein which promotes assembly of nucleosomes by binding histones and transferring them to DNA (2). Recently Germond, Brutlag, Yaniv and Rouvière-Yaniv (3) have shown that highly purified nicking-closing can also convert histones and DNA into nucleosomes at physiological strength.

This paper considers the evidence that the two pathways for nucleosome assembly can occur independently *in vitro* and it considers the possibility that they could occur together *in vivo* as complementary components of the cellular mechanism of nucleosome assembly.

PROPERTIES OF THE THERMOSTABLE PROTEIN FROM EGGS OF XENOPUS LAEVIS

The thermostable protein isolated from eggs of *Xenopus laevis* (2) has an isoelectric point of 5 and consists of subunits with an apparent molecular weight on SDS gels of 29,000. Gel filtration, sedimentation and cross-linking with dimethyl suberimidate (J.O. Thomas unpublished) suggest that

ISBN 0-12-068350-4

the native form of the protein contains at least four subunits. Two types of subunit can be resolved on two-dimensional gels. Not only is the protein thermostable, but it is relatively resistant to trypsin, chymotrypsin, and proteinase K, though it is inactivated by digestion with proteinase K at pH values more alkaline than pH 8.0.

Another unusual property of the thermostable protein is solubility in acetic acid at pH 3.6 or in 2% trichloroacetic acid (TCA), a property which it shares with the HMG (high mobility group) proteins extracted from chromatin by Goodwin and Johns (4). This similarity is emphasized sharply when the amino acid composition of the protein is considered (Table 1).

TABLE 1
AMINO ACID COMPOSITION OF THE THERMOSTABLE PROTEIN

Amino acid	Moles %
Asp (+Asn)	10.74
Thr	6.65
Ser	5.55
Glu (+Gln)	18.72
Pro	6.23
Gly	7.98
Ala	7.42
Val	9.20
Cys	not detectable
Met	1.68
Ile	3.18
Leu	6.15
Tyr	1.50
Phe	2.00
His	0.89
Lys	10.87
Arg	1.21
Trp	not estimated

The thermostable protein from Xenopus eggs differs from calf thymus HMG 1 and HMG 2 in having only 10.9 moles % lysine compared to 19.3 moles % HMG 1 and HMG 2 (5) and in having significantly higher threonine, valine and leucine contents. However the remainder of the amino acid composition is conspicuously similar, the most notable features being that all three proteins contain 18 - 19 moles % glutamic acid (including glutamine) and 10 - 11 moles % aspartic acid (including asparagine). Although the apparent molecular weight of the subunits of the thermostable protein is roughly similar to that of the HMG proteins, HMG 1 and 2 appear to exist as single subunits unlike the multi subunit structure of the thermostable protein.

THE THERMOSTABLE PROTEIN IS NOT NICKING-CLOSING ENZYME

The thermostable protein separates from the egg's nicking-closing enzyme during chromatography on DEAE cellulose, phosphocellulose, CM cellulose or ECTEOLA cellulose. Its sedimentation, gel filtration and cross-linking properties indicate that it has a much higher molecular weight in solution, more subunits and a smaller subunit size than nicking-closing enzyme from other sources. We have been unable to detect nicking-closing activity in preparations of the thermostable protein, prepared with or without heating, and assayed under conditions which include those used for nucleosome assembly. We conclude that the thermostable protein is not nicking-closing enzyme.

THE THERMOSTABLE PROTEIN AND NICKING-CLOSING ENZYME CAN BOTH ASSEMBLE NUCLEOSOMES INDEPENDENTLY IN VITRO

Germond _et al_. (3) have demonstrated that 0.4 μg of highly purified nicking-closing enzyme can convert 1 μg DNA and 1 μg histones into nucleosomes. The purity of the nicking-closing enzyme and the low concentration required (3) strongly suggest that the assembly process mediated by nicking-closing enzyme _in vitro_ under these conditions does not require the type of factor purified from _Xenopus_ eggs.

Similarly as shown previously (2) and in Fig. 1 the thermostable protein from _Xenopus_ eggs can assemble a beaded nucleoprotein complex in the absence of nicking-closing enzyme. In addition incubation of DNA with the core histones and the purified thermostable protein results in protection of a characteristic pattern of discrete subnucleosomal DNA fragments from digestion by micrococcal nuclease (unpublished observation).

DO THE THERMOSTABLE PROTEIN AND NICKING-CLOSING ENZYME BOTH FULFIL THE CRITERIA EXPECTED OF THE CELLULAR MECHANISM OF NUCLEOSOME ASSEMBLY?

Both nicking-closing enzyme and the thermostable protein promote nucleosome assembly at physiological ionic strength and both reactions are inhibited above 150 mM NaCl (2,3). Approximately 20 times more purified nicking-closing enzyme is required for assembly than is required for relaxation of

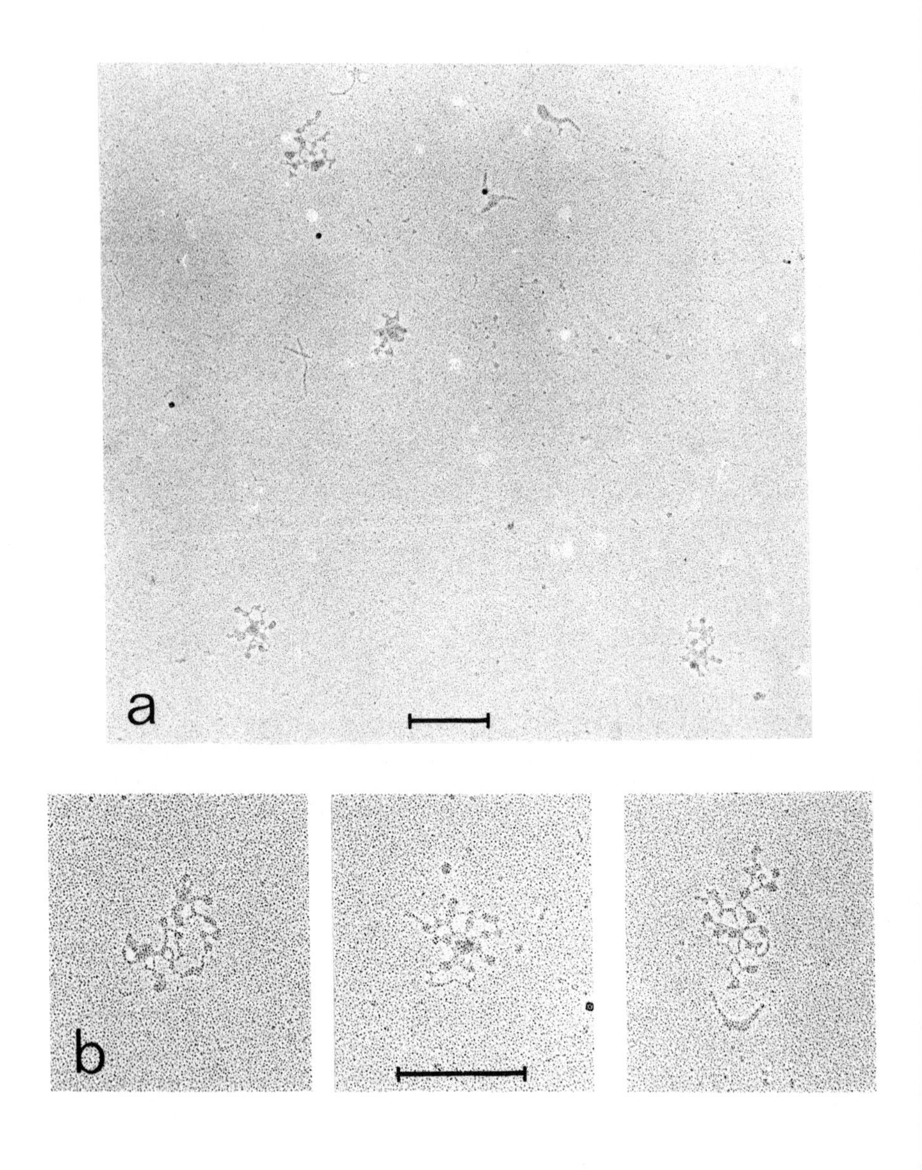

FIGURE 1. Nucleoprotein complexes obtained by incubating 0.5 μg form I PMB9 plasmid DNA with 0.9 μg of core histones and 1.25 μg thermostable protein without nicking-closing enzyme for 2 h at 21°C. Complexes were isolated by gradient centrifugation as described previously (2) and spread by the procedure of Thoma and Koller (11) without dilution. The scale bar represents 0.2 microns.

the same amount of DNA (3). Nevertheless the concentrations which assemble *in vitro* are roughly similar to those which occur within the cell nucleus (3).

The amount of the thermostable protein required to assemble 1 μg of DNA is 3-4 μg, i.e. 7-10 times more than the concentration of nicking-closing enzyme required to perform the same function. Therefore the thermostable protein remains a viable candidate for the function of nucleosome assembly only if it occurs in the nucleus at high concentrations. Ovulated unfertilized eggs of *Xenopus laevis* are in arrested meiotic metaphase and do not contain a nucleus. Therefore in collaboration with E.M. De Robertis and P. Black we have examined full-grown ovarian oocytes to determine if their nuclei contain the thermostable protein which can be extracted from eggs. We have looked for the protein using the criteria of its position and characteristic shape on two-dimensional gels and its selective solubility in TCA. We shall show elsewhere that a protein fulfilling these criteria is present in the oocyte nucleus, where it is one of the most abundant components. We have not detected it in oocyte cytoplasm. Its intranuclear concentration is approximately 5 mg/ml. The amount of the protein which can be extracted from the oocyte nucleus by TCA is similar to the amount which can be extracted by other methods from total egg homogenates (2). The intranuclear location of the thermostable protein in oocytes has been confirmed by micro-injecting the iodinated pure protein into oocyte cytoplasm and observing that it becomes highly concentrated within the nucleus (E.M. De Robertis and P. Black, unpublished).

We have found previously that high concentrations of nicking-closing enzyme are present in eggs and in oocyte nuclei of *Xenopus laevis* (2,6). In view of the finding of Germond *et al.* (3) our estimates of nicking-closing enzyme activity suggest that both nicking-closing enzyme and the thermostable protein are present in sufficient amounts to account for all the observed assembly activity independently of each other. However, neither factor alone exhibits all the main features of the unfractionated assembly system. In both cases the concentration range of histones and DNA which can be assembled is more limited that we observe for unfractionated egg homogenates. In particular Germond *et al.* found that ratios of histones to DNA greater than 1:1 inhibit assembly. However egg homogenates contain a histone pool sufficient for at least 6,000 diploid nuclei per egg (1,7,8) and this can assemble a wide range of DNA concentrations (1). Furthermore the capacity for assembly can be increased by adding exogenous histones and then the optimum histone to DNA ratio is 2:1. The same optimum ratio

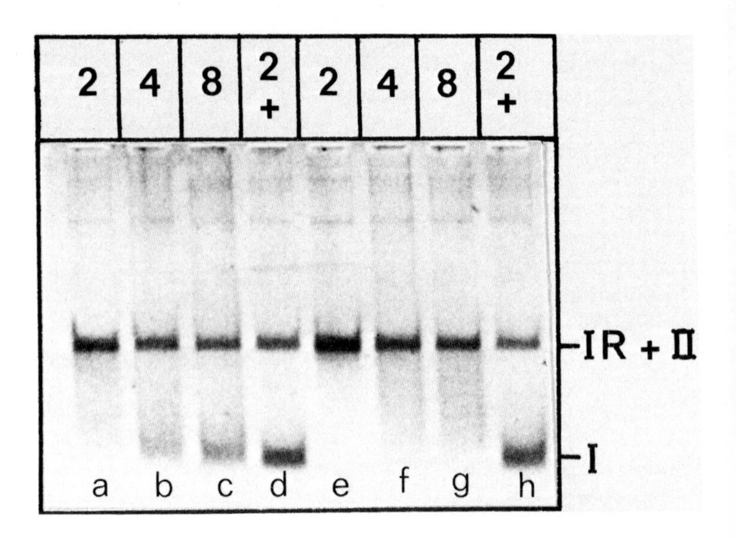

FIGURE 2. Effect of the thermostable protein on the insertion of superhelical turns into DNA by nicking-closing enzyme in the presence of histones. 0.1 μg DNA and 0.2 μg histones were incubated (a-c) with 2, 4 or 8 μl of an unfractionated nicking-closing enzyme preparation (9) from 3T6 mouse fibroblasts or with a partially purified nicking-closing enzyme from Xenopus eggs (2). (d) and (h) are the same as (a) and (e) except that 0.3 μg of the purified thermostable protein were added. DNA was added last to each incubation. Conditions of incubation and electrophoresis in a 1% agarose gel were described previously (2).

is observed for the thermostable protein. Although we have observed assembly with crude unfractionated preparations of nicking-closing enzyme (Fig. 2) the thermostable protein stimulates this assembly at histone to DNA ratios of both 2:1 and 1:1, though the optimum concentration of the thermostable protein is lower at histone to DNA ratios of 1:1.

The ability of the thermostable protein to promote nucleosome assembly in the presence of excess histones could indicate that it is an adaptation involved in the accumulation and mobilization of the egg's large histone pool. We know of no evidence against this possibility at present. We have isolated an acidic thermostable protein from wheat embryos which also assembles nucleosomes, but the subunits of this protein are larger than those of the Xenopus protein and we do not yet know if the two proteins are structurally related. Mouse fibroblast nuclei also contain an assembly activity which resists 10 min at 60°C (but is decreased by 10 min at 80°C), though we have not identified it yet, nor shown that its action is independent of nicking-closing enzyme. The possibility that these activities are related remains to be tested.

In view of the extensive literature on proteins which can be extracted from somatic nuclei by TCA we feel it is unlikely that an identical form of the thermostable protein exists in high concentrations in somatic nuclei. However we suggest that further structural and functional relationships of this protein to the HMG proteins deserve to be investigated.

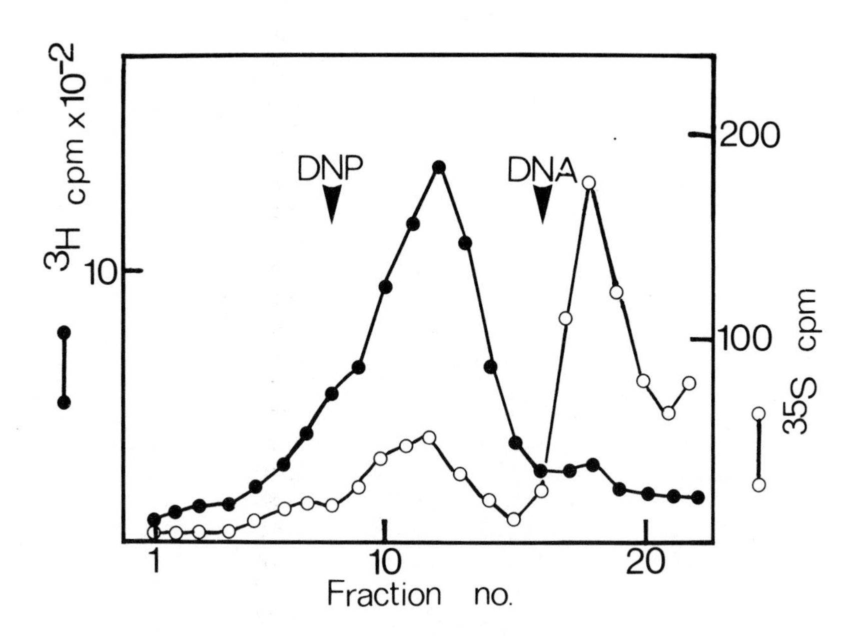

FIGURE 3. Transfer of ^{35}S histones from a complex with the thermostable protein to DNA. A complex between ^{35}S histones and the thermostable protein was isolated by sucrose gradient centrifugation as described previously (2). An aliquot of 100 μl from the sucrose gradient peak (made 100 mM in NaCl) containing approximately 0.3 μg histones, O — O (7,000 cpm/μg) and 0.6 μg thermostable protein was incubated 3 hr at 20°C with 0.15 μg SV40 ^{3}H-DNA form I ● — ● (10^5 cpm/μg). The nucleoprotein product was analysed by centrifugation as described previously (2). Arrows indicate the positions of free DNA and the nucleoprotein complex assembled by unfractionated egg homogenates.

POSSIBLE MECHANISMS OF NUCLEOSOME ASSEMBLY

The thermostable protein from *Xenopus* eggs binds histones and causes them to cosediment with it in sucrose gradients (2). Fig. 3 shows that the bound histones can be transferred to DNA with the formation of a nucleoprotein complex. The transfer does not require nicking-closing enzyme but addition of low concentrations of partially purified nicking-closing enzyme reveals insertion of superhelical turns. The specificity of the interaction between histones and the thermostable protein requires further investigation, but preliminary results obtained in collaboration with J.O. Thomas indicate that the thermostable protein induces cross-linkable interactions between the core histones. We have not observed any affinity of the thermostable protein for DNA and labelled preparations of this protein do not associate with the chromatin product detectably. Because nucleosomes can be reconstituted by dialysis of DNA and histones from 2 M NaCl, we previously proposed (2) that the thermostable protein might act simply by preventing random ionic interactions between histones and DNA, and allowing only certain specific interactions to occur.

Germond *et al*. (3) suggest that nicking-closing enzyme promotes assembly by interacting directly with DNA since they observe better assembly when these components are preincubated before addition of histones. Since neither the thermostable protein nor nicking-closing enzyme is essential for nucleosome assembly at physiological ionic strength, it appears to us to be most likely that they both act by minimizing non-specific electrostatic interactions between histones and DNA while allowing more specific interactions to occur. This interpretation is also one of those favoured by Germond *et al*. (3).

When exogenous DNA is introduced into intact or disrupted eggs and oocytes of *Xenopus laevis* it is immediately acted on by high levels of nicking-closing enzyme (1, 6 and unpublished observations). In addition, at least part of the egg's histone pool and exogenous histones added to homogenates appear to be bound to a component with similar properties to the thermostable protein (2,10). Therefore, in *Xenopus* eggs it appears likely that both the DNA and the histones are 'masked' by association with other components before they interact to form nucleosomes. If so, it will be interesting to determine whether this is a specialised adaptation for pre-accumulating a histone pool or a general feature of nucleosome assembly.

Acknowledgements

We are grateful to Drs. Germond, Brutlag, Yaniv, and Rouvière-Yaniv for permission to discuss their work before its publication.

References

1. Laskey, R.A., Mills, A.D., and Morris, N.R. (1977). Cell 10, 237.
2. Laskey, R.A., Honda, B.M., Mills, A.D., and Finch, J.T. (1978). Nature. 275, 416.
3. Germond, J-E., Brutlag, D., Yaniv, M., and Rouvière-Yaniv, J. Manuscript submitted.
4. Goodwin, G.H. and Johns, E.W. (1973). Eur. J. Biochem. 40, 215.
5. Walker, J.M., Goodwin, G.H., and Johns, E.W. (1976). Eur. J. Biochem. 62, 461.
6. Wyllie, A.H., Laskey, R.A., Finch, J.T., and Gurdon, J.B. (1978) Devl. Biol. 64, 178.
7. Adamson, E.D. and Woodland, H.R. (1974). J. Mol. Biol. 88, 263.
8. Woodland, H.R. and Adamson, E.D. (1977). Devl. Biol. 57, 118.
9. Germond, J-E., Hirt, B., Oudet, P., Gross-Bellard, M., and Chambon, P. (1975). P.N.A.S. 72, 1843.
10. Laskey, R.A., Honda, B.M., Mills, A.D., Morris, N.R., Wyllie, A.H., Mertz, J.E., De Robertis, E.M., and Gurdon, J.B. (1978). Cold Spring Harbor Symp. Quant. Biol., 42, 171.
11. Thoma, F. and Koller, Th. (1977). Cell 12, 101.

CHROMATIN STRUCTURE OF A SPECIFIC GENE IN ACTIVE AND INACTIVE STATES: A BRIEF REVIEW [1]

Carl Wu and Sarah CR Elgin

Harvard University, The Biological Laboratories,
16 Divinity Avenue, Cambridge, MA 02138

ABSTRACT We have studied the chromatin structure of specific genes in Drosophila by using the nucleolytic enzymes DNase I and micrococcal nuclease as probes of DNA accessibility in chromatin. ^{32}P-labelled, cloned Drosophila DNA has been used as sequence-specific "stains" of the resultant DNA fragments fractionated on agarose gels and transferred onto nitrocellulose sheets by the Southern blotting technique. The DNase I cleavage patterns of specific chromosomal regions so examined reveal discrete DNA fragments ranging in size from approximately 2 kilobase pairs to greater than 20 kilobase pairs of DNA. Different chromosomal regions yield uniquely different cleavage patterns. From restriction enzyme analyses, the cleavage sites are shown to be position-specific with respect to DNA sequence. We refer to the chromosomal regions bounded by the position-specific, preferential DNase I cleavage sites as supranucleosomal or higher order domains of defined DNA sequence. These "domains" may potentially define a new level of structural organization in the eukaryotic chromosome. In addition to the typical oligonucleosome pattern generated by micrococcal nuclease digestion of nuclear chromatin, the micrococcal nuclease cleavage patterns of specific chromosomal regions also reveal larger discrete DNA fragments. At least some of the micrococcal nuclease cleavage sites are position-specific. The discrete DNA fragment patterns generated by DNase I and micrococcal nuclease digestion of chromosomal regions encoding the major heat shock

[1]This work was supported by grants GM 20779 and GM 24069 from the National Institutes of Health to S.C.R.E. C.W. is also supported by a scholarship from the Harvard Faculty of Arts and Sciences. S.C.R.E. is supported by an NIH Research Career Development Award (GM 00234).

ISBN 0-12-068350-4

protein in uninduced Drosophila tissue culture cells may be taken as a criterion of structural integrity in chromatin. Both specific cleavage patterns are smeared to a considerable extent in experiments using heat shocked tissue culture cells, indicating a disruption in nucleosomal and possibly in higher order organization during gene activity. The discrete cleavage patterns can be generated again when the cells are removed from heat shock and allowed to recover.

INTRODUCTION

Since the establishment of the nucleosome as a paradigm for the first level of DNA packaging in chromatin, a problem of some interest has been whether the chromatin structure of active genes is organized in a similar manner. Initial evidence had shown that transcribed sequences could be recovered in 11S nucleoprotein particles (1), indicating that at least some semblance of nucleosome structure is maintained during gene expression. However, Weintraub and Groudine reported that the active globin gene is preferentially susceptible to DNase I, suggesting an alteration in chromatin structure during gene activity (2). This observation has been confirmed in the globin case as well as in other systems in several laboratories (3, 4, 5). One may then consider whether or not nucleosome organization as judged by the characteristic pattern of cleavage of nuclear chromatin by micrococcal nuclease is preserved in active genes. Several recent advances in technology, as well as characterization of the heat shock response in Drosophila tissue culture cells, make the question amenable to detailed experimental analysis. The DNA fragment pattern generated from chromatin in general after micrococcal nuclease digestion may be observed by agarose gel electrophoresis and ethidium bromide staining. The pattern of cleavage at specific chromosomal regions can then be visualized by making a "print" of the general DNA fragment pattern onto a nitrocellulose sheet by the Southern blotting technique followed by filter hybridization with radioactively labelled, cloned DNA sequences from loci of interest. In Drosophila, elevated environmental temperatures (heat shock) result in the new expression of a limited number of genes; several of these genes have now been cloned (6, 7, 8). One may therefore analyze the micrococcal nuclease patterns of chromosomal regions encoding the major heat shock protein, and of other control loci, from a Drosophila tissue culture cell line in the normal or heat shock induced state. The recombinant plasmid probes used in this study were

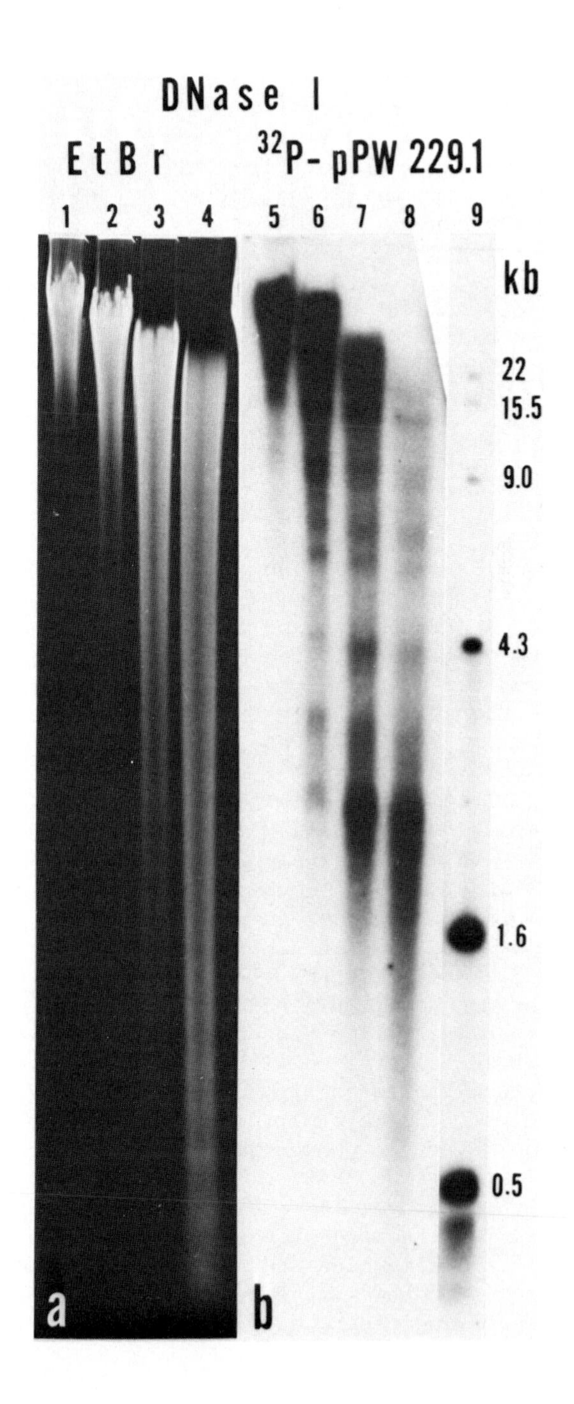
DNase I
EtBr
^{32}P- pPW 229.1
1
2
3
4
5
6
7
8
9
kb
22
15.5
9.0
4.3
1.6
0.5
a
b

obtained and characterized by M. S. Meselson and his colleagues in collaboration with P. Wensink. In addition to the micrococcal nuclease cleavage patterns, the DNase I patterns of cleavage at specific loci were also examined as a further means of studying the preferential susceptibility of active chromatin to DNase I. The results obtained from the latter studies were surprising; DNase I, a non-specific nuclease, was observed in limited digestions to generate a specific DNA fragment pattern of discrete bands from individual loci, instead of the anticipated random fragment pattern. The results of these studies are reported in detail in two recent communications (9, 10). We present here a summary of the major findings.

RESULTS

Nuclei were isolated from Drosophila tissue culture cells (Schneider Line 2) maintained under normal conditions (25° C) and digested with increasing concentrations of DNase I (0% to approximately 4% acid-solubility of the DNA). The DNA was purified and electrophoresed on agarose gels. After staining with ethidium bromide, a photographic record of the pattern of cleavage of the total genome was made (Figure 1a). DNA fragments in the gel were then transferred onto a sheet of nitrocellulose by the method of Southern (11). The specific cleavage pattern of chromosomal regions homologous to pPW 229.1 is visualized (Figure 1b) by filter hybridization and autoradiography, using ^{32}P-labelled pPW 229.1 as sequence probe. pPW 229.1 contains an 800 bp Drosophila DNA insert, localized entirely within sequences encoding the major heat shock protein. It is repeated approximately 5 times in the

---

Figure 1. Comparison of the General DNase I Digestion Pattern with the DNase I Digestion Pattern of Sequences Homologous to pPW 229.1 in Drosophila Tissue Culture Cells. (a) 0.8% agarose gel stained with ethidium bromide. Electrophoresis is from top to bottom. The sample load is 10 μg DNA/slot. Digestions were carried out for 3 minutes (25° C) at the following enzyme concentrations: slot 1,6 units/ml; slot 2,8 units/ml; slot 3,12 units/ml; slot 4,16 units/ml. (b) Southern blot of gel (a) probed with ^{32}P-pPW 229.1. Slot 9 is a track blotted from the same gel containing marker restriction fragments which share sequence homology to the labelled plasmid probe. The sizes of the bands seen in slot 7 are, in increasing order, ca. 2.4 (doublet), 3.2, 4.2, 5.8, 6.7, 9.0 (doublet) and 14.4 kb respectively. From reference 9.

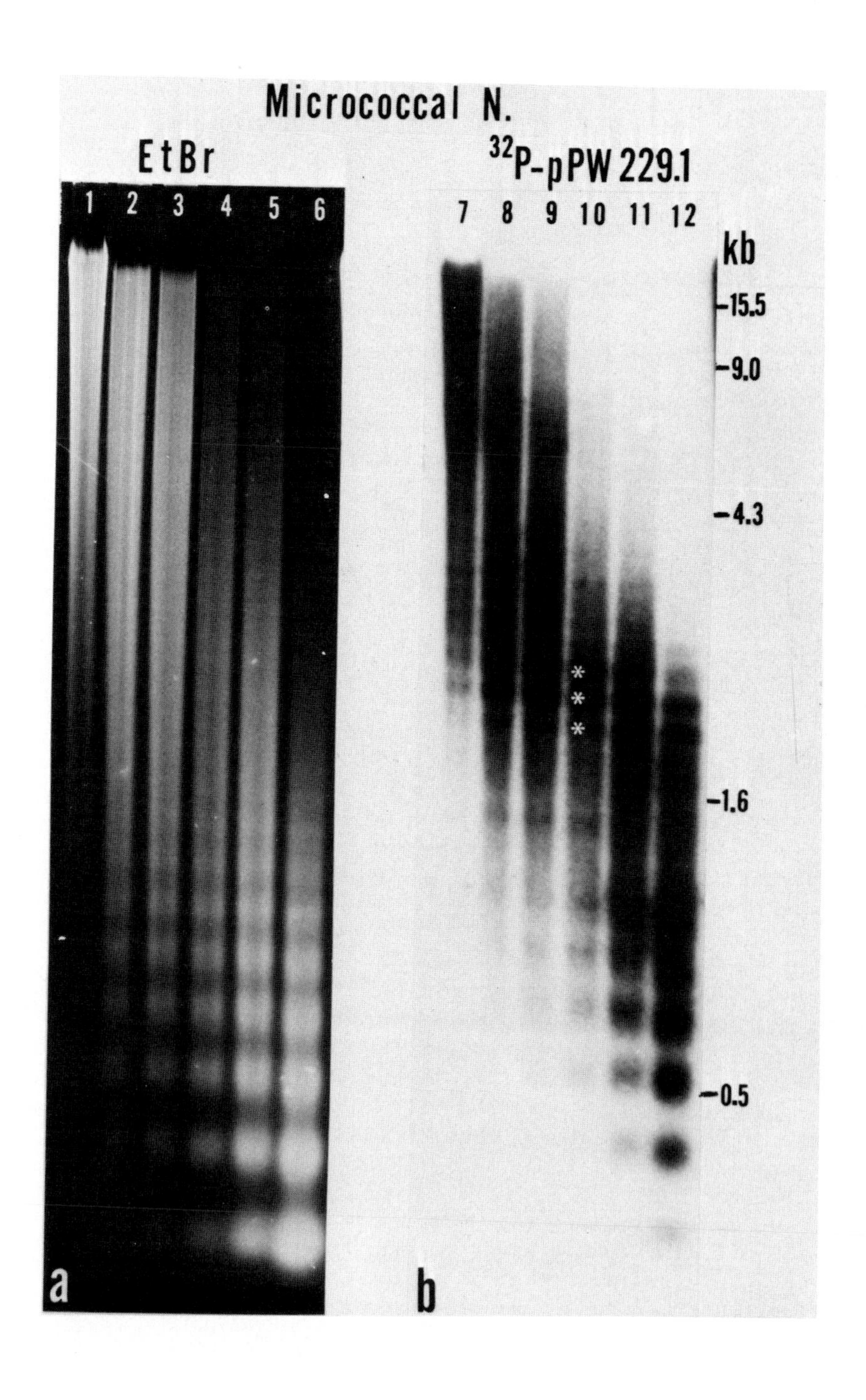
Micrococcal N.
EtBr
1 2 3 4 5 6
32P-pPW 229.1
7 8 9 10 11 12
kb
15.5
9.0
4.3
1.6
0.5
a
b

genome, and maps in situ to loci 87A and 87C on the Drosophila polytene chromosomes (8). A striking difference between general and specific cleavage patterns is observed. 7 discrete bands are evident in Figure 1b, ranging from approximately 2 kb to 14.4 kb in size; an additional band of approximately 20 kb can be seen in gels of higher resolution (12). A similar pattern is observed using nuclei isolated from Drosophila embryos (ref. 12; see also Figure 3). This unique pattern is not observed when naked Drosophila DNA is digested with DNase I and subjected to a similar analysis; only a smear of fragments of decreasing mean size is obtained (9). Thus the specificity of cleavage by DNase I must reflect the properties of the chromatin complex. Different Drosophila DNA clones employed in studying the cleavage patterns from other chromosomal regions yield uniquely different cleavage patterns. Four other loci have been investigated thus far (9). It appears, therefore, that the general smeared pattern seen with ethidium bromide staining may be a collection of unique discrete cleavage patterns from the entire genome.

A comparison of the general pattern of cleavage using increasing concentrations of micrococcal nuclease (0% to approx. 6% acid-solubility of the DNA) on isolated nuclei from normal tissue culture cells with the specific cleavage pattern of chromosomal regions homologous to pPW 229.1 is shown in Figure 2. A pattern typical of oligonucleosomal organization is observed in both cases; additionally, in the region of the filter around 2 kb, where the general nucleosomal pattern loses resolution, a set of sharp bands are prominently observed with pPW 229.1. A different unique band pattern above the oligonucleosomal pattern is observed using different clones as sequence probe (e.g. see Figure 4).

Figure 2. Comparison of the General Micrococcal Nuclease Digestion Pattern with the Pattern from Sequences Homologous to pPW 229.1 in Drosophila Tissue Culture Cells. (a) 1% agarose gel stained with ethidium bromide. Digestions were carried out for 3 minutes (25° C) at the following enzyme concentrations: slot 1 9 units/ml; slot 2, 12 units/ml; slot 3, 17 units/ml; slot 4, 23 units/ml; slot 5, 35 units/ml; slot 6, 71 units/ml. (b) Southern blot of gel (a) probed with ^{32}P pPW 229.1. The bands marked by asterisks in slot 10 are ca. 2.0, 2.2 and 2.4 kb in size. The nucleosome monomer fragments are hardly visible on the autoradiogram because DNA fragments in this size range do not stick to nitrocellulose very well. From reference 9.

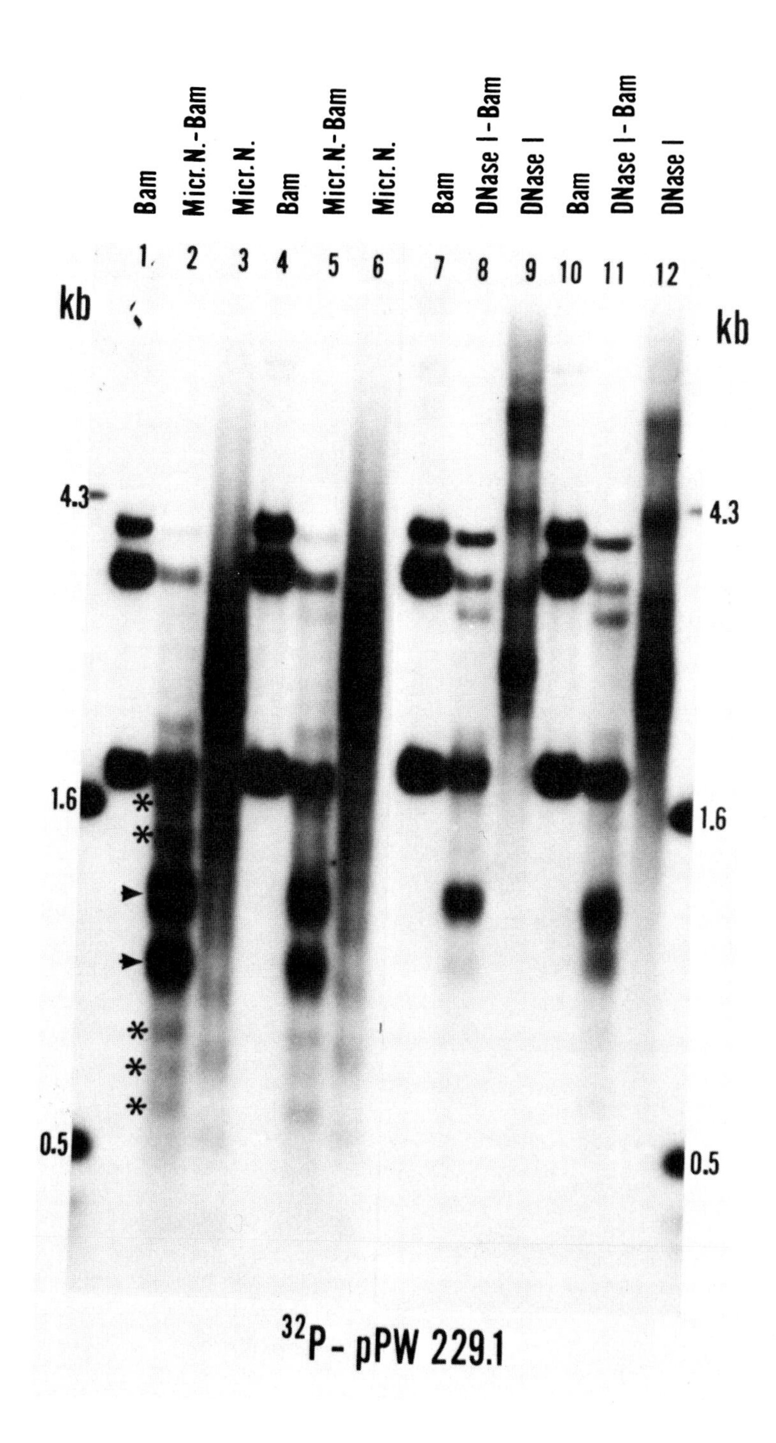
Bam
Micr. N.-Bam
Micr. N.
Bam
Micr. N.-Bam
Micr. N.
Bam
DNase I-Bam
DNase I
Bam
DNase I-Bam
DNase I
1 2 3 4 5 6 7 8 9 10 11 12
kb
kb
4.3
4.3
1.6
1.6
0.5
0.5
³²P-pPW 229.1

The presence of a unique and discrete DNA fragment pattern generated by cleavage with DNase I, and to a lesser extent by micrococcal nuclease, at particular chromosomal loci suggests that the cleavage sites are position-specific with respect to DNA sequence. To demonstrate this directly, purified DNA samples from DNase I digests and micrococcal nuclease digests of nuclear chromatin were subjected to restriction enzyme cleavage followed by Southern analysis, using pPW 229.1 as sequence probe The results are shown in Figure 3. In both cases, new fragments are generated in the double-digested samples which are not produced by DNase I digestion of chromatin nor by restriction cleavage of naked genomic DNA alone. The data indicate that at least some of the nucleolytic cleavage sites are position-specific. We refer to the chromosomal regions bounded by the preferential DNase I cleavage sites as supranucleosomal or higher order domains. The preferential DNase I cleavage sites may reflect the boundaries of higher order structures or they may simply be especially susceptible regions along an otherwise continuous chromatin fiber. Further work will be required to distinguish between these alternatives. The specificity of cleavage shown by micrococcal nuclease is intriguing. If one assumes that this nuclease cleaves only at internucleosomal regions, then the results suggest a non-random positioning of nucleosomes with respect to DNA sequence. This point requires further detailed analysis.

Employing the discrete DNase I and micrococcal nuclease cleavage patterns at sequences homologous to pPW 229.1 as criteria of integrity in the chromatin structure of those sequences, one may examine the respective cleavage patterns generated by digestion of isolated nuclei from cells subjected to a 15-minute heat shock stimulus. The discrete DNase I pattern is observed to be smeared considerably in the heat shock induced cells. In addition, the active loci were more

---

Figure 3. Comparative Restriction Enzyme Analysis of Higher Order Bands Produced by Micrococcal Nuclease and DNase I Cleavage. All samples (10 μg DNA/slot) were electrophoresed on the same 1.1% agarose gel, blotted and hybridized with ^{32}P-pPW 229.1. Slots 1-3 and 7-9 are DNA from Drosophila tissue culture cells; 4-6 and 10-12 are DNA from embryos. Slots 1, 4, 7 and 10 are Bam H1 digests of purified Drosophila DNA. Slots 2, 5, 8 and 11 are Bam H1 digests of DNA purified after micrococcal nuclease (2, 5) or DNase I (8, 11) digestion of nuclei. Slots 3, 6, 9 and 12 micrococcal (3, 6) or DNase I (9, 12) digests of nuclei. From reference 9.

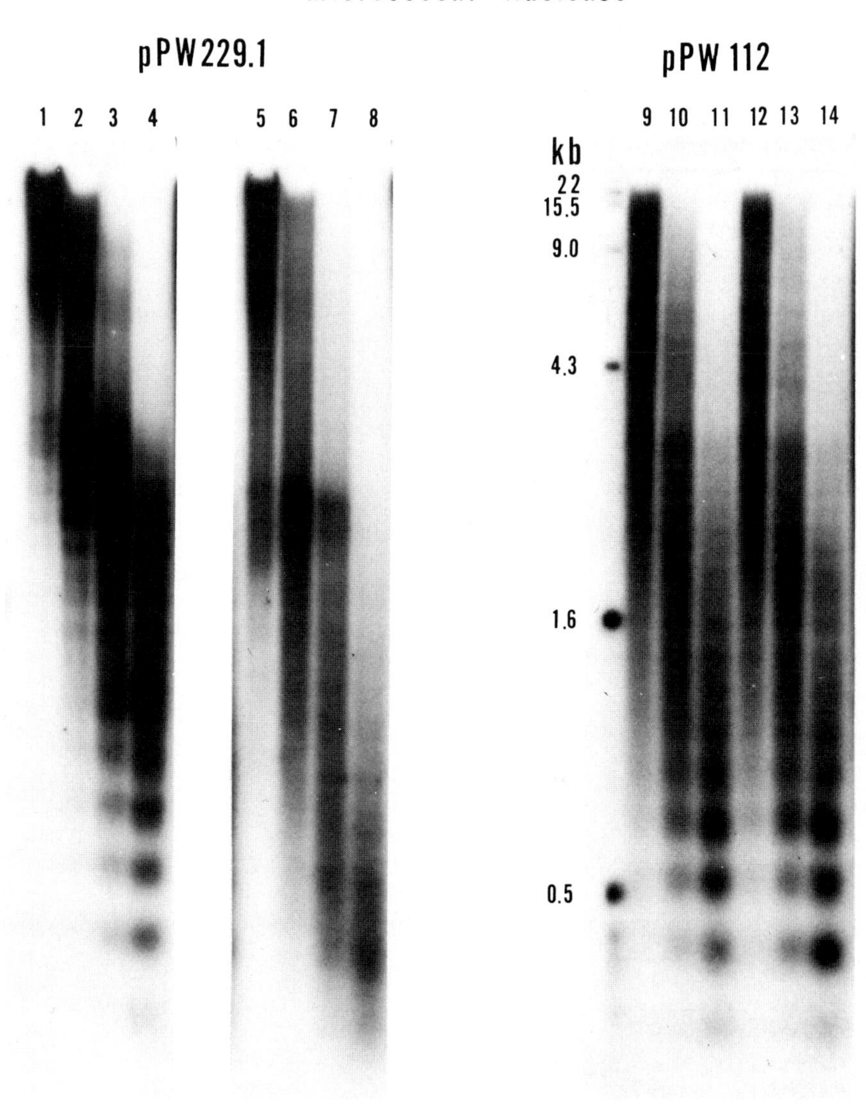

Micrococcal Nuclease
pPW229.1
pPW 112
1 2 3 4
5 6 7 8
9 10 11 12 13 14
kb
22
15.5
9.0
4.3
1.6
0.5

rapidly cleaved to small DNA fragments, confirming in this system the results of Weintraub and Groudine (2). The discrete DNase I cleavage pattern unique to a locus not activated by heat shock is retained in heat shocked cells (10).

The micrococcal nuclease cleavage pattern of active sequences homologous to pPW 229.1 is shown in Figure 4. Here the cleavage pattern indicative of oligonucleosomal organization is also considerably smeared upon heat shock, while a locus not activated by heat shock (pPW 112: 78A) retains its structural integrity during heat shock.

The smearing of the micrococcal nuclease cleavage pattern could be due to an alteration in the relative susceptibility to cleavage between nucleosome core and linker regions and/or to a random sliding of intact nucleosome cores during transcription. The smearing of the discrete DNase I cleavage pattern might reflect the increased DNase I susceptibility of the transcribed regions and/or a generalized disruption of the higher order domains. Further work is required to clarify these points. The DNase I and micrococcal nuclease cleavage patterns obtained from cells allowed to recover for 3 hours at 25° C after a 15-minute heat shock stimulus are similar to those from uninduced tissue culture cells (10).

DISCUSSION

The experiments summarized here extend our knowledge of chromatin structure in relation to gene activity. However, they also raise many intriguing questions. Is the organization of chromatin into higher order domains revealed by the small number of clones used here representative of the whole Drosophila genome and of other eukaryotes? What is the physical basis for the observed locus-specific DNase I

Figure 4. The Disruption of Chromatin Structure upon Gene Activation. All samples (10 μg DNA/slot) were electrophoresed on the same 1.1% agarose gel, blotted and hybridized with ^{32}P-pPW 229.1 (slots 1-8) or ^{32}P-pPW 112 (slots 9-14). Samples were either from tissue culture cells maintained at 25° C (slots 1-4 and 9-11) or heat shocked at 35° for 15 minutes (slots 5-8 and 12-14). Digestions (3 min., 25° C) were carried out with micrococcal nuclease at the following concentrations: slots 1 and 5- 12 units/ml; slots 2, 6, 9, 12- 23 units/ml; slots 3, 7, 10, 13- 47 units/ml; slots 4, 8, 11, 14- 94 units/ml.

cleavage patterns? What is the molecular basis of the smearing of both cleavage patterns in active genes and what are the spatial limits of the underlying alterations in chromatin structure? Solutions to these and other related questions will help determine the role of chromatin structure in the process and control of gene activation in higher organisms.

ACKNOWLEDGEMENTS

We thank Cell, the M.I.T. Press, for permission to reprint figures 1, 2 and 3.

REFERENCES

1. Lacy, E. and Axel, R. (1975). Proc. Natl. Acad. Sci. USA 72, 3978.
2. Weintraub, H. and Groudine, M. (1976). Science 193, 848.
3. Garel, A. and Axel, R. (1977). Proc. Natl. Acad. Sci. USA 73, 3966.
4. Flint, S. J. and Weintraub, H. M. (1977) Cell 12, 783.
5. Young, N. S., Benz, E. J. Jr., Kantor, J. A., Kretschmer, P. and Nienhuis, A. W. (1978). Proc. Natl. Acad. Sci. USA 75, 5884.
6. Lis, J. T., Prestidge, L. and Hogness, D. S. (1978). Cell 14, 901.
7. Schedl, P., Artavanis-Tsakonas, S., Steward, R., Gehring, W., Goldschmidt-Clermont, M., Moran, L. and Tissieres, A. (1978). Cell 14, 921.
8. Livak, K. J., Freund, R., Schweber, M., Wensink, P. C. and Meselson, M. (1978). Proc. Natl. Acad. Sci. USA 75, 5613.
9. Wu, C., Bingham, P. M., Livak, K. J., Holmgren, R. and Elgin, S.C.R. (1979). Cell, in press.
10. Wu, C., Wong, Y.-C. and Elgin, S.C.R. (1979). Cell, in press.
11. Southern, E. M. (1975). J. Molec. Biol. 98, 503.
12. Wu, C. (1979). Ph.D. Thesis, Harvard University.

ABSENCE OF NUCLEOSOMES IN THE ACTIVELY TRANSCRIBED rDNA OF CALLIPHORA ERYTHROCEPHALA LARVAL TISSUE AND OF QUAIL MYOBLASTS[1]

Milan Jamrich,[2] Ellen Clark, and O. L. Miller, Jr.

Department of Biology, University of Virginia,
Charlottesville, Virginia 22901

ABSTRACT. Chromatin of Calliphora larval tissue and of tissue cultured quail myoblasts was prepared for electron microscopic analysis using modifications of the technique developed by Miller and Beatty (1969). The chromatin of actively transcribed rDNA was compared to transcriptionally inactive chromatin. The actively transcribed rRNA genes are present in an altered, non-nucleosomal configuration. Nucleosome-free regions are also present beyond the putative termination sites of the transcription units. These regions vary in length and occasionally include an entire non-transcribed spacer.

INTRODUCTION

Electron microscopic analysis of spread chromatin provides an approach to the study of structural changes in chromatin associated with transcriptional activity. Transcriptionally inactive chromatin is present in a condensed, nucleosomal configuration (1,4,12). Nuclease digestion experiments (6,13) have suggested that ribosomal chromatin is organized into nucleosomal structures. More recently, however, the same authors (7,14) demonstrated that there is a biochemically measurable alteration in ribosomal chromatin structure coincident with changes in transcriptional activity. Electron microscopic observations of nucleolar chromatin (2,3,5,9,15,16,17,18) support the hypothesis that the actively transcribed rDNA occurs in an extended, non-nucleosomal configuration. We have examined the morphology of actively transcribed rRNA genes from Calliphora larval tissue and from quail myoblasts. Examination

[1] This work was supported by NIH grant number 5RO1 GM21020 to O.L.M. M.J. is a Fellow of the Jane Coffin Childs Memorial Fund for Medical Research.

[2] Present address: Department of Biology, Kline Biology Tower, Yale University, New Haven, Conn., 06520.

ISBN 0-12-068350-4

of Calliphora larval rRNA genes allows a correlation of transcription frequency with changes in chromatin morphology, since RNA polymerase density changes significantly during larval development. Quail myoblast rRNA genes are seldom fully loaded with RNA polymerase molecules, permitting observation of chromatin segments between polymerases.

METHODS

Chromatin of Calliphora larval tissues and of tissue cultured quail myoblasts was prepared for electron microscopic observation using modifications of the technique of Miller and Beatty (1969; also see Miller and Bakken, 1972). Spreading conditions were varied in pH (7.5-9.0), in detergent concentration (0-0.1% Joy, 0-0.1% Digitonin), in presence of tRNA (0-100 μg/ml), and in degree of agitation.

RESULTS AND DISCUSSION

Two types of chromatin structure are observed in spreads of Calliphora larval tissue and quail myoblasts. The bulk of the chromatin has a beaded, nucleosomal morphology (Fig. 1). The chromatin segments visible between RNA polymerase molecules on actively transcribed rRNA genes are usually in a relatively smooth, non-nucleosomal configuration (Fig. 1,2,3). Smooth chromatin is also frequently observed in the non-transcribed spacers of Calliphora rDNA. As shown in Figures 2 and 4, a spacer may be partially or almost entirely free of nucleosomes; the length of this smooth region, when present, varies considerably from one spacer to another.

Smooth chromatin associated with active transcription is observed under a variety of spreading conditions (described in Methods). Though an artefactual change in chromatin morphology by these preparative methods cannot be ruled out, it is clear that rRNA chromatin is not organized in the nucleosomal configuration of inactive chromatin, since the active chromatin does behave differently under the applied spreading conditions. Furthermore, it is apparent that this change is not limited to the immediate vicinity of polymerase molecules (see Fig. 1).

In attempting to establish a functional relationship between transcription and chromatin morphology, the possibility that the structure of rDNA chromatin is dependent on the number of associated RNA polymerases was ruled out by the observation of V. Foe et al (2) that in Oncopeltus rRNA genes, smooth chromatin is observed with a low density of transcribing polymerases or even in the absence of polymerases. We have made similar observations of smooth chromatin in rRNA genes of Calliphora larval tissue transcribed by as few as 4 polymerase

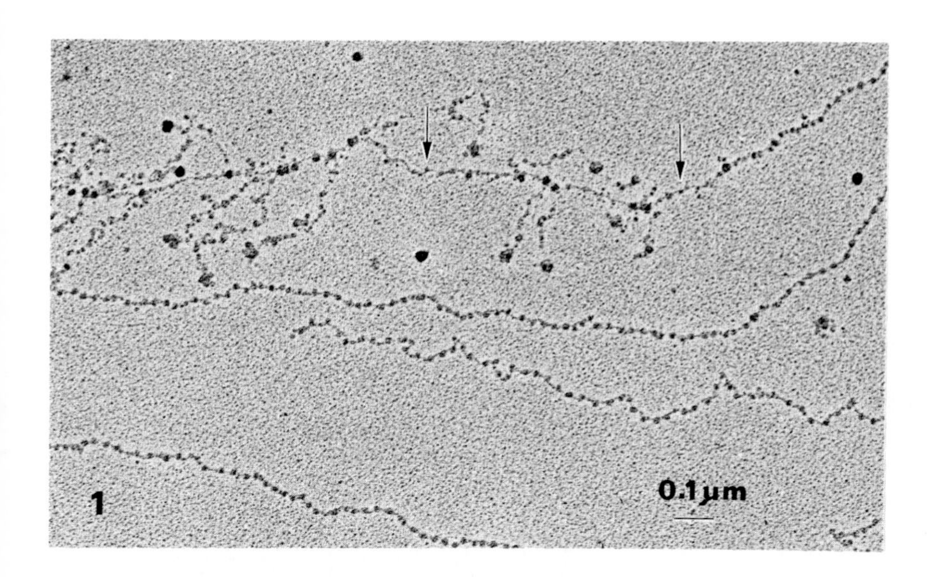

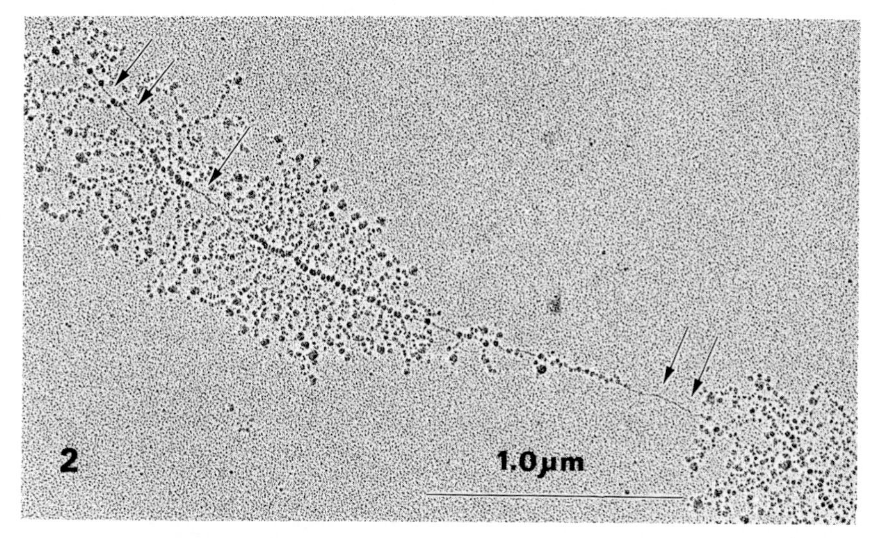

FIGURE 1. rRNA gene with low RNA polymerase density. Inactive chromatin is present in nucleosomal configuration, chromatin segments between RNA polymerase molecules are present in an altered, non-nucleosomal configuration. Arrows indicate possible transient structures (Calliphora fat body, one hour old larva).

Figure 2. rRNA genes from 6 hour old larval fat body. Arrows indicate smooth chromatin in the non-transcribed spacer and between RNA polymerases.

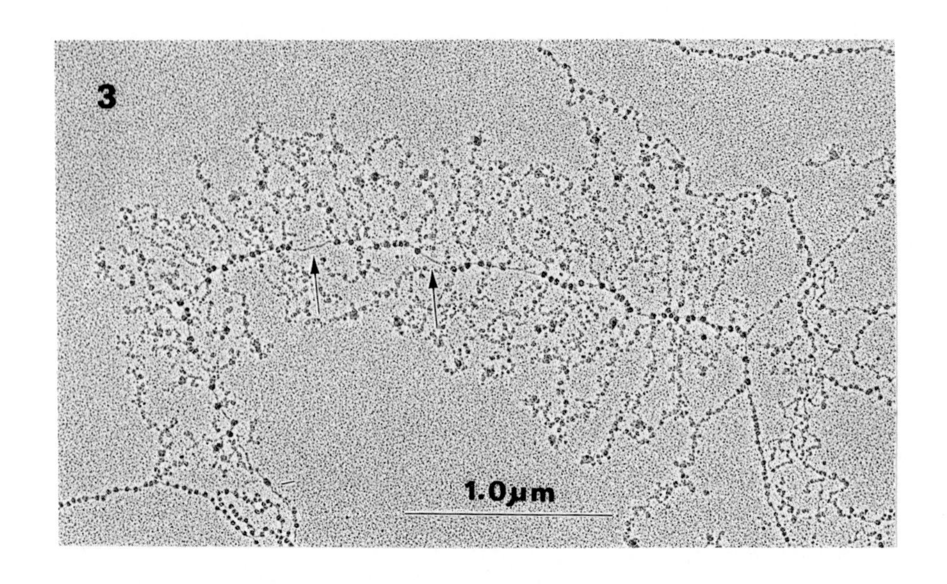

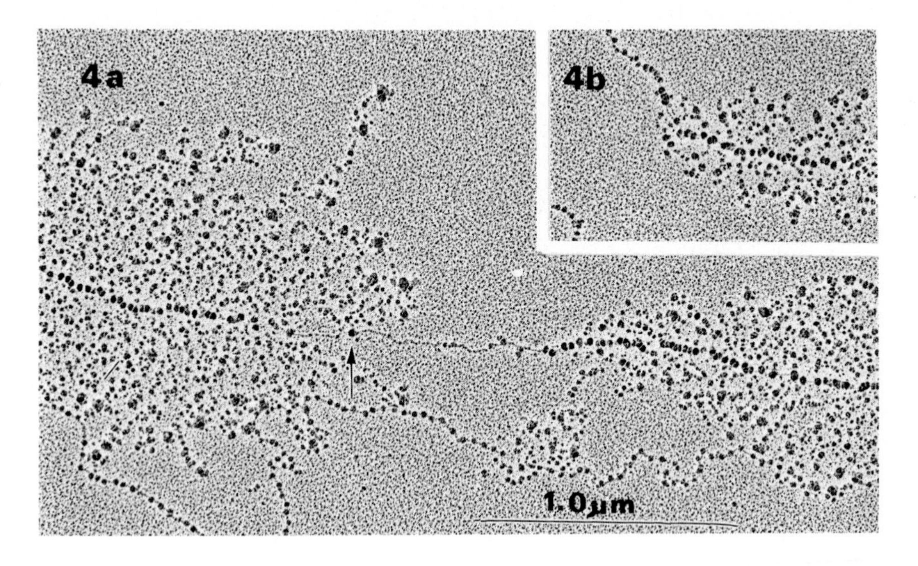

FIGURE 3. rRNA gene from quail myoblast. Arrows indicate the smooth chromatin in the transcribed region.

FIGURE 4a. Smooth spacer visible in spread chromatin from 2 day old larval fat body. Arrow indicates an incorrectly terminated RNA polymerase molecule.

FIGURE 4b. Beaded chromatin before initiation site (see text).

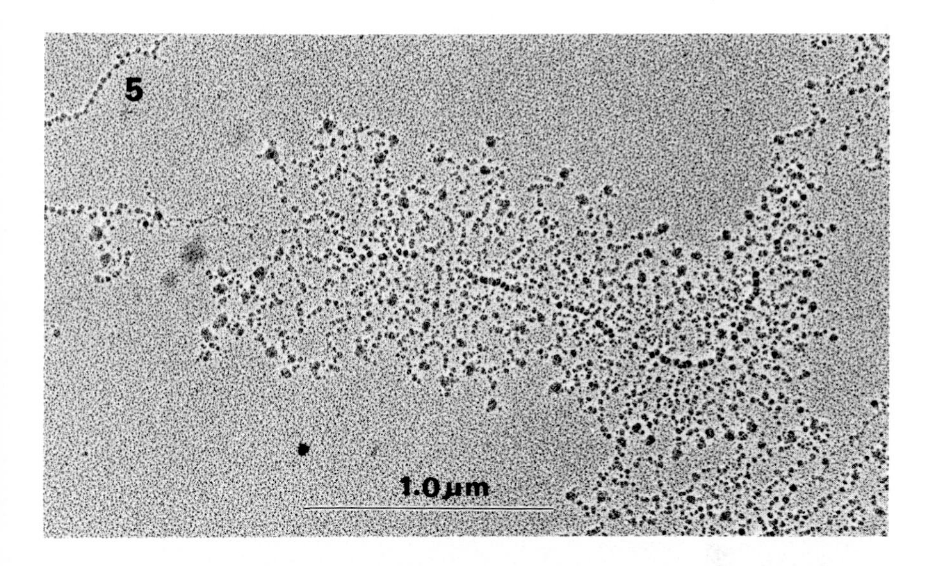

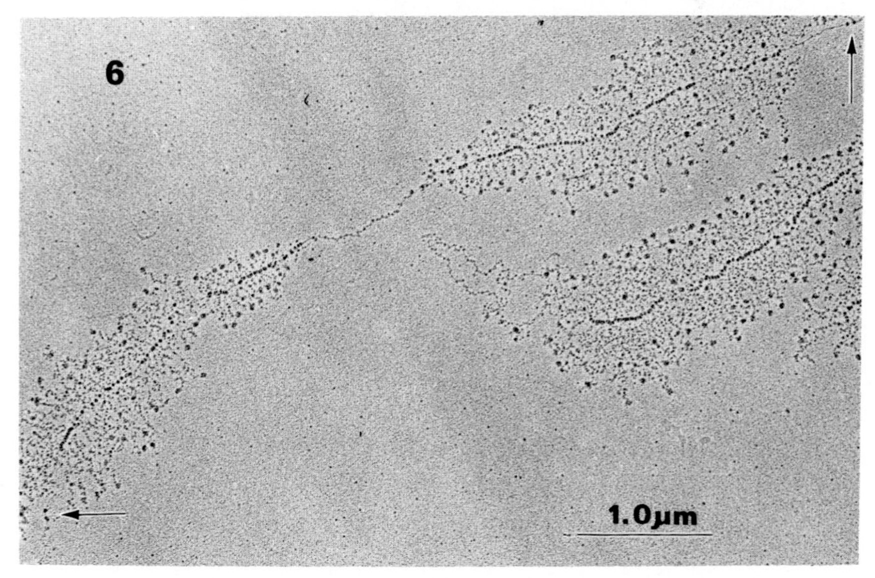

FIGURE 5. Smooth chromatin beyond rRNA gene termination site (6 hour old larval fat body).

FIGURE 6. rRNA genes of divergent polarity. Note the chromatin with a beaded morphology between the two initiation sites, the smooth chromatin beyond the termination sites. Arrows indicate incorrectly terminated RNA polymerase molecules (3 day old larval fat body).

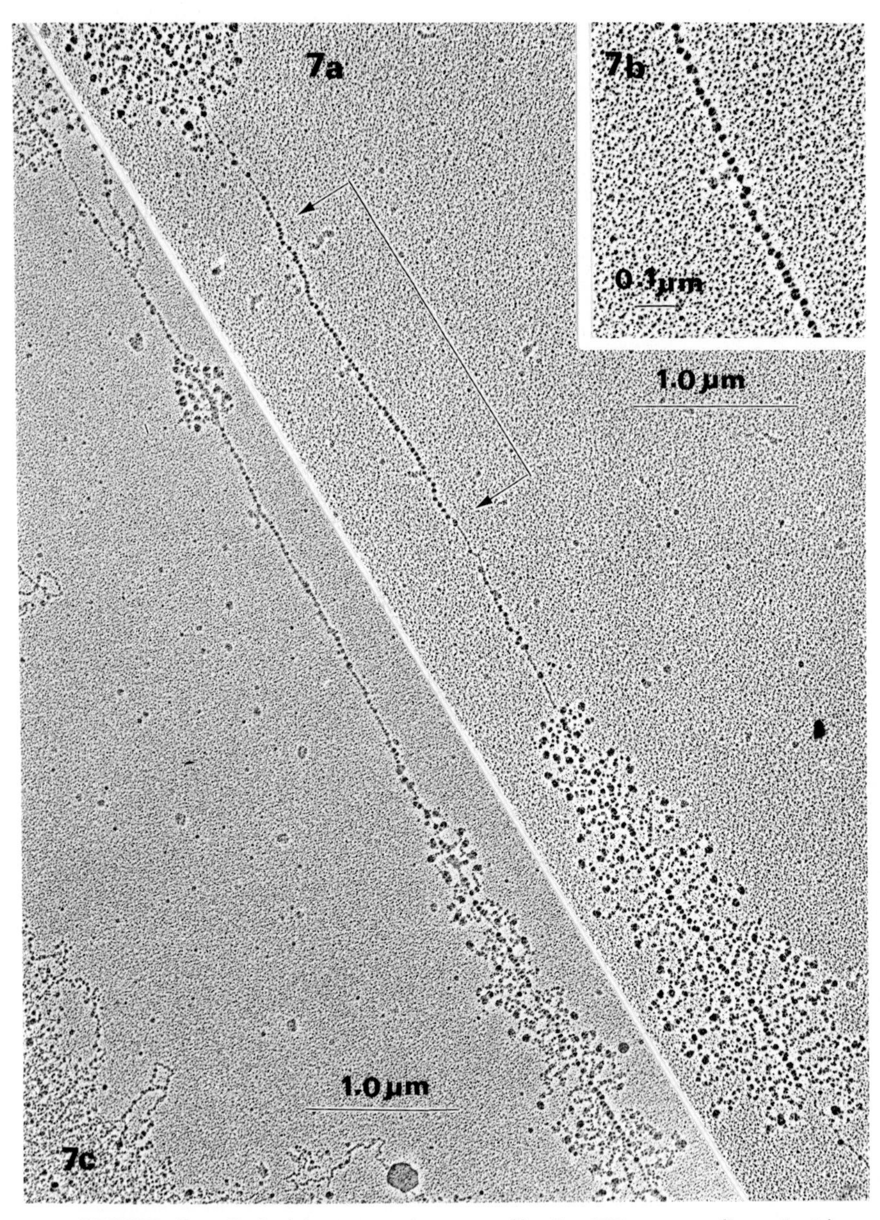

FIGURE 7a. Putative non-transcribed rRNA gene (bracket) has a beaded morphology (larval midgut, 24 hours).

FIGURE 7b. Enlargement of the putative non-transcribed gene.

FIGURE 7c. Newly initiated transcription of an rRNA gene. Yet-to-be-transcribed region has a beaded morphology (fat body, one hour).

molecules. In newly activated rRNA genes of *Calliphora*, we do not usually observe a nucleosome-free region preceding the first transcribing polymerases (Fig. 7a,b,c). A similar observation was reported for *D. melanogaster* rRNA genes (8). We therefore conclude that the smooth chromatin morphology may be a result of transcription. These observations are in conflict with the conclusion of Foe et al (2) that a smooth chromatin transition occurs in rRNA genes before transcription is initiated.

The smooth chromatin seen in non-transcribed spacers may be a result of incorrect termination. Of 60 *Calliphora* rRNA genes examined, 40 appeared to have at least one RNA polymerase present in the smooth spacer region beyond the putative termination site. The smooth configuration of this chromatin is apparently not the result of incorrect initiation. Figure 4 shows chromatin segments preceding the initiation sites of two genes within the same nucleolus. In one case (4a), the chromatin is part of a non-transcribed spacer which immediately follows another active rRNA gene; the chromatin has a smooth morphology. In the other case (4b), the spacer segment does not immediately follow a transcribed gene and is in a beaded configuration. Figure 5 shows a chromatin segment beyond the termination site of an rRNA gene which is not followed by another rRNA gene. The chromatin immediately adjacent to the termination site is smooth, whereas that more distal is beaded. Further evidence that smooth chromatin does not result from abnormal polymerase initiation is given in figure 6, where genes of divergent polarity are separated by a non-transcribed spacer which has a beaded morphology. The spacers following the termination points of these genes are smooth and have incorrectly terminated RNA polymerase molecules associated with them.

Again, although the possibility of artefactual changes in chromatin morphology by our preparative methods cannot be ruled out, our observations are consistent with the possibility that the altered chromatin morphology within rRNA genes of *Calliphora* larval tissue and quail myoblasts may be generated by the passage of actively transcribing RNA polymerases.

REFERENCES

1. Felsenfeld, G. (1978). Nature 271, 115.
2. Foe, V. E., Wilkinson, L. E., and Laird, C. D. (1976). Cell 9, 131.
3. Franke, W. W., Scheer, U., Trendelenburg, M. F., Spring, H. and Zentgraf, H. (1976). Cytobiologie 13, 401.
4. Kornberg, R. D. (1974). Science 184, 868.
5. Laird, C. D., Wilkinson, L. E., Foe, V. E., and Chooi, W. Y. (1976). Chromosoma 58, 169.

6. Mathis, D. J., and Gorovsky, M. A. (1976). Biochem. 15, 750.
7. Mathis, D. J., and Gorovsky, M. A. (1977). Cold Spring Harbor Symp. Quant. Biol. 42, 773.
8. McKnight, S. L., and Miller, O. L., Jr. (1976). Cell 8, 305.
9. McKnight, S. L., Bustin, M., and Miller, O. L., Jr. (1977). Cold Spring Harbor Symp. Quant. Biol. 42, 749.
10. Miller, O. L. Jr., and Bakken, A. H. (1972). Acta Endocrinol. Suppl. 168, 155.
11. Miller, O. L. Jr., and Beatty, B. R. (1969). Science 164, 955.
12. Olins, A. L., and Olins, D. E. (1974). Science 183, 330.
13. Reeves, R., and Jones, A. (1976). Nature 260, 495.
14. Reeves, R. (1977). Cold Spring Harbor Symp. Quant. Biol. 42, 709.
15. Scheer, U. (1978). Cell 13, 535.
16. Trendelenburg, M. F., Scheer, U., Zentgraf, H., and Franke, W. W. (1976). J. Mol. Biol. 108, 453.
17. Trendelenburg, M. F., and Gurdon, J. B. (1979). Nature 276, 292.
18. Woodcock, C. L. F., Frado, L. L. Y., Hatch, C. L., and Ricciardiello, L. (1976). Chromosoma 58, 33.

REGULATION OF EARLY ADENOVIRUS mRNA SYNTHESIS[1]

Frank Lee[2], Arnold J. Berk[3], Tim Harrison[4], J. Williams[5], and Phillip A. Sharp

Center for Cancer Research and Department of Biology
Massachusetts Institute of Technology
Cambridge, Massachusetts 02139

ABSTRACT In mammalian cells few examples are known where one gene product controls the expression of mRNAs from several other genes. We have recently shown that an adenovirus gene product, defective in the Ad5 host range 1 mutant, is required for expression of viral mRNAs from four widely separated transcription units. These transcription units are normally expressed before the onset of viral DNA replication, i.e., early in the lytic phase of infection. We report here that four other adenovirus 5 host range mutants in the same complementation group have a similar phenotype for the synthesis of early mRNAs. These data show that synthesis of all but two adenovirus mRNAs are under control of viral gene products. Hopefully, this biochemically and genetically defined system will provide a good model for studying regulation of mRNA synthesis in mammalian cells.

INTRODUCTION

The adenovirus genome is a large (35 kilobases) viral genome with a complex organization of transcriptional units whose expression is both quantitatively and temporally regulated during the course of a lytic infection. We expect that

[1]This work was supported by an American Cancer Society grant to P.A.S., two Helen Hay Whitney Fellowships to A.J.B. and F.L., respectively, and a Damon Runyon Walter Winchell Fellowship to T.H.

[2]Present address: Department of Pharmacology, Stanford University.

[3]Present address: Department of Microbiology, University of California at Los Angeles.

[4]Present address: MRC Virology Unit, Virology Institute.

[5]Present address: The Mellon Institute, Carnegie-Mellon University.

ISBN 0-12-068350-4

in elucidating the mechanisms evolved by the virus to regulate the expression of its genes we will also reveal mechanisms used by the mammalian host cell in controlling its own genes. In the early phase of infection prior to the replication of viral DNA the genome is transcribed from four regions (1,2,3). The structure of the early cytoplasmic transcripts and their locations on the viral genome have been well characterized (4,5) (see Figure 1). These analyses have shown that the mature cytoplasmic transcripts are in general the products of post-transcriptional processing from a precursor molecule, and consist of RNA segments spliced together following the removal of intervening regions.

We have recently shown that the expression of most of the early mRNAs requires pre-early viral gene products and that an Ad5 host-range mutant is defective in one of these pre-early products (6). Ad5 host-range mutants were isolated by Harrison, Graham and Williams (7) by the property that the mutants grew to high titers on Ad5 transformed human embryonic cells (293 cell line) but were defective for growth on HeLa cells. Hypothetically, gene products encoded by the integrated viral DNA in the 293 cells would complement defects in the mutants. Approximately 18 mutants were isolated and these were divided into two complementation classes. The biological properties of mutants in these two complementation classes are listed in Table I. Mutants in both complementation classes are defective for transformation of embryonic hamster cells and stimulation of T antigen in HeLa cells. Ad5 host-range mutants in complementation class I by marker rescue between 0-4.4 units while class II mutants map between 6.1 and 9.0 units (8).

TABLE I
PROPERTIES OF ADENOVIRUS 5 HOST RANGE MUTANTS[a]

		Complementation groups I	Complementation groups II
293 cells	host range	+	+
HeLa cells	host range	−	−
HEK[b] cells	host range	−	+
Transformation[c]		−	−
T antigen		−	−
Viral DNA replication		−	±
Map position[d]		0-4.4	6.1-9.0

[a]Graham, Harrison and Williams (12)
[b]HEK: primary human embryonic kidney cells
[c]Assayed with primary rat embyro or brain cells (some abortive transformation was seen in baby rat kidney cells with class I mutants)
[d]Frost and Williams (8)

Our previous studies of the synthesis of early mRNAs after infection with Ad5 hr 1, the prototype complementation class I mutant, have shown that this variant was defective for transcription for mRNA C and the mRNAs from early regions II, III and IV. Thus infection of HeLa cells with Ad5 hr 1 stimulates the synthesis of only mRNAs *A* and *B*. These conclusions were based on S1 nuclease-gel analysis (see below) of cytoplasmic and nuclear RNA produced in HeLa cells 8 hours post infection. As the mutation in Ad5 hr 1 lies in the 0-4.4 region, proteins translated from either *A* or *B* mRNAs are probably necessary for synthesis of the other early mRNAs. That this gene product acts at the level of either initiation of transcription or stabilization of nuclear RNA is suggested by the absence of nuclear RNA processing intermediates complementary to early regions II, III and IV in Ad5 hr 1 infected HeLa cells. Nuclei prepared from w.t. Ad5-infected cells contain significant amounts of non-spliced precursor RNAs which are destined to have intervening sequences excised during maturation to cytoplasmic mRNA. These precursors have 5' and 3' termini in common with mRNAs and can readily be detected and mapped by S1 nuclease-gel procedures.

An independent set of host range mutants has recently been isolated by Jones and Shenk (1979, Cell, in press). These mutants were selected for resistance for cleavage by Xba 1 restriction endonuclease at position 4.5 on the genome. Variants of Ad5 altered in this cleavage site were propagated on the 293 cell line. Seven of the eight mutants selected were deletions of viral sequences spanning the 4.5 sites. Some of these deletions extended leftward into the 1.5-4.5 region and others rightward from 4.5-10 m.u. As with the host range mutants of Harrison, Graham and Williams (7), these deletions are defective for transformation and viral DNA replication. Jones and Shenk (1979, Proc. Natl. Acad. of Sci., USA, in press) have also shown that these deletion host-range mutants are defective for synthesis of mRNAs from early regions II, III and IV after infection of HeLa cells.

We describe here two extensions of these studies on the requirement of a pre-early gene product for the synthesis of early viral mRNAs. First, we show that four other Ad5 complementation group I host-range mutants have a phenotype similar to that of Ad5 hr 1 for the synthesis of early mRNAs. Second, by following the synthesis of early mRNAs after infection of HeLa and 293 cells with w.t. Ad5 in the absence of protein synthesis, we suggest that all viral gene products required for transcription of mRNAs from regions II, III and IV are encoded in region I.

METHODS

I. Cells and Viruses

Wild type Ad5 and the Ad5 group I host-range mutants were grown either on HeLa cells in suspension culture or on 293 cells growing as monolayers.

Stocks of the host-range mutants 2, 3, 4, and 5 were generously provided by Jim Williams. These mutants were isolated by Harrison, Graham and Williams (7) and fall into complementation group I as described previously.

II. Analysis of viral-induced RNAs

HeLa cells or 293 cells were infected with mutant or wild type virus at a m.o.i. of 10 for all experiments.

Early cytoplasmic RNA was isolated from cells 8 hours after virus infection according to the NP40 lysis procedure described previously (5).

The analysis of viral RNA hybridization to ^{32}P-labeled viral restriction fragments followed by S1-nuclease digestion and agarose gel electrophoresis has been previously described (9). Briefly, this method involves hybridization of an excess of ^{32}P-labeled viral DNA with total unlabeled cytoplasmic RNA in a solution containing 80% formamide. The hybridization is conducted at a temperature above the Tm of DNA but below that of RNA-DNA hybrids (10). Hybridization is allowed to proceed until greater than 90% of the complementary RNA is annealed to the excess DNA. The hybridization mixture is diluted and hybrids are treated with S1 nuclease, degrading single-stranded DNA and RNA. In the experiments presented here, all S1-resistant products were resolved by electrophoresis on alkaline 1.4% or 2.0% agarose gels followed by autoradiography of the dried gel (11).

RESULTS

I. Analysis of viral RNAs induced by Ad5 host range mutants

We wished to examine the pattern of early viral transcription produced by the Ad5 host-range mutants when grown in the non-permissive HeLa cell line. The early transcription pattern of the closely related Ad2 virus has previously been determined by the S1-nuclease-gel procedure (5). As in Ad2, transcription early after infection with Ad5 occurs from four regions of the genome; the wild type Ad5 transcription map closely resembles that of Ad2 and is diagrammed in Figure 1.

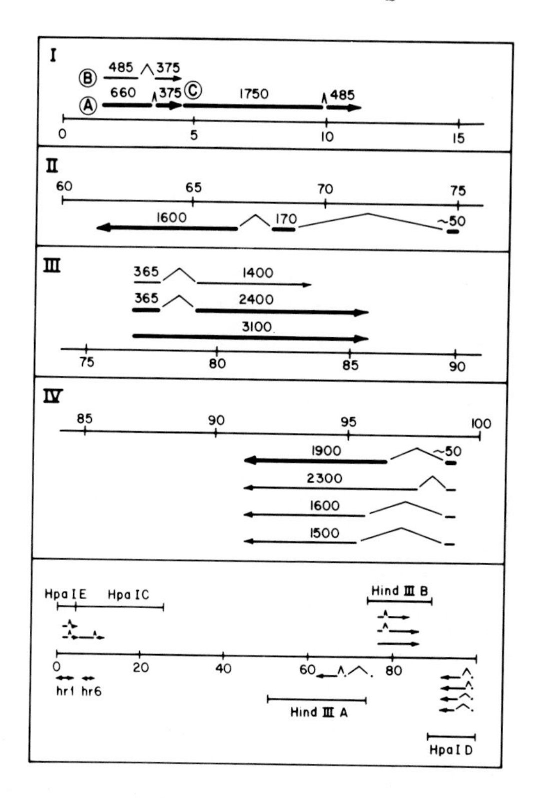

FIGURE 1. *Ad5 early transcription map. Panels I-IV show diagrammatically the locations of the transcripts originating from each of the four early transcription units. Genome map coordinates are given on the lower lines in each case. The particular restriction fragments of viral DNA used to detect the RNAs from the different regions are shown in the lower panel. The arrows represent the cytoplasmic transcripts that have been mapped with the arrow indicating the direction of transcription. Interruptions in the arrows indicate the location of intervening sequences which are removed in the mature transcripts, and the numbers give the lengths in nucleotides of the segments spliced together. Thick lines designate major transcripts while thin lines represent less abundant species. A detailed description of the Ad5 transcripts can be found in Berk* et al. *(6).*

This transcription map may be derived in part from the data shown in Figure 2 and is discussed in detail elsewhere (6).

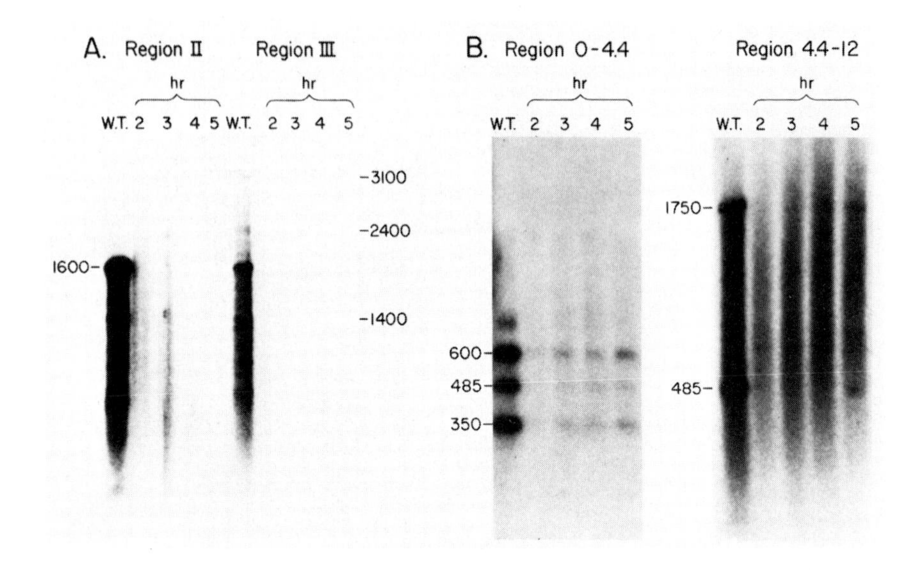

FIGURE 2. *A. S1-endonuclease-gel analysis of early region II and III from wild type Ad5 and Ad5-host range mutants. The RNAs were detected by hybridization to the following restriction fragments; region II: Hind III A fragment; region III: Hind III B fragment; region IV: Hpa I D fragment. Hybridizations between DNA and total cytoplasmic RNA in 80% formamide and S1 digestion of RNA-DNA hybrids were performed as described previously (9). S1-resistant products were resolved by electrophoresis on an alkaline 1.4% agarose gel.*

B. Analysis of RNAs from early region I. Hybridizations were performed using either Hpa I E fragment (0-4.4) or Hpa I C fragment (4.4-25.5) to detect region I mRNAs. Hpa I E is complementary to mRNAs A and B (see Fig. 1 and text) while Hpa I C hybridizes to mRNA C. The S1-nuclease digested products were resolved on a 2% alkaline agarose gel.

In analyzing the early mRNAs expressed from the four Ad5 host-range mutants, non-permissive HeLa cells were infected with 10 PFU/cell of either mutant or wt Ad5. Eight hours after infection, cytoplasmic RNAs were extracted and the amount of each early RNA was quantitated by hybridization of equal quantities of total cell RNA with an excess of ^{32}P-labeled viral DNA restriction fragment. After digestion of such hybrids with S1 nuclease, the amount of each mRNA is equal to the number of ^{32}P counts migrating in the corresponding band on the alkaline agarose gel. As a control, the synthesis of early RNAs following infection of the permissive cells, 293, with the same virus stocks was analyzed in parallel.

The results for early regions II, III and IV may be discussed together. As shown in Figure 2A, infection of HeLa cells by wt Ad5 stimulates the synthesis of abundant RNAs complementary to 1600 nt and 2400 nt segments of viral DNA from early regions II and III, respectively. However, infection of HeLa cells with equal m.o.i. of any of the Ad5 group 1 host range mutants fails to stimulate the synthesis of these mRNAs. Comparable results were obtained for early region IV (data not shown). The amount of mRNA synthesized after infection with the host range mutant can be quantitated by densitometer scanning of gels such as those shown in Fig. 2. In general, mutants induced from these regions 1/300 the level of mRNA found after wt Ad5 infection. These results parallel those for Ad5 hr 1, the mutant in host range complementation group 1 which was analyzed in our previous study (6).

The expression of region I mRNAs in the host range mutant-infected cells differs from that of the other early regions. Three mRNAs are encoded in region I; these are designated *A*, *B* and *C* in Fig. 1 (top). When the RNA from wild type infected cells is hybridized to Hpa I E DNA, mRNAs *A* and *B* are detected, and the S1-resistant products are resolved into three bands with lengths 660, 485 and 375 nucleotides, respectively (Fig. 2B). The two longest bands are produced by mRNAs *A* and *B*, respectively, while the 375 nucleotide band is produced by both mRNAs (see Fig. 1). mRNA *C* is detected by hybridization to Hpa I C and gives rise to two alkaline gel bands of 1750 and 485 nt. The results in Fig. 2B show that mRNAs *A* and *B* from region I are produced at somewhat reduced levels by all the host range mutants. However, mRNA *C* is produced in greatly reduced amounts, comparable to the reduction in mRNAs from early regions II, III and IV. Thus within early region I only two of the mRNAs are synthesized in significant levels, the third is markedly reduced.

II. Experiments with protein synthesis inhibitors

The results obtained with the host-range mutants would be consistent with the hypothesis that a viral protein encoded by one of the region I mRNAs can act in trans to affect the production of mRNAs from the other early regions as well as one mRNA from region I.

The 293 cell line offers a possibility of testing this hypothesis. If the function defective in the host-range mutants was in fact a protein, one should be able to reproduce the block in transcription of the mutants during wt virus infection of HeLa cells under conditions where protein synthesis was inhibited. No viral proteins would be produced, and one would predict that transcription from early regions II, III and IV would be suppressed. 293 cells on the other hand complement the host range mutants and would be expected to contain the viral functions defective in the host-range mutants. If the proposed viral gene product were stable, inhibiting protein synthesis in 293 cells during infection with wild type virus should not inhibit transcription of early regions II, III and IV or mRNA C from region I.

Our initial experiments were performed with the inhibitor of protein synthesis, cycloheximide. In experiments where HeLa cells were treated with 25 μg/ml of this drug we found that there was no effect on the ability of infecting wild type virus to express the mRNAs which rely on the host-range functions (data not shown). We determined that cycloheximide reduces the level of protein synthesis in HeLa cells to about 3% of the untreated level. Since it is conceivable that this residual level of protein synthesis produces sufficient quantities of the required product, we used another more effective inhibitor of protein synthesis, emetine. This drug reduces protein synthesis to less than 1% of the untreated level and most of this residual synthesis should be due to the mitochondria. When wild type virus was infected into HeLa cells pretreated with emetine, we obtained the results shown in Fig. 3. Under these conditions no transcription from regions II, III and IV was observed. The bands observed after hybridization to RNA from untreated Ad5-infected cells are not detected or are present in markedly reduced levels after similar analysis of RNA from emetine pretreated cells infected with wild type Ad5. The situation in region I with emetine-blocked cells is different, both from the control and from the host-range mutants. The host-range mutants are defective in the production of mRNA *C* but still synthesize mRNAs *A* and *B*. The untreated control cells synthesize as expected all three mRNAs (Fig. 3, Hpa I E and Hpa I C). The cells treated with emetine produced greatly reduced levels of both mRNA *B* and *C*. This effect on the accumulation of mRNA *B* is not a general effect of addition of emetine because if we add this drug 3 hours post infection, thus allowing a period for viral protein synthesis, then the full complement of wild type RNA is observed (data not shown).

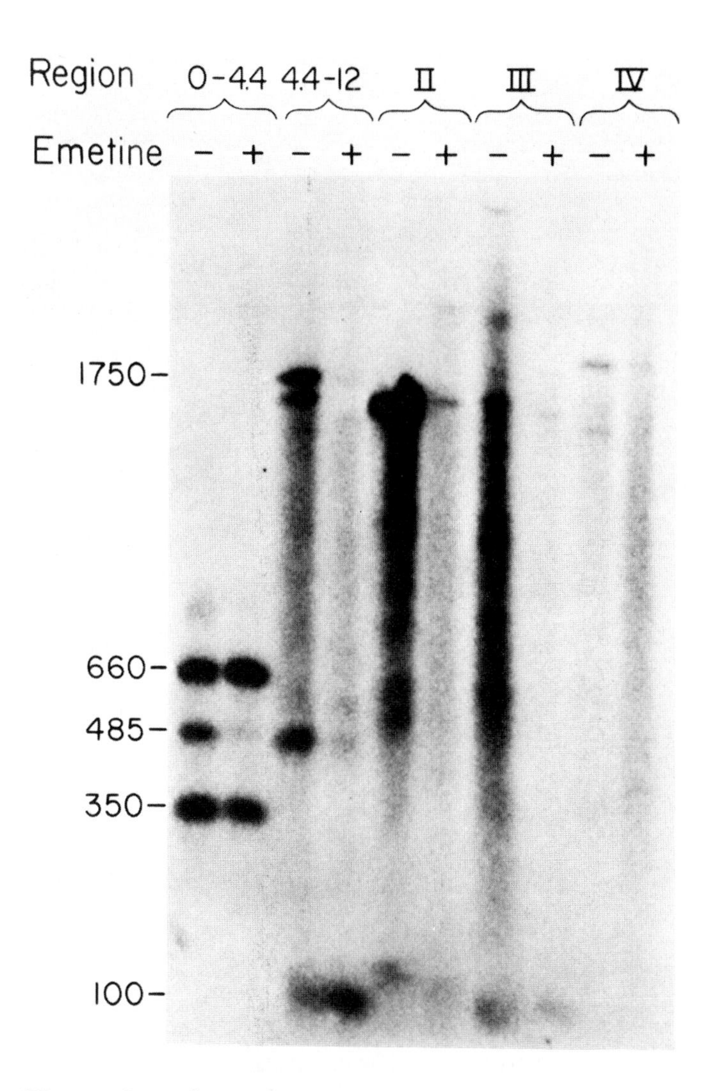

FIGURE 3. *The effect of emetine on expression of early RNAs during infection of HeLa cells with wild type Ad5. Cytoplasmic RNA was isolated 8 hours post infection from a control culture with no emetine added and a culture which was treated with 50 μg/ml emetine beginning 10 min prior to virus infection. RNA was hybridized to the following restriction fragments to detect transcripts from each of the four regions (left to right); region I: Hpa I E (0-4.4); Hpa I C (4.4-25.5). Region II: Hind III A. The S1-resistant products were run on an alkaline 1.4% agarose gel.*

The 293 cell line contains a portion of the left end of the Ad5 genome including all of region I integrated into the cellular genome. This cell line expressed continuously the mRNAs from region I (6) and we would expect it to produce all the encoded proteins. When it is treated with emetine starting 10 min before infection with wild type virus, all the region I mRNAs are processed as well as the mRNAs from regions II, III and IV (Fig. 4). Thus, in a cell line in which the

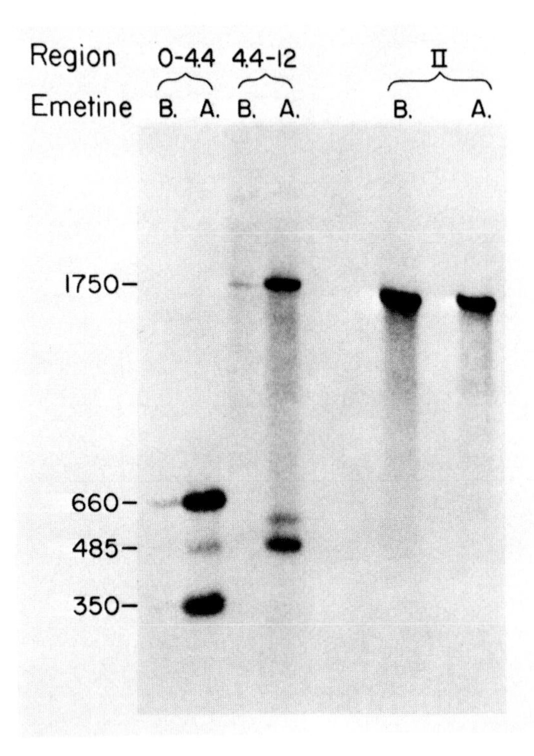

FIGURE 4. *The effect of emetine on expression of early RNAs during infection of 293 cells with wild type Ad5. Emetine was added to 50 μg/ml either 10 min before (denoted B) or 3 hours after (denoted A) virus infection. Cytoplasmic RNA extracted at 8 hr post infection was hybridized to restriction fragment probes to detect mRNAs from early regions I and II (left to right). The S1-nuclease resistant products were run on an alkaline 1.4% agarose gel.*

gene products encoded by region I are already present, inhibiting protein synthesis by addition of emetine does not prevent the expression of viral transcripts encoded in early regions II, III and IV. If emetine is not added until 3 hours post infection, all the RNAs are again found. Under these conditions there is an enhancement of the level of transcription from some of the early regions, particularly early region I. This enhancement is also observed when this experiment is done with HeLa cells (data not shown).

DISCUSSION

We have presented here a characterization of the transcription properties of a set of group I host range mutants of Ad5. These mutants are unable to replicate on HeLa cells but are complemented and grow in Ad5 transformed 293 cell line. These host range mutants have been shown to map in the region from 0-4.5 units (8).

The most striking property of all the host range mutants in this complementation group is that when infecting the nonpermissive HeLa cell line, they are defective in the production of all but two mRNAs. The transcripts normally expressed from early regions II, III and IV are drastically reduced or are not detectable; in addition, only two of the three mRNAs encoded in early region I are produced at significant levels. Infection of the complementing 293 cell line with the host range mutants results in the full wild type transcription pattern. Because the host range mutants map within early region I, these results demonstrate that a viral gene product encoded in this region is required for accumulation of viral RNAs from other regions of the genome.

Only mRNAs *A* and *B*, both of which appear to be processed from a common precursor, are expressed at significant levels in host range group I mutant infected HeLa cells. However, these mRNAs are consistently present at reduced levels when compared to the amounts induced by wild type virus. It is possible then that the host range gene product, or other viral gene products dependent on the host range function, is required for high level expression of the precursor to these two mRNAs.

Ad5 infection of HeLa cells pretreated for 10 minutes with emetine (50 μg/ml) results in the accumulation of only one early mRNA, mRNA *A* (Figure 1). This RNA is present 8 hours after infection of pretreated cells at levels comparable to that found after infection of untreated HeLa cells. There are two possible explanations for inhibition of protein synthesis blocking the lytic cycle at the stage of synthesis of mRNA *A*. First, the protein product of mRNA *A* is required for synthesis and/or accumulation of mRNAs from the other early regions. In this case, the absence of protein synthesis arrests the lytic cycle at a different stage than the defect in Ad5 hr 1. Cells infected with this mutant accumulate both mRNAs *A* and *B*. Second, the addition of emetine at 50 μg/ml to a culture essentially stops RNA metabolism of the cells shortly thereafter and the accumulation of only mRNA *A* might reflect the window of mRNA synthesis shortly after infection. In this case, the block of the lytic cycle does not reflect the absence of a viral protein but a non-specific effect of the drug on RNA metabolism in these cells.

Cellular poly A(+) RNA synthesis is reduced by approximately 90% within 1.5 hours after addition of 50 μg/ml emetine (D. Spector, personal communication; J. Nevin, personal communication; H. Handa, personal communication). Thus, the viral RNA found 8 hours post infection of cells pretreated with this drug probably reflects RNA synthesis within 1-2 hours post infection. Several controls suggest that viral RNAs are stable in emetine arrested cells and that the RNAs found 8 hours post infection reflect the earlier transcriptionally active state. The most interesting control is the analysis of cytoplasmic RNA from 293 cells which had been pretreated with emetine 10 minutes before infection with Ad5 (see Figure 4). mRNAs, complementary to all four early regions, are present within these cells 8 hours post infection. If emetine did arrest RNA synthesis by 1.5 hr post infection, then these RNAs were not degraded during the 6 hour interim before extraction. Viral RNAs are also stable when emetine is added 3 hours post infection with w.t. Ad5. This is also shown in Figure 4 where hybridizations were done with cytoplasmic RNA extracted 8 hours post infection, 5 hours after emetine addition to Ad5-infected 293 cells. Here mRNAs reflecting enhanced levels of all early regions were detected; again suggesting that viral RNAs are stable in emetine arrested cells. Another indication of the stability of viral RNAs in emetine treated cells comes from the analysis of nuclear RNA 5 hours post addition of the drug. In experiments not presented here HeLa cells were infected with w.t. Ad5 and 3 hours later emetine at 50 μg/ml was added. Nuclear RNA was extracted from these cells 5 hours later and analyzed by the S1 nuclease-gel procedure. Nuclei from these emetine-treated cells contained levels of unspliced precursors to viral mRNAs in comparable amounts to that found in equivalent untreated cultures. If transcription stops 1-2 hours post emetine addition, then these nuclear RNAs have survived 3 hours of incubation within the cell.

The observation that w.t. Ad5 infection of emetine pretreated HeLa cells only permits accumulation of mRNA *A*, while a similar experiment using 293 cells results in synthesis of all early adenovirus mRNAs, might reflect the relative kinetics of synthesis of early mRNAs in each cell type. Early mRNAs from regions II, III and IV might be synthesized immediately after infection of 293 cells because the required region I products are already present in this cell. Hence, these emetine pretreatment experiments cannot be interpreted as proving that the region I product, defective in Ad5 hr 1 and required for expression of mRNAs from the other early regions, is a viral coded protein. Settling this issue must await further experiments. However, the emetine results further confirm the existence of a pre-early stage of the lytic cycle of Ad5. Elucidating the biochemical mechanism by which pre-early gene products control the expression of other early mRNAs will unveil a mechanism for gene regulation in mammalian cells.

We have previously analyzed transcription of Ad5 hr 7, one member of the host range mutants in complementation group II (6). When infected into either HeLa or 293 cells, this mutant induces the full complement of wild type mRNAs.

REFERENCES

1. Sharp, P. A., Gallimore, P. H., and Flint, S. J. (1974). Cold Spring Harbor Symp. Quant. Biol. *39*, 457.
2. Philipson, L., Pettersson, U., Lindberg, U., Tibbets, C., VennStrom, B., and Parsson, T. (1974). Cold Spring Harbor Symp. Quant. Biol. *39*, 447.
3. Flint, S. J., Berget, S. M., and Sharp, P. A. (1976). Virology *72*, 443.
4. Kitchingman, G. R., Lai, S.-Pu, and Westphal, H. (1977). Proc. Natl. Acad. Sci. USA *74*, 4392.
5. Berk, A. J., and Sharp, P. A. (1978). Cell *14*, 695.
6. Berk, A. J., Lee, F., Harrison, T., Williams, J., and Sharp, P. A. (1979). Cell, in press.
7. Harrison, T., Graham, F., and Williams, J. (1977). Virology *77*, 319.
8. Frost, E., and Williams, J. (1978). Virology *91*, 39.
9. Berk, A. J., and Sharp, P. A. (1977). Cell *12*, 45.

Berk, A. J., and Sharp, P. A. (1977). Cell *12*, 721.
10. Casey, J., and Davidson, N. (1977). Nucl. Acids Res. *4*, 1539.
11. McDonell, M. W., Simon, M. N., and Studier, F. W. (1977). J. Mol. Biol. *110*, 119.
12. Graham, F. L., Harrison, T., and Williams, J. (1978). Virology *86*, 10.

SYNTHESIS AND PROCESSING OF ADENOVIRUS 2 RNA IN VITRO[1]

James L. Manley , Phillip A. Sharp, and Malcolm L. Gefter

Department of Biology, Massachusetts Institute of Technology, Cambridge, Mass. 02139

ABSTRACT Nuclei isolated from HeLa cells during the late phase of infection by adenovirus serotype 2 (Ad2) synthesize high levels of RNA *in vitro*. The average chain elongation rate, when nuclei are incubated at 25°C, is 320 nucleotides per minute at the start of the reaction, and approximately 150 nucleotides per minute after 90 min of *in vitro* incubation. This continued rate of transcription is reflected in the observation that the promoter distal portion of the Ad2 major late transcription unit (25 kb in length) is transcribed almost as efficiently after 2 hrs of *in vitro* incubation as at the start of the reaction.

RNA synthesized in isolated nuclei, as well as the endogenous, steady-state Ad2 specific RNA, is extremely stable during *in vitro* incubation. Non-sepcific degradation of the primary transcript, presumptive processing intermediates or mature mRNA can not be detected after 2 hrs of incubation. The Ad2 specific RNA synthesized *in vitro* undergoes many of the processing reactions known to occur *in vivo*. The RNA is accurately capped and methylated at its 5' end. RNA molecules are synthesized which have discrete 3' ends corresponding to the 3' ends found on mature mRNAs *in vivo*. In particular, the 3' ends of mRNAs for the abundant hexon and 100K protein are formed very efficiently, and are polyadenylated shortly after their creation.

INTRODUCTION

In order to understand the mechanisms by which eukaryotic cells regulate the expression of their genetic information, *in vitro* systems that accurately carry out the various reactions

[1]This work was supported by grants to MLG from the American Cancer Society (NP-6) and the National Institutes of Health (AI13357). JLM was supported by the training grant in immunology to the Dept. of Biology, M.I.T., by NIH (CA09255).

ISBN 0-12-068350-4

involved in mRNA synthesis will be required. In vitro systems which make use of isolated nuclei have been widely used (1-3). One problem with such systems is that very little information has been available concerning the nature of the in vitro product. We have recently begun to characterize the RNA synthesized in vitro by nuclei isolated from adenovirus 2 (Ad2) infected HeLa cells. We have taken advantage of the detailed knowledge concerning the pathways by which Ad2 specific mRNA is synthesized during the late phase of infection.

The vast majority of late viral mRNAs are apparently derived from one (approximately 25 kb) primary transcript (4,5). During the pulse-label, 40-50% of teh total radioactivity incorporated into RNA are incorporated into this transcript (5). Transcription is initiated at a promoter located at 16.5 map units on the viral genome, and continues rightwards towards the end of the genome (6,7). All mRNAs derived from this transcriptional unit contain a common 5' tripartite leader segment, encoded at coordinates 16.5, 19.5 and 26.5 which is covalently joined to the coding region of the mRNA presumably by RNA splicing (8-10). Since the three leader segments appear to be encoded only once in the genome, each primary transcript, although containing the coding information necessary for at least 13 separate mRNAs (11), gives rise to only one mRNA. This suggests that the bulk of each primary transcript is not processed into mRNA. Consistent with this are observations which show that as much as 80% of Ad2 specific RNA is turned over within the nucleus (12).

The cytoplasmic mRNAs encoded in this transcription unit can be grouped in 5 "families"; mRNAs within a family have common 3' termini (11-14). Thus, two mRNAs with a common 3' termini may differ with regard to the coordinates at which the tripartite leader sequence is attached to the "main body" sequences. Figure 1 outlines the structure of these mRNAs.

We have previously described an isolated nuclei system which synthesizes viral specific RNA at a near linear rate for several hours. We showed that this system supports in vitro initiation of the major late transcription unit and that the transcripts are accurately capped and methylated at their 5' ends (15). Furthermore, presumptive splicing intermediates are formed in vitro, and these transcripts are accurately terminated and polyadenylated (16). Here we present some additional properties of the transcription system, as well as summarize our current view of the structure of in vitro synthesized RNA.

RESULTS AND DISCUSSION

Nuclei isolated from HeLa cells between 16 and 24 hrs after infection by Ad2 synthesize RNA in vitro at a nearly

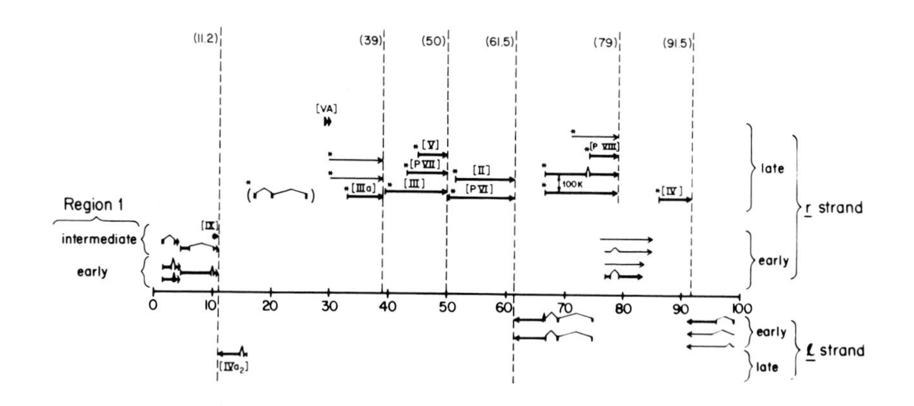

FIGURE 1. Transcription map of adenovirus 2. Colinear segments of mRNAs are represented by bars which span the length of complementary DNA sequences. Two segments of RNA which are joined together in a mRNA are shown by caret symbols. Segments with arrows pointing to the right are transcribed from the r strand. All late mRNAs from the r strand have a tripartite set of leaders spliced to their 5' terminus. The presence of this tripartite leader set is denoted by the asterick (*). The polypeptide specified by each mRNA is denoted above the bar.

linear rate for 2.5 hrs. This synthesis consists of elongation of molecules initiated *in vitro* (15). Specifically, the bulk of transcription originates at the major late promoter and continues rightwards towards the end of the genome, 25 kb away. After 2 hrs, approximately equal amounts of *in vitro* synthesized RNA have accumulated complementary to regions of the genome between position 17 and 80 (16). For reasons not yet clear, RNA from the region 80-100 map units, which is also apparently part of the major late transcription unit (7,8), is much less abundant after a similar incubation. The promoter proximal 4,000 nucleotides of the major late transcription unit is transcribed efficiently throughout a 2 hour *in vitro* incubation (15). This is also shown in Figure 2, lanes 1-4. Nuclei were pulse-labeled at different 20 minute intervals, over a two hour time period. Aliquots of RNA purified from these nuclei were hybridized to a restriction fragment containing the promoter region (BamHI B, 0-29.1 map units), non-hybridized RNA was digested with nuclease S1, and the resistant duplexes were analyzed by agarose gel electrophoresis (16,17). Two bands of ^{32}P-labeled RNA/DNA hybrids were detected in samples from each 20 min period. An *in vitro* synthesized primary transcript (with a 5' terminus of colinear sequences at 16.5) gives rise to a

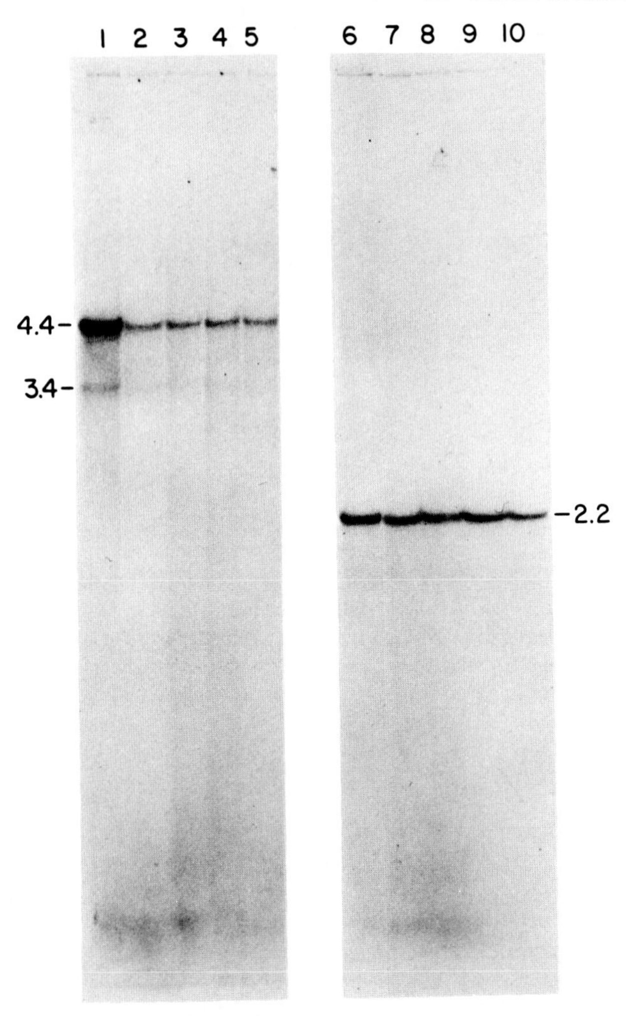

FIGURE 2. Synthesis in vitro of the Ad2 major late transcription unit for long periods of time. Nuclei (8 x 10^6) were incubated in 100 μl reaction mixtures as described (15,16). At the times indicated below, 15 μCi of α-^{32}P UTP were added to each reaction mixture. After an additional 20 min, RNA was extracted, and aliquots hybridized to either the BamHI B (0-29.1 map units; lanes 1-5) or HindIII H (73.6-79.9 map units; lanes 6-10) restriction fragment (16). Following nuclease S1 treatment and agarose gel electrophoresis, RNA-DNA hybrids were visualized by autoradiography. The size of the RNA fragments are expressed in kilobases. Lanes 1 and 6 show the results of hybridization with RNA extracted from nuclei labeled from 0-20 min; lanes 2 and 7, 20-40 min; lanes 3 and 8, from 40-60 min; lanes 4 and 9, from 80-100 min; lanes 5 and 10, from 100-120 min.

duplex 4.4 kb in length (16.5 to 29.1 map units, from the promoter to the end of the restriction fragment). Another RNA fragment (3.4 kb), which probably represents an intermediate in the RNA splicing process (16) can also be detected. The intensity of the promoter proximal fragment of the primary transcript is three-four fold more intense with RNA from nuclei labeled during the first 20 minutes of incubation than with RNA from nuclei labeled during the second 20 minutes. However, RNA from nuclei labeled during the second through the fifth 20 minute periods give rise to equally intense bands. This continuous synthesis in the promoter proximal region of the transcription unit is consistent with transcription initiation occurring throughout the incubation period. The three-fold decrease in transcription rate after the first 20 minutes most likely arises from the movement of in vivo initiated RNA polymerase molecules. The synthesis which occurs in the promoter proximal region during the remainder of the incubation is likely the result of elongation of transcripts initiated in vitro. This is consistent with our estimate that only 12% of the chains elongated in vitro during a two hour incubation were actually initiated in vitro (15).

If this model is correct, the levels of transcription across regions of the genome lying downstream from the promoter should not decrease sharply after the initial 20 minute period. To test this, we analyzed RNA complementary to a restriction fragment with map coordinates 73.6-79.9 (HindIII H). RNA transcribed from this fragment is roughly 20 kb downstream from the promoter. Assuming a roughly equal distribution of polymerase molecules throughout the transcription unit at the time the nuclei were isolated, we should see approximately equal amounts of RNA complementary to this fragment synthesized in vitro during each 20 minute period until all in vivo initiated polymerase molecules have transcribed through this region, or the system begins to "die".

We hybridized RNA purified from nuclei labeled at 20 minute intervals as described above to the HindIII H restriction fragment, digested single strand nucleic acid with nuclease S1, and resolved the resistant duplexes by agarose gel electrophoresis. The results (Figure 2, lanes 6-10) show that a RNA/DNA fragment 6.3 Ad units (2,200 base pairs) in size is present in all samples. This is the length of the full-sized, intact restriction fragment, and reflects hybridization to a primary transcript spanning the fragment (discussed in [16]). The intensity of this band is virtually identical (within 25%) in all five samples. Therefore, approximately equal amounts of transcription occur in this region during the first 20 min of in vitro incubation as do in the interval between 100 and 120 min of incubation. This suggests that the entire transcription unit (at least the region from 16.5 to approximately

80) is transcribed at a reasonable rate throughout the in vitro reaction.

Assuming that the three-fold decrease in transcription rate we observed after the first 20 min in the promoter proximal region is indeed due to the movement of "pre-initiated" polymerase molecules out of this region, we can estimate the average chain elongation rate during this period. The length of the transcribed region is approximately 4,400 nucleotides (16.5-29.1 Ad2 units). In order for nascent polymerase molecules near the promoter to transcribe through this region in 20 minutes, the transcription rate must be greater than 200 nucleotides per minute (npm). If we assume that the absence of a sharp decrease in the intensities of the bands shown in Figure 2 occurs after 120 minutes of incubation is that all nascent RNA polymerase molecules have not passed through this region, we can again calculate the transcription rate. That is, the distance from the promoter to the end of the HindIII H restriction fragment (16.5-79.9 Ad2 units) is approximately 22,200 nucleotides. Since, after 120 minutes the nascent polymerase molecules originally located in the promoter proximal part of the transcription unit have not passed position 79.9, the average elongation rate during two hours of transcription must be 185 nucleotides per minute or less. Although these two numbers agree quite well, we cecided to measure the chain elongation rate more directly. For this, we adapted a procedure similar to the one described by Marzluff et al. (18). We labeled nuclei with ^{3}H-GTP for one minute, after either 5 or 90 minutes of incubation, then extracted RNA, and isolated molecules greater than 18S by sucrose gradient centrifugation. This RNA was digested to mononucleotides with ribonucleases T2, T1 and A, and the digestion products resolved by chromatography on Whatman #1 paper. The positions of guanosine monophosphate and guanosine were determined by ultraviolet illumination of marker compounds. Regions of the chromatograph corresponding to these molecules were cut out and quantitated by liquid scintillation counting. The ratio of cpm in GMP to the cpm in guanosine gives the chain elongation rate in nucleotides per minute. The results in Table 1 show that the chain elongation rate after five minutes of in vitro incubation is approximately 320 npm. After 90 minutes of incubation, the rate has decreased to about 150 npm. These values agree well with those presented above. Several points are suggested by this data. First, the elongation rate is sufficiently high so that polymerase molecules which initiate transcription early in the reaction should be able to completely transcribe the entire transcription unit. Although this property is most likely not necessary to study the various reactions of RNA processing, it is reassuring that transcription continues at a high enough rate so that complete primary transcripts can be synthesized

TABLE I

RATE OF CHAIN ELONGATION IN ISOLATED NUCLEI[a]

Time	GMP	G-OH	Rate (npm)
5 min	12,200	38	320
90 min	5,200	35	150

[a]100 μl reaction mixtures containing 8 x 10^6 nuclei were incubated *in vitro* for the times indicated. At these times, 20 μCi ^{3}H-GTP was added to each reaction. After an additional minute of incubation, the reactions were terminated, and RNA was extracted as described (16). The ratio of ^{3}H-GMP to ^{3}H-guanosine in high molecular weight RNA was determined by the method of Marzluff *et al.* (18). This ratio is the transcription rate in nucleotides per minute. The above numbers are the average of values obtained in two separate experiments.

in vitro. It would be of interest to compare the chain elongation rate *in vitro* to that observed *in vivo*. However, precise estimates of the *in vivo* chain elongation rate of RNA polymerase II are not available. A number of estimates, though, suggest a value of approximately 3,000 nucleotides per minute (*e.g.*, those referenced in references 12 or 18). If we assume that this estimate is correct, and also that the elongation rate is 3-4 fold lower at 25° than at 37°, then the rate after five minutes of *in vitro* incubation is approximately 40% the comparable *in vivo* rate. After 90 minutes of *in vitro* synthesis, the elongation rate is still almost 20% of the *in vivo* rate.

Stability of Nuclear RNA in vitro. We previously showed that the bulk of RNA synthesized *in vitro* after two hours of incubation is of very high molecular weight (*i.e.*, greater than 10 kb) (15). This suggests that degradation of RNA by nonspecific nucleases is not a major problem during *in vitro* synthesis. To test this more directly, we examined the fate of several specific high molecular RNA species during the course of a two hour incubation. For this, we made use of the hybridization-nuclease S1-gel electrophoresis technique described above. We also took advantage of the fact that, during the late phase of viral infection, the concentrations of viral specific nuclear RNAs are sufficiently high that many of these molecules can be detected simply by ethidium bromide staining of agarose gels. For example, if nuclear RNA is hybridized to the HindIII A restriction fragment (50.1-73.6 map units), several discrete RNA-DNA duplexes can be detected following nuclease S1 digestion and agarose gel electrophoresis. In Figure 3, we show the results of an experiment in which RNA extracted from nuclei incubated _in vitro_ for varying lengths of time was fractioanted on the basis of its ability to bind to oligo dT cellulose, hybridized to the HindIII A fragment, digested with nuclease S1, and the resistant duplexes separated by gel electrophoresis. Bands were visualized by ethidium bromide staining and photographed under ultraviolet illumination. Four bands, which we have previously identified, are readily detectable (16,19). The largest band (8.2 kb) is the size of the intact restriction fragment. It is formed by hybridization of an RNA species with both 5' and 3' ends located outside the boundaries of the restriction fragment. This RNA corresponds to long primary transcripts. The RNA/DNA hybrid 4,000 bp long results from an RNA species with a 3' end at position 61.5 and 5' ends to the left of position 50.1. These RNA molecules are a mixture of unspliced transcripts of presumptive processing intermediates. The remaining two bands are formed as a result of hybridization to mature mRNA. They represent the "amin-body" sequences of the complete hexon mRNA (3,500 bp) and a 5' fragment of 100K mRNA (2,200 bp). The 5' leader segment and poly A, as well as the 3' coding region of 100K mRNA, are not complementary to the HindIII A fragment, and are therefore removed by nuclease S1 treatment. As we showed previously (16), the presumptive hexon mRNA precursor, as well as the mature mRNAs, are detectable only as polyadenylated molecules. (The low level of 100K protein mRNA in the poly A^- fraction in this figure likely results from incomplete selection on the oligo dT cellulose columns. In other experiments, we have not been able to detect any of this mRNA in a poly A^- form [_e.g._, see ref. 16].) These molecules are extremely stable _in vitro_. There is virtually no change in the intensity of the bands produced after either 0 (lane 1) or 120 min (lane 4) of incubation. This is also true of the polyadenylated 8.2 kb primary transcript. The site(s) at which these molecules are polyadenylated is (are) not clear from this experiment. However, it is likely that poly A is added at least at position 78.5, which corresponds to the 3' end of the 100K protein on RNA. The non-polyadenylated material is less stable and a substantial fraction of the non-polyadenylated primary transcript is apparently degraded during the final hour of incubation (lanes 5 and 6). If nuclei are incubated at 37°, rather than 25°, and the RNA extracted from them analyzed by the same method, quite different results are obtained. Polyadenylated RNA, as well as non-adenylated RNA, begins to be degraded after 20 minutes of _in vitro_ incubation. By the time two hours have elapsed, bands are no longer detectable. We do not know why the RNA is so much more susceptible

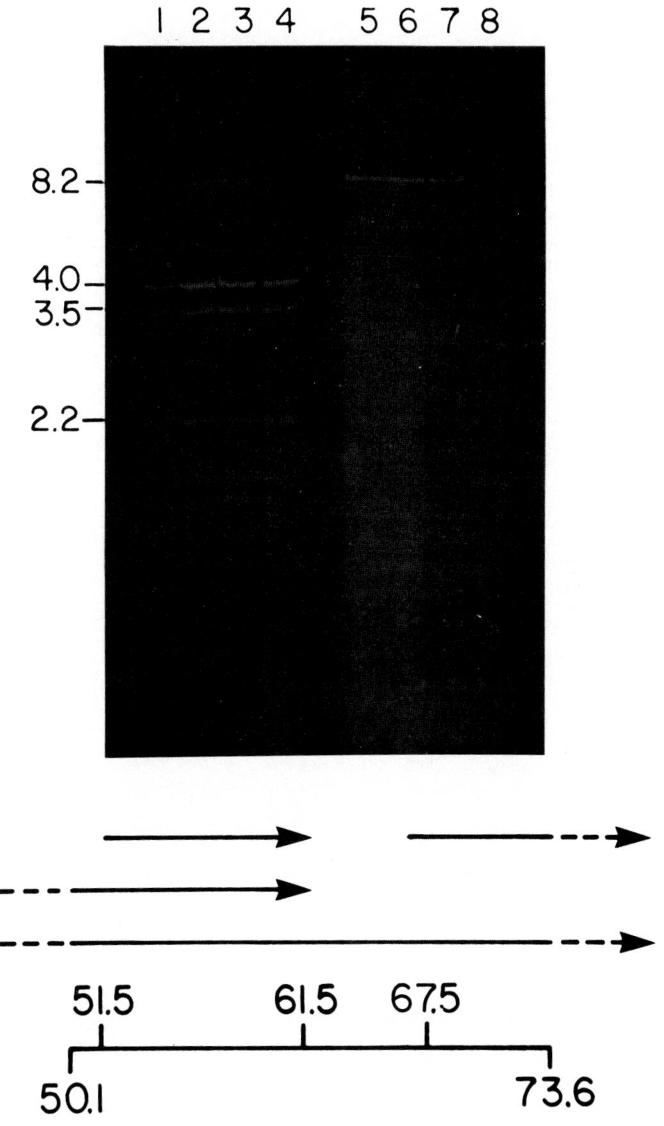

FIGURE 3. Stability of Ad2 nuclear RNA species during *in vitro* incubation. RNA was extracted from 100 μl reaction mixtures containing 8 x 10^6 nuclei after 0 (lanes 1 and 5), 20 (lanes 2 and 6), 60 (lanes 3 and 7), or 120 (lanes 4 and 8) minutes of incubation. RNA was passed over columns of oligo dT cellulose and aliquots of the poly A^+ (lanes 1-4) and poly A^- (lanes 5-8) fractions hybridized to the HindIII A restriction fragment (map units 60.1-73.6). The sizes of the RNA fragments which are formed by nuclease S1 (depicted schematically below the figure) are expressed in kilobases.

to RNAse at 37° than at 25°. One possibility is that the RNAse activity is compartmentalized, and not accessible to the RNA at 25° but that at 37°, the compartmentalization breaks down.

Properties of in vitro Synthesized RNA. The data presented above show that isolated nuclei synthesize large amounts of RNA from throughout the Ad2 late transcription unit, and that this RNA is quite stable. With this background, we have begun to characterize the RNA synthesized *in vitro*, and to study the processing reactions which occur in isolated nuclei. A sample of the results we have obtained to date is presented below.

We have recently shown that RNA molecules with specific 5' and 3' ends are synthesized *in vitro* (16). For example, Figure 2, lane 1, shows the primary transcript, which is initiated at position 16.5, as well as an RNA species colinear with viral DNA with a 5' end at position 19.5. This is the structure expected of a species which has undergone a single splice in the pathway which ultimately results in the creation of the tripartite leader sequence (19). Figure 4 shows that RNA species with discrete 3' ends are created during *in vitro* incubation and that these molecules, which correspond to species found *in vivo*, are efficiently polyadenylated *in vitro*. Aliquots of RNA extracted from nuclei which had been incubated for 2 hrs *in vitro* in the presence of α-^{32}P UTP were hybridized to either EcoRI B (map units 58.5-70.7, lanes 1-3) or EcoRI D (map units 75.9-83.4, lanes 5-6), and the nuclease S1 resistant duplexes were then analyzed by agarose gel electrophoresis. In such case, an RNA-DNA duplex indentical in size to the full-length restriction fragment can be detected. These represent RNA species with both 5' and 3' end points which lie outside the boundaries of the restriction fragment. In addition, bands which correspond to smaller RNA-DNA hybrids are apparent. The RNA fragment which gives rise to the hybrid approximately 3.0 Ad2 units (1,000 bp) in length (lanes 1 and 3) is the 3' end of the large precursor RNA which terminates at position 61.5. The 5' end of this RNA is presumably encoded at the major late promoter, at position 16.6. Likewise, the 2.0 Ad2 unit long fragment in lanes 4 and 5 is the 3' end of a large presumptive precursor RNA with 5' end at position 16.5 and 3' end at 78.5. These coordinates were determined by comparing the results of hybridizations to fragments of viral DNA produced by digestion with different restriction endonucleases (16, unpublished results). The results shown in Figure 4 do not, by themselves, allow us to determine whether these fragments are the 3' ends of mature mRNAs (*e.g.*, hexon and 100K mRNAs) or the longer precursors of these mRNAs. Two lines of evidence argue that they represent the 3' ends of precursor RNAs. First, if the ^{32}P is hybridized to larger DNA restriction fragments which extend

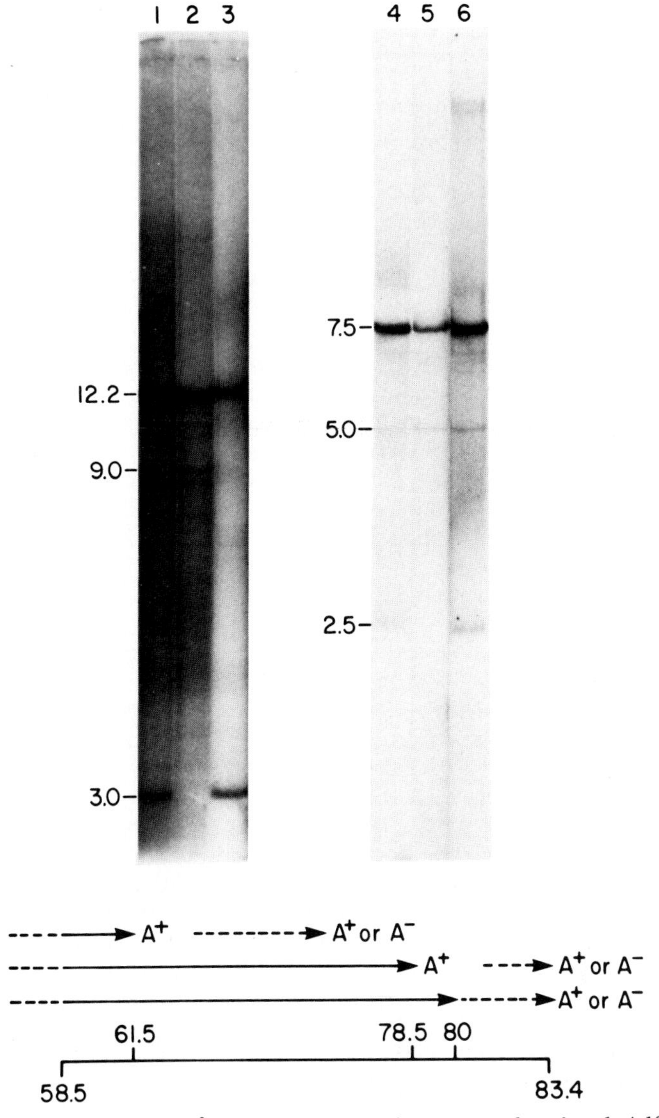

FIGURE 4. The 3' ends of in vitro synthesized Ad2 specific RNA. RNA was extracted from 100 μl reaction mixtures containing 8 x 10^6 nuclei after 120 min of incubation in the presence of 15 μCi α-^{32}P UTP. After selection on columns of oligo dT cellulose, aliquots of poly A^+ (lanes 1 and 4), poly A^- (lanes 2 and 5) and unselected (lanes 3 and 6) RNA were hybridized to either EcoRI B (58.5-70.7 map units, lanes 1-3), or EcoRI D (75.9-83.4 units, lanes 4-6) restriction fragments. In this figure, sizes of the RNA fragments produced by nuclease S1 digestion are expressed in Ad2 map units (1.0 map unit = 350 base pairs).

beyond the 5' ends of the mRNAs from these families, the only RNA-DNA hybrids produced are larger than would be produced from mRNA. For example, hybridization to BamHI A (59-100) gives rise to an RNA-DNA duplex after nuclease S1 digestion with its 3' end at 78.5 and its 5' end at 59 (results not shown). This is much longer than hybrids which would be produced by mRNA. Second, although we find molecules longer than mature mRNA, we find, at most, only very low levels of RNA-DNA hybrids of the size expected from mature mRNA. For example, when ^{32}P-RNA is hybridized to the HindIII A restriction fragment, which encodes the entire "main-body" sequence of hexon mRNA (*e.g.*, Figure 3), we can detect no RNA-DNA hybrid of the size expected.

As is the case with their *in vivo* counterparts, the *in vitro* synthesized RNA molecules with 3' ends at 61.5 or 78.5 can be detected only in a polyadenylated form. The material which does not bind to oligo dT cellulose contains no RNA molecules with mature 3' ends (Figure 4, lanes 2 and 5). Thus these molecules are polyadenylated with a high efficiency *in vitro*.

3' Ends can be Created by Cleavage. An important question is quether the 3' ends of eukaryotic mRNAs are created by transcription termination, or by cleavage of larger precursors. Current indirect evidence suggests that the Ad2 late mRNA 3' ends are created by cleavage (12). The *in vitro* nuclei system we have described here is ideal for asking this question in a more direct way. Two experiments we have done demonstrate this. If 3' ends are created by cleavage, this suggests that, in addition to the 3' mRNA precursor, a 5', promoter distal fragment should be formed. Such a fragment should be detected by the hybridization-ncuelase S1-agarose gel technique used above, if the fragment is stable, and if transcription continues beyond the boundary of the DNA restriction fragment used in the assay. Such RNA fragments have not been detected in analysis of nuclear steady-state viral specifid RNA (19, unpublished data). However, the analysis of *in vitro* synthesized RNA shown in Figure 4 reveals the presence of RNA molecules of the expected size. These are the duplex 9.0 Ad2 units in length which is produced following hybridization to EciRI B, and the hybrid 5.0 Ad2 units obtained from EcoRI D. As might be expected, if they are indeed by-products of the cleavage reaction which gives rise to mRNA 3' ends, these molecules are predominantly non-polyadenylated. Both of the fragments, but especially the one produced by the presumptive cleavage at 61.5 (lanes 1 and 2), are present in less than equimolar amounts compared with the mRNA 3' ends. This could be the result of either degradation or transcription termination by a fraction of the RNA polymerase molecules before reaching the boundary of the restriction fragment.

We have taken advantage of the fact that pulse-chase experiments can be easily performed with isolated nuclei to demonstrate directly that the 3' end of the 100K protein mRNA is created by cleavage. To do this, we pulse-labeled nuclei with α-^{32}P UTP for 5 min, and then diluted the label by adding a 25-fold excess of unlabeled triphosphate. (We previously calculated, by isotope dilution, the endogenous pool of UTP to be approximately 30 μM. Measurement of TCA-precipitable radioactivity showed that the chase was effective immediately, and that after 90 min, the amount of precipitable counts had increased by only about 10%.) Samples were taken immediately after the pulse, and at several times during a 90 min chase. RNA extracted from these samples was selected on columns of oligo dT cellulose, and aliquots of both poly A$^+$ and poly A$^-$ fractions were hybridized to the HindIII H restriction fragment (73.6-79.9). The results obtained following nuclease S1 digestion and agarose gel electrophoresis are shown in Figure 5. Several points are apparent: First, a significiant increase in the intensity of the "rimary transcript" (the 2.2 kb fragment in both the poly A$^+$ and poly A$^-$ fractions) occurs between the pulse and the first chase point (lanes 1 and 2, and 5 and 6). This results from movement of polymerase molecules which were within the boundaries of the restriction fragment during the pulse, but which did not transcribe beyond the promoter distal end of the fragment (coordinate 79.9) until the chase. Only RNA molecules which have moved beyond this point can give rise to a band. Second, while the intensity of the poly A$^+$ primary transcript remains constant during the chase, the poly A$^-$ material decreases in intensity. This is most likely due to a combination of degradation (e.g., see Figure 3) and specific cleavage which produces the polyadenylated mature 3' end at 78.5. Third, the RNA species terminated at position 78.5 increases dramatically during the chase. RNA extracted from nuclei immediately after the pulse does not give rise to a detectable band of the size expected (1.8 kb, lane 1) but a band of increasing intensity is seen with RNA extracted from "chased" nuclei (lanes 2-4). The band shows the greatest intensity increase during the time interval from 45-90 min, indicating the cleaving activity can function for extended times *in vitro*. Also the cleaved material is exclusively polyadenylated. We have never been able to detect mature, cleaved 3' ends in poly A$^-$ RNA. This suggests that the cleaving activity and the poly A adding enzyme might function in a tightly coupled reaction.

We have recently shown that the precursor to hexon mRNA (3' end at 61.5 units) is also created by cleavage. In addition we have shown that this cleavage, as well as the cleavage at 78.5, occurs equally efficiently whether or not RNA synthesis is allowed to continue during the "chase" (manuscript in preparation).

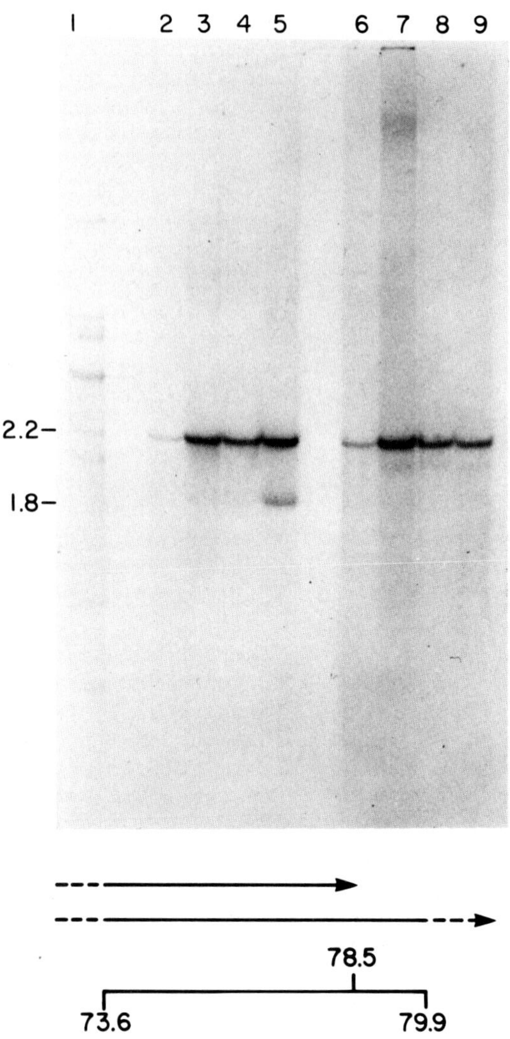

FIGURE 5. Formation of the 3' end of 100K in RNA in vitro by cleavage of a larger precursor. 100 μl reaction mixtures containing 8 x 10^6 nuclei were labeled with 50 μCi α-^{32}P UTP (50 Ci/mmole) for 5 min. After this time, 100 nmoles each of unlabeled UTP and $MgCl_2$ were added to each reaction, and incubation continued for 0 (lanes 2 and 6), 10 (lanes 3 and 7), 30 (lanes 4 and 8) or 90 (lanes 5 and 9) minutes. RNA was extracted, selected on oligo dT cellulose columns, and aliquots of poly A^+ (lanes 2-5) or poly A^- (lanes 3-9) were hybridized to the HindIII H (map units 73.6-79.9) restriction fragment. Nuclease S1 resistant hybrids were resolved by electrophoresis through agarose gel. The sizes of the RNA fragments are shown in likobases. Lane 1 shows a HindIII digest of ^{32}P Ad2 DNA, which was used to calibrate the gel.

SUMMARY

We have developed an *in vitro* transcription system for the synthesis of late adenovirus 2 RNA (15,16). In this system, viral RNA synthesis is followed during *in vitro* incubation of isolated nuclei. A majority of the transcription we observe during a two hour incubation *in vitro* is elongation of chains which had been initiated *in vivo*. However, approximately 10-20% of the chains elongated *in vitro* are also initiated *in vitro*. These molecules are initiated at the correct promoter, and are capped and methylated correctly. In fact, our data suggest that capping occurs very early during transcription, perhaps simultaneously with initiation. Long transcripts which are colinear with the viral DNA extend towards the righthand end of the genome. Approximately 50% of these molecules have a 5' end point at 16.5, the position of the major late promoter. These are presumably unspliced, primary transcripts. The remaining 50% have 5' ends at several other sites, the most abundant of which is at 19.5. The most likely origin of these species is that they are splicing intermediates in the pathway which leads to the formation of the tripartite leader sequence. For example, the species with its 5' end located at 19.5 by S1 mapping would appear to have undergone one splice, joining the approximately 45 nucleotides encoded at 16.5 (7) to sequences encoded at 19.5. Whether these molecules were spliced *in vivo* as nascent chains, and then elongated *in vitro*, or are the products of *in vitro* splicing, is not clear.

We have also shown that correct polyadenylated 3' ends are synthesized in *in vitro* nuclei. In particular, we have followed the time course of synthesis of 3' ends at sites in common with mature mRNAs for hexon and 100K polypeptide. These termini are created by cleavage of a larger precursor, as inhibition of transcription does not affect their rate of synthesis. Cleaved molecules are polyadenylated immediately, perhaps in a concerted reaction of cleavage and polymerization of A.

These findings, that initiation of transcription, extensive elongation of chains and cleavage and polyadenylation occur during incubation of isolated nuclei, encourage us to work on development of a soluble transcription system. Only when the factors involved in the process of mRNA synthesis are biochemically defined will we understand gene regulation in mammalian cells.

REFERENCES

1. Price, R., and Penman, S. (1972). J. Virol. 9, 621-626.
2. Marzluff, W.F., Jr., Murphy, E.C., Jr., and Huang, R.C.C. (1973). Biochemistry 12, 3440-3446.

3. Mory, Y.Y., and Gefter, M.L. (1978). Nucl. Acids Res. 5, 3899-3912.
4. Backenheimer, S., and Darnell, J.E. (1975). Proc. Nat. Acad. Sci. USA 72, 4445-4449.
5. Philipson, L., Pettersson, U., Lindberg, U., Tibbetts, C., Vennstrom, U., and Persson, T. (1974). Cold Spring Harb. Symp. Quant. Biol. 39, 447-456.
6. Goldberg, S., Weber, J., and Darnell, J.E. (1977). Cell 10, 617-621.
7. Ziff, E., and Evans, R. (1978). Cell 15, 1463-1475.
8. Chow, L.T., Gelinas, R.E., Broker, T.R., and Roberts, R.J. (1977). Cell 12, 1-8.
9. Klessig, D.F. (1977). Cell 12, 9-21.
10. Berget, S.M., Moore, C., and Sharp, P.A. (1977). Proc. Nat. Acad. Sci. USA 74, 3171-3175.
11. Chow, L.T., and Broker, T.R. (1978). Cell 15, 497-510.
12. Nevins, J.R., and Darnell, J.E. (1978). Cell 15, 1477-1493.
13. McGrogan, M., and Raskas, H.J. (1978). Proc. Nat. Acad. Sci. USA 75, 625-629.
14. Ziff, E., and Frazer, N. (1978). J. Virol. 25, 897-906.
15. Manley, J.L., Sharp, P.A., and Gefter, M.L. (1979). Proc. Nat. Acad. Sci. USA 76, 160-164.
16. Manley, J.L., Sharp, P.A., and Gefter, M.L. (1979), submitted to J. Mol. Biol.
17. Berk, A.J., and Sharp, P.A. (1977). Cell 12, 721-132.
18. Marzluff, W.F., Jr., Pan, C-J., and Cooper, D.L. (1978). Nucl. Acids Res. 5, 4177-4193.
19. Berget, S.M., and Sharp, P.A. (1979). J. Mol. Biol., in press.

ALTERNATIVE RNA SPLICING PATTERNS AND THE CLUSTERED TRANSCRIPTION AND SPLICING SIGNALS OF HUMAN ADENOVIRUS-2[1]

Thomas R. Broker and Louise T. Chow

Cold Spring Harbor Laboratory
Cold Spring Harbor, New York 11724

ABSTRACT We have used electron microscopic heteroduplex analyses to identify and map the cytoplasmic RNAs generated by human adenovirus serotype 2 (Ad2) at early and late times after permissive infection of cultured HeLa cells. We draw specific attention to the large number of nearly coincident transcription signals and processing sites for different RNA molecules and to the alternative splicing fates of precursor transcripts. Some of the principles governing Ad2 gene expression and the biological consequences of RNA splicing have been inferred from the structures and relative abundances of the RNAs.

INTRODUCTION

Adenovirus-2 grows lytically in cultures of human HeLa cells and establishes a complex program of RNA transcription and processing. The approximate map coordinates of Ad2 nuclear transcripts and messenger RNAs have been determined in several laboratories by hybridization of the RNAs to the separated strands of DNA restriction fragments (1-9). The map was refined by electron microscopic examination of RNA loops (10) formed in double-stranded DNA (11-13). Five regions of the Ad2 chromosome are expressed at early times (2-8 hr post infection), before the onset of DNA replication (Fig. 1). Regions 1A, 1B, and 3 are transcribed from the r-strand and regions 2 and 4 from the ℓ-strand. Regions 1B, 2, and 3 continue to be expressed at intermediate times (8-22 hr), but transcription of regions 1A and 4 declines (6, 14). Two additional transcription units, one on each strand, are active at late times from 8 hr until cell death after 40 hr.

---

[1] This research was sponsored by NIH-NCI Cancer Center Grant 13106 to Cold Spring Harbor Laboratory.

ISBN 0-12-068350-4

Many of the cytoplasmic mRNAs have been correlated with the encoded proteins, which were identified by cell-free translation of fractionated RNAs and comparison of the resulting polypeptides with viral-induced proteins and virion components (15-18, R. Ricciardi, J. Miller, B. Roberts, B. Paterson, and M. Mathews, personal communication).

We and others discovered the phenomenon of RNA splicing in the mature late Ad2 mRNAs (19-24). Shortly thereafter, splicing was also recognized in the early Ad2 RNAs (25, 26) and in a variety of other eukaryotic transcripts. The nucleotide sequences comprising most adenoviral RNAs found in the cytoplasm are derived from two or more noncontiguous segments of the chromosome. The segments become covalently linked via internal deletions and re-ligations that take place in the primary nuclear transcripts.

There is a different promoter for each of the early regions (27-29), and each transcription unit gives rise to a family of RNA products with common 5' and 3' ends but alternative internal deletions (14, 25, 26) (Fig. 1). Sets of proteins that share common peptides are produced from these messages (17), establishing the existence of overlapping, nested genes in adenoviruses.

A different type of polycistronic RNA is observed at late times (Fig. 1). Transcription of the r-strand is initiated at a single promoter at coordinate 16.5 (27, 30) and the long primary transcript can extend through all genetic regions located downstream (31). Only some transcripts continue uninterrupted through the distal gene (fiber: IV) in the unit and these become polyadenylated at coordinate 91.3. The remainder become polyadenylated at one of four other preferred 3' terminal sites at coordinates 38.8, 49.5, 61.3, or 78.3 (7, 8, 9, 11, 32) or at one of several minor sites at coordinates 45.1, 74.0, 82.7, or 86.0 (14, 32). Each of these various nuclear transcripts then undergoes a series of internal deletions that join the 5' terminal sequences (16.5-16.6) and two additional leader segments (19.5-19.7 and 26.5-26.8) to the 5' end of different coding regions. As a result, only one RNA is derived from each primary transcript and the RNAs fall into families with overlapping members which have common 3' proximal sequences. The longer members of the 3' coterminous families of cytoplasmic RNAs are physically polycistronic (cf., Fig. 1) but functionally monocistronic because only the 5' proximal message is translatable (reviewed in ref. 33). Unlike most of the early mRNA families, the overlapping late r-strand transcripts code for unrelated proteins (with the exception of the 33K protein, which is the carboxy-terminal portion of the 100K protein (34)). Comparisons of the map sites of promoters with the 5' ends of mature RNAs and of the polyadenylation sites in nuclear RNAs with the 3' ends of

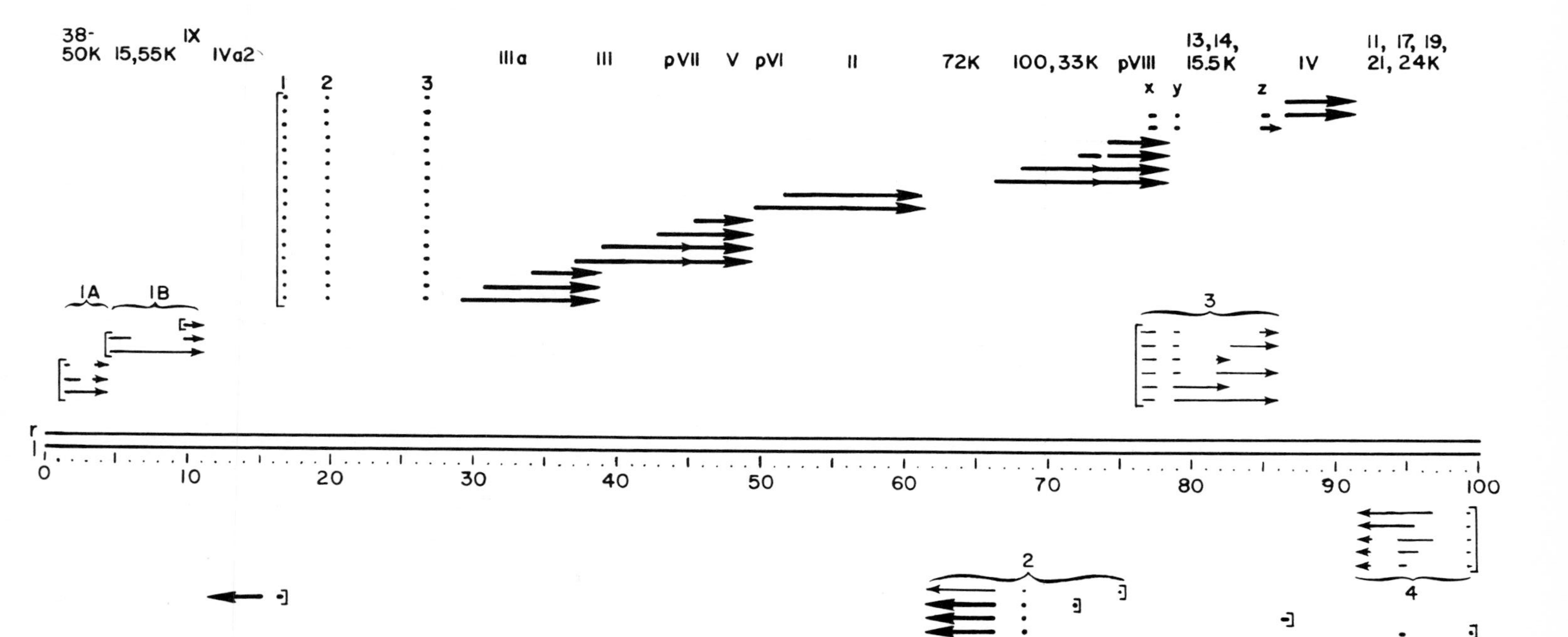

Figure 1: The electron microscopical map of adenovirus-2 cytoplasmic RNAs. The conserved segments constituting early RNAs (14) are depicted by thin arrows and those in late RNAs (19, 32) by thick arrows. Vertical brackets mark mature 5' ends and arrowheads correspond to the 3' ends. Gaps in arrows represent deletions in the RNAs resulting from splicing. At intermediate to late times, region 2 is expressed from additional promoters, as shown, and region 3 can be transcribed under the direction of the major r-strand late promoter. All derivatives of the late r-strand transcript have the same 3-part leader. Precursors to the fiber (IV) protein can also contain some combination of ancillary leaders x, y, z (Figure 3). The correlation of mRNAs with encoded proteins was based on cell-free translation of RNA selected by hyridization to DNA restriction fragments (15-18).

mature cytoplasmic RNAs have revealed the important principle that the original ends are conserved during RNA processing.

MATERIALS AND METHODS

The experimental procedures employed for the isolation of RNA, its hybridization to single-stranded or double-stranded DNA, the preparation of heteroduplexes for electron microscopy, and EM data gathering and analysis have been described previously (11, 14, 19, 32).

RESULTS AND DISCUSSION

Using electron microscopic heteroduplex analysis, we have mapped the abundant and the minor RNA species synthesized at early and late times after infection of HeLa cells with Ad2 (11, 14, 19, 32). These have been aligned and are summarized in Figure 1. Examination of this complete map has led to the identification of a large number of short intervals in the Ad2 chromosome where various combinations of promoters, terminators, and upstream or downstream splicing sites for different RNAs (derived from either the same strand or from opposite strands) are nearly coincidental. Table 1 lists these clustered sites and Figures 2 and 3 provide graphic representations of two of the more complex regions. We have adopted a system of numerical notation to represent each type of 5' end, 3' end and splice junction. When different mRNAs have ends or splice junctions that coincide, the site is labeled by a pair of numbers designating the two types of boundaries involved and whether the RNAs are complementary to the same or opposite strands. Of the thirty-six types of alignment theoretically possible, sixteen are redundant in that they are symmetrical with sixteen others except in strand assignment. Consequently, there are twenty unique combinations, of which representative RNAs have been found in sixteen (Table 1). The groups of RNAs in coincidence classes 1-1, 2-2, 3-3, and 4-4 represent alternative transcription and processing patterns observed in certain genetic regions. Many of these features are discussed below.

Identical promoters for different RNAs synthesized from the same strand (coincidence class 1-1) give rise to nested or to serially polycistronic messages. They exist for early regions 1A (coordinate 1.3), 1B (4.6), 3 (76.6), and 4 (99.3) (14, 25-29) and for the major r-strand transcription unit (coordinate 16.5) (19, 30, 32). Other promoters do not fit in this class and give rise to messenger RNAs for component IX

Table 1: Coincidences of adenovirus-2 transcription signals and RNA splice junctions: catalogue of chromosomal sites used in multiple ways.

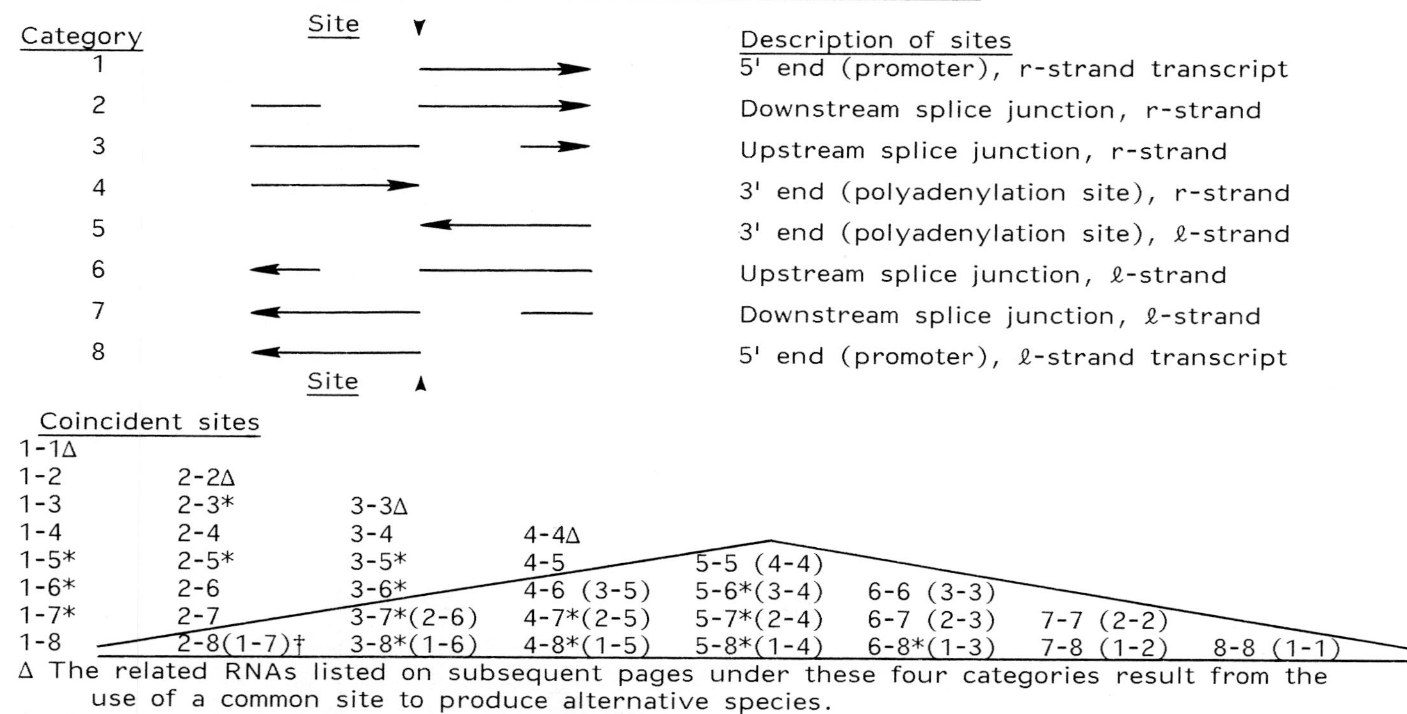

Category	Site	Description of sites
1		5' end (promoter), r-strand transcript
2		Downstream splice junction, r-strand
3		Upstream splice junction, r-strand
4		3' end (polyadenylation site), r-strand
5		3' end (polyadenylation site), ℓ-strand
6		Upstream splice junction, ℓ-strand
7		Downstream splice junction, ℓ-strand
8		5' end (promoter), ℓ-strand transcript

Coincident sites

1-1Δ							
1-2	2-2Δ						
1-3	2-3*	3-3Δ					
1-4	2-4	3-4	4-4Δ				
1-5*	2-5*	3-5*	4-5	5-5 (4-4)			
1-6*	2-6	3-6*	4-6 (3-5)	5-6*(3-4)	6-6 (3-3)		
1-7*	2-7	3-7*(2-6)	4-7*(2-5)	5-7*(2-4)	6-7 (2-3)	7-7 (2-2)	
1-8	2-8(1-7)†	3-8*(1-6)	4-8*(1-5)	5-8*(1-4)	6-8*(1-3)	7-8 (1-2)	8-8 (1-1)

Δ The related RNAs listed on subsequent pages under these four categories result from the use of a common site to produce alternative species.

* These combinations have no observed examples.

† Parentheses indicate redundant combinations (enclosed in the triangle) that differ only in the strand designation. Specific examples are listed on subsequent pages.

Conventions in RNA designations:

→ , ← : rightward or leftward direction of transcription

[,] : r- or ℓ-strand promoter (5' end)

.. : polyadenylation site (3' end)

\- : RNA is continuous between the adjacent coordinates

/ : splice junction

-- : numbers flanking "--" are 5' and 3' coordinates of a family of RNAs with different internal splice patterns.

__ : underlined coordinates are the coincident or nearly coincident sites

; : separates the description of different RNAs.

In the catalog below and in Figures 2 and 3, the symmetrical combinations indicated on the previous page are grouped together irrespective of the r- or ℓ-strand origins and are listed under one pair of numbers. The strand origins are, however, clear from the directional arrow at the 5' side of each indicated RNA species or segment of RNA.

To conserve space, the description of RNAs is generally abbreviated to include only the segments adjacent to the indicated site but, in most cases, the RNA can be identified in Figure 1, 2, or 3.

Differences in map coordinates of 0.1-0.2 units (35-70 nucleotides) are not necessarily significant and fall within the 95% confidence limits of the mapping.

Coincident sites	Description of sites
1-1 (8-8)	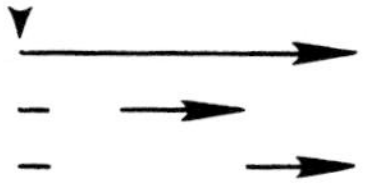 Alternative transcripts that share a common promoter and have the same 5' end.
(a) →[1.3 -- 4.4..	(Early region 1A)
(b) →[4.6 -- 11.1..	(Early region 1B)
(c) →[76.6 -- 86.0..	(Early region 3)
(d) ..91.3 -- 99.3]←	(Early region 4)
(e) →[16.5 -- 91.3..	(Late r-strand unit)
1-2 (7-8)	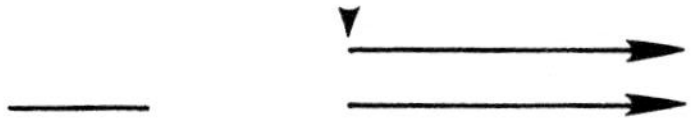 Coincidence of a 5' end and a downstream splice junction on the same strand
(a) →[9.8 - 11.1.. ; →[4.6 - 6.1 / 9.8 - 11.1..	
(b) /71.9 - 72.0]← ; / 71.9 - 72.0 / 75.1 - 75.2]←	
1-3 (6-8)*	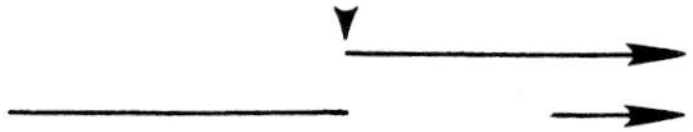 Coincidence of a 5' end and an upstream splice junction on the same strand
(a) →[4.6 - 11.1.. ; → /3.3 - 4.4/ 9.8 - 11.1..	
1-4 (5-8)*	 Coincidence of a 5' end and a 3' end on the same strand
(a) →[4.6 - 11.1.. ; →[1.3 - 4.4..	
1-5*(4-8)*	Coincidence of a 5' end and a 3' end on opposite strands
(no examples)	

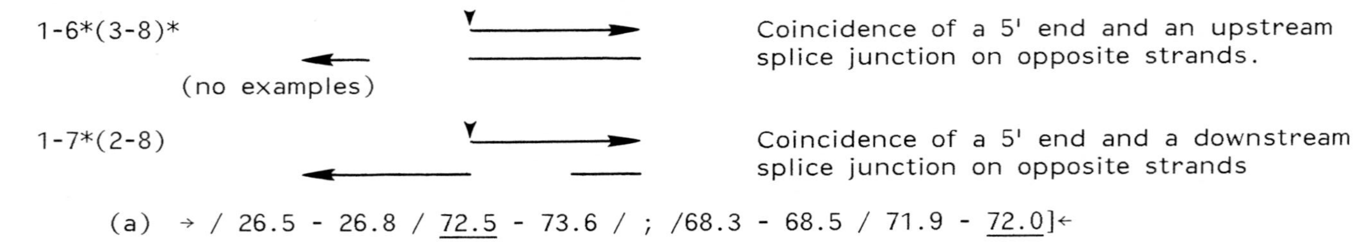

1-6*(3-8)* — Coincidence of a 5' end and an upstream splice junction on opposite strands.

(no examples)

1-7*(2-8) — Coincidence of a 5' end and a downstream splice junction on opposite strands

(a) → / 26.5 - 26.8 / 72.5 - 73.6 / ; /68.3 - 68.5 / 71.9 - 72.0]←

1-8 — Coincidence of promoters that initiate divergent transcription.

(a) ..11.1 -- 16.1]← ; →[16.5 -- 91.3..
(b) ..61.5 -- 75.2]← ; →[76.6 -- 86.0.. (not very close)

2-2 (7-7) — Alternative upstream splice junctions and the same downstream splice junction on the same strand

(a) →[1.3 - 3.1 / 3.3 - 4.4.. ; → [1.3 - 2.6 / 3.3 - 4.4.. ;
→[1.3 - 1.7 / 3.3 - 4.4..
(b) / 68.3 - 68.5 / 94.4 - 94.8/ ← ; / 68.3 - 68.5 / 86.2 - 86.7]← ;
/ 68.3 - 68.5 / 75.1 - 75.2] ← ; / 68.3 - 68.5 / 71.9 - 72.0]←
(c) → / 19.5 - 19.7 / 26.5 - 26.8/ ; → / 22.0 - 23.2 / 26.5 - 26.8 /
(d) → / 26.5 - 26.8 / 74.0 - 78.3.. ; → / 72.5 - 73.6 / 74.0 - 78.3..
(e) → / 26.5 - 26.8 / 78.6 - 82.7.. ; → [76.6 - 77.6 / 78.6 - 82.7..
(f) → / 26.5 - 26.8 / 86.2 - 91.3.. ; → / 77.3 - 77.6 / 86.2 - 91.3.. ;
→ / 78.6 - 79.1 / 86.2 - 91.3.. ; → / 84.7 - 85.1 / 86.2 - 91.3.. ;
(g) (numerous other examples are found among late RNA processing intermediates in which a short RNA segment from the 5' end of an upstream transcript remains between the tripartite leader and the main body of a downstream transcript)

2-3*(6-7)

Coincidence of downstream and upstream splice junctions on the same strand

(a) ..91.3 - 96.8 / 99.2 - 99.3]← ; / 94.4 - 94.8 / 96.8 - 97.1 / ←

2-4 (5-7)*

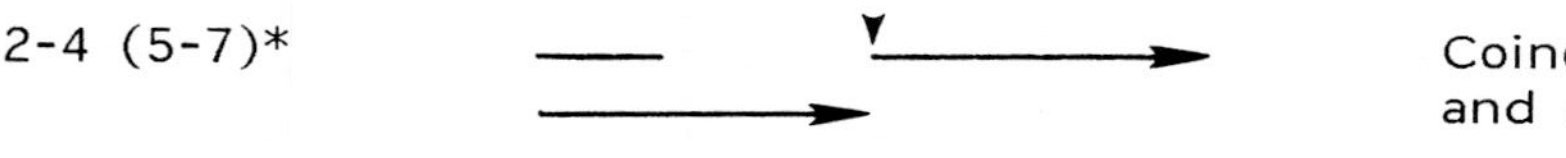

Coincidence of a downstream splice junction and a 3' end on the same strand

(a) → / 26.5 - 26.8 / 38.8 - 49.5.. ; → / 29.1 - 38.8..
(b) → / 26.5 - 26.8 / 45.1 - 49.5.. ; → / 38.8 - 45.1..
(c) → / 26.5 - 26.8 / 49.5 - 61.3.. ; → / 38.8 - 49.5..
(d) → / 26.5 - 26.8 / 74.0 - 78.3.. ; → / 66.1 - 74.0..
(e) → / 26.5 - 26.8 / 78.6 - 86.0.. ; → / 66.1 - 78.3..
(f) → / 78.6 - 79.1 / 82.7 - 86.0.. ; → / 78.6 - 82.7..
(g) → / 26.5 - 26.8 / 86.2 - 91.3.. ; → / 78.6 - 86.0..

2-5*(4-7)*

Coincidence of a downstream splice junction on one strand and a 3' end on the opposite strand.

(no examples)

2-6 (3-7)*

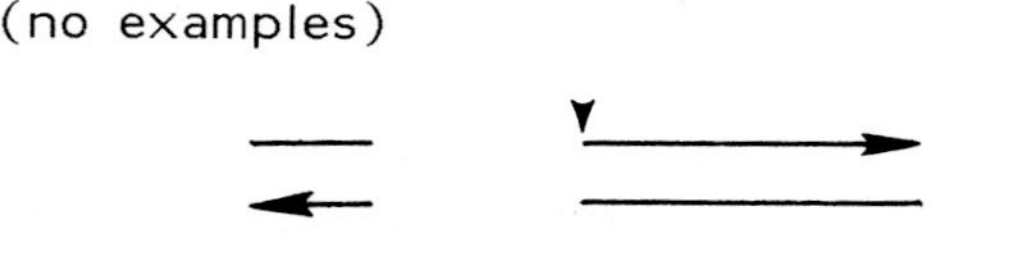

Coincidence of a downstream splice junction on one strand and an upstream splice junction on the opposite strand.

(a) → / 26.5 - 26.8 / 86.2 - 91.3.. ; /68.3 - 68.5 / 86.2 - 86.7]←
(b) → / 26.5 - 26.8 / 68.0 - 78.3.. ; ..61.6 - 66.5 / 68.3 - 68.5/←

2-7 — Coincidence of downstream splice junctions on opposite strands

(a) → / 26.5 - 26.8 / 66.1 - 78.3.. ; ..61.6 - 66.2 / 66.3 - 66.5 / ←
(b) → / 26.5 - 26.8 / 72.5 - 73.6 / ; / 71.9 - 72.0 / 75.1 - 75.2] ←

3-3 (6-6) — The same upstream splice junction and alternative downstream splice junctions on the same strand.

(a) / 71.9 - 72.0 / 75.1 - 75.2] ← ; / 68.3 - 68.5 / 75.1 - 75.2] ←
(b) → [76.6 - 77.6 / 78.6 - 82.7.. ; → / 77.3 - 77.6 / 86.2 - 91.3..
(c) → / 78.6 - 79.1 / 81.5 - 82.7.. ; → / 78.6 - 79.1 / 82.7 - 86.0..;
→ / 78.6 - 79.1 / 84.7 - 86.0..
(d) / 96.8 - 97.1 / 99.2 - 99.3] ← ; ..91.3 - 96.8 / 99.2 - 99.3] ←
..91.3 - 95.7 / 99.2 - 99.3] ← ; / 99.4 - 94.8 / 99.2 - 99.3] ←
(e) ..91.3 - 92.4 / 94.4 - 94.8 / ← ; / 68.3 - 68.5 / 94.4 - 94.8 / ←
(f) → / 19.5 - 19.7 / 22.0 - 23.2 / ; → / 19.5 - 19.7 / 26.5 - 26.8 /
(g) → / 26.5 - 26.8 / to many different downstream sites: 29.1 ; 30.5; 33.9; 37.0; 38.8; 42.8; 45.1; 49.5; 51.2; 66.1; 68.0; 72.5; 74.0; 77.3; 78.6; 84.7; 86.2 (see Figure 1 for examples)

3-4 (5-6)* — Coincidence of an upstream splice junction and a 3' end on the same strand

(a) → [1.3 - 4.4.. ; → /3.3 - 4.4 / 9.8 - 11.1..

3-5*(4-6) — Coincidence of an upstream splice junction on one strand and a 3' end on the opposite strand.

(a) → / 78.3 - 86.0.. ; / 68.3 - 68.5 / 86.2 - 86.7]←

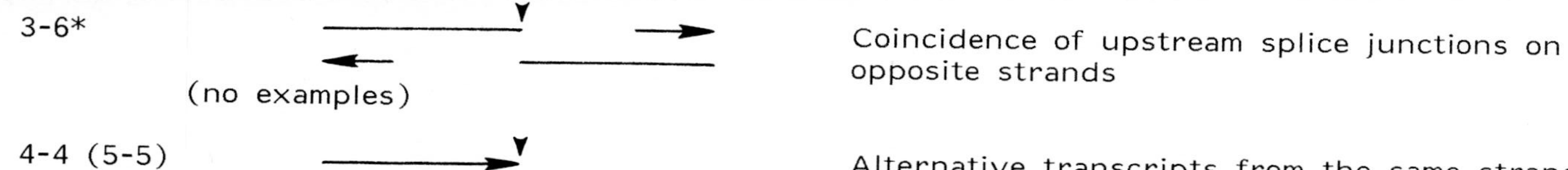

3-6* — Coincidence of upstream splice junctions on opposite strands

(no examples)

4-4 (5-5) — Alternative transcripts from the same strand that have the same 3' ends

(see Figure 1 for complete coordinates of 3' co-terminous transcripts)

```
(a)  →[1.3              --    4.4..                                  (Early region 1A)
(b)  →[4.6              --   11.1..                                  (Early region 1B)
(c)                    ..   61.6   -- 72.0]←, 75.2]←, 86.7]←        (Early and late region 2)
(d)  →[16.5, →[76.6 --      82.7..                                  (Early and late region 3)
(e)  →[16.5, →[76.6 --      86.0..                                  (Early and late region 3)
(f)                    ..   91.3   -- 99.3]←                         (Early region 4)
(g)  →[16.5             --   38.8..                                  (Late)
(h)  →[16.5             --   45.1..                                  (Late)
(i)  →[16.5             --   49.5..                                  (Late)
(j)  →[16.5             --   61.3..                                  (Late)
(k)  →[16.5             --   74.0..                                  (Late)
(l)  →[16.5             --   78.3..                                  (Late)
(m)  →[16.5             --   91.3..                                  (Late)
```

4-5 — Convergent 3' ends (on opposite strands)

```
(a)  → [  4.6 - 11.1.. ; .. 11.1 - 15.1 / 15.7 - 16.1] ←
(b)  → / 49.5 - 61.3.. ; .. 61.6 - 66.5 / ←
(c)  → / 86.2 - 91.3.. ; .. 91.3 - 96.8 / ←
```

Figure 2

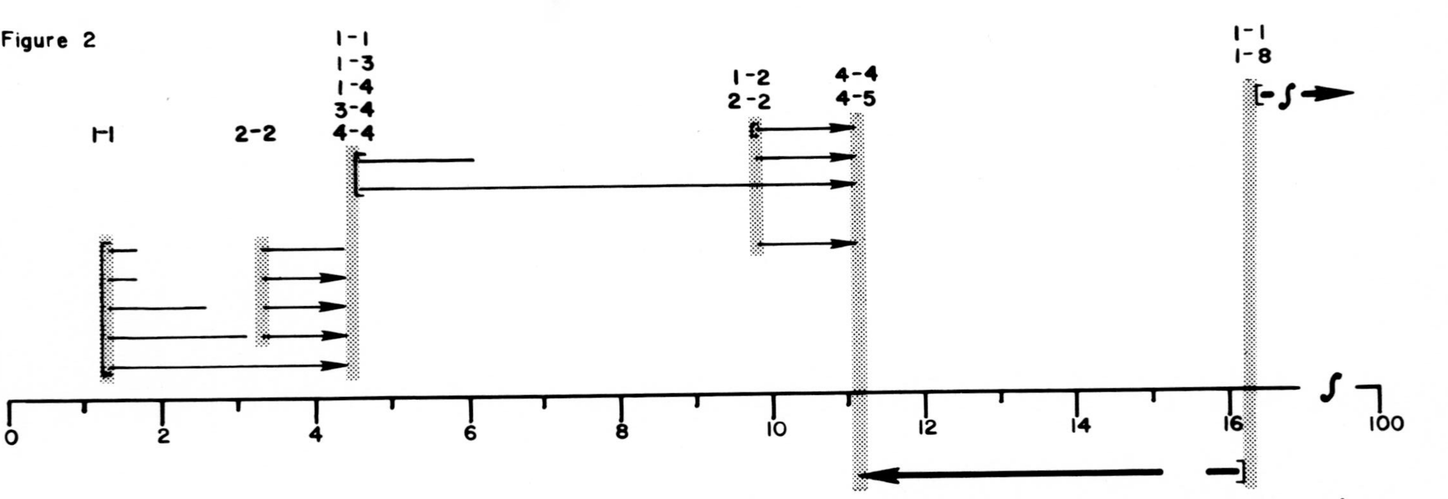

Figure 2. <u>Coincident RNA processing sites in the chromosome region 0.0-17.0</u>. The conventions used in Figure 1 to represent spliced early RNAs (thin arrows) and late RNAs (thick arrows) are maintained. Short DNA or RNA regions that serve multiple roles in the synthesis and maturation of different mRNAs are indicated by stippled bars. The pairs of numbers above the bars designate the nature of the coincident sites (cf. Table 1). Symmetrical combinations are indicated by one pair of numbers, irrespective of the r- or ℓ-strand origin of the RNAs.

Figure 3. <u>Coincident RNA processing events in the chromosome region 60.0-100.0</u>. The conventions are described in the legends to Figures 1 and 2 and in Table 1. All late r-strand RNAs have the tripartite leader (3) derived from coordinates 16.5-16.6/19.5-19.7/26.5-26.8/. The fiber RNAs (86.2-91.3) have multiple forms, some with ancillary leaders, but probably encode a single protein. The DBP RNAs (61.6-66.5/68.3-68.5/) result from four different promoters but probably encode the same protein. Each DBP RNA can be observed in an alternative form with a microsplice from 66.2-66.3 (14); only two examples are depicted.

Figure 3

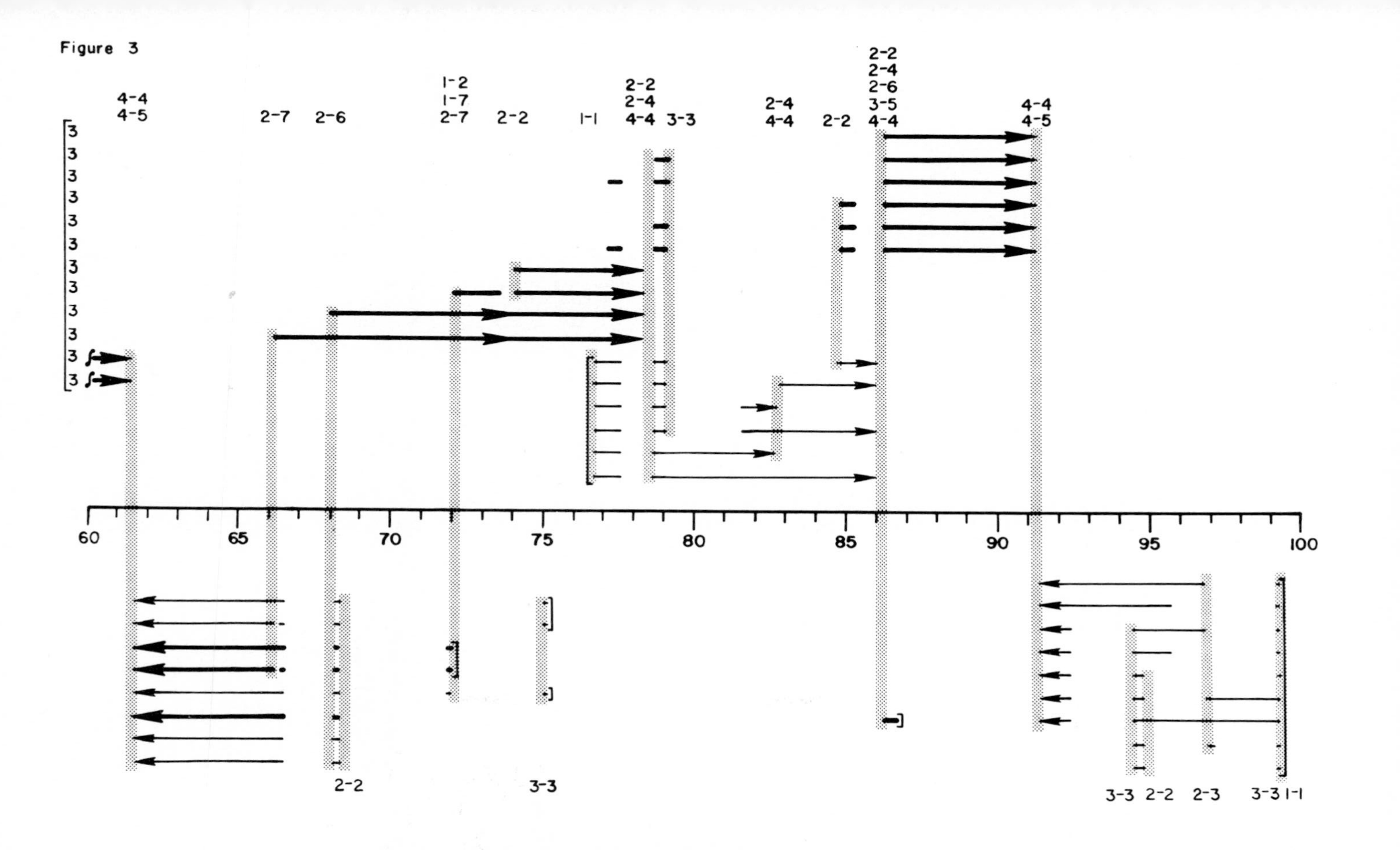

4-4
4-5
2-7
2-6
1-2
1-7
2-7
2-2
1-1
2-2
2-4
4-4
3-3
2-4
4-4
2-2
2-2
2-4
2-6
3-5
4-4
4-4
4-5
3
3
3
3
3
3
3
3
3
3
3
3
60
65
70
75
80
85
90
95
100
2-2
3-3
3-3
2-2
2-3
3-3
1-1

(coordinate 9.8, r-strand) (11, 18), IVa2 protein (coordinate 16.1, ℓ-strand) (21, 23, 32), and the single-stranded DNA binding protein (DBP) (coordinates 72.0, 75.2, or 86.7, ℓ-strand) (14, 16, 25, 26). All product mRNAs from early regions 3 and 4 and from the major r-strand unit have, within each family, consanguinous 5' leader segments. The related messages are distinguished by the alternative coordinates to which the leaders are spliced and are therefore examples of class 3-3 coincidences of the same upstream splice site recombining with different downstream sites.

Alternative promoters for the same message result in common downstream splice junctions (coincidence class 2-2). During the transition from early to late transcription, control of the expression of region 2 is gradually transferred from the early promoter (75.2) (27, 28) to several alternative late promoters at 72.0 (major), 86.7 (minor), and 99.3 (rare) (14). The DBP from region 2 is necessary for replication at intermediate to late times and the activation of late promoters assures its continued production when early transcription declines. The structures of the DBP mRNAs reveal additional information about the factors modulating RNA splicing. At early times, the splice 68.5-71.9 is not achieved, although it occurs readily at late times when the 5' end is at 72.0. Similarly, when the 5' leader segment comes from 86.2-86.7, both of the potential leader segments from 75 and from 72 are bypassed. And when the rare leader from region 4 is coupled to the DBP mRNA, each of the potential leader segments from 86, 75, and 72 is excluded from the spliced RNA product. In RNA samples prepared at intermediate times, we have detected a few examples of the DBP mRNA early leader from 75.1-75.2 joined to the 5' boundary of the late leader at 72.0 and then to the common segments (..61.6-66.5/ 68.3-68.5/71.9-72.0/75.1-75.2]←), a class 1-2 use of coincident sequences. In summary, the presence of a nearby 5' end seems necessary to potentiate recognition of the leader splice junctions or to confer stability to their inclusion.

Region 3 is the only early region embedded within a late transcription unit. At early times it is expressed under control of its own promoter at 76.6 (14, 25-28). At late times some region 3 RNAs originate, instead, from the late promoter at 16.5 and have the tripartite leader; it is spliced to a site just downstream of the early 5' end (to coordinate 77.3) or, alternatively, to the beginning of the second early RNA segment (y) at 78.6 or to one of the alternative third segments at 84.7 (z junction) (14) (Fig. 4). Despite the different 5' proximal sequences, the variety of alternative splicing patterns and 3' ends seen at early times are observed in these late forms.

Alternative 5' leader components were found for some fiber region RNAs. They are short optional segments termed x (coordinate 77), y (coordinate 79), and z (coordinate 85), and one or more can be interposed between the tripartite leader and the main body of the message (23, 32, 35) (Figs. 1, 3 and 4). We have recently found that one abundant form of early region 3 RNAs consists of three segments (14). The x leader corresponds with the downstream portion of the first early RNA segment, y is identical to the middle segment, and z is the upstream portion of the third segment. Thus, the x/y and y/z splice junctions are the same at early and late times and the ancillary leaders in some fiber RNAs represent the vestigial recognition of the early splicing signals. All seven possible combinations of x, y, and z leaders and the eighth option, their complete absence in mature mRNAs, have been observed in fiber RNAs (32; D.F. Klessig and L.T. Chow, unpublished results). The coupling of either the x segment or the late tripartite leader to y, and the coupling of either the y segment or the tripartite leader to z, and the coupling of either the x segment, the y segment, the z segment, or the tripartite leader to the main body of fiber mRNA all represent class 2-2 coincidences of alternative upstream splice junctions recombining with a common downstream splice site. Conversely, the x/y and x/fiber, and the y/fiber and the y/81.5, y/82.7 and y/84.7 (z) splices observed in region 3 RNAs all are class 3-3 coincidences of the same upstream splice junction and alternative downstream junctions.

Alternative splicing patterns can result in the production of related mRNAs from the same precursor. The presence of a potential splice junction in a transcript does not obligatorily result in its utilization. Indeed, one of the alternative RNA species in each of the nested polycistronic sets from early regions 1A, 1B, 3, and 4 is the unspliced form. The distribution of species in such a set can be very sensitive to time after infection and to the presence of metabolic inhibitors (14).

In early region 1A, several spliced messages have been observed (Fig. 2). They all have the same 5' end and the same 3' (downstream) conserved segment but differ in the length of the deletion and in the upstream splice site (coincidence class 2-2) (14, 26). Alignment of the Ad2 RNA splicing maps with the DNA sequence of the closely related serotype Ad5 (36) suggests that these splices delete protein termination codons and probably also result in a frame shift to allow the synthesis of larger proteins. These alternative RNAs could be expected to give rise to a set of related proteins sharing amino- and carboxy-terminal peptides but differing in some internal peptides. The relative abundance of the

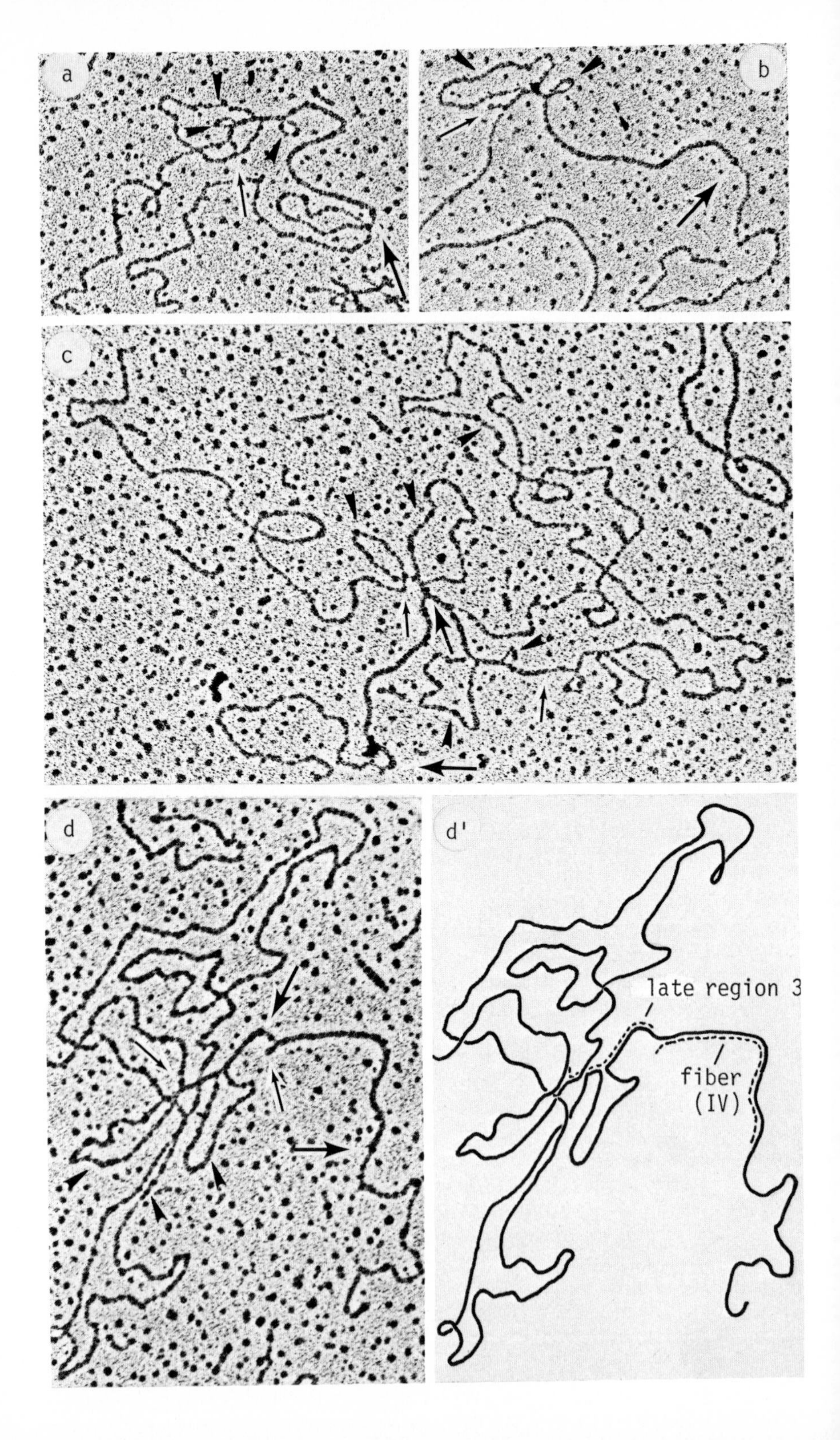
a
b
c
d
d'
late region 3
fiber
(IV)

species with the longest deletion increases with time after infection and the species with no splice or with the shortest splice correspondingly decrease. This transition does not occur in the presence of cycloheximide, an inhibitor of protein synthesis, suggesting either that some stage in RNA translation or that an early adenoviral protein is involved in the region 1A RNA splicing. This same effect of cycloheximide in decreasing the relative regional abundances of RNAs with long deletions or with two deletions is also observed in RNAs from regions 1B and 4 (14).

The 3' end of early region 1A and the 5' end of region 1B are separated by no more than 70 nucleotides (class 1-4 coincidence) (14, 26). In region 1B, two messages share 5' and 3' termini but differ by the presence or absence of an internal splice, another kind of alternative processing (14, 19, 25) (Fig. 2). The region encodes proteins with apparent molecular weights of 15,000 daltons and 55,000 daltons, the smaller coming from the spliced RNA (16, 37; A.J. van der Eb, J.H. Lupker, A. Davis-Olivier, and H. Jochemsen, personal communication). Examination of the DNA sequences of Ad5 reveals a continuous open reading frame sufficient to encode the larger protein (38). The splice may well place a termination codon present in the segment 9.8-11.1 into the open frame, resulting in the truncation of the 55K protein to a 15K protein. The downstream segment 9.8-11.1 in the region 1B spliced RNA also exists as an independent RNA (6, 11) which has its own promoter near coordinate 9.8 (29) and

---

Figure 4. Coincidence of the x, y, and z leader segments in fiber RNA and conserved segments in early region 3 RNAs. Small arrows point to the 5' ends of RNAs, large arrows to the 3' ends, and arrowheads to DNA deletion loops indicative of RNA splicing. (a) fiber IV RNA (86.2-91.3) with an unhybridized tripartite leader and x (77.3-77.6), y (78.6-79.1), and z (84.7-85.3) ancillary leaders. (b) fiber RNA with an unhybridized tripartite leader and y and z leaders. (c) fiber RNA with a hybridized tripartite leader (16.6/19.6/26.6) forming three DNA deletion loops. In the deletion loop between the third leader segment and fiber RNA, a spliced region 3 RNA (76.6-77.6/78.6-79.1/84.7-86.0) has co-annealed. The 3' end of the early RNA and the 5' end of the main body of fiber RNA are nearly coincident. The deletion loops formed by the early RNA are identical to those formed by the x, y, and z leaders in fiber RNAs shown in panels (a) and (b). (d and d') The late form of one region 3 RNA consists of the tripartite leader (16.6 [unpaired]/19.6/26.6) coupled to RNA from coordinates 78.6-79.1/84.7-86.0 (cf., panel b). The proximity of the 3' end of this RNA and the 5' end of the main body of the fiber RNA is evident.

encodes the virion component IX (18). Thus, the promoter for peptide IX RNA and the downstream splice junction in region 1B nearly coincide (class 1-2).

Alternative upstream and downstream splice junctions (classes 2-2 and 3-3) also occur during the maturation of the tripartite leaders in late RNAs. A group of apparent processing intermediates are observed with additional leader sequences interposed between the second and third leaders (39; L. Chow, D. Klessig, T. Grodzicker, J. Sambrook, J. Lewis, and T. Broker, unpublished). These are particularly prevalent when RNA samples are prepared at intermediate times after infection of cells grown in monolayers. The most frequently seen extra leader has coordinates 22.0-23.2; other extra sequences are observed in the form of long third leader segments which extend upstream (e.g., 22.0-26.8 or 24.2-26.8). In most cases these "immature" leader combinations are already spliced to the normal set of downstream coding sequences.

Alternative orders of splicing have been observed during the maturation of RNA leader and main body sequences. Certain splices almost always preceed others. Some series of deletions, however, can occur in variable combinations indicative of alternative pathways. Furthermore, downstream splices can, and in some cases usually, preceed upstream splices in the same transcript. Therefore, splicing is not necessarily processive. Prominent examples include (a) the incorporation and removal of x, y, and z ancillary leaders from fiber region RNAs (32, 35); (b) the coupling of the leader segment to the main bodies of some region 4 RNAs, which can sometimes follow the downstream deletion of the intervening segment 92.4-94.4 (14); (c) the coupling of the tripartite leader to downstream messages, which proceeds by multi-step deletions and involves the elimination of temporary additional leader components (Chow et al., unpublished) and (d) the multistep joining of the second to the third leader segment of the r-strand late transcripts, just described. Leader sequences can be spliced into their mature configuration prior to or following maturation of the main body coding sequences. In all cases, the conserved segments in spliced RNAs are in the same relative order as they exist in the primary transcript: permutations have never been observed.

Coincident 3' ends (class 4-4) have been observed in the alternative mRNAs derived from most transcription units (Fig. 1). In early regions 1A, 1B and 4, the 5' ends of all their alternative species also coincide. As discussed previously, all DBP (region 2) mRNAs have a common 3' end but possess alternative 5' ends. Both the region 3 and the late r-strand

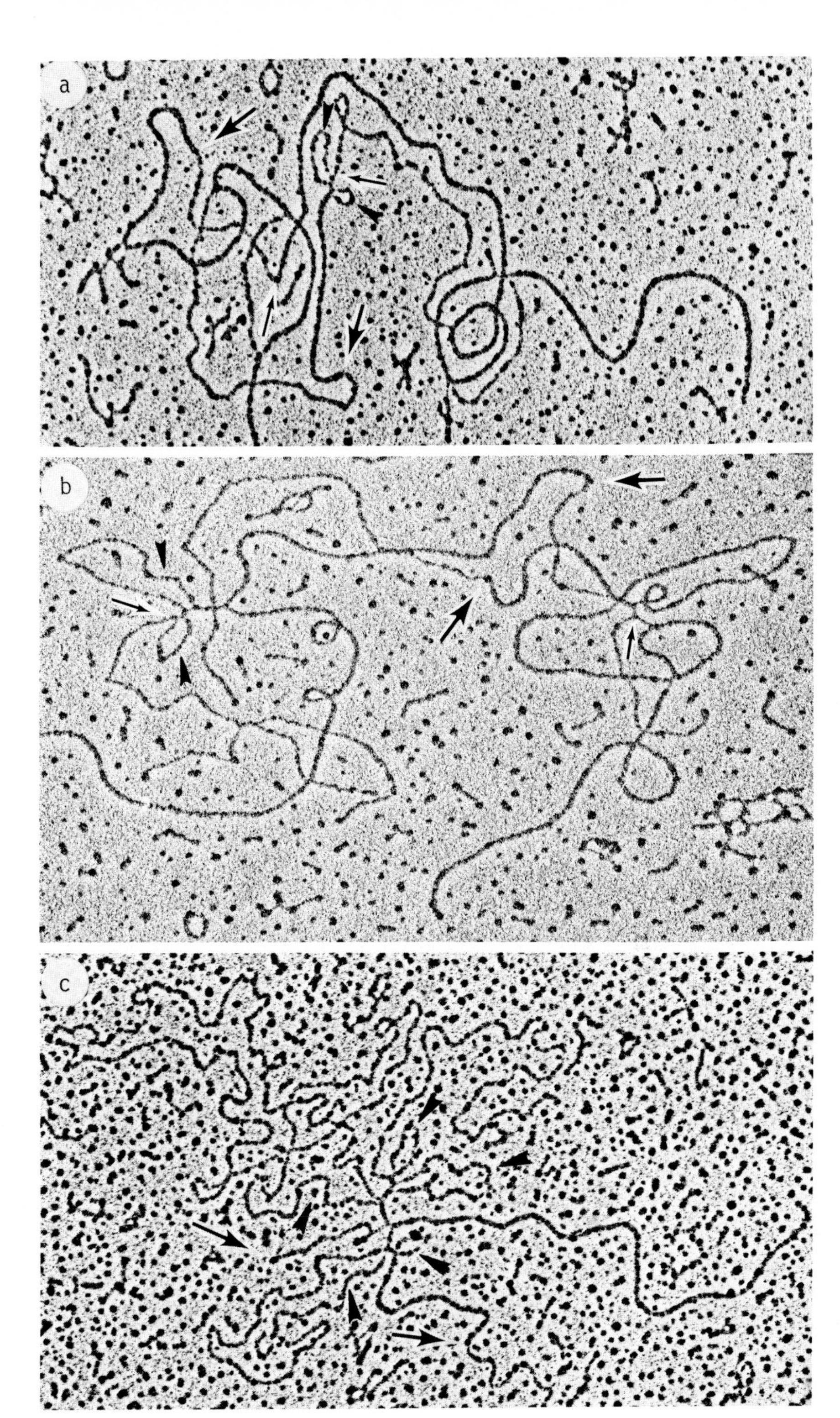
a
b
c

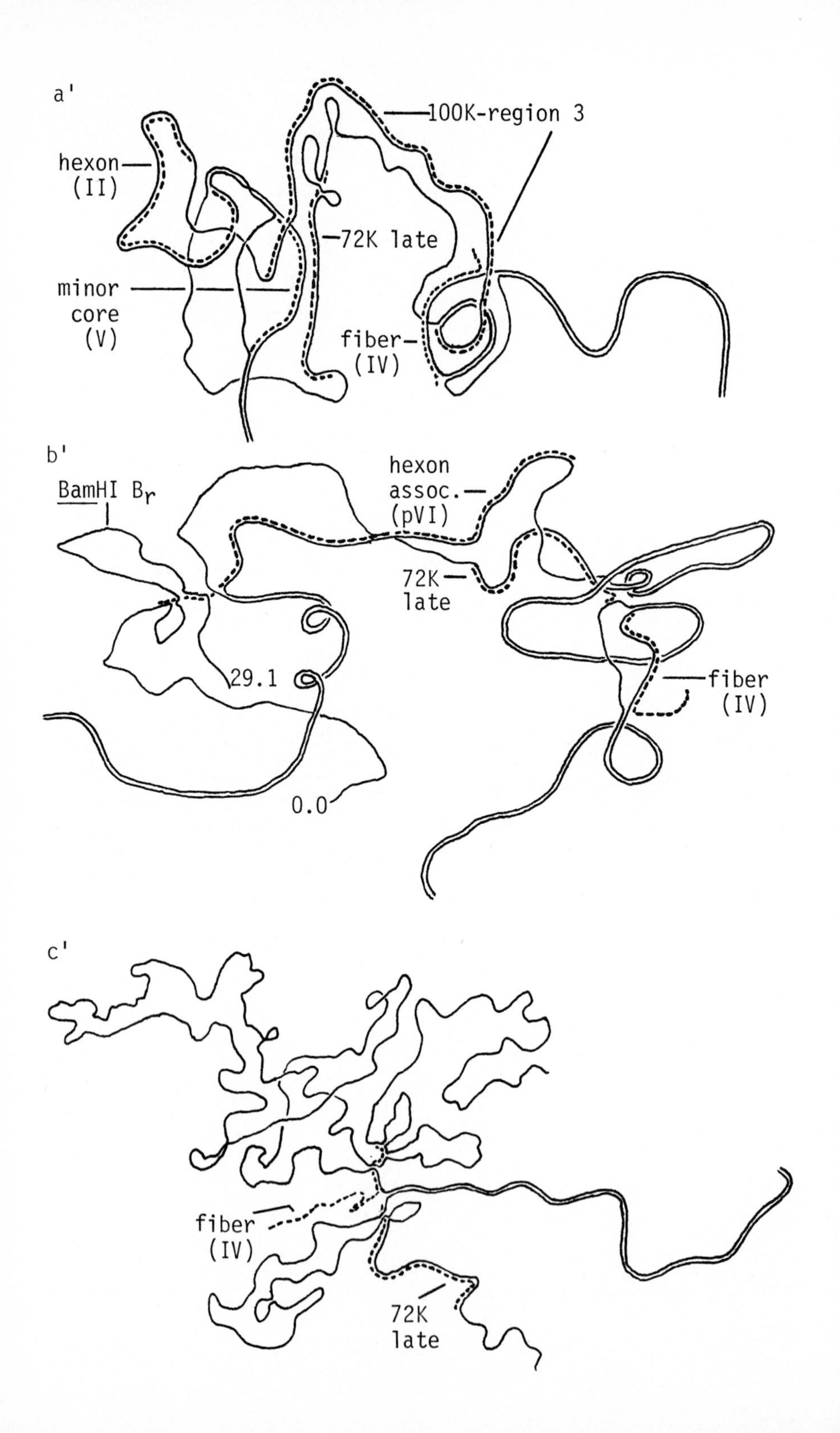

a'
100K-region 3
hexon
(II)
72K late
minor
core
(V)
fiber
(IV)
b'
BamHI B_r
hexon
assoc.
(pVI)
72K
late
29.1
fiber
(IV)
0.0
c'
fiber
(IV)
72K
late

Figure 5. Late form RNAs for the single-stranded DNA binding protein: multi-component RNA:DNA heteroduplexes that reveal 5' leaders, strand switches, and coincident sites. In all panels small arrows point to the 5' ends of RNAs, large arrows to the 3' ends, and arrowheads to DNA deletion loops constrained in the heteroduplexes by the spliced RNAs. (a,a') Convergent R loops in double-stranded Ad2 DNA. r-strand RNAs: hexon (II) coordinates (51.2-61.3); 100K region-3 (66.0-82.7), a continuous RNA in which no region 3-specific splices are evident. ℓ-strand RNA: DBP (72K), major late form (..61.6-66.5/68.3-68.5/71.9-72.0]←), with the splices revealed by two DNA deletion loops. The convergence of the 3' ends of hexon and DBP RNAs at 61.3-61.6 and the divergence of the 5' ends of the main bodies of 100K and DBP RNAs from 66.0-66.5 result in an expanded R loop extending from 51.2 to 82.7. R loops of r-strand transcripts for the minor core protein (V) (45.1-49.5) and for fiber (IV) (86.2-91.3) are also present. (b,b') A convergent R loop formed between Ad2 DNA and the r-strand transcript for the precursor to hexon-associated protein (pVI) (49.5-61.3) and the minor late form of the ℓ-strand mRNA for the DBP (..61.6-66.5/68.3-68.5/86.2-86.7-←). The r-strand mRNA for fiber (IV) (86.2-91.3) formed a separate R loop. The 5' end of the pVI RNA paired with the r-strand of the BamHI B restriction fragment of Ad2 DNA (coordinates 0.0-29.1). The fragment was constrained into two deletion loops diagnostic of the spliced tripartite leader (16.6/19.6/26.6) (19). The 5' leader of the DBP RNA was able to hybridize back to its coding sequences on the ℓ-strand at 86.7 because this region was exposed in the fiber R loop. (c,c') Late RNA was hybridized to denatured DNA. The minor form of late DBP mRNA (..61.6-66.5/68.3-68.5/86.2-←) annealed to an ℓ-strand, forming two DNA deletion loops due to the splices. An r-strand of DNA then paired with the ℓ-strand between coordinates 86.7 and 100. Annealing of the two DNA strands prevented the full body of a fiber mRNA from hybridizing to the r-strand. Only its tripartite leader and the 5' end of the message body (86.2-86.7) annealed. The rest of the fiber transcript appears as an RNA tail. The overlap on opposite strands of the 5' leader of the DBP RNA and the 5' portion of the main body of the fiber RNA is evident.

transcription units give rise to families of alternative mRNAs with common 3' ends located at several different positions.

The polyadenylate probably plays an important directive role in message maturation. The portion of the primary transcript next to the poly(A) is the particular group of message bodies to be kept in the mature mRNA. In the late r-strand transcriptional unit, the five major polyadenylation sites result in five major families of transcripts. The splicing of the tripartite leader to several alternative sites then provides translation accessibility to the different genetic regions within each family.

This directive role of poly(A) can also be inferred from some unusual transcripts which result from the extension of transcription past normal termination sites to sites further downstream. The joining of region 4 leader (and possible coding sequences) to region 2 RNA segments (..61.6-66.5/68.3-68.5/94.4-94.8/(±96.8-97.1)/99.2-99.3]←) is an example (14). It is instructive that the normal conserved segment at the 3' end of region 4 (91.3-92.4) is deleted when the poly(A) is located much further downstream. The very rare spliced form →[1.3-1.7/3.3-4.4/ 9.8-11.1.. resulted from the joining of two RNA segments that normally constitute a region 1A species to the 3' proximal segment of a region 1B species; the generally conserved segment of the region 1B RNAs (4.6-6.1) was deleted (14). Thus, splice junctions in separate transcription units can recombine faithfully if transcription read-through puts them in the same precursor RNA molecule.

Convergent 3' ends (class 4-5) of transcripts from opposite strands were observed at three sites (coordinates 11.1, 61.5, and 91.3) (Fig. 5). These boundaries separate blocks of early and late genes (1, 2, 11, 12).

Overlapping or nearly overlapping transcripts on opposite strands have also been found. The early and late forms of ℓ-strand transcripts for the single-stranded DNA binding protein contain sequences that are complementary to those in some r-strand transcripts. Several nearly coincidental processing sites for the r- and ℓ-strand transcripts in this region are observed (Fig. 5). The 3' ends of DBP and hexon (II) RNAs converge at 61.5. The microsplice found in some DBP RNAs at 66.2-66.3 (14) is close to the downstream splice site where the tripartite leader is joined to the main body of 100K mRNAs (class 2-7). The upstream junction for the common leader of DBP RNAs at 68.3 is close to an alternative downstream splice for 100K region RNAs (class 2-6). The 5' end of the late form of DBP RNA at 72.0 can also be a downstream splice target for the leader of the early DBP RNA (class 2-6) on the same strand and the same site nearly coincides with a downstream splice target for the tripartite leader in a late r-strand transcript (classes 1-7 and 2-7). In the interval 86.0-86.2, we observed a multitude of coincidences of the 3' ends of early and late region 3 RNAs (r-strand), the downstream splice junctions for ancillary leaders or the tripartite leader for fiber RNA (r-strand), and the upstream splice junction of the minor late leader for DBP mRNA from the ℓ-strand (classes 2-2, 2-4, 2-6, 3-5, and 4-4) (14).

The promoter and both leader segments of the major late form of DNA binding protein mRNA at coordinates /68.3-68.5/ 71.9-72.0]← on the ℓ-strand are complementary to two portions of the coding region for the 100 K protein from the r-strand. Similarly, the promoter and 5' leader for the minor late form of DBP mRNA at ℓ-strand coordinates /86.2-86.7] are complementary to coding sequences for the r-strand mRNA for fiber (IV) protein (Fig. 5) (40).

Other examples of RNA processing signals within protein encoding sequences occur on the same strand. The downstream target for the late tripartite leader at 38.8 and the polyadenylation site for IIIa mRNA, also at 38.8, are almost certainly within the penton base gene (8, 9, 11, 32, 41).

CONCLUSIONS

The maps of adenovirus-2 cytoplasmic RNAs observed early and late after infection reveal the existence of short DNA sequences, generally of less than 100 nucleotides, that serve different functions during RNA synthesis and processing.

By cataloging these sites, we have provided much of the information necessary for computer searches of DNA sequences to identify segments that can be implicated as promoters, terminators, and signals for splicing.

ACKNOWLEDGEMENTS

We wish to thank our colleagues, James B. Lewis, Richard E. Gelinas, Daniel F. Klessig, Terri Grodzicker, Joe Sambrook, and Jeffrey A. Engler for the adenovirus RNA and DNA samples used in the investigations. The manuscript was skillfully prepared by Marie Moschitta.

REFERENCES

1. Sharp, P.A., Gallimore, P.H., and Flint, S.J. (1974). Cold Spring Harbor Symp. Quant. Biol. 39, 457.
2. Pettersson, U., Tibbetts, C. and Philipson, L. (1976). J. Mol. Biol. 101, 479.
3. Büttner, W., Veres-Molnár, Z., and Green, M. (1976). J. Mol. Biol. 197, 93.
4. Flint, S.J. (1977). J. Virol. 23, 44.
5. Craig, E.A., Sayavedra, M., and Raskas, H.J. (1977). Virol. 77, 545.
6. Spector, D.J., McGrogan, M., and Raskas, H.J. (1978). J. Mol. Biol. 126, 395.
7. McGrogan, M. and Raskas, H.J. (1978). Proc. Nat. Acad. Sci. USA 75, 625.
8. Nevins, J.R. and Darnell, J.E. (1978). J. Virol. 25, 811.
9. Fraser, N. and Ziff, E. (1978). J. Mol. Biol. 124, 27.
10. Thomas, M., White, R.L., and Davis, R.W. (1976). Proc. Nat. Acad. Sci. USA 73, 2294.
11. Chow, L.T., Roberts, J.M., Lewis, J.B., and Broker, T.R. (1977). Cell 11, 819.
12. Neuwald, P.D., Meyer, J., Maizel, J.V., Jr. and Westphal, H. (1977). J. Virol. 21, 1019.
13. Westphal, H. and Lai, S.P. (1977). J. Mol. Biol. 116, 525.
14. Chow, L.T., Broker, T.R. and Lewis, J.B. (1979). (submitted for publication).
15. Lewis, J.B., Atkins, J.F., Anderson, C.W., Baum, P.R. and Gesteland, R.F. (1975). Proc. Nat. Acad. Sci. USA 72, 1344.
16. Lewis, J.B., Atkins, J.F., Baum, P.R., Solem, R., Gesteland, R.F. and Anderson, C.W. (1976). Cell 7, 141.
17. Harter, M.L. and Lewis, J.B. (1978). J. Virol. 26, 736.
18. Pettersson, U. and Mathews, M.B. (1977). Cell 12, 741.
19. Chow, L.T., Gelinas, R.E., Broker, T.R. and Roberts, R.J. (1977a). Cell 12, 1.
20. Klessig, D.F. (1977). Cell 12, 9.
21. Lewis, J.B., Anderson, C.W. and Atkins, J.F. (1977). Cell 12, 37.
22. Dunn, A.R. and Hassell, J.A. (1977). Cell 12, 23.
23. Broker, T.R., Chow, L.T., Dunn, A.R., Gelinas, R.E., Hassell, J.A., Klessig, D.F., Lewis, J.B., Roberts, R.J. and Zain, B.S. (1977). Cold Spring Harbor Symp. Quant. Biol. 42, 531.

24. Berget, S.M., Moore, C. and Sharp, P.A. (1977). Proc. Nat. Acad. Sci. USA 74, 3171.
25. Kitchingman, G.R., Lai, S-P. and Westphal, H. (1977). Proc. Nat. Acad. Sci. USA 74, 4392.
26. Berk, A.J. and Sharp, P.A. (1978). Cell 14, 695.
27. Evans, R.M., Fraser, N., Ziff, E., Weber, J., Wilson, M. and Darnell, J.E. (1977). Cell 12, 733.
28. Berk, A.J. and Sharp, P.A. (1977). Cell 12, 45.
29. Wilson, M.C., Fraser, N.W. and Darnell, J.E. Jr. (1979). Virology 94, 175.
30. Ziff, E.B. and Evans, R.M. (1978). Cell 15, 1463.
31. Goldberg, S., Nevins, J. and Darnell, J.E. (1978). J. Virol. 25, 806.
32. Chow, L.T. and Broker, T.R. (1978). Cell 15, 497.
33. Kozak, M. (1978). Cell 15, 1109.
34. Axelrod, N. (1978). Virology 87, 366.
35. Dunn, A.R., Mathews, M.B., Chow, L.T., Sambrook, J. and Keller, W. (1978). Cell 15, 511.
36. van Ormondt, H., Maat, J., deWaard, A. and van der Eb, A.J. (1978). Gene 4, 309.
37. Halbert, D.N., Spector, D.J. and Raskas, H.J. (1979). J. Virol. (in press).
38. Maat, J. and van Ormondt, H. (1979). Gene 5 (in press).
39. Kilpatrick, B.A., Gelinas, R.E., Broker, T.R. and Chow, L.T. (1979). J. Virol. (in press).
40. Zain, S., and Roberts, R.J. (1979). J. Mol. Biol. (in press).
41. Patterson, B.M., Roberts, B.E. and Kuff, E.L. (1977). Proc. Nat. Acad. Sci. USA 74, 4370.

INHIBITION OF SV40 T ANTIGEN ENZYMATIC ACTIVITY BY A MONOSPECIFIC ANTIBODY

R. Tjian, A. Robbins and D. Lane*

Department of Biochemistry, University of California, Berkeley, California 94720

*Cold Spring Harbor Lab, Cold Spring Harbor, New York 11724

ABSTRACT Antibody directed against the tumor antigen of SV40 was prepared by immunizing rabbits with an SV40 T-antigen related protein, the D2 hybrid protein, purified to homogeneity by column chromatography and sodium dodecyl sulfate gel electrophoresis. Antisera to the D2 hybrid protein specifically reacts with SV40 T antigen from monkey cells lytically infected with SV40 as well as the D2 hybrid protein from Hela cells infected with the defective adenovirus SV40 hybrid virus, Ad2 + D2. Unlike the sera from hamsters bearing tumors induced by SV40, the anti-D2 sera does not react with little t. This monospecific antibody was subsequently used to test the specificity of enzymatic activities carried out by purified D2 hybrid protein. Anti-D2 antibody efficiently inhibited the ATPase activity of the D2 hybrid protein but did not directly affect the T-antigen associated protein kinase activity.

INTRODUCTION

Until recently, the only biological activities which had been demonstrated for a purified A gene product of SV40 were its antigenic and immunogenic reactivities, its capacity to bind the origin of SV40 DNA replication and its ability to induce cellular DNA replication and helper function when injected into cells in culture (1-6). Our most recent studies of the SV40 T-antigen related protein, the D2 hybrid protein, (7) indicates that there may be enzymatic activities catalyzed by the A gene product of SV40. Our first attempts at detecting enzymatic activity revealed that both an ATPase and protein kinase activity co-purified with the D2 hybrid protein. Upon more rigorous purification, the ATPase activity remained

ISBN 0-12-068350-4

with the viral antigen but the protein kinase activity was resolved by column chromatography under high ionic conditions. Thus, it was unclear whether these enzymatic activities are intrinsic to the A gene of SV40 or merely associated with T antigen. One way to distinguish these possibilities is to test the sensitivity of the enzyme activity to antibody specificity directed against T antigen. Here, we report the specific inhibition of the T antigen associated ATPase activity by monospecific antibody raised against D2 hybrid protein isolated by conventional chromatography and further purified by sodium dodecyl sulfate gel electrophoresis.

RESULTS AND DISCUSSION

Antibody Specificity. The D2 hybrid protein was purified from Hela cells infected with Ad2 + D2 as described previously (7). The antigen was further isolated by gel electrophoresis on a 25 cm long slab gel formed with a gradient of acrylamide (7-15%) containing sodium dodecyl sulfate (8). Approximately 50 μg of D2 hybrid protein was used in each gel and after extraction from the gel, the 107,000 dalton D2 hybrid protein was used directly to immunize rabbits. The sera from immunized animals was subsequently used to prepare anti-D2 gamma globulin (IgG).

A comparison of the monospecific anti-D2 IgG with hamster anti-T IgG indicates that both antibodies are capable of specifically immunoprecipitating the 96,000 dalton T antigen from cells lytically infected with SV40 as well as the 107,000 dalton D2 hybrid protein from cells infected with Ad2 + D2 (Fig. 1). By contrast, only the hamster anti-T IgG reacted with small t from monkey cells infected with SV40. Extracts from Hela cells infected with Ad2 + D2 failed to yield any detectable small t protein when treated with either hamster anti-T IgG or anti-D2 IgG. These results are consistent with the finding (9) that the genome of Ad2 + D2 lacks sequences which code for small t and the amino terminus of big T (10-12). Although there are other proteins which can be detected in the immunoprecipitates of Figure 1, the 107,000, 96,000 dalton and their partial proteolytic degradation products appear to be the only polypeptides which are specifically immunoreactive to anti-D2 and anti-T and not reactive to IgG from pre-immune control sera. It was expected that perhaps anti-D2 IgG would react specifically to an Ad2 coded protein because the antigen, D2 hybrid protein, is thought to carry some sequences coded by either the 100,000 or pVIII protein of Ad2 (7). However, there does not appear to be any Ad2 proteins which is specifically immunoprecipitated although the 100,000 dalton

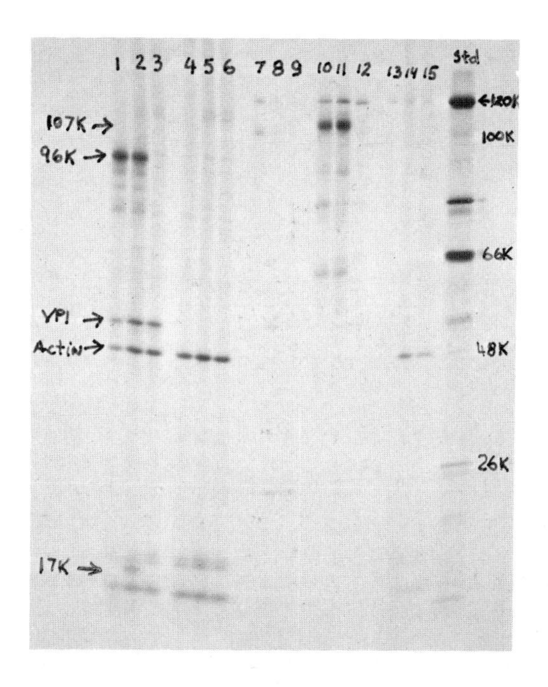

FIGURE 1. Immunoprecipitation of the SV40 T antigen and D2 hybrid protein. Tissue culture plates (100 mm) containing either CV1 monkey cells or Hela cells were infected with SV40 and Ad2 + D2 (MOI = 10) respectively. After 48 hrs, the cells were labeled with 250 μCi of ^{35}S Met for 3 hrs at 37° C and harvested. Extracts of the labeled cells were prepared for immunoprecipitate as described previously (14,15). Antibody precipitation was carried out by adding 20 μl of either anti-D2 IgG, anti-T IgG or control rabbit IgG to the cell lysate. After incubating at 0° C for 4 hrs, 50 μl of a 10% solution of formalin treated staphylococcus aureus was added to precipitate immune complexes. The precipitate was subsequently washed 4 times with Buffer W (0.1 M Tris, .1 mM EDTA, .4 M $LiCl_2$) and once with Buffer W + 2 M urea. Samples were then prepared for sodium dodecyl sulfate gel electrophoresis by adding 50 μl of sample buffer and boiling for 2'. Immunoprecipitates were analyzed by electrophoresis on a 25 cm long gel formed with a gradient of acrylamide (7-15%) containing sodium dodecyl sulfate (16). Extracts from CV1 monkey cells infected with SV40 776 (1,2,3), uninfected CV1 cells (4,5,6), Hela cells infected with Ad2 (7,8,9), Hela cells infected with Ad2 + D2 (10,11,12) and uninfected Hela cells (13,14,15) were immunoprecipitated with anti-D2 IgG (1,4,7,10,13), anti-T IgG (2,5,8,11,14) and control IgG (3,6,9,12,15).

species seems slightly enriched (Fig . 1-7). Other major viral and cellular proteins such as VP1, Hexon, 100,000 dalton Ad2 protein and actin appear to be non-specifically trapped and brought down by the immunoprecipitation procedure irrespective of IgG specificity. Thus, it appears that the anti-D2 IgG is a highly specific antibody which is directed exclusively against the big T portion of the SV40 early gene products and does not cross react with small t.

Effect of Anti-D2 IgG on ATPase and Protein Kinase Activities. We had shown previously that the ATPase activity associated with purified D2 hybrid protein could be efficiently inhibited by antibody isolated from the sera of hamsters bearing SV40-induced tumors (7). As further evidence for a specific catalytic activity of the SV40 A gene product, we now demonstrate that monospecific antibody directed against the D2 hybrid protein also inhibits the ATPase activity. Increasing amounts of anti-D2 IgG rapidly inhibits the hydrolysis of ATP by purified D2 hybrid protein (Fig. 2A) while IgG from control rabbit serum has no effect. Similarly, the *in vitro* phosphorylation of D2 hybrid protein by an associated protein kinase is efficiently blocked by anti D2 IgG (Fig. 2B). In marked contrast, phosphorylation of an exogenously added phospho-acceptor protein, phosvitin, was not inhibited by anti-D2 IgG (Fig. 2C). The effects of the anti-D2 IgG on the T antigen associated protein kinase activity is most easily explained if the phospho-transferase activity were carried out by a tightly associated protein kinase and not the D2 hybrid protein itself (7). In addition, it is likely that inhibition of the phosphorylation of D2 hybrid protein is a result of antibody molecules binding to the phosphate acceptor site of the substrate rather than at the catalytic site of the protein kinase. We cannot, however, exclude the possibility that the anti-D2 IgG, in fact, reacts directly with the kinase subunit but fails to inhibit activity. To test the possibility that the protein kinase associated with the purified D2 hybrid protein is actually an enzyme which interacts directly with and binds tightly to the viral antigen, we performed an immuno absorption experiment designed to remove D2 hybrid protein from the protein kinase reaction mixture. Increasing amounts of anti-D2 IgG were first incubated with the D2 hybrid protein. The antigen-antibody complex was then removed by absorption to sepharose beads coupled to purified staphylococcus aureus A protein (13). The supernatant was subsequently assayed for phospho-transferase activity by adding phosvitin as the phospho acceptor. The results indicate that when the D2 hybrid protein is partitioned from the reaction by virtue of its binding to anti-D2 IgG, an

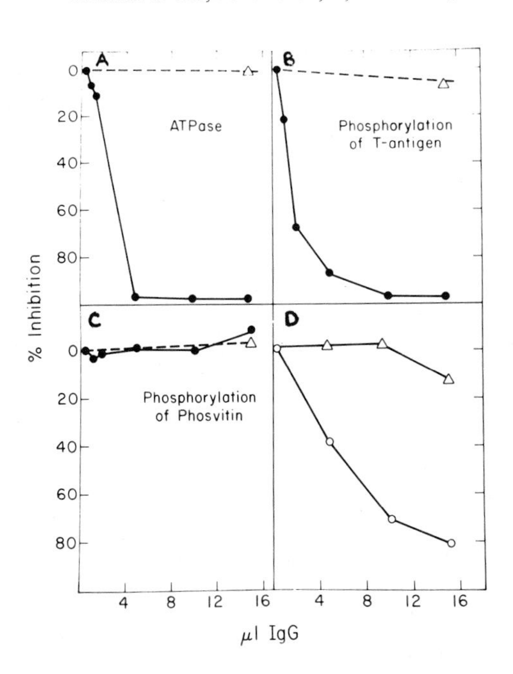

FIGURE 2. Inhibition of enzymatic activities by anti-D2 IgG. The D2 hybrid protein was purified to greater than 95% homogeneity by the procedure recently described (7). The hydrolysis of ATP and phospho-transferase reactions were performed according to the previously described procedures (7). (A) Increasing amounts of anti-D2 IgG (●—●) or preimmune IgG (Δ—Δ) were added to an ATPase reaction mixture containing approximately 1.0 μg of phosphocellulose-purified Form I (7) D2 hybrid protein. (B) The indicated amounts of anti-T or preimmune IgG were added to the protein kinase reaction mixture containing 1.0 μg of Form II (7) D2 hybrid protein.

associated protein kinase activity is also removed and the remaining supernatant is incapable of carrying out the phosphorylation of an exogenously added substitute. When preimmuned sera is used, no such inhibition is observed. These results, taken together, provide further evidence that the D2 hybrid protein is an ATPase which is associated with a protein kinase.

REFERENCES

1. Koch, M. A., and Sabin, A. B. (1963) Proc. Soc. Exp. Biol. Med. 113, 4-12.
2. Tevethia, S. S., and Rapp, F. (1966) Proc. Soc. Exp. Biol. Med. 123, 612-615.
3. Girardi, A. J., and Defendi, V. (1970) Virology 42, 688-698.
4. Reed, S. I., Furguson, J., Davis, R. W., and Stark, G. R. (1975) Proc. Natl Acad. Sci. USA 72, 1605-1609.
5. Tjian, R. (1978) Cell 13, 165-179.
6. Tjian, R., Fey, G., and Graessmann, A. (1978) Proc. Natl Acad. Sci. USA 75, 1279-1283.
7. Tjian, R., and Robbins, A. (1979) Proc. Natl Acad. Sci. USA 76, 610.
8. Tjian, R., Stinchcomb, D., and Losick, R. (1976) J. Biol. Chem. 250, 8824-8828.
9. Hassel, J. A., Lukanidin, E., Fey, G., and Sambrook, J. (1978) J. Mol. Biol. 120, 209.
10. Crawford, L. V., Cole, C. N., Smith, A. E., Paucha, E., Togtmeyer, P., Rundell, K., and Berg, P. (1978) Proc. Natl Acad. Sci. USA 75, 117-121.
11. Sleigh, M. J., Topp, W. C., Hanich, R., and Sambrook, J. (1978) Cell 14, 79-88.
12. Paucha, E., Mellor, A., Harvey, R., Smith, A. E., Hewick, R. W., and Waterfield, M. D. (1978b) Proc. Natl Acad. Sci. USA 15, 2165-2169.
13. Kessler, W. W. (1975) J. Immunol. 115, 1617-1624.
14. Rundell, K., Collins, J. K., Tegtmeyer, D., Ozer, H. L., Lai, C., and Nathans, D. (1977) J. Virol. 21, 636-646.
15. Lane, D., and Robbins, A. (1978) Virology 87, 182-193.

RABBIT β GLOBIN SYNTHESIS IN CULTURED MONKEY KIDNEY CELLS INFECTED WITH SV40 β GLOBIN RECOMBINANT GENOMES

Richard C. Mulligan, Bruce H. Howard, and Paul Berg

Department of Biochemistry, Stanford University
Stanford, California 94305

ABSTRACT Rabbit β globin complementary DNA (cDNA) has been inserted in various locations of the late region of SV40 DNA. The resulting recombinant genomes multiply efficiently in monkey kidney cell cultures in the presence of helper virus. Cells propagating two of the recombinants (SVGT5-βG and SVGT7-βG) produce substantial quantities of mRNA containing the β globin cDNA sequence and rabbit β globin polypeptide. Cells infected with another (SVGT9-βG), produce a hybrid protein with antigenic determinants present in rabbit β globin and SV40's major structural protein VP1.

INTRODUCTION

SV40 provides a useful vector to study the expression of cloned genes in mammalian cells (1-3). It is particularly attractive as a transducing vector because the entire nucleotide sequence of its small, circular genome has been determined (4,5), the regions of the DNA responsible for almost all known viral functions have been accurately mapped (4-7), and extensive information is available about replication, expression and regulation of the viral genome in different host cells.

Although initial attempts in our own and other laboratories to obtain expression of exogenous DNA segments joined to SV40 DNA were discouraging (1-3), more recent experiments demonstrated that insertion of a rabbit β globin cDNA segment in place of SV40's VP1 coding sequence (SVGT5-βG contains the β globin cDNA segment at SV40 map position 0.94 to 0.145) led to the formation of β globin cytoplasmic polyadenylated mRNAs and authentic β globin peptide (8). In this paper we summarize experiments concerning the construction and expression of two additional SV40 β globin cDNA recombinants: One, which contains the β globin cDNA sequence interposed between SV40 map coordinates 0.86 and 0.145 (SVGT7-βG), produces substantial amounts of β globin

ISBN 0-12-068350-4

after infection; the other, which contains its β globin cDNA sequence between map coordinates 0.99 and 0.145 (SVGT9-βG), induces a hybrid protein containing VP1 and β globin antigenic determinants.

RESULTS

Construction of SVGT-RaβG Genomes. The mRNAs which code for SV40's three capsid proteins, VP1, VP2 and VP3 are composite structures; that is, the mRNAs contain nucleotide sequences transcribed from non-adjacent DNA segments (9, 10). Each mRNA contains a 5'-terminal segment (the leader) transcribed from the l-strand between map coordinates 0.69 and 0.76 (see Fig. 1). Joined to the leader sequence, the VP2 and VP3 mRNAs (the 19S class) have a segment of RNA homologous to the l-strand between map coordinates 0.77 and 0.175 (the 19S body); the body of the VP1 mRNA (16S class) is transcribed from the l-strand of the DNA between map coordinates 0.94 and 0.175.

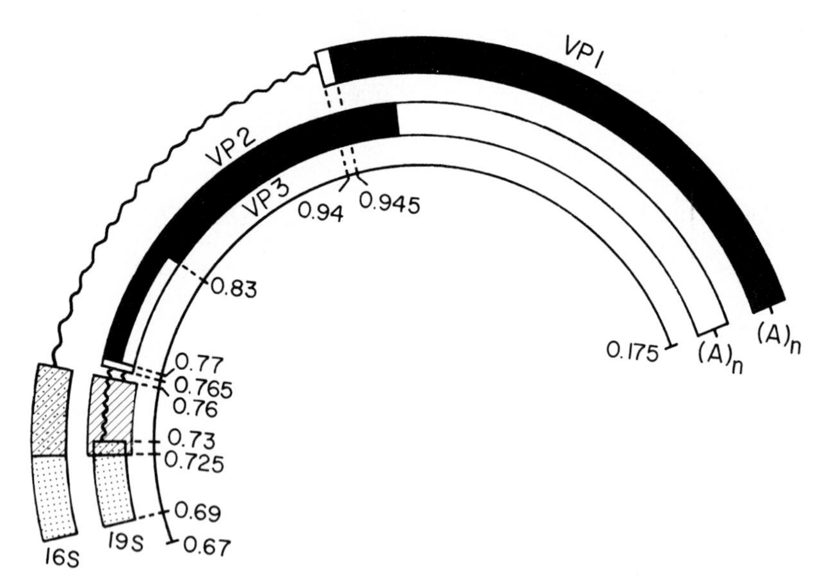

FIGURE 1. The composite structure of SV40 late mRNAs. The map of SV40 DNA from the origin of DNA replication (0.67) to the end of the late region (0.175) is shown on the inner circle. The regions coding for the structure of VP1, VP2, and VP3 appear as shaded regions within bars that define the 'bodies' of the 19S and 16S mRNAs. The 'leader' segments are shown as bars spanning map coordinates 0.69-0.76. The cross-hatched and stippled regions of the 16S mRNA leader, for example, are intended as symbolic, rather than literal

representations of more than one type of leader segment; one leader segment spans the region from 0.725 to 0.76 and another the region from 0.69 to 0.73 and is joined to the body at 0.76; another class of leaders extends from 0.725 to 0.76.

Our objective in this work was to construct a series of SV40 vectors which would permit the transcription and translation of inserted segments of DNA. One approach which seemed feasible was to retain in the vector those SV40 sequences needed for initiating and terminating transcription, polyadenylation and splicing. Then, appropriate DNAs coding for specific proteins could be inserted in such vectors to assess the efficiency of their expression as mRNAs or proteins.

Inspection of the SV40 restriction map suggested several restriction sites which could be utilized to generate vectors fulfilling the above requirements. SV40 DNA is cleaved by Hind III in three positions in the late region: 0.86, 0.94 and 0.99 map units (Fig. 2). The Hind III cleavage site

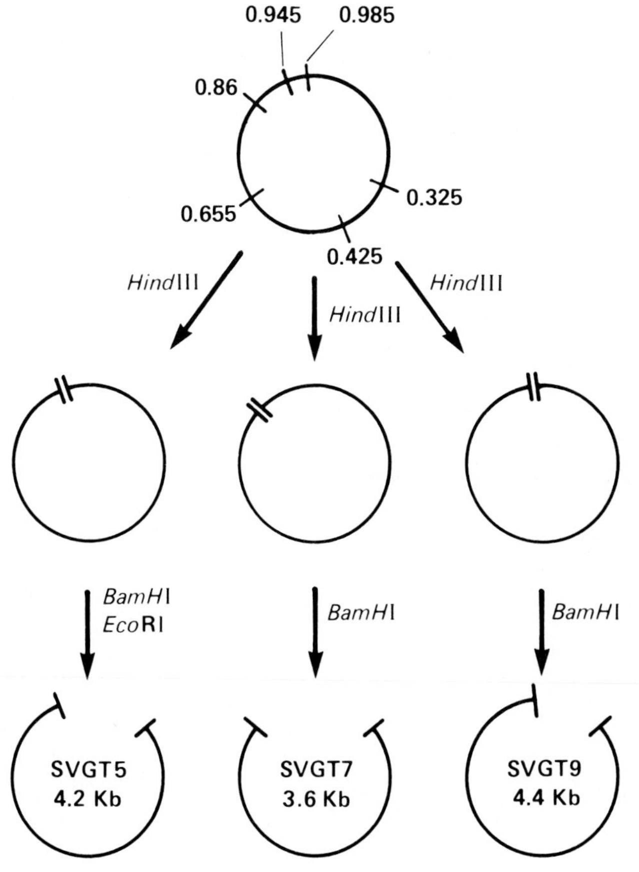

FIGURE 2. Construction of SVGT5, 7 and 9 vectors.

at map position 0.86 lies within the protein coding sequence for VP2 and VP3, approximately 300 and 500 bases beyond the initiator codons for VP3 and VP2, respectively, and thus, within the coding sequences of the 19S RNA. The Hind III cleavage site at 0.94 is six nucleotides before the initiator codon for VP1 and approximately 50 nucleotides beyond the 5' end of the body portion of the VP1 mRNA (11, 12). The Hind III cleavage site at 0.99 lies within the coding sequence of VP1, approximately 200 bp beyond the initiator codon for VP1. Bam H1 restriction endonuclease cleaves SV40 DNA at map coordinate 0.145, 50 nucleotides before the termination codon for VP1 and 150 nucleotides before the poly A sequence at the 3' end of the VP1 (and VP2/3) mRNAs (4).

Partial cleavage of SV40 DNA with Hind III and complete digestion with Bam H1, generated the three vectors used in this study: SVGT5, a 4.18 kb segment bounded by the Hind III site at 0.94 and the Bam H1 site; SVGT7, a 3.7 kb segment bounded by the 0.86 Hind III site and the Bam H1 site, and SVGT9, a 4.4 kb segment bounded by the Hind III site at 0.99 and the Bam H1 site (see Fig. 2). SVGT5 and SVGT9 contain the regions involved in the splicing of both 16S and 19S viral mRNAs while SVGT7 contains only the regions implicated in splicing of 19S mRNAs (see Fig. 1). All three vectors retain the origin of DNA replication and the entire early region; hence, they can be propagated by complementation with an appropriate tsA mutant (13).

The rabbit β globin cDNA sequence was selected as the transducible gene because it is small enough to be accommodated in these vectors and the protein can be isolated and characterized readily. Moreover, the complete coding sequence for β globin as well as 43 of 56 5'-and all of the 3'-terminal nucleotide sequences present in β globin mRNA have been cloned in E. coli by Maniatis et al. (14, 15).

To obtain the β globin cDNA segment suitable for insertion into the SV40 vectors, the β globin cDNA segment in the bacterial plasmid PβG1 (14) was recloned at the Hind III restriction site of pBR322 as already described (8). Digestion of this recombinant with Hind III and Bgl II endonuclease yielded a β globin cDNA segment with a Hind III cohesive end 37 nucleotides before the β globin initiator codon and a Bgl II cohesive end 3 nucleotides beyond the terminator codon of the β globin cDNA. Even though the Bam H1 (GGATCC) and Bgl II (AGATCT) endonuclease recognition sites differ, the cohesive ends generated by the two endonucleases are identical (GATC). Accordingly, the 485 bp β globin cDNA fragment was ligated to SVGT5, 7 and 9 with T4 ligase and the ensuing SVGT-βglobin recombinants were cloned and propagated at 41°C with tsA58 in CV1P cells. (tsA58 is a thermosensitive mutant of the SV40 early function and can not multiply at 41°C (16)).

Plaques (about 2.5 x 10^4 per microgram of ligated DNA) were picked, screened for virus containing the β globin cDNA sequence (90% of the plaques contained DNA that hybridized specifically to the β globin cDNA (17)), purified by plaque isolation and both high-titre virus stocks and purified viral DNA were prepared. Judging from the relative yields of the two DNAs from such mixed infections, we estimate that the recombinant and helper genomes multiply equally well. The DNAs of independent isolates of the three kinds of recombinants were found to have the expected structures by a battery of restriction endonuclease digestions (data not shown).

Formation of β Globin Protein. To determine if the infection with various SVGT-βglobin recombinant genomes results in the production of β globin, CV1 cells were infected with each of the three recombinants plus tsA58, and after 48 hours at 37° the cultures were labelled with ^{3}H-labeled leucine and valine for 30 minutes; SV40 and mock-infected cultures were labeled in a comparable way. Extracts of each culture were prepared by sonication and treatment with NP40 DOC and aliquots were electrophoresed on SDS 15-20% gradient gels (Fig. 3). Although the autoradiograms of each electrophoresed extract have many labeled protein bands, the extracts from cells infected with SVGT5-βG (track c) and SVGT7-βG (track d) contain protein bands migrating precisely at the position of added authentic rabbit β globin. However, the extract from SVGT9-βG infected cells does not contain a discernible band at the β globin position; instead, a new band with a mobility corresponding to a protein of 25,000 daltons is visible (track e). None of these bands is discernible in either the SV40 or mock-infected extracts (tracks a and b, respectively).

Each of these extracts was also immunoprecipitated with goat anti-rabbit β globin antibody and the precipitated samples electrophoresed as above (tracks f-j). Extracts from cells infected with SVGT5-βG (track h) and SVGT7-βG (track i) contain an immunoprecipitable protein that coelectrophoreses with authentic rabbit β globin. Immunoprecipitates of SV40 (track f) or mock-infected (track g) cell extracts did not contain such a β globin like protein. By contrast, immunoprecipitates of SVGT9-βG infected cell extract (track j) contain the 25,000 dalton protein as well as a family of proteins which migrate a bit more slowly than authentic rabbit β globin. The 25,000 dalton protein, seems to contain anti-

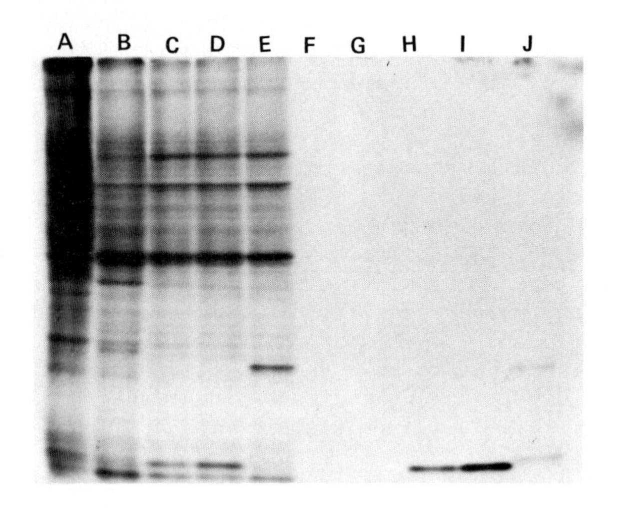

FIGURE 3. Production of rabbit β globin in cells infected with SV40 RaβG recombinants. Tracks a-e are untreated cell extracts. Tracks f-j are extracts immunoprecipitated with goat anti-rabbit β globin antiserum. Tracks a, f: uninfected cell extract. b, g: SV40-infected cell extract, c,h: SVGT5-βG infected cell extract. d,i: SVGT7-βG infected cell extract, e,j: SVGT9-βG infected cell extract.

genic determinants from VP1 as well since the same band appears in the immunoprecipitates with anti-VP1 antibody (data not shown). Immunoprecipitation of the infected cell extracts with anti-VP1 antibody did not bring down any proteins with electrophoretic mobilities corresponding to β globin.

The putative β globin polypeptide synthesized after infection with SVGT5-βG yields at least 8 of the 10 expected rabbit β globin peptides after digestion with trypsin and two dimensional electrophoretic-chromatographic analysis (unpublished experiment done in collaboration with Tony Hunter, Salk Institute). Moreover, we have also recently shown that the N-terminal tryptic peptide of the β globin made in SVGT5-βG infected cells co-migrates with the corresponding peptide from authentic rabbit β globin.

Using quantitative densitometry of the autoradiographed bands we estimate that the amounts of β globin and VP1 synthesized in cells infected with SVGT5-βG and SVGT7-βG are about equal. Taking into account the different number of labeled leucine and valine residues in VP1 and β globin and the ratio of recombinant to helper genomes at the time of labeling, it appears that β globin and VP1 are being synthesized at nearly equal rates.

Formation of SV40-RaβG Hybrid mRNAs. It is possible, on the basis of present ideas about SV40 late mRNA formation, to anticipate the structures of the hybrid mRNAs that would be produced during infections with each of the three recombinant genomes (Fig. 4). Since the regions believed to be involved in splicing of the 16S and 19S mRNAs are intact in SVGT5 and SVGT9, infection with either SVGT5-βG or SVGT9-βG should produce corresponding, but smaller, hybrid mRNAs: SVGT5-βG and SVGT9-βG mRNAs should be about 0.5 and 0.3 kb, respectively, smaller than those formed by wild-type SV40. Since SVGT7-βG lacks some of the sequences needed for splicing 16S mRNA, only one hybrid mRNA, approximately 1 kb smaller than the SV40 19S RNA, should be formed.

So far only the transcription of the SVGT5-βG recombinant has been examined in detail. CV1 cells were infected with tsA58 and two independent SVGT-βG recombinant viruses (MOI, 10-50 PFU per cell) and after 48 h at 37°C, the cytoplasmic, poly A containing RNA (freed from DNA) was isolated; analogous RNA preparations were isolated from SV40 and mock-infected cultures. Each RNA preparation was denatured with glyoxal (18,19), electrophoresed on 1.5% agarose gel and transferred to diazotized benzyloxmethyl paper (20). The imprints were annealed with ^{32}P-DNA to reveal the positions of the separated RNAs (Fig. 5).

With [^{32}P]-labelled pBR322-βG DNA as hybridization probe, two discrete β globin mRNA species can be detected in the RNA from cells infected with both SVGT5-βG recombinants (Fig. 5, tracks c and d). These RNAs are missing in the RNA mock- or SV40-infected cells (tracks a and b). Using DNA fragments of known size as molecular length standards (tracks e and f), we estimate that these mRNAs are 1.8 and 1 kb in size. When duplicate tracks were hybridized with [^{32}P]-DNA containing only the SV40 late mRNA leader sequences, two labeled bands are visible in the SV40-infected cell RNA (track i), but not in the RNA from mock-infected cells (track j). The faster migrating, more heavily labeled band is the 16S mRNA (1.5 kilobase) and the slower, less intensely labeled band, the 19S mRNA (2.3 kilobase). There are four labeled bands when RNA from cells infected with the SVGT5-βG recombinants is hybridized with the labeled SV40 leader probe (tracks g and h). Two (the 1.5 and 2.3 kilobase species) are the 16S and 19S mRNAs produced by the tsA58 helper genome. Another has the mobility of the 1.0

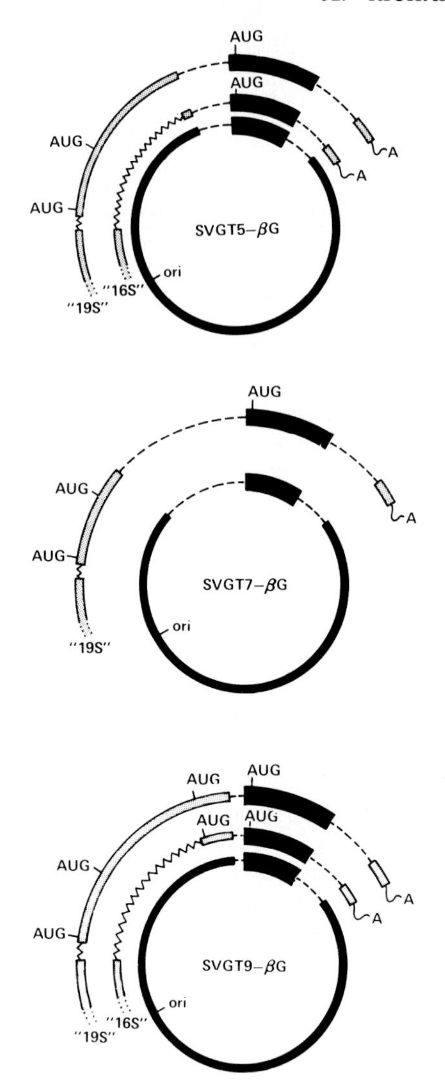

FIGURE 4. Predicted mRNAs produced in cells infected with SVGT5, 7 and 9 βG recombinants. The boxed AUGs denotes the position of the initiation codons for the late viral proteins within the particular messages.

kilobase RNA species that hybridized to the β globin cDNA probe. The fourth, which is only barely discernible in this exposure, has a mobility very close to that of the 1.8 kilobase RNA detected with the β globin cDNA probe. Thus the 1 and 1.8 kilobase polyadenylated RNAs contain the β globin cDNA sequence and the leader segment characteristic of SV40

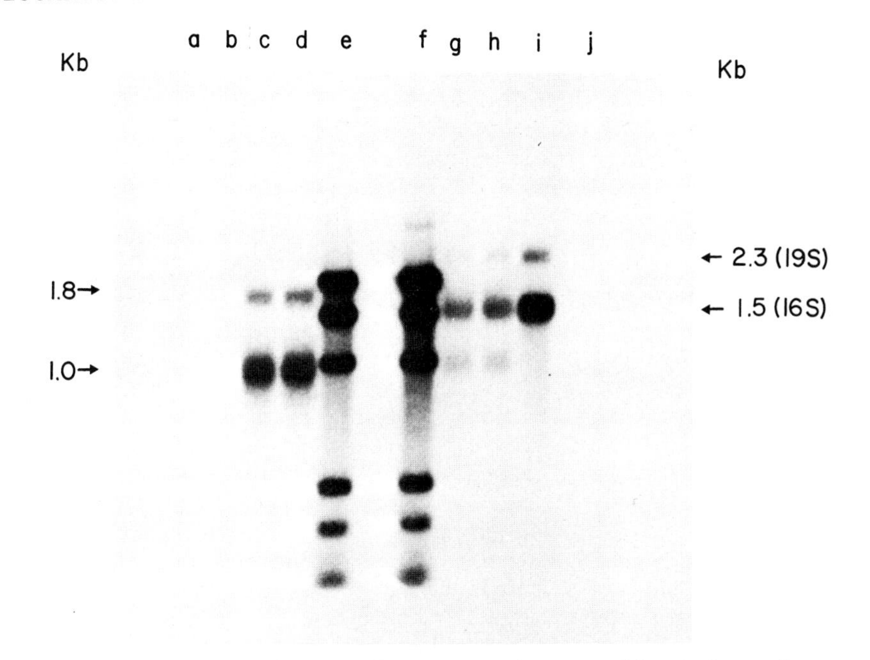

FIGURE 5. Analysis of mRNAs made in cells infected with SVGT5-βG. ^{32}P-labelled pBR322-βG2 DNA was the hybridization probe for tracks a-d and a ^{32}P-SV40 DNA fragment (from map position 0.67-0.76, produced by cleavages with Bgl I and Hpa I endonucleases) was the probe for tracks g-j. Tracks e and f contained a Hind III digest of [^{32}P]-SV40 DNA as size standards. The lengths of the fragments (kilobases) in decreasing order are: 1.96, 1.54, 1.07, 0.369, 0.240. Tracks a, j; RNA from non-infected cells; b, i SV40 RNA; c, h SVGT5-βG10A RNA; d, g SVGT5-β12A RNA.

late mRNAs. Comparing the labeling intensities of the 1 kilobase β globin mRNA and 1.5 kilobase SV40 mRNA (in Fig. 5 tracks g and h) and taking account of the ratio of recombinant to helper DNA in these infected cultures (1:3), we surmise that the steady state levels of the two mRNAs are within a factor of two of each other.

DISCUSSION

These experiments establish that an exogenous gene, in this case the coding segment for rabbit β globin, can be recombined with SV40 DNA *in vitro* at several positions in the late region to yield hybrid genomes that express the incorporated gene.

At present we do not know which mRNAs are being translated to yield the β globin or fused VP1-β globin peptides. VP1 is believed to be translated exclusively from the 16S mRNA although the protein's coding sequence is also contained within the 19S mRNA (21). Therefore, we assume that the production of β globin after infection with SVGT5-βG results from the translation of the smaller 16S-like mRNA; in this case, the β globin initiator AUG is within 30-35 nucleotides of the corresponding position of the VP1 initiator AUG. With SVGT9-βG, however, the smaller mRNA retains the VP1 initiator AUG and therefore, protein synthesis probably begins at that position. Since the β globin coding sequence is in phase with the VP1 coding sequence,it seems likely that translation of that hybrid mRNA proceeds into the β globin coding sequence, terminating at the β globin terminator codon; this would account for the 25K protein with antigenic determinants of both VP1 and β globin. Moreover, we surmise that translation probably does not begin at the AUG that initiates the β globin coding sequence because little or no β globin peptide seems to be produced. Probably, the β globin-like peptides are formed by the proteolysis of the 25K hybrid protein, but further studies are needed to explore that possibility.

The nature of the mRNA responsible for formation of β globin after infection with SVGT7-βG is more obscure. Since the sequence needed for splicing the 16S-like mRNA is absent in this recombinant, the logical candidate for the mRNA is the 19S analogue (see figure 4). But this would place the β globin initiator AUG codon distal to the AUG's that initiate the translation of VP2 and/or VP3. According to the "dogma" (22) the internal β globin initiator AUG should not function. Hence, either the "dogma" is wrong or a novel β globin containing mRNA is generated and translated in SVGT7-βG infected cells. Preliminary experiments suggest the latter, but further work is needed to confirm and characterize the structure of the putative novel mRNA.

Expression of the β globin gene in SVGT-βG recombinant genomes permits us to explore the requirements for proper transcription and processing of SV40 mRNAs. Such studies are more readily performed because of the ability to monitor the formation and structure of β globin containing mRNAs even in the presence of analogous SV40 mRNAs generated by the helper.

The ability to obtain expression of the β globin cDNA sequence in, for example, SVGT5-βG also permits us to test the effect of mutational alterations in the β globin segment itself; for example, in collaboration with Charles Weissmann's laboratory we are testing cDNAs with base changes in the β globin coding region or deletions of 5'-proximal non-translated nucleotides for their relative efficiencies in directing β globin synthesis *in vivo*.

Other variations on the basic strategy outlined here include the insertion of the β globin cDNA segment at different locations of the early region to explore the requirements for obtaining proper and efficient expression from such recombinant genomes. Also of particular interest is whether segments of DNA coding for a variety of proteins, prokaryote or eukaryote in origin, can also be expressed if inserted properly in SVGT vectors.

ACKNOWLEDGEMENTS

We wish to express our sincerest appreciation to Dr. T. Maniatis for providing PBG-1 and to S. Boyer for his gracious gifts of goat anti-rabbit β globin antiserum. All experiments involving the construction and propagation of recombinant genomes were performed according to the stipulations in the existing NIH guidelines: P2 for PβG1 and pBR322-βG and P3 for SVGT-βG. This research was supported by research grants NIH GM 13235-14 and NCI CA 15513-05. Richard C. Mulligan is a NSF Pre-doctoral fellow and Bruce H. Howard is an Investigator in the USPHS.

REFERENCES

1. Goff, S.P., and Berg, P. (1976). Cell 9, 695.
2. Ganem, D., Nussbaum, A. L., Davoli, D., and Fareed, G.C. (1976). Cell 7, 349.
3. Hamer, D. (1977). In "Recombinant Molecules: Impact on Science and Society (R. G. Beers, Jr., and E. G. Bassett, eds.), pp. 317-335. Raven Press, New York.
4. Reddy, V. B., *et al*. (1978). Science 200, 494.
5. Fiers, W., *et al*. (1978). Nature 273, 113.
6. Cole, C. N., *et al*., (1977). J. Virol. 24, 277.
7. Lai, C. J., and Nathans, D. (1974). Virology 60, 466.
8. Mulligan, R. C., Howard, B. H., and Berg, P. (1979). Nature 277, 108.
9. Aloni, Y., *et al*. (1977). Proc. Nat. Acad. Sci. U.S.A. 74, 3686.
10. Lavi, S., and Groner, Y. (1977). Proc. Nat. Acad. Sci. U.S.A. 74, 3523.
11. Celma, L., Dhar, R., Pan, J., and Weissmann, S. M. (1977). Nucleic Acids Res. 4, 2549.
12. Haegeman, G., and Fiers, W. (1978). Nature 273, 70.
13. Mertz, J. E., and Berg, P. (1974). Virology 62, 112.

14. Maniatis, T., Sim Gek Kee, Efstratiadis, A., and Kafatos, F. C. (1976). Cell 10, 571.
15. Efstratiadis, A., Kafatos, F. C., and Maniatis, T. (1976). Cell 8, 163.
16. Tegtmeyer, P. (1972). J. Virol. 10, 591.
17. Villarreal, L. P., and Berg, P. (1977). Science 196, 183.
18. McMaster, G. K., and Carmichael, G. G. (1977). Proc. Nat. Acad. Sci. U.S.A. 74, 4835.
19. Villarreal, L. P., White, R. T., and Berg, P. (1979). J. Virol. 29, 209.
20. Alwine, J. C., Kemp, D. J., and Stark, G. R. (1977). Proc. Nat. Acad. Sci. U.S.A. 74, 5350.
21. Prives, C. L. *et al.* (1974). Cold Spring Harbor Symp. Quant. Biol. 39, 309-316.
22. Jacobson, M. F. and Baltimore, D. (1968). Proc. Nat. Acad. Sci. 61, 77-84.

INDEX

D

E

F

G

H

N

O

P

R

S

T

V

Y

X